OCEAN WORLD
ENCYCLOPEDIA

OCEAN WORLD ENCYCLOPEDIA

DONALD G. GROVES

and

LEE M. HUNT

Professional Staff, National Academy of Sciences

McGRAW-HILL BOOK COMPANY

New York St. Louis San Francisco Auckland Bogotá
Hamburg Johannesburg London Madrid Mexico
Montreal New Delhi Panama Paris São Paulo
Singapore Sydney Tokyo Toronto

Library of Congress Cataloging in Publication Data

Groves, Donald G
Ocean world encyclopedia.

Includes index.
1. Oceanography—Dictionaries. I. Hunt, Lee M., joint author. II. Title.
GC9.G76 551.4′6′003 79-21093
ISBN 0-07-025010-3

2 3 4 5 6 7 8 9 0 VHVH 8 9 8 7 6 5 4 3 2 1

The editors for this book were Robert A. Rosenbaum and Ruth L. Weine, the designer was Elliot Epstein, and the production supervisor was Thomas G. Kowalczyk. It was set in Baskerville by Progressive Typographers, Inc.

Printed and bound by Von Hoffmann Press, Inc.

CONTENTS

PREFACE

There is no credible evidence that modern human beings are, in any fundamental sense, intellectually superior to their counterparts at the dawn of recorded history. And yet, we have left far behind the period when transportation and construction relied solely on puny human muscles and those of domesticated animals; when disease—such as the Great Plague that laid waste to 33 million Europeans in 1348—repeatedly rode the land and when ignorance and superstition combined to shackle the questing mind. Instead, we have tapped and are learning to use the fabulous energy of the atom; we have brought to heel the ancient diseases and tripled life expectancy; and we have walked the boulder-strewn surface of the moon and explored the deepest, darkest depths of the ocean. For those who would seek the answer to this disparity in intellectual achievement, it can be found in a single word: accumulation. Each generation has the decisive advantage of being able to draw on the accumulated knowledge of all previous generations.

Beginning with Denis Diderot's great 35-volume *Encyclopédie* of 1772, it is the traditional objective of an encyclopedia to translate and summarize the accumulated knowledge of a given field for the benefit of those not specialized in that field. In keeping with established tradition, McGraw-Hill's *Ocean World Encyclopedia* is the first encyclopedia of oceanography to be written specifically for the nonspecialist and the first to address, in a single volume, all the major divisions of oceanography along with several categories of related subjects. Included are articles on

- Physical Oceanography
- Geological Oceanography
- Chemical Oceanography
- Biological Oceanography
- Oceanographic Instrumentation
- Hurricane

- International Marine Sciences Organizations
- Individual Famous Oceanographers

By their writing style (largely nontechnical) and by the selection of articles, the authors have attempted to produce an encyclopedia both useful and interesting to high school and college students as well as to interested nonspecialists from many walks of life. For easy accessibility, considerable care has been taken in selecting the more obvious titles for articles, and the articles themselves have been arranged in alphabetical order. Liberal use has been made of *See* and *See also* to indicate either additional articles which expand on a given subject or related articles of possible interest to the reader. Words and phrases in small capital letters have also been used to indicate a separate article under that title. And, wherever appropriate, broad survey articles have been provided for those interested in a quick overview, followed (alphabetically) by specialized articles for those interested in a more detailed discussion of some aspect of the general subject.

The plant and animal life of the ocean, or that related to it (e.g., MARINE BIRDS), has been covered quite extensively by both general and specific articles. For the benefit of those with more than a casual interest in the subject, taxonomy has been emphasized so that members of common groups may be more easily related.

For those interested in the geography of the coeans, there is a separate article on each of the five oceans and the 60-odd seas describing the physical and political boundaries, water characteristics, bottom configuration and, in most cases, early exploration in the area. In like manner, all the major ocean currents are discussed under individual headings.

In discussing the physics, geology, geophysics, chemistry, and meteorology of the oceans the authors have concentrated on those subjects thought to be of greatest interest and use to the intended audience. Broad articles are often used to explain specific phenomena and processes: e.g., CORIOLIS EFFECT to discuss the mechanics of ocean and atmospheric circulation; SUBMARINE CANYON to discuss the processes of sediment transport and erosion in the ocean; and the anatomy of a HURRICANE to discuss the interaction between ocean and atmosphere. In all cases the authors have attempted to use an uncomplicated approach to the discussion and have avoided the use of formulas wherever possible.

Ocean engineering and oceanographic instrumentation, both vital to the study and utilization of the ocean, have not been addressed by way of numerous specialized articles. Rather, both have been treated extensively under a single, broad article. Acoustics, or underwater sound, has been similarly treated.

Perhaps the most difficult task faced by the authors was the selection of those scientists whose biographies were to be included in the encyclopedia. Space limitations dictated that only 25 could be chosen. Final selection was made on the basis of a questionnaire completed by many members of the oceanographic community. The authors take full responsibility for the final selection, but are indebted to those who responded with their recommendations.

In completing a task that has spanned three years the authors are, first and foremost, indebted to that long line of explorers, adventurers, scientists, engineers, technicians, and sailors whose inquisitive minds, exhaustive labors and lonely voyages produced the knowledge summarized in this volume.

Special acknowledgement and appreciation is also due the many institutions (credited in the text) that were so helpful in providing photographs. And finally, we are deeply indebted to those friends, colleagues and family members who provided support, encouragement and inspiration throughout the effort.

Donald G. Groves
Lee M. Hunt

SI UNITS

SI UNITS are values used in the International System for measurements (Système International). SI represents a streamlined version of the metric system in which units of length, time, and mass are the meter, second, and kilogram, respectively, or decimal multiples and submultiples thereof. The United States is a signatory to the General Conference on Weights and Measures which gave official status to the metric SI units in 1960.

Metric System The metric system has been used by oceanographers and other scientists for many years in varying degrees. Now it is used almost exclusively worldwide in all scientific work. However, the United States is the last industrialized nation of the world to adopt the system in any widespread way. United States engineers, including ocean engineers, and the American public have traditionally used a system of measurement that consists of the inch, foot, pound, pounds per square inch, and so forth.

This customary system of measurement traces its ancestry back to Anglo-Saxon days and is probably based on the ancient Egyptian practice of measuring things by using parts of the human body as gauges. For instance, in Egypt the standard of length for the construction of pyramids, the cubit, was the distance from the elbow to the end of the middle finger [about 18 in (46 cm)]. The hands and arms were used to measure short lengths, while a man's stride served as a measure for longer distances. The Anglo-Saxons adopted this scheme based on physical standards. They used the cubit as well as the inch (the distance from the knuckle to the end of the thumb), the foot (the length of four palms or sixteen fingers), and the yard (the distance from King Edgar's nose to the end of the middle finger of his extended arm). A fathom was the length of a Viking's embrace.

When, some 1000 years ago, the Anglo-Saxons realized that not everyone's thumb, nose, or embrace was quite the same, more meaningful standards began to evolve. These were eventually sophisticated by basing various measurement units on standardized bars, metal weights, and all the types of measures that have existed, especially in the American-used system.

In the late eighteenth century, the French initiated the metric system and based its measurement units on various natural constants. For instance, the meter was defined first as that length equal to one ten-millionths of the quadrant of the earth measured on the meridian passing through Dunkirk and Barcelona. The liter was the volume occupied by a unit mass of one cubic decimeter of water at the TEMPERATURE of its maximum DENSITY. While there are some built-in inaccuracies in these particular definitions, the metric system, based on natural standards, is excellent in almost every respect and perhaps much sounder than the customary one that has been used in the United States.

While there is considerable resistance to metric use in the United States, its usage is rather substantial. Since international activity in oceanographic research and engineering promises to increase at a rapid rate as do other international efforts, it is logical and practical to utilize some compatible internationally agreed-upon standard of measurements in these efforts. The SI units are a step in this direction. Some conversion factors for measurements are listed in the following tables.

When You Know	You Can Find	If You Multiply By
Length		
inches	millimeters	25
feet	centimeters	30
yards	meters	0.9
miles (statute)	kilometers	1.6
miles (nautical)	kilometers	1.85
millimeters	inches	0.04
centimeters	inches	0.4
meters	yards	1.1
kilometers	miles (statute)	0.6
Area		
square inches	square centimeters	6.5
square feet	square meters	0.09
square yards	square meters	0.8
square miles	square kilometers	2.6
acres	square hectometers (hectares)	0.4
square centimeters	square inches	0.16
square meters	square yards	1.2
square kilometers	square miles	0.4
square hectometers (hectares)	acres	2.5
Mass		
ounces	grams	28
pounds	kilograms	0.45
short tons	megagrams (metric tons)	0.9
grams	ounces	0.035
kilograms	pounds	2.2
megagrams (metric tons)	short tons	1.1
Liquid Volume		
ounces	milliliters	30
pints	liters	0.47
quarts	liters	0.95
gallons	liters	3.8
milliliters	ounces	0.034
liters	pints	2.1
liters	quarts	1.06
liters	gallons	0.26
Temperature		
degrees Fahrenheit	degrees Celsius	$\frac{5}{9}$ (after subtracting 32)
degrees Celsius	degrees Fahrenheit	$\frac{9}{5}$ (then add 32)
degrees Celsius	kelvins (SI temperature unit, K)	by adding 273
Pressure Stress		
pounds per square inch	newtons per square meter	6895 (= 1 pascal)
pounds per square foot	newtons per square meter	47.88 pascal

Equivalents

The following is a partial list of equivalents:

10 millimeters	=	1 centimeter
10 centimeters	=	1 decimeter
10 decimeters	=	1 meter
10 meters	=	1 dekameter
10 dekameters	=	1 hectometer
10 hectometers	=	1 kilometer

1 centimeter	=	0.3937 inch
1 inch	=	2.54 centimeters
1 decimeter	=	3.937 inches
1 foot	=	3.048 decimeters
1 meter	=	39.37 inches
1 yard	=	0.9144 meter
1 dekameter	=	1.9884 rods
1 rod	=	0.5029 dekameter
1 kilometer	=	0.62137 mile (statute)
1 mile (statute)	=	1.6093 kilometers
1 fathom	=	1.828 meters (6 feet)

1 square centimeter	=	0.1550 square inch
1 square inch	=	6.452 square centimeters
1 square decimeter	=	0.1076 square foot
1 square foot	=	9.2903 square decimeters
1 square meter	=	1.196 square yards
1 square yard	=	0.8361 square meter
1 acre	=	3.954 square rods
1 square rod	=	0.2529 acre
1 hectare	=	2.47 acres
1 acre	=	0.4047 hectare
1 square kilometer	=	0.386 square mile
1 square mile	=	2.59 square kilometers

1 cubic centimeter	=	0.061 cubic inch
1 cubic inch	=	16.39 cubic centimeters
1 cubic decimeter	=	0.0353 cubic foot
1 cubic foot	=	28.347 cubic decimeters
1 cubic meter	=	1.308 cubic yards
1 cubic yard	=	0.7646 cubic meter

1 gram	=	0.03527 ounce
1 ounce	=	28.35 grams
1 kilogram	=	2.2046 pounds
1 pound	=	0.4536 kilogram
1 metric ton	=	1.1023 USCS tons
1 USCS ton	=	0.9027 metric ton

1 liter	=	0.908 quart dry
1 liter	≐	1.0567 quarts liquid
1 quart dry	=	1.101 liters
1 quart liquid	=	0.9436 liter
1 dekaliter	=	2.6417 gallons
1 gallon	=	0.3785 dekaliter

10 milliliters	=	1 centiliter
10 centiliters	=	1 deciliter
10 deciliters	=	1 liter
10 liters	=	1 dekaliter
10 dekaliters	=	1 hectoliter
10 hectoliters	=	1 kiloliter

Thermal Conductivity, κ

1 cal (thermochemical)/cm · s · °C = 418.40* watt/meter · kelvin

Specific Heat, C

1 cal (thermochemical)/g · °C = 4184.00* joule/kilogram · kelvin

The following table makes it easy to do conversions from the old system to the new one. The letter f as in lbf stands for "force."

Conversion of U.S. Customary Units to Equivalent Values in SI Units

1 ft	= 0.3048 m
1 in	= 25.4 mm = 2.54 cm
1 milli-inch	= 25.4 μm
1 ft^2	= 0.0929030 m^2 = 920.030 cm^2
1 in^2	= 645.16 mm^2 = 6.4516 cm^2
1 yd^3	= 0.764555 m^3
1 ft^3	= 28.3168 dm^3
1 in^3	= 16.3871 cm^3
1 Imp. gal	= 4.54609 dm^3 = 4.546 L
1 U.S. gal	= 3.78541 dm^3 = 3.785 L
1 qt	= 1.13652 dm^3 = 1.137 L
1 pt	= 0.568261 dm^3 = 0.568 L
1 ft/s^2	= 0.3048 m s^{-2}
1 ton	= 1016.05 kg = 1.01605 metric ton
1 cwt	= 50.8023 kg
1 lb	= 0.45359237 kg
1 oz	= 28.3495 g
1 lb/ft^3	= 16.0185 kg m^{-3}
1 lb/in^3	= 27.6799 g cm^{-3} = 27.6799 Mg m^{-3}
1 lb/gal	= 0.0997763 kg dm^{-3} = 0.09978 kg l^{-1}
°F	$= \frac{°C \times 9}{5} + 32$
°C	= K − 273
1 tonf (loosely 1 ton)	= 9.96402 kN
1 lbf (loosely 1 lb)	= 4.44822 N
1 $tonf/in^2$	= 15.4443 MN m^{-2}
1 lbf/in^2	= 6894.76 N m^{-2} = 68.9476 mbar
1 ft H_2O	= 2989.07 N m^{-2}
1 in H_2O	= 249.089 N m^{-2}
1 in Hg	= 3386.39 N m^{-2} = 33.8639 mbar
(1 torr	= 1 mm Hg)
1 therm	= 105.506 MJ
1 hph	= 2.68452 MJ
1 kWh	= 3.6 MJ
1 Btu	= 1.05506 kJ
1 ft·lbf (loosely 1 ft·lb)	= 1.35582 J
1 hp	= 745.700 W (J s^{-1}) = 0.745700 kW
1 ft lbf/s	= 1.35582 W
1 Btu/h	= 0.293071 W (J s^{-1})
1 Btu/ft^2 h	= 3.15459 W m^{-2} (J m^{-2} s^{-1})

Decimal multiples and submultiples of the SI units are formed by means of the prefixes given below:

Factor by Which the Unit Is Multiplied	Prefix	Symbol*
10^{12}	tera	T
10^{9}	giga	G
10^{6}	mega	M
10^{3}	kilo	k
10^{2}	hecto	h
10	deka	da
10^{-1}	deci	d
10^{-2}	centi	c
10^{-3}	milli	m
10^{-6}	micro	μ
10^{-9}	nano	n
10^{-12}	pico	p
10^{-15}	femto	f
10^{-18}	atto	a

* The symbol of a prefix is considered to be combined with the symbol to which it is directly attached, forming with it a new unit symbol which can be raised to a positive or negative power and which can be combined with other unit symbols to form symbols for compound units.

ABBREVIATIONS USED IN THIS BOOK

~	approximately
Btu	British thermal unit
°C	degrees Celsius
cm	centimeter
eH	redox potential
°F	degrees Fahrenheit
ft	feet
g	gram
h	hour
in	inches
K	kelvin
kg	kilogram
kHz	kilohertz
km	kilometer
kn	knot (kts in some places)
L	liter
lb	pound
m	meter
mg	milligram
mL	milliliter
mm	millimeter
mph	miles per hour
%	percent
pH	hydrogen ion concentration
ppm	parts per million
ppt	parts per thousand
‰	parts per thousand (salinity)
psi	pounds per square inch

ABALONE is a gastropod mollusk composing the single genus *Haliotis* of the family Haliotidae. (See GASTROPODS, MOLLUSK.) Also known as ear shell, ormer, or paua, the abalones (over a hundred species are known) are found in temperate and tropical oceans and are fished on the Pacific Coast of North America, as well as in southern Peru, China, Japan, Korea, and the Republic of South Africa. These animals tightly adhere to a solid substrate such as rocks, especially in waters from the low-tide mark to depths of about 120 ft (36.6 m). The attachment to a substrate is made by a large and muscular foot or disk, which is edible and highly esteemed as a food by many peoples. Abalones are vegetarians and feed primarily on ALGAE, which they take from the substrate by means of the radula, a ribbon-shaped organ, studded with chitinous teeth, that is found in the mouth of these mollusks.

The shell of the abalone resembles a valve of a large clam except for the spiral whorl (see CLAMS). Typically, this flattened shell is perforated by a series of small pores or natural holes. SEAWATER is drawn in under the edges of the shell, passes over the gills, and exits from the mantle cavity through the pores. The nacreous lining of the shell, composed mostly of calcium carbonate and held together by a tenuous network of organic conchiolin, is used commercially as mother-of-pearl.

Reproduction takes place by separate abalone sexes, and fertilization is external. The single gonad empties into the kidney, thus permitting transport of the gametes to the outside. The larvae are pelagic and swim among the PLANKTON in coastal waters for about 2 days. They then settle and develop to the adult stage.

ABYSSAL PLAIN is an unusually flat area of the deep ocean basin in which the slope of the bottom is no more than 1:1000, and may be as little as 1:10 000. This smoothness is due to an evenly distributed sediment cover which has masked the original irregularities of the bottom. These sediments appear to be derived from the CONTINENTAL SHELF and transported downslope by turbidity currents. A TURBIDITY CURRENT tends to seek the lowest elevation on the bottom across which it flows to deposit the greater part of its sediment load. Therefore, depressions on the ocean floor tend to be erased early.

The existence of abyssal plains was not determined until the Mid-Atlantic Ridge Expedition of 1947 in which extensive use was made of the newly developed continuous depth recorder. (See UNDERWATER SOUND.) In 1948, the Swedish Deep-Sea Expedition, using the same technique, discovered abyssal plains in the INDIAN OCEAN. Since then such features have been discovered and mapped in all oceans.

Since abyssal plains owe their existence to an ample supply of eroding sediment and a slope sufficient to spawn and nourish turbidity currents, they are most commonly found just seaward of the CONTINENTAL RISE that borders the continents. Further seaward the smooth abyssal plain terminates in hilly terrain composed of abyssal hills. Such hills lie outside the reach of the turbidity currents with their masking sediment loads and represent the true nature of the original seafloor. These hills range from small mounds to hills of a few hundred feet in height and a few miles in width. Little or no sediment overlies the surface of these hills, whereas the sediment cover overlying abyssal plains averages about 0.6 mi (1 km).

Where continental margins terminate in deep-sea TRENCHES, as off the west coast of Central and South America, trench abyssal plains often occupy the floor. In other words, the trench serves to trap the turbidity current flowing down the slope and its floor is smoothed by the resulting sediment deposition. Where trenches exist, no abyssal plain will be found beyond their seaward rims. In some cases an abyssal plain will be interrupted by a rough area called an abyssal gap. This is thought to be the long-used path of turbidity currents which lies short of the point at which sediment deposition begins to take place. An abyssal plain which lies off an oceanic island or group of such islands is called an *archipelagic plain.*

ACOUSTICS See INSTRUMENTATION; SEA NOISE; SONAR; UNDERWATER SOUND.

ADSORPTION is the property of a substance to retain or concentrate at its surface one or more components from another substance in contact with the surface. Adsorption is a fundamental physiochemical property of solids and liquids. The greatest importance of adsorption is in colloidal systems (see COLLOID). In such systems the particle sizes are small, but the very large surface area of all PARTICULATE MATTER results in many binding sites for adsorption. Many ocean-life processes depend upon adsorption.

ADVECTION FOG See FOG.

AGASSIZ, LOUIS (1807–1873), a Swiss naturalist, made many fundamental contributions to geology and marine biology.

Like his fellow countryman and friend Arnold Guyot (see GUYOT, ARNOLD), Agassiz spent his boyhood in an impressive geographical area that doubtless first stimulated his interest in a study of the mysteries of nature.

At the age of 17, Agassiz entered the Medical School of Zurich. Two years later he enrolled at the University of Heidelberg to study physiology, anatomy, zoology, and botany. He finished his academic training with a degree of Doctor of Philosophy at the University of Erlangen in 1829 and a Doctor of Medicine at Munich in 1830.

Concurrent to his university studies. Agassiz was chosen by von Martius; the Brazilian explorer, to describe the fishes collected during the latter's expedition. Agassiz' outstanding work in this regard ranked him among the best naturalists of the time. Following this accomplishment, and while continuing his preparations in 1830 for the publication of a natural history of the freshwater fishes of Europe and a dissertation on fossil fishes of Europe, Agassiz visited Paris and Vienna to study the museum collections there.

In 1832, Agassiz accepted the position of professor of natural history in the College of Neuchâtel where he served for 14 years. In this capacity, he published his research works in a most accurate and scholarly manner. For example, his *Recherches sur les Poissons Fossiles,* plus other books related to echinoderms and studies of fossil mollusks, (see ECHINODERM; MOLLUSK) greatly enhanced his reputation as an outstanding scientist of the time.

Agassiz, in this period, conducted several investigations relative to glaciers and made important contributions by his lectures and publications (e.g., *Système Glaciaire*) on the subject. However, important as these particular research areas were, his friend Humboldt (see HUMBOLDT, ALEXANDER VON) thought Agassiz should not be diverted from other natural history investigations. Accordingly, Humboldt convinced the King of Prussia to send Agassiz on a scientific mission for the comparison of the fauna of temperate Europe and the United States. At the same time Agassiz received an invitation to lecture before the Lowell Institute in Boston, Massachusetts. He accepted the invitation and made an extraordinary impression both in scientific circles and on the United States public in general.

Because of this, Alexander Bache, Superintendent of the U.S. Coast Survey, funded Agassiz in his investigations of marine life on the U.S. Atlantic coast and among the Florida reefs. Also, an expedition to Brazil and the Amazon was arranged, as well as the means provided for Agassiz' publication *Contributions to the Natural History of the United States,* for the establishment of a biological laboratory, and for the organization of a scientific school and museum of comparative zoology at Harvard University which largely embodied and displayed Agassiz' ideas. Such ideas were reinforced by Louis Agassiz' son, Alexander Agassiz (1835–1910), who made many outstanding contributions to physical and biological oceanography.

AGE OF OCEAN WATER is the time that has elapsed since a given WATER mass was last at the surface of the sea. In many areas of the world's oceans the surface water increases in density—due to decreasing temperature or increasing salinity, or both—and sinks to great depths. Water movement at these depths is very slow, and conditions tend to be rather stable. Therefore, it becomes of interest to know how long a mass of water has been away from the surface. Such knowledge is useful in determining the rate at which overturn occurs and deep nutrient-rich water is brought to the surface. It is also useful in reaching decisions regarding the disposal of toxic chemicals and radioactive waste in the deep sea.

The most effective technique thus far for determining the age of seawater is to measure the depletion that has taken place through radioactive decay in carbon 14. The technique assumes that the only source of carbon 14 available to the ocean is the atmosphere. It further assumes that, once away from the surface, the water receives no further carbon 14. The technique indicates the following ages for different water masses, with a reported accuracy of ± 100 years: North Atlantic Central water, 600 years; North Atlantic Bottom water, 900 years; North Atlantic Deep water, 700 years; Antarctic Intermediate and Bottom water in the South Atlantic, less than 350 years. Measurements of South Pacific Deep water have given ages ranging from 650 to 900 years.

AGE OF THE OCEAN is, for the present, unknown. As yet no evidence has been uncovered in the geologic record which would indicate that the ocean did not exist prior to the event responsible for creating that evidence. The best that we can do is bracket the probable age with two statements: If we assume that WATER is essential to the creation of life on this planet, then the ocean is at least as old as the earliest fossils. Obviously, the ocean cannot be older than the Earth itself.

Based largely on the study and dating of meteorites, the Earth is now taken to be about 4.5 billion (4.5×10^9) years old. The oldest fossils are those of algae of the blue-green variety, fungus colonies, and worm burrows found in some pre-Cambrian rocks. The pre-Cambrian era ended about 600 million (600×10^6) years ago. Therefore, we may consider the ocean to be between 600 million and 4.5 billion years old, but we must realize that these numbers are likely to increase as new evidence is uncovered and more sophisticated analytical tools are brought to bear.

Attempts have been made, of course, to measure the age of the ocean. In 1715, Edmund Halley, the English astronomer after whom Halley's comet is named, suggested that the age of the ocean might be determined by dividing the total salt content of the world's ocean by the annual amount of salts added to the ocean by all the rivers emptying into it. But reliable data upon which to estimate the salt content of the ocean did not exist then, and would not until the *Challenger* expedition of 1872–1876—the first attempt to systematically examine the world's oceans from the chemical, physical, and biological points of view. Subsequently a number of calculations, beginning with those of John Joly in 1899, were made as a result of Halley's suggestion. These calculations yielded an "age" for the ocean which ranged from about 80 million to around 150 million years. (See CHALLENGER EXPEDITION.)

Clearly, as indicated by the fossil evidence noted earlier, the salt calculations yielded an age far too small. In part, this was due to factors now considered obvious. For instance, account was not taken of the large volumes of salts extracted by evaporation from arms of the sea cut off from normal circulation for long periods of time. And, it has only recently been known that tiny salt crystals are continually being transferred from the ocean to the atmosphere by the evaporation of water droplets ejected by spray and bursting bubbles. These salt crystals, swept upward by wind currents, serve the important function of acting as a nucleus about which raindrops form. By this mechanism, known as the "salt cycle," enough salts are transferred to the land by rain to quantitatively account for the salt found in rivers. And finally, the calculations were biased in the opposite direction by failing to account for the salts added to the ocean by volcanic activity on the seafloor.

AGUAJE See EL NIÑO

AGULHAS CURRENT, one of the swiftest of ocean currents, flows southward along the east coast of Africa. As the South Equatorial Current (see EQUATORIAL CURRENT SYSTEM), flowing from east to west in the vicinity of the equator, approaches the African coast, it is deflected to the left by the CORIOLIS EFFECT and flows south. Part of this current flows between Madagascar and the coast to form the Mozambique Current, while the remainder passes to the east of Madagascar to feed the Agulhas Current. South of 30° south LATITUDE the Agulhas becomes a narrow, well-defined current that extends less than 62 mi (100 km) from the coast. Reaching the tip of Africa, part of the current apparently enters the ATLANTIC OCEAN to join the northward-flowing BENGUELA CURRENT. However, most of the WATER turns sharply south and then eastward to join the WEST WIND DRIFT CURRENT as it flows west to east across the southern INDIAN OCEAN.

Being fed by water warmed by a slow drift across the Indian Ocean in the vicinity of the equator, the Agulhas water is warm, although the temperature drops gradually as it flows southward toward Antarctica. A narrow tongue of water—called Warm Agulhas Water—extends to a depth of about 492 ft (150 m) and has a temperature of about 68° F (20° C). Below and to either side of the tongue of Warm Agulhas Water, and extending down to a depth of about 1300 ft (400 m), is another identifiable envelope of water with a temperature of around 62.6° F (17° C). The temperature of the water to either side and below this second envelope drops rather sharply. The boundary between the warm and cool Agulhas water is marked by a salinity of around 35.6 ppt.

The velocity at which the Agulhas Current flows varies with location, depth, and season. A velocity range of 7.8–24 in/s (20–60 cm/s) is representative of the main southerly current.

ALBACORE See TUNA.

ALBATROSS See MARINE BIRDS.

ALEUTIAN CURRENT is an eastward-flowing current in the north PACIFIC OCEAN. Also called the Subarctic Current, it flows between the North Pacific Current to the south and the Aleutian Island chain to the north. An early branch of the current turns northward to flow into the BERING SEA. Further along, as the current nears the North Pacific coast of North America, one branch turns north to flow into the Gulf of Alaska while another turns south to become the CALIFORNIA CURRENT which flows south along the coast of California.

The Aleutian Current originates as a mixture of water from the KUROSHIO CURRENT and the OYASHIO CURRENT. Where it is best developed, the current flows at the rate of 530 million (530×10^6) ft³/s (15 million m³/s) above a depth of 2000 ft (610 m).

ALGAE is a general term for autotrophic organisms which includes both prokaryotes and eukaryotes; that is, which includes those organisms with a primitive type of nucleus lacking a clearly defined membrane (prokaryotes) and those with a well-defined nuclear membrane, chromosomes, and mitotic cell division (eukaryotes); approximately 8000 of the known 18 000 species are marine forms. Algae are the predominant "plant" forms in the world's oceans and the producer organisms in the marine ecosystem.

Some algae belong to the subkingdom Thallobionta (Thallophyta) of the kingdom Metaphyta; others are members of the kingdom Protista (see TAXONOMY). These organisms vary in size from microscopic acellular species to massive seaweeds, (see KELP), which may attain a length of 200–300 ft (60–91 m). They dominate in the oceans not only in their number of species (about 8000), but also in their number of individual organisms.

Like land plants, they possess chlorophyll by which they utilize sunlight in the process of PHOTOSYNTHESIS to manufacture their own food. Organisms with this ability are called autotrophic. Unlike most terrestrial plants, algae never form true roots, stems, or leaves. Because of their lack of structural complexity, algae are characterized as primitive or "low" forms of plant life, or thallus plants.

However, some thallus plants display more complex developmental patterns than flowering plants. For example, in the algae, the plant body varies greatly in both size and shape, and the methods of reproduction are quite diverse. Three general types of reproduction are used: vegetative, which includes cell division and fragmentation of the thallus; asexual, by motile (zoospores) or nonmotile spores; and sexual, by gametes which may be isogamous (undifferentiated in respect to maleness and femaleness), anisogamous (with a degree of differentiation), heterogamous (differentiated as egg and sperm), or oogamous (differentiated, with small male gametes and large female gametes).

Vegetative reproduction is accomplished primarily by fragmentation of the plant. This occurs when the mature nonfilamentous parts are split up into two or more segments, or when filaments (or fragments) of the plant are broken apart. In the larger brown and red algae, relatively small portions of the plant bodies often become detached to form entire new plants. The extent to which such dispersion and propagation occurs is largely unknown. However, it is known that the plant resulting from a fragment contains the same chromosomal composition as the fragment. Thus, each new plant is not a different generation but an identical twin.

Reproduction may also take place in some species by simple cell division. By this division, two daughter cells are formed, each becoming a new individual. These cells also may separate immediately after formation, or they may stay together for a time so as to resemble an algal colony.

Asexual reproduction by zoospores is common. These zoospores, or animallike unicellular reproductive units, possess a cell membrane rather than a true cell wall. They are free-swimming and move by means of whiplike threads, or flagella, which sometimes grow laterally on the zoospores or (more frequently) on the ends of them. Since these zoospores yield spores, plants of this generation are called sporophytes. Sporophytes are diploid (that is, their genetic complement of chromosomes is duplicated), with pairs of each kind of chromosome in every cell. The spores, however, are monoploid (that is, they contain only one of each kind of chromosome), and the settled spores give rise to male and female plants (gametophytes). The sexual gametophyte plants, as in the higher plants, produce reproductive cells, or gametes. (A gamete is a cell that grows into a new individual only after fusion with another gamete.) The gametes are male (sperm, antherozoids, or spermatia) and female (eggs, or ova). Union of the sperm and egg produces a diploid single-cell zygote which then develops into the sporophytic generation (green, brown, and some red algae) or the carposporophytic generation (most red algae).

Most "sexual" algae deposit all or a portion of the reproductive cells (spores and gametes) into the water. The cells are mesoplanktonic (floating) for varying periods, and chance usually determines whether spores settle on a suitable substrate. However, chemical attractants may assist sperm of some species in locating mature ova. The spermatia of red algae lack flagellar swimming mechanisms so that fertilization apparently depends on random contacts.

Algae do not have, nor do they require, true roots, stems, or leaves. The stem of a typical plant is designed to transport water and food and provide structural support, but because most algae have the ability to absorb food and materials for sustenance without such stems, there is no need for these types of root systems or stem appendages. However, some algae, notably brown and red algae, have what is termed a *holdfast,* a structure that holds the plant in place. Such a structure is, by definition, not a root since it does not absorb water or nutrients from the soil. On the other hand, some algae have blades resembling leaves, which are extensions of the plant body. These blades act to increase the surface area of the plant body itself and make absorption and photosynthesis more efficient.

In all, there are some 18 000 varieties of algae which exist on land, in fresh water, and in SEAWATER. Many of these are microscopic one-celled organisms; others are land forms as large as bushes and still others, the massive sea KELPS (brown seaweed), are really huge. They grow in the tropics and in the Arctic regions—on ice-locked mountains, in hot geysers, and as floating plants in oceans, lakes, ponds, rivers, and creeks.

Most authorities agree that the algae comprise seven different divisions. These phyla, which vary according to body structure, reproductive organs, the types of pigments produced, and the kinds of stored food in each, are as follows:

- *Green Algae* (*Chlorophyta*). Some 10 percent are marine, and many are small, with the largest being the *Ulva* or sea lettuce.
- *Blue-Green Algae* (*Schizophyta—prokaryotes*). Common both in salt and fresh water; reproduce asexually.
- *Yellow-Green Algae* (*Chrysophyta*). Mostly found in fresh water, although marine DIATOMS (class Bacillariophyceae) are included in this group (Chrysophyta) as well as the golden-brown algae (class Crysophyceae) which are mostly found in fresh water but also occur as marine forms.
- DINOFLAGELLATES (*Pyrrhophyta*). More common in marine forms than freshwater forms.
- *Euglenoids* (*Euglenophyta*). Mostly freshwater forms. Most species are autotrophic; others such as *Astasia* have no chlorophyll.
- *Brown Algae* (*Phaeophyta*). Primarily marine.
- *Red Algae* (*Rhodophyta*). Primarily marine.

Some green or blue-green algae may become commensal with, or parasitized by, different species of fungi (see FUNGUS) to produce different species of LICHENS. Many marine algae, especially Rhodophyta, are epiphytic or parasitic on other larger algal forms which sometimes are close relatives. A few species occur as endozoophytes within the cells of small animals (PROTOZOA, SPONGES, and Hydra) or in the digestive tract of mammals; some are epizoic (externally attached) on skin, hair, or scales.

Some red algae, such as the coralline algae of the family Corallinaceae, have the ability to secrete lime within and between the cell walls so that the fossils commonly show the cellular structure of the tissue. These algae make important contributions to the building of limestones and coral reefs. (See CORAL REEF.) In a number of the atolls of the PACIFIC OCEAN, coralline algae have contributed significantly to CORAL in the formation of the reefs. (See ATOLL.)

Marine algae show distinct zonations along coasts and in the marginal waters of continental shelves. (See CONTINENTAL SHELF.) Littoral algae (*Fucus, Ulva,* and *Chondrus*) occupy the intertidal zone. In the supralittoral zone are forms especially adapted for existing during intermittent moist and long dry periods (the splash zone), whereas the infralittoral is characterized by those genera which must live submersed, occuring below the lowest low-tide (see TIDES) level and on out into deeper waters.

In this regard, the particular algae that are found in the oceans are usually further divided into two main ecological classifications or groups: (1) the small drifting phytoplankton; and (2) the larger holdfast, or attached, plants called macrophytes. However, some of the latter marine plants grow while drifting, and attachment to a solid base or substrate is not necessarily a requirement among the macrophytes.

The tiny phytoplanktonic life forms of the ocean world are represented chiefly by the diatoms (the Chrysophyta division of algae) and dinoflagellates (the Pyrrhophyta division). Both have high surface-to-volume ratios, which make them quite efficient in feeding upon the dissolved nutrients in ocean waters. They live in the sunlight (photic) and epipelagic HABITABLE ZONES of the oceans and are incalculably abundant, prodigally self-renewing, and beautifully structured. When examined under a microscope, diatoms and dinoflagellates exhibit fabulous designs resembling strings of jewelry of great beauty. In addition to such beauty, together with their approximate total of some 6000 varied and diverse species (5000 diatoms and 1000 dinoflagellates), these algae provide the basis of aquatic food chains. In that sense, they are really the giants and wealth of the ocean (planktonic) flora. It has been estimated that in 1 ft^3 (0.028 m^3, or 28.3 L) of coastal seawater off the British Isles, there are on an average more than 20 000 of these minute plants, plus millions of even smaller plant forms. In a corresponding volume of water, there are probably some 120 minute animal forms, primarily COPEPODS or tiny (1–10 mm long) CRUSTACEANS, medusae, and larvae which later develop into decapods, echinoderms, cirripedes, and GASTROPODS (see LARVA; ECHINODERM). These copepods are the chief link in the food chain between the phytoplankton and higher forms of animal life of the ocean.

Some other kinds of more complex attached marine algae are commonly called seaweeds. This group of plants, contrary to the connotation of their namesake—weeds—yields a number of products having many important uses. These larger (macrophytic) plants are members of three of the groups of algae:

1. Brown (Phaeophyta)
2. Red (Rhodophyta)
3. Green (Chlorophyta)

Although most species of seaweeds grow in the ocean's intertidal zone where they are partially or totally exposed at low tide, a considerable number also grow below the surface [approximately 50–100 ft (15 to 30 m)] in abundant beds. It is in deep water that the large weeds are found. Agarum, a species of KELP, commonly known as the sea colander, is an abundant seaweed on the ATLANTIC OCEAN coast of Nova Scotia that grows in these depths.

In southern California many seaweed species are found at deeper levels along the offshore islands, as compared with their maximum depths on the mainland shelves. The difference probably results from clearer waters offshore, permitting higher light intensities at a given depth. In some coastal areas, the large kelps, including *Macrocystis*, become sparse at 50–60 ft (15–18 m). Their range may extend to 90–100 ft (28–31 m) in clearest coastal situations. At San Clemente Island, the lower limits can be as great as 200 ft (61 m) for some brown algae and as deep as 130 ft (40 m) for *Macrocystis*.

Another species of seaweed is the sargassum weed which floats in the surface layers of the SARGASSO SEA. It is a gold and olive-colored alga that was once land-based but became pelagic in the oceans millions of years ago when parts drifted out into the ocean surface from coastal areas. The plant reproduces asexually by breaking off fragments that then thrive separately. The sargasso weed comprises two species of brown algae, *Sargassum natans* and *S. fluitans*. These plants normally live in the shallow tropical waters of the Atlantic coasts of North and South America. When they are detached by the action of WAVES, they float because of their gas-filled pea- and grape-sized bladders and follow the current to the Sargasso Sea in the North Atlantic Ocean. (See CURRENTS.) Here, the weed is trapped and joins other patches of accumulated sargassum that make up a huge egg-shaped area two-thirds as large as the contiguous United States.

A wide variety of animal life lives on and among this ponderous, widespread mass of floating weeds that itself is a mystery to science and has long been a subject of legend and folklore. (See MYTHS AND LEGENDS.) Most species of fauna are similar to species found in coastal waters, and it is said that when Columbus saw a sargassum CRAB (*Planes minutus*) there, many hundreds of miles from North America, he believed himself to be near land.

One of the most interesting forms of Sargasso life is the sargassum fish (e.g., *Histrio histrio*) that can devour prey almost as large as itself and whose mottled coloration blends perfectly with the new and old plant life. The small toad fish (*Antennarius marmoratus*) with its gold and brown marbled body melts into its surroundings and is hardly distinguishable from the seaweeds.

Perhaps the most fantastic of all the inhabitants are the American and European eels, temporary visitors to the area. (See EEL.) These eels come annually from the rivers of Europe and the United States to spawn. The larvae they produce develop into young eels called elvers, and they make their way back somehow to the very same rivers that their parents came from. (See LARVA.)

Although thousands of different kinds of seaweeds have been identified, only about a dozen are used commercially. However, their many potential applications are yet to be found. These highly useful plants are capable of yielding a number of products having many important uses, since the major constituents in seaweeds are the carbohydrates, while minor constituents include protein, fat, minerals, and vitamins. Interestingly, some seaweeds have long been used as fodder for domesticated animals (especially around the NORTH SEA areas) and as food for humans (particularly in the Orient). These plants were also considered to be of considerable medicinal value as far back in history as 3000 B.C., and the Japanese and Chinese peoples used them in the treatment of goiter and other glandular diseases. The early Romans also used them for the healing of burns, rashes, and wounds; the British employed *Porphyra* (an edible red alga) to prevent scurvy during long sea voyages; and other peoples of the past used various of these plants as laxatives, as a means to reduce fever, as anticoagulants, as aids in childbirth by dilating the uterus during labor, for the treatment of various stomach and intestinal disorders, and for other applications.

In Japan, the green algae *Monostroma* (Heteogusa), *Enteromorpha* (Aonori), and *Ulva* (Aosa) are currently used as foods. *Monostroma,* the most popular of these plants, is grown in the intertidal zone of shallow seas and estuaries. The U.S. Bureau of Commercial Fisheries reports that although seaweeds are not popular as food in the United States, the Japanese eat certain seaweeds as regularly as Americans eat tomatoes or lettuce.

In general, the brown algae, that grow best in the colder waters of the ocean, and the red algae, that flourish in warmer and usually deeper waters than the brown, represent the most important ocean plant classes as far as economic usage is concerned. For instance, agar, algin, and carrageenin, three colloids of invaluable assistance to human beings, can be extracted from them.

Agar, used widely in biomedical laboratories as an all-purpose culture medium as well as, like algin and carrageenin, an emulsifier, stabilizer, gelling agent, or thickener in food processing, can be extracted from at least 28 species of red algae. Several red algae are also used as source materials to extract carrageenin, which is also a polysaccharide. One such source is *Chondrus crispus* or "Irish moss." Incidentally, carrageenin derives its name from the town Carragheen in Ireland which is a major source of Irish moss. The red algae dulse (*Rhodymenia palmata*) and Irish moss are both used as a food in a variety of ways, either raw or cooked, in parts of western Europe, Canada, and New England (U.S.).

Algin (common name for alginic acid and its derivatives) is extracted from the large brown algae such as the giant kelps, *Macrocystis* and *Laminaria.* Algin can absorb large quantities of water, and the addition of algin to ice cream prevents the water in the ice cream from forming coarse ice crystals while being frozen. Algin also holds moisture and prevents icing on cupcakes from sticking to wrappers.

The production and sale of these three colloids, agar, algin, and carrageenin, derived from sea plants represent a multimillion dollar business in the United States and other parts of the world. In the United States alone, they have a wide and varied number of applications in the following products: clarifiers for wine and beer, die lubricants for tungsten wire, oil-well drilling muds, battery-plate separators, plastics, water-based paints, bacterial inoculum, pill coatings, surgical dressings, dental impression compounds, low-calorie and other diet foods, toothpaste, hair dressing, hair sprays, cold creams, and shaving creams—plus a number of foods: canned meat, sauces, syrups, fruit juices, pie fillings, ice cream, chocolate milk, fruit cakes, and many others.

These uses of ocean plant life may well be augmented in the future, for it has been recognized that kelp beds, when protected and cultivated, serve not only as the source for valuable products such as algin and seaweed meal but also as an essential link in the marine food chain; they also function as a habitat for a number of marine organisms, including many valuable ocean fish and SHELLFISH. In addition, several investigators today envision large oceanic algae farms associated with fish pens, livestock pens, a power plant, and a chemical-conversion facility. In this arrangement, photosynthesis may be utilized more efficiently than it is utilized for dry land agriculture.

See also MARICULTURE; PLANT LIFE IN THE OCEANS; SEAGRASS.

AMA is the name applied to the "diving women" of Japan who for at least 2000 years have dived in the coastal WATER of the PACIFIC OCEAN and the Sea of Japan to gather shellfish and ALGAE from the seafloor. In the ancient Japanese language the word *ama* meant the ocean or sky. Later it was applied to both men and women who used any of several techniques to gather animals and plants from the sea. The Chinese characters which distinguish male from female ama translate to "sea warrior" and "sea woman," respectively. The Korean equivalents of the ama are called hae-nyo. Synonyms for the ama are "suijin," "kazuki," "osaiso," "umibito," "mogurime," and "umiko."

Over the centuries the technique of diving to the seafloor to gather SHELLFISH and algae fell increasingly to the women because, at least in part, of their greater capacity to withstand long hours of submersion in cold water. Traditionally the ama who dived in the Pacific waters along the Japanese coast wore only a cotton shirt to cover their bodies. In recent years the wet suit has become the preferred garment. On the other hand, the ama who dived in the Sea of Japan wore nothing but a cord tied around the waist from which to hang ballast, in the form of lead beads, and tools, such as a rope basket to hold the catch and a metal pallet to help pry shells from the rock. The northern limit of the ama is the line representing the average air temperature of 53.6° F (12° C). Interestingly this is also the northern limit of cultivated tangerines and tea plants. South of this line the water temperature from May to October remains above 68° F (20° C). In these temperatures the ama commonly make 100 dives a day to depths of around 60–75 ft (18–23 m). The dives and the harvesting require that they hold their breath for 50–60 s. The hae-nyo of Korea dive year-round but make only about 50 dives a day during the winter months. A fire pit in the support boat or on the beach is used to warm the divers when they become uncomfortably chilled.

The equipment and techniques used by the ama are simple and practical. To protect their eyes and to provide better underwater vision, goggles are worn. However, goggles have a tendency to collapse at depth as the water pressure builds up above that inside the goggles; this could cause bleeding and irritation of the eyes. Therefore, for over a hundred years, the ama have used a bulb of rubber or soft leather attached to their goggles. The bulb collapses at depth, forcing air into the goggles and equalizing the pressure. A cord about the waist carries lead beads to speed the ama's descent to the bottom. A variation on this technique is to use two lines, one tied to a weight and the other around the ama's waist. Both lines are managed from the boat by a father, brother, or husband. In practice the ama allows the weight to pull her to the bottom. Releasing the weighted line, she collects shellfish (she uses a small metal pallet called a "kaigane" to pry the shells loose from the rock) and algae. When she feels the need to breathe, she tugs on the line about her waist and is pulled rapidly to the surface by her assistant in the boat.

The ama fall into three classes defined by both age and diving depth. The koisodo are the young girls from 15–20 years of age who are in training (and those over 60). The koisodo use a tub floating on the surface to deposit their catch and dive only in depths of 6.5–13 ft (2–4 m). The nakaisoso are the intermediate class, aged 18–25, and dive to depths of 13–23 ft (4–7 m). The ooisodo are the fully trained divers, aged 20–50, and dive to depths as great as 82 ft (25 m). Some women work in the ooisodo class until they are well past 50 years of age.

Based on the kind of harvest in which they specialize, the ama may be further classed as either kusa-ama or kai-ama. The kusa-ama collect marine plants (*kusa* means "grass" or "marine plants") from which agar is made. The more common of these plants are tengusa (*Gelidium amansii, G. pacificum, G. linoides*) hirakusa (*G. subcostatum*), onikusa (*G. japonicum*), dorakusa (*Perocladia capillacea*), toriashi (*Acanthopeltis japonicum*), and ego (*Geramium hypnaeoides*). The more important animals collected by the kai-ama (*kai* means "shellfish") are awabi (abalone; *Haliotis gigantea*), sazae (top shell), uni (sea urchin; *Echinoidea*), namako (sea cucumber; *Holothuroidea*), and igai (sea mussel; *Mytilus erassitesta*).

Since the life of the ama is hard and often extremely uncomfortable, young girls are finding the lure of the city with its office and factory jobs more to their liking. As a result, the number of ama has steadily declined over recent years. Of those who remain, many find that the tourists who come to watch them work are more lucrative than the harvest.

Paumotan Divers The ama are not actually the most spectacular working breath-hold divers of the Pacific: that distinction goes to the Paumotan pearl divers of the Tuamotu Archipelago. These stocky, muscular men typically dive to depths of 120 ft (36.5 m) and hold their breath up to 2 min and 35 s. They, too, operate from a small boat on the surface and use an assistant (tete) to manage the boat and to pull them up if they become disabled.

The Paumotan divers also use a faceplate and a weighted line to help them reach the bottom faster. A period of 5–7 min of hyperventilation precedes each dive. After hyperventilating, the diver grasps the weight with his feet and the line with his hand and sinks feet first to the bottom. During the descent he holds his nose to help with pressure equalization. The descent typically takes from 30 to 50 s. The diver, on reaching the bottom, swims off in search of shells, for a period lasting from 30 to 60 s. On the return to the surface, the diver usually requires only about 20 s to pull himself up hand-over-hand.

Pearl divers commonly suffer from an ailment known as taravana (*tara,* "to fall"; *vana,* "crazily"). The onset of taravana symptoms are quite sudden and consist of vertigo, nausea, anxiety, paralysis, and, occasionally, death. On occasion the symptoms may be preceded by a "sparkling before the eyes." Paralysis and unconsciousness may follow rapidly after the symptoms begin, and divers are often pulled unconscious from the water. The symptoms usually disappear within hours or days, but some divers are paralyzed for life. And, many of those

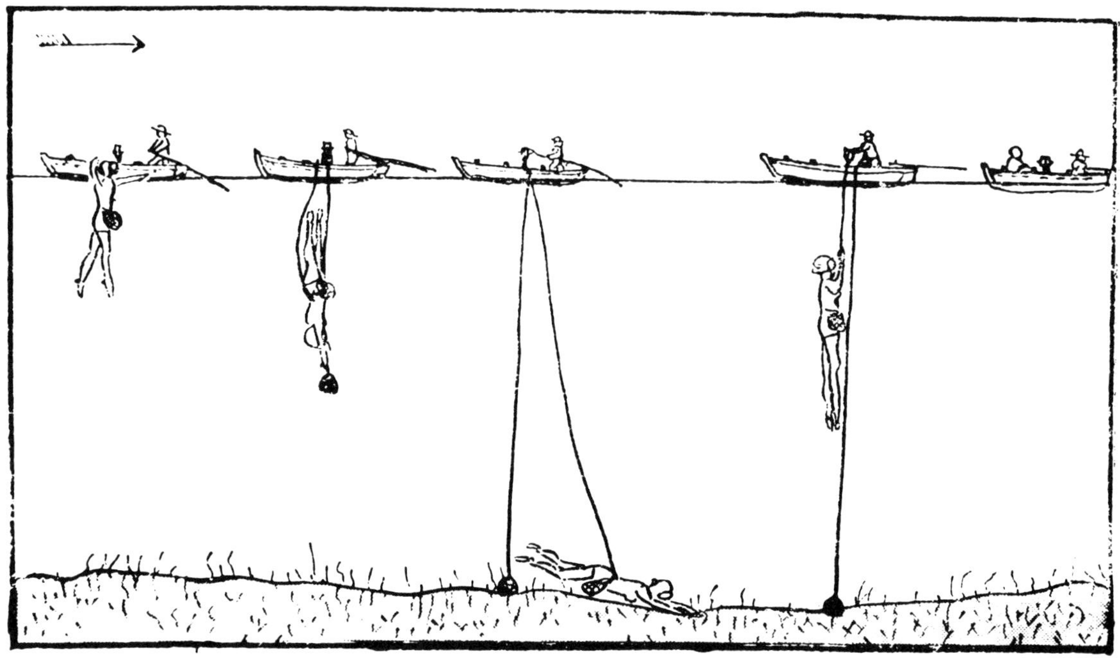

AMA. (*Top*) Ama, wearing goggles, drop to the coastal seafloor to gather shellfish and algae. In shallow water, they attach lead beads to their waists to speed the descent, and they return unaided to the surface. (*Bottom*) In deeper waters, a family member remains in the boat and maneuvers two lines: one attached to a weight and the other to the ama's waist.

who have dived for years and had taravana several times appear punch-drunk. That is, they are forgetful, clumsy, and make silly mistakes.

AMOEBA is the generic name for a number of species of rhizopod protozoans in the order Amoebida. Amoebas are microscopic, acellular animals, having a thin protective membrane (plasmalemma) and an irregular false foot (lobopodium) consisting of a temporary extrusion of cytoplasm used for movement. About 75 marine species are known. (See PROTOZOA.)

AMPULLAE OF LORENZINI is the name applied to the cutaneous receptors in the head region of the ELASMOBRANCHS (the sharks and RAYS). The ampullae of Lorenzini are highly modified sensory organs consisting of very deep canals filled with a jellylike material which apparently performs a thermosensory function. Concentrated on the heads and faces of both sharks and rays are also tiny sense organs, corresponding to the lateral-line organs in the higher BONY FISH since they continue along a line back to the tail in a fine tube under the skin, which opens at intervals to the outside through minute canals. These sense organs on the sharks and rays transmit vibrations and changes in the flow of water.

See also FISH; LATERAL LINE; SHARK.

AMUNDSEN SEA is the marginal sea of the South PACIFIC OCEAN and is located off the Walgreen Coast of Antarctica. It is bounded on the west by the BELLINGSHAUSEN SEA in the vicinity of Thurston Island, and on the east by the ROSS SEA at Cape Dart. The northern boundary lies at about the ANTARCTIC CIRCLE. Peacock Bay and Burke Island lie within its boundaries. As is characteristic of the Antarctic, the CONTINENTAL SHELF is depressed below the normal 600 ft (183 m) and breaks at a depth of about 1500 ft (457 m). This is thought to be due to the shelf sinking under the weight of ice during the last Ice Age. The seafloor is covered by poorly sorted terrigenous materials typical of a glacial origin. The outer limit of the sea lies at depths as great as 13 123 ft (4000 m). The sea is perpetually covered by ice out to about 100 mi (160 km) from the coast.

The Amundsen Sea is named for the great Norwegian explorer Roald Amundsen who was on the Norwegian whaler *Belgica* in 1897–1898 when it became the first ship to winter over in the Antarctic (Bellinghausen Sea). Amundsen was also the first to reach the South Pole in December 1911.

ANADROMOUS FISH refers to any marine fish that spends most of its feeding life in the ocean and moves into fresh waters to spawn. Among the fishes having an anadromous behavior are the SALMON, the sturgeons, the shads, the basses, and the sea and river LAMPREY.

The most important of the anadromous commercial fishes of the world's oceans is the salmon (see COMMERCIAL FISHING). Typical of the anadromous fish, the salmon lives for a period of years in the ocean before it returns to fresh water to spawn in the same river in which it was hatched.

Another unique feature of these fishes is their ability to survive the change from SEAWATER to fresh water. In the case of the salmon, the young of ocean-living parents must live in fresh water until such time as salt-secreting cells have developed in their gills. Once this is accomplished, they are equipped with a kidney structure that enables them to enter the ocean environment as well as the freshwater areas. Therefore, there is no problem when the mature fishes return to the rivers to spawn. However, there are differences between the Atlantic and Pacific salmon. When the latter return to fresh water, there is a decrease in the osmotic concentration of their blood. Thus, with this irreversible chemical change, the Pacific salmon waste away and die in fresh water after breeding.

See also FISH; OSMOSIS.

ANDAMAN SEA, sometimes called the Burma Sea, is located in the northeastern corner of the INDIAN OCEAN and lies off the coasts of Burma, Sumatra, and the Malay Peninsula. The BAY OF BENGAL lies to the west, and the STRAIT OF MALACCA, between Sumatra and Malaysia, to the southeast. The official boundary is a line running southward from Cape Negrais in Burma, through the Andaman and Nicobar Islands, to Oejong Raja in Sumatra. The sea is separated from the Strait of Malacca by a line running from Pedropunt in Sumatra to Lem Voalan on the Malay Peninsula. The Andaman Sea covers an area of 232 372 mi^2 (602 000 km^2), occupies a volume of 158 340 mi^3 (660 000 km^3), and has a mean depth of 3596 ft (1096 m). The maximum depth is 13 714 ft (4180 m). Two great rivers—the Irrawaddy and the Salween of Burma—flow into the Andaman Sea from the north. The mouths of these two rivers lie on either side of the Gulf of Martaban.

The Andaman-Nicobar Ridge forms a graceful curve between the tips of Burma and Sumatra. The ridge breaks the surface in several places to form the Andaman and Nicobar island groups, both of which are controlled by India. The islands have an area of 3202 mi^2 (8293 km^2) and a total population (1975 estimate) of 148 000. The native population of the Andaman Islands is made up of Negritos, whereas that of the Nicobar Islands is of mixed Burmese and Malaysian blood. The major islands of the Andaman group are North, Middle, South, and Little Andaman; those of the Nicobar group are Car, Little, and Great Nicobar. The Andaman Sea is one of the world's great shipping lanes because of the strategic Strait of Malacca which connects the INDIAN OCEAN with the PACIFIC OCEAN. Shipping channels across the Andaman-Nicobar Ridge are the North and South Preparis Channels near the edge of the Irrawaddy-Salween Delta which has a depth of 656 ft (200 m), the Ten Degree Channel between the Andaman and Nicobar Islands which has a depth of 2625 ft (800 m), and the Great Passage south of Great Nicobar with a depth of 5905 ft (1800 m).

The CONTINENTAL SHELF of the Andaman Sea may be considered a continuation (through the Strait of Malacca) of the great Sunda Shelf to the southeast. The shelf runs up the Malay and Burma coasts as the Mergui Platform and terminates in the sprawling Irrawaddy-Salween Delta. The shelf consists of fine sands derived from the granitic islands of the Mergui Archipelago that lies off the Burma Coast, and sands, silts, and clays deposited in the sea by the two great river systems. The Irrawaddy alone dumps approximately 400 million (4.00×10^8) tons (363 million metric tons) of sediment a year into the sea, and its delta is growing at the rate of about 3 mi (5 km) per century. The thickness of the sediment in the delta is estimated at around 9842 ft (3000 m). By way of a network of submerged channels and submarine canyons, sediment is fed from the shelf to the deep basin between the shelf and the Andaman-Nicobar Ridge. See SUBMARINE CANYON.

The deep basin of the Andaman Sea is cut by an inner arc of seamounts and active volcanoes. (See SEAMOUNT.) Barren and Narcondam Islands are part of this arc. The basin on either side rests at depths as great as 13 714 ft (4180 m) and is covered with sand, silt, and clay fed into it from the bordering shelf. The basin is filling more rapidly from the north because of material brought in by the rivers.

Since it lies within 15° of the equator, the surface WATER temperature of the Andaman Sea has a very small seasonal range—82.4–86° F (28–30° C). The water circulation is strongly influenced by the seasonal monsoons and a current that flows into the sea through the Strait of Malacca. During the spring and summer the southwest monsoon creates a current that flows into the sea from the Bay of Bengal. With the reversal of the monsoon during the fall and winter, the current also reverses and flows southwest. The enormous volume of water entering the sea through the Irrawaddy and Salween Rivers during the southwest monsoon (wet period) drops the salinity in the northern part of the sea to around 20 ppt, whereas that in the southern part remains at around 33.5 ppt.

ANGELFISH (sometimes erroneously called the butterfly fish), is the name used for any of about 150 species of bony marine and freshwater fishes of the order Perciformes and family Chaetodontidae (bristle-toothed); these beautifully colored and distinctly arrowhead-shaped fishes have enlarged dorsal and ventral winglike fins.

Marine angelfishes are found primarily in and among the coral reefs of tropical seas as well as in shallow parts of the ocean. (See CORAL REEF.)

These fish possess several interesting features. They display fantastic colors and designs over a body that can attain a 2-ft (61-cm) length. For example, the queen angelfish, *Holacanthus ciliaris,* has entirely yellow caudal (tail) and pectoral fins. The spot on the nape is ringed and spotted with bright blue. The blue angelfish, *H. bermudensis,* lacks the spotting, and only the outer edges of the caudal and pectoral fins are yellow. These particular species also hybridize so that color variations take place. Angelfishes have small trunk-shaped mouths equipped with many small teeth which they use to catch the small invertebrates on which they feed. (See INVERTEBRATE.) Some species possess an elongated snout, and these angelfishes are especially able to pick up their food from crevices of the branched CORAL of their environment. In spite of the fact that they are poor swimmers, few carnivorous species ever harm them, apparently because the meat of the angelfish is bad-tasting. In addition, the conspicuous color patterns that angelfishes possess may serve as a warning to predators that the wearer of these colors is poisonous or has a sting. However, they are neither poisonous nor do they sting.

Some angelfishes live alone while young, but later, when they find a mate, they are mated for life. They are strongly territorial and use their colors to warn off intruders.

ANNELID See SEAWORMS.

ANTARCTIC CIRCLE is the line of latitude 66°32′ S (often taken as 66°30′ S). Along this line the sun does not set on the day of the summer solstice, about December 22 (Southerm Hemisphere), and does not rise on the day of the winter solstice, about June 21. Southward from the Antarctic Circle the number of 24-h periods of continuous day or continuous night increases to about 6 months at the South Pole. The period of daylight at the Pole lasts from September 21 to March 21; and the period of darkness, from March 21 to September 21. The Antarctic Circle separates the South Frigid Zone from the South Temperate Zone. It was first crossed by James Cook on January 17, 1773.

ANTARCTIC CIRCUMPOLAR CURRENT is the largest current in all the oceans, with a flow rate estimated to be between 3.2 and 3.9 billion (3.9×10^9) ft^3/s [90 and 110 million (110×10^6) m^3/s]. Even the lower figure is 400 times the rate of the world's largest river, the Amazon. The current flows from west to east around the Antarctic continent and, therefore, through the southern extremities of the ATLANTIC OCEAN, the PACIFIC OCEAN, and the INDIAN OCEAN.

In its course around the Antarctic continent, the circumpolar current is locally deflected by projections of land, by submarine topography, and by other currents along its northern boundary. On the northern edge, the current is continuous with the SOUTH ATLANTIC CURRENT, the SOUTH PACIFIC CURRENT, and the eastward-flowing extension of the AGULHAS CURRENT in the Indian Ocean.

ANTARCTIC OCEAN See SOUTHERN OCEAN.

ANTICYCLONE See HURRICANE.

ANTILLES CURRENT is the northern branch of the North Equatorial Current (see EQUATORIAL CURRENT SYSTEM) which flows along the northern side of the Greater Antilles, carrying WATER that is identical with that of the SARGASSO SEA. The current eventually joins the northern-flowing FLORIDA CURRENT in the vicinity of the Bahamas. In the vicinity of Cape Hatteras, these currents, along with waters from the Sargasso Sea, form the GULF STREAM.

APHOTIC refers to the absence of light, as below the photic zone of the world's oceans, the zone where light intensity is sufficient for PHOTOSYNTHESIS; the aphotic zone hence is that lower region of water bodies not reached by sunlight.

See also HABITABLE ZONES; MARINE OPTICS.

AQUACULTURE refers to the production of food from managed aquatic systems. See MARICULTURE.

ARABIAN SEA is located between Africa and India in the northern INDIAN OCEAN. Its official boundaries are as follows: on the west, the eastern limit of the Gulf of Aden (between Somalia and Yemen); on the north, a line joining Ràs al Hadd, the east point of Arabia (Oman), and Ràs Jiyuni on the coast of Pakistan; on the south, a line joining Addu Atoll (Maldives) to Ràs Hafun (Somalia); and on the east, the western limit of the LACCADIVE SEA. Arms of the Arabian Sea include the Gulf of Aden and the RED SEA between Africa and Iran, and the Gulfs of Kutch and Cambay off the west coast of India. The Indus and the Narbada Rivers of India are the major sources of fresh WATER flowing into the sea. The Arabian Sea covers an area of 1 491 118 mi^2 (3 863 000 km^2), occupies a total volume of some 2 533 690 mi^3 (10 561 000 km^3), and has a mean depth of 8970 ft (2734 m). Most oceanographers consider the Laccadive Sea on the east to be a part of the Arabian Sea. The two are discussed as a single unit here.

The CONTINENTAL SHELF of the Arabian and Laccadive seas is best-developed between Sri Lanka and Pakistan, where it ranges from 75 to 219 mi (120 to 353 km) in width. Most of the shelf off Iran, Arabia, and Africa is less than 25 mi (40 km) in width. The shelf lies at a depth which ranges from 722 ft (220 m) off India to 121 ft (37 m) off Iran, and is primarily covered by mud. The shelf off the Indus River is cut by a SUBMARINE CANYON which carries sediment into the Arabian Basin to form the great Indus Cone. This immense fan of sediments is, in turn, cut by numerous channels (fan valleys) down which sediment is distributed to the deeper basin.

The most impressive feature of the deep basin of the Arabian Sea is the mid-Indian Ocean Ridge which curves westward from the vicinity of the Chagos Archipelago to enter the basin of the Gulf of Aden. The ridge is part of the 47 000-mi- (75 623-km-) long Mid-Ocean Ridge that circles the globe and is found in all oceans. The ridge represents a zone from which the seafloor is spreading as new material wells up to replace that moving laterally away from the ridge. (See CONTINENTAL DRIFT.) In the Arabian Sea the top of the ridge lies at a depth of about 16 000–20 000 ft (5000–6000 m). Just before entering the Gulf of Aden, the ridge is offset by a fault (Owen Fracture Zone) running north-south from the deep seafloor off Iran to the vicinity of the Amirande Isles. To the south of the ridge is the Samoli Basin, with depths of nearly 17 000 ft (5182 m). To the north lies the Arabian Basin, with depths as great as 14 826 ft (4519 m). The floor of the basin, except along the southeastern edge, is covered by sediment deposited by the Indus River in the form of a great alluvial fan (Indus Cone). Along the eastern edge of the Arabian Sea is the

Chagos-Laccadive Plateau which runs north-south from the vicinity of the Laccadive Islands off the coast of India to the Chagos Archipelago in the south. The Maldive Islands also rest on this plateau. The western edge of the plateau provides the boundary between the Arabian and Laccadive seas. The eastern boundary of the Laccadive is a line running from the Maldive Islands to Sri Lanka (Ceylon). The sediment in the deep basin of these two seas is primarily red clay and globigerina ooze. See MARINE SEDIMENTS.

The circulation of the surface water of the Arabian and Laccadive seas is under the influence of the seasonal monsoon. From April to November the monsoon blows from the southwest in response to the heating of the air over the landmass to the north. As a result, part of the South Equatorial Current (See EQUATORIAL CURRENT SYSTEM) turns north and flows along the African coast to become the swift Somali Current flowing around the northern rim of the Arabian Sea to join the Monsoon Drift Current and flow eastward past Sri Lanka. Since the southwest monsoon is moisture-laden, this is the period of high precipitation. From November to March the monsoon reverses and blows fitfully from the northeast. This dry air causes the current to reverse. Now, much weakened, one branch flows into the Gulf of Aden, while the remainder flows southward along the Somali Coast. The temperature of the surface water ranges from 75.2 to 82.4° F (24 to 28° C).

ARAFURA SEA lies between West Irian (West New Guinea) and the northern coast of Australia. The boundaries are defined as the southern limits of the CERAM SEA and the BANDA SEA on the north; the southwest coast of West Irian and the western limits of the CORAL SEA at the Torres Straits on the east; the coast of Australia on the south; and a line from Cape Don to the Tanimbar Islands on the west. The Arafura Sea extends over an area of 398 352 mi² (1 032 000 km²), occupies a volume of 48 942 mi³ (204 000 km³), and has a mean depth of 646 ft (197 m).

The floor of the Arafura Sea is very similar to that of the TIMOR SEA to the west; a CONTINENTAL SHELF, the Arafura Shelf, extends northward for much of the total area and is abruptly terminated by an elongate trough called the Aru Trough. The trough has depths in excess of 11 975 ft (3650 m) and merges further west with the Timor Trough. The Arafura Shelf, which also underlies the Gulf of Carpentaria, is 164–262 ft (50–80 m) in depth and is covered by glauconitic sands and calcareous muds. The Aru Trough is filling with terrigenous muds and globigerina ooze.

The water of the Arafura Sea is probably renewed by both the PACIFIC OCEAN and the INDIAN OCEAN, with Pacific WATER entering southward through the Banda Sea, and Indian water flowing in from the southwest through the Timor Trough. Surface currents in the area are rather random during the summer (Southern Hemisphere) but flow generally westward during the winter. Surface temperatures range from 82 to 79° F (28 to 26° C).

ARAL SEA See INLAND SEA.

ARCHIPELAGIC PLAIN See ABYSSAL PLAIN.

ARCTIC CIRCLE is the imaginary line of latitude 66°32′ N (often taken as 66°30′ N). Along this line the sun does not set on the day of the summer solstice, about June 21, and does not rise on the day of the winter solstice, about December 22. Northward from the Arctic Circle the number of annual 24-h periods of continuous day or continuous night increases to about 6 months at the North Pole. The Arctic Circle, which separates the Arctic region from the North Temperate Zone, runs through northern Russia, the Scandinavian Peninsula, Greenland, northern Canada, and Alaska and touches the northern tip of Iceland. In 1778 James Cook, the famous English explorer, crossed the Arctic Circle on his voyage through the Bering Strait to become the first to cross both the Arctic and Antarctic Circles. (He had crossed the ANTARCTIC CIRCLE in 1773.) See COOK, JAMES. William Bligh, later Captain of the *Bounty,* accompanied Cook on this voyage as master of the *Resolution.*

ARCTIC OCEAN, located at the top of the world and nearly landlocked by the North American and Eurasian landmasses, is the smallest of the world's oceans with an area of only 5 404 000 mi² (14 000 000 km²). Its extensive continental shelves (the widest in the world) and deeply indented coastlines harbor such peripheral seas as the GREENLAND SEA, BARENTS SEA, WHITE SEA, KARA SEA, LAPTEV SEA, EAST SIBERIAN SEA, CHUKCHI SEA, and BEAUFORT SEA. Numerous northward-flowing rivers, notably the Mackenzie, Anderson, and Colville of Canada, and the Kolyma, Indigirka, Lena, Yana, Ob, and Yenisei of the U.S.S.R., add their silt-laden WATER to that contributed by the PACIFIC OCEAN and the ATLANTIC OCEAN. The surface of the Arctic Ocean is covered by a perennial ice sheet—the polar ice pack—which extends to the encircling landmasses during winter and averages 10 ft (3 m) in thickness. The ocean floor, with depths as great as 16 995 ft (5180 m), is divided into several major basins by three submarine ridges—Lomonosov, Alpha, and mid-Arctic Ocean—which partition the ocean basin. One of the most hostile regions on earth, with water temperatures of 29° F (−1.7° C) and air temperatures as low as −76° F (−60° C), the Arctic Ocean supports relatively less life than the more southerly regions where the waters of the Arctic Ocean mix with those of the Atlantic and Pacific.

Early Exploration Lured by better fishing and whaling grounds, the acquisition of new territory, and the search for a short route across the Atlantic to the Orient, explorers began to arrive at the Arctic region as early, perhaps, as 320 B.C. Exploration continued with increasing intensity until the navigation of the Northwest Passage by Amundsen in 1903–1906 and the conquest of the Pole by Peary in 1909. The more notable of the explorations during this period of nearly 2000 years are listed below in chronological order:

320 B.C. Pytheas of Massalia sailed out through the Strait of Gibraltar to circumnavigate the British Isles and then explore northward, perhaps into the Norwegian Sea.

ca. 870 The Norseman Ottar (Othere) claimed to have sailed around northern Norway, along the Murman Coast, and into the White Sea as far as the Kola Peninsula, in search of WALRUS.

ca. 875 During the latter half of the ninth century Irish monks settled in Iceland in search of solitude.

877 Gunnbjorn Ulfsson, driven westward from Iceland, sighted Greenland.

982–985 Eric the Red, outlawed from Iceland, founded a colony in western Greenland.

1000–1006 Leif Ericson and Thorfinn Karlsefni, sailing from Greenland, explored and tried to settle the coast of North America.

1194 Iceland annals record the discovery of Spitsbergen by hunters of SEALS, WHALES, and WALRUS. These hunters explored the Greenland, Barents, and White Seas and discovered Novaya Zemlya.

1553–1554 Sir Hugh Willoughby and Richard Chancellor sailed around the northern coast of Norway and reached the present site of Archangel.

1556 Stephen Burrough reached Novaya Zemlya and the Kara Strait.

1594–1597 Willem Barents and Cornelis Nay discovered Bear Island, rounded Novaya Zemlya and wintered at Ice Haven (the first expedition to successfully winter in the Arctic), and explored the Barents and Kara seas as far as the Yaimal Peninsula.

1607–1611 Henry Hudson discovered Jan Mayen Island and sailed through Hudson Strait to explore Hudson Bay.

1610–1648 The Russian Cossacks in conquering Siberia reached the mouths of the Yenisei (1610), Lena, Yana (1636), and Kolyma (1644) rivers. There is strong evidence that Simon Dezhnev led an expedition from the Kolyma River, through the Bering Strait, to the Gulf of Anadyr.

1725–1741 Vitus Bering, in a series of voyages, discovered the Bering Strait, explored the Aleutian Islands, and discovered and named Mt. St. Elias on the American side.

1819–1820 Edward Parry sailed through Lancaster Sound and Barrow Strait, and discovered Wellington Channel, Prince Regent Inlet, North Somerset Island, Melville Sound, and Melville Island.

1819–1826 John Franklin traveled down the Coppermine River of Canada to the Arctic Ocean and explored eastward for 550 mi (885 km) to Cape Turnagain.

1827 W. Edward Parry tried to reach the North Pole from northern Spitsbergen, using sledgeboats, but the southern drift of the floe was too fast. He did not get beyond 82°45′ north LATITUDE.

1831 James Ross located the north magnetic pole and planted the British flag there.

1845–1848 Sir John Franklin reached Cape Herschel to complete the discovery of the Northwest Passage, although ice prevented his navigating it.

1850–1854 Richard Collinson and Robert McClure wintered in Prince of Wales Strait from which they could see Banks Strait. Thus, they too, confirmed the final links in the intricate Northwest Passage.

1871–1874 Julius Payer and Carl Weyprecht discovered Franz Josef Land.

1878–1879 N.A.E. Nordenskjöld navigated the Northwest Passage for the first time from west to east.

1879–1881 G.W. De Long in the *Jeannette* attempted to explore the coast of Siberia from the Bering Strait. The ship was crushed by the ice, but the surviving party discovered Jeannette and Henrietta islands in the New Siberian group.

1881 Robert M. Berry, in search of De Long, discovered Wrangel Island which, in turn,

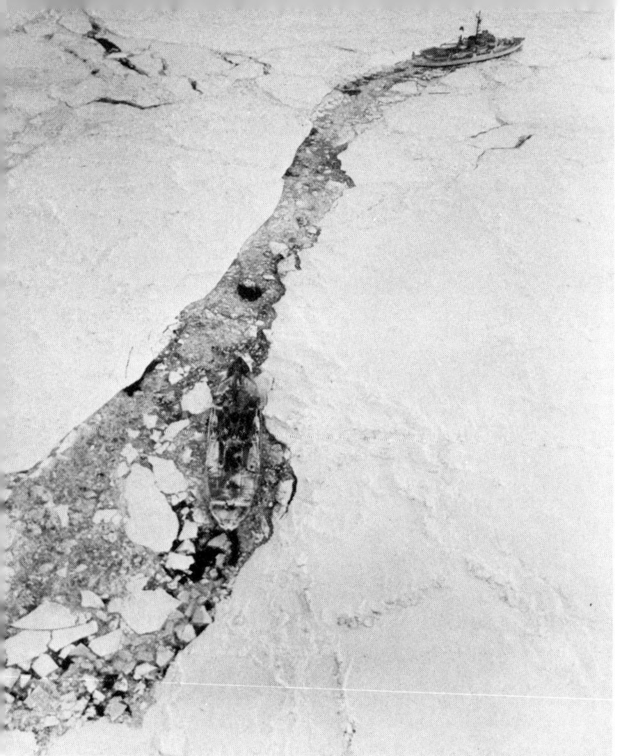

ARCTIC OCEAN. (*Left*) An ice cutter forces a path through the Polar Ice Pack in the white-coated Beaufort Sea, and another vessel takes advantage of the new trail. (*National Archives*) (*Right*) Icebergs float freely in the Arctic Ocean. (*National Archives*)

had been sighted and named by Kellett in 1849.

1888 Fridtjof Nansen became the first man to cross Greenland.

1893–1896 Fridtjof Nansen and Otto Sverdrup, aboard the *Fram,* made the first successful drift across the Arctic Ocean. They entered the pack off the New Siberian Islands and finally broke free near Spitsbergen after a drift of 3 years. During the drift, Nansen, using skis and dog sledges, managed to reach 86°14′ north latitude—a record.

1903–1906 Roald Amundsen navigated the Northwest Passage by way of the east coast of King William Land and out through the Bering Strait.

1909 Robert E. Peary claimed to have reached the Pole on this date.

1926 Richard E. Byrd and Floyd Bennett flew from King Bay and reached the Pole in a Fokker monoplane named *Josephine Ford.*

The Ice Cover The Arctic Ocean has been characterized as the only ocean that can be walked across. This is true, at least in winter, because of an ice cover averaging 10 ft (3 m) in thickness that forms a solid bridge between the North American and Eurasian coastlines. In summer the ice in the peripheral seas begins to melt and break up, except for that off the north coast of Greenland, where the pattern of circulation keeps ice close by for the entire year. The aerial extent of the ice cover recedes by 15–20 percent in summer over its winter maximum. The central polar ice pack, of course, is permanent, though always in motion and always being renewed. The ice cover is a mixture of SEA ICE and icebergs contributed by glaciers on Greenland and along the Siberian coast. The ice also contains numerous ice islands, some measuring 231 mi^2 (600 km^2) in aerial extent and standing as much as 33 ft (10 m) above the surrounding ice. Ice islands are generally formed from the more stagnant ice of the Beaufort Sea.

Normal seawater of 35 ppt (parts per thousand) salinity (see SEAWATER) freezes at 28.6° F (−1.9° C). In the process of forming ice crystals, the salt is excluded to form brine pockets. Since there is little precipitation in the Arctic, new ice is added from the bottom of the ice pack. In summer perhaps one-third of the ice thickness melts, even on the polar ice pack, only to be replaced in the winter. Although the ice is only about 10-ft (3-m) thick on the average, pressure ridges created by the collision of large ice floes may thicken it to as much as 40 ft (12 m).

Under the influence of polar wind patterns and ocean currents, the ice cover is in constant motion. Three major circulatory systems have been identified. One system drifts westward from the East Siberian and Laptev seas toward the Greenland Sea. Two loops spin off from the main drift: one is confined largely to the Laptev Sea, and the other flows around the islands in the vicinity of Franz Josef Land and to the north of Novaya Zemlya. Both loops turn in a counterclockwise pattern, and both are augmented by young ice from the Kara Sea.

A second ice drift turns clockwise within the Beaufort Sea to form a zone of stagnation. Because it is confined to a closed pattern and flows neither into the Pacific nor the Atlantic Ocean, the ice in this system is thicker than in other areas.

The third drift system originates in the East Siberian and Chukchi seas and drifts directly toward the Pole, where it joins the first system on its drift toward the Greenland Sea. Each year a part of the ice from the first and third drifts passes through the Greenland Sea into the North Atlantic, and through the straits of the Arctic Archipelago into Baffin Bay.

Water Masses and Currents As a result of the stratification and circulation of the waters of the Arctic Ocean, the marine life on the Pacific side is different from that on the Atlantic side, and those species found in one layer are not found in another. Pacific water enters through the Bering Strait; that of the Atlantic enters through the area between Greenland and Spitsbergen. The major outflow is by way of the GREENLAND CURRENT. The Atlantic water, with its higher temperature [37.4–39.2° F (3–4° C)] and salinity, flows into the Arctic Basin in a 2000-ft (609-m) layer which amounts to nearly twice that contributed by the Pacific. Rivers flowing northward over the North American and Eurasian landmasses contribute fresh water amounting to about one-tenth that flowing through the Bering Strait. Precipitation amounts to only about 0.01 percent of the total.

The Arctic water is divided into several distinctive layers. The surface water is about 28.7° F (−1.8° C) in winter and 29.3° F (−1.5° C) in summer; it extends to a depth of approximately 150 feet (45.7 m). Between 150 and 600 ft (45.7 and 182.8 m) lies the intermediate layer, with a temperature range of 28.6–29.1° F (−1.9 to −1.6° C). On the Pacific side a layer of somewhat warmer water is interleaved between the surface and intermediate layers at a depth of 250–350 ft (76.2–106.6 m). The Atlantic water flows across the Arctic Basin, becoming colder and thinner as it approaches the Beaufort Sea. It lies just beneath the surface layer, with a depth ranging to 2952 ft (900 m). Beneath the Atlantic water and extending to the bottom is a water mass with near-uniform temperature [31.4–30.5° F (−0.3 to −0.8° C)] and salinity (34.9–34.99 ppt). The major contributor to the bottom water is the Atlantic water which cools and sinks.

The Continental Margin The CONTINENTAL SHELF bordering the Arctic Ocean Basin covers approximately two-thirds of the 5 404 000 mi² (14 000 000 km²) which constitute the total area. It is irregular in both shape and width—ranging from 12.4 to 24.8 mi (20 to 40 km) wide north of Alaska and Canada in the Beaufort Sea, and from 310.6 to 745.5 mi (500 to 1200 km) wide (the widest in the world) in the Barents, East Siberian, and Chukchi seas. In many areas the shelf is deeply cut by submarine canyons, notably Herald Canyon, which heads up in the Chukchi Sea and runs northward to the continental margin with canyon depths ranging to 295 ft (90 m), and Barrow Canyon, which begins 93 mi (150 km) west of Point Barrow, Alaska, and runs northeastward into the Beaufort Sea with canyon depths of up to 328 ft (100 m). (See SUBMARINE CANYON.) The break in the outer edge of the continental shelf is at a depth of 656 ft (200 m)—considered normal worldwide—except that north of Greenland which is at a depth of 984 ft (300 m), probably due to downwarping caused by the Greenland ice cap. The slope of the shelf is also considered normal at 1½–4°.

The Ridges The Arctic Ocean Basin is partitioned by three submarine ridges: the Alpha Ridge, the Lomonosov Ridge, and the trans-Arctic Ocean extension of the Mid-Atlantic Ridge, or the Arctic Mid-Oceanic Ridge. These ridges play an important role in influencing the circulation of both the deep water masses and the perennial ice cover, and in establishing and maintaining the main provinces of ocean life.

The Alpha Ridge (named after drift station Alpha) joins the continental shelf of North America from the vicinity of Ellesmere Island to that part of the Siberian Shelf covered by the East Siberian Sea, a distance of about 559 mi (900 km). The ridge rises to a minimum depth of 4593 ft (1400 m) and is divided by a broad shallow trough which lies at depths of around 6561 ft (2000 m). The sides of the ridge do not grade smoothly into the deep ocean floor on either side, but drop abruptly by means of escarpments nearly 1968 ft (600 m) in height. The rugged topography of the ridge itself is thought to be due to repeated faulting.

MAP 1. The Arctic Ocean.

The Lomonosov Ridge, nearly parallel to the Alpha Ridge, runs for some 1118 mi (1800 km) across the Arctic Basin to connect the continental shelf in the vicinity of Ellesmere Island to that near the New Siberian Islands. The Lomonosov Ridge is narrower than the other Arctic ridges, being only 24.8 mi (40 km) wide at its narrowest point and 124 mi (200 m) wide at its widest. The crest of the ridge rises to a depth ranging from 2789 to 3937 ft (850 to 1200 m), and rises some 9842 ft (3000 m) from the adjacent seafloor. The crest of the ridge is relatively smooth and measures approximately 16 mi (26 km) in width. Recent evidence indicates that the ridge is a fragment of the former Eurasian continental shelf which split away from the continental block when the Arctic Basin was formed.

Between the Lomonosov and Alpha ridges, a low ridge known as the Marvin Ridge projects into the Makarov Deep. These three ridges join on the North American side of the basin to form a broad shelf.

The Mid-Atlantic Ridge enters the Arctic Basin between Greenland and Spitsbergen, where it becomes the Arctic Mid-Oceanic Ridge. This ridge crosses the basin and intersects the Eurasian Shelf in the vicinity of the Lena River delta. The width of the ridge over its entire length is about 124 miles (200 km). The physiography of the ridge differs somewhat from that of the Mid-Atlantic Ridge, but the associated earthquake belt is continuous. Practically all earthquakes in the Arctic Basin fall in a narrow band which coincides with the Arctic Mid-Oceanic Ridge. The ridge is made up of peaks, ridges, and deep rifts whose relief above the ocean floor is from 3280 to 4921 ft (1000 to 1500 m).

The Deeps The Alpha, Lomonosov, and Arctic Mid-Oceanic ridges partition the Arctic Basin into four deeps: the Canadian Deep, the Makarov Deep, the Eurasian Deep, and the Fram Deep. These deep sub-basins, along with their confining ridges and continental margins, provide a measure of stability and uniformity to the deep water masses and act as catchment basins for the sediment which flows and creeps down their steeply sloping sides.

The Canada Deep lies between the continental shelf of the Beaufort Sea and the Alpha Ridge, a distance of about 683 mi (1100 km). The remarkably smooth floor of the Canada Deep lies at a depth of about 12 926 ft (3940 m). Extending under the Beaufort Sea is a smaller deep—the Beaufort Deep—which is separated from the Canada Deep by a broad SILL rising some 1148 ft (350 m) from the ocean floor. However, the name Canada Deep is usually applied to both because the degree of their separation is still uncertain.

The Makarov Deep, enclosed by the Alpha, Marvin, and Lomonosov ridges, lies at a depth of 13 221 ft (4030 m). The smooth floor of the Makarov Deep butts abruptly against the sides of its confining ridges, indicating a rapid buildup of sediment throughout its basin.

The Eurasian Deep and the Fram Deep were referred to as the Nansen Deep until it was determined that they were actually separated by the Arctic Mid-Oceanic Ridge. The floor of the Eurasian Deep, like those of the other deeps, is featureless and lies at a depth of 14 074 ft (4290 m). The depth of the floor is some 1345 ft (410 m) greater on the side near the Lomonosov Ridge than that near the Arctic Mid-Oceanic Ridge.

The Fram Deep is the smallest of the four deeps—590 by 217 mi (950 by 350 km)—and the deepest, at 16 995 ft (5180 m). It also is the most irregular with a change in elevation of 14 600 ft (4450 m) over a distance of 50 mi (80 km), and two seamounts which rise to within 2395 and 3281 ft (730 and 1000 m) of the surface. (See SEAMOUNT.) At present very little is known of the Fram Deep, and new features may be found at any time.

ARISTOTLE'S LANTERN See SEA URCHIN.

ATLANTIC OCEAN, named by the Romans for the Atlas Mountains on the northwestern rim of Africa, is the second largest (after the PACIFIC OCEAN) of the world's oceans. It also appears to be the youngest ocean, emerging only within the past 100 million (100×10^6) years when the Americas separated from Europe and Africa as a result of CONTINENTAL DRIFT. The east-west boundaries of the Atlantic—North and South America on the west and Europe and Africa on the east—are well known. However, the north-south boundaries are a matter of some interpretation. Some have argued that the ARCTIC OCEAN should properly be classified as a sea of the Atlantic, in which case the northern boundary of the Atlantic would be the Bering Strait between North America and Siberia. But the Arctic has been officially designated a separate ocean. Therefore, the generally accepted northern boundary of the Atlantic is the LATITUDE of northern Spitsbergen, and includes the GREENLAND SEA and the NORWEGIAN SEA.

Similarly the southern boundary of the Atlantic Ocean is a subject of some debate. The International Hydrographic Bureau does not officially recognize a SOUTHERN OCEAN encircling Antarctica. Therefore, the southern boundary of the Atlantic extends to the shores of that frozen land. In this case, the eastern boundary between Africa and Antarctica is taken as the 20° E meridian, and the western boundary between South America and Antarctica is a line representing the shortest distance between the two. However, some authorities would prefer the western boundary to follow the eastward-looping shallow-water contour that runs through the Falkland, South Sandwich, South Orkney, and South Shetland Islands. In this case, the SCOTIA SEA would belong to the Pacific Ocean. If, on the other hand, a Southern Ocean is accepted—as many oceanographers would prefer—then the southern boundary lies at 55° S latitude. It should be noted that the northern boundary of the Southern Ocean is open to some debate, with 60, 55, 40, and 35° S being used by various authorities.

Under the more restricted interpretation, the Atlantic covers a total area of 32 476 496 mi² (84 136 000 km²), occupies in all a volume of 73 505 360 mi³ (306 400 000 km³), and has a mean depth of 10 777 ft (3285 m) for the North Atlantic and 13 421 ft (4091 m) for the South Atlantic. The maximum depth of 30 183 ft (9200 m) is found in

the Puerto Rican Trench along the northern rim of Puerto Rico. If the CARIBBEAN SEA, MEDITERRANEAN SEA, and other marginal seas are included, the area of the Atlantic is 40 916 000 mi² (106 000 000 km²).

The climatic interrelationship between the Atlantic and the huge landmasses on either side, together with certain topographic controls (i.e., the position of the Andes Mountains, etc.), results in the Atlantic receiving about 4 times the land drainage of either the Pacific or the Indian Oceans. As a consequence, the less saline waters of the Atlantic can be found to

ATLANTIC OCEAN. North America lies in the upper left of this satellite photo, South America dead center, and the west coast of Africa toward the upper right. Much of the Atlantic Ocean rests under swirls of clouds. *(NASA)*

flow into all the major oceans. The two longest marginal seas are the Caribbean–Gulf of Mexico system and the Mediterranean Sea. Through both these seas the Atlantic connects with another ocean by means of an artificial canal. Other marginal seas include the Greenland, Norwegian, and Celtic, as well as the NORTH SEA, BALTIC SEA, IRISH SEA, BLACK SEA, IRMINGER SEA, LABRADOR SEA, SCOTIA SEA, and WEDDELL SEA.

Early Exploration The greatest distance across the Atlantic is 4500 mi (7240 km) between the coasts of Mexico and Africa. The shortest distance is 1500 mi (2413 km) between Cape São Roque on the coast of Brazil and Sierra Leone on the West African coast. Undoubtedly, the relative narrowness of the Atlantic, along with the close cultural and commercial ties between Europe and the New World, had much to do with the fact that today the Atlantic is the best-known and most completely explored of the world's oceans. In the beginning, however, the existence of the New World was unknown, and the first major landfalls west of the Pillars of Hercules were thought to be Japan, China, and India. The distance to these lands was estimated at about 7000 mi (11 263 km) if one agreed with Ptolemy (as most did) and accepted the 18 000-mi (28 962-km) circumference of the earth arrived at by Posidonius rather than the 24 000-mi (38 616-km) circumference estimated by Eratosthenes. While this distance probably acted as a deterrent to early explorers, it later served to make an Atlantic trade route to the Orient appear practical.

The earliest recorded voyage of discovery in the Atlantic was reported by the Greek historian, Herodotus. According to Herodotus, King Necho of Egypt, in the seventh century B.C., sent Phoenician ships southward through the RED SEA to circumnavigate Africa and return through the Strait of Gibraltar. The voyage was reported to have taken 4 years. Hanno, the Carthaginian admiral, reportedly sailed out through the Strait of Gibraltar around 500 B.C. and turned southward along the African Coast. The records indicate that he may have reached the Gulf of Guinea before turning back. It is quite possible that all the islands that lie off the European and African coasts were known to those early Phoenician sailors. To learn the trading secrets of the Phoenicians may have been the inspiration behind the voyage of Pytheas of Massilia who, according to the accounts of Strabo, Diodorus, and Pliny, sailed out through the Strait of Gibraltar in 325 B.C. to explore northward. Pytheas is believed to have crossed the Bay of Biscay, circumnavigated the British Isles, and then sailed 6 days north to Thule on the edge of the frozen sea. In a second voyage, it is speculated, Pytheas sailed along the coast of the NORTH SEA and may have entered the BALTIC SEA.

By the middle of the third century B.C., Rome emerged as the dominant Mediterranean power as far west as the British Isles. But Rome contributed little to the exploration of the Atlantic. It remained for the northern Europeans, at the close of the Dark Ages, to revive the interest in westward exploration. During the period 565–573 A.D., the Irish monk St. Brendan is said to have sailed in search of "the promised land of the Saints" later recorded as "St. Brendan's Isle" on early maps. In the ninth century, when the Norsemen were raiding southward along the coast of Europe, Irish monks, in search of solitude, settled in Iceland. During this same period Othere of Helgeland sailed north around the tip of Norway (North Cape) and into the WHITE SEA. A Viking ship blown off course discovered Iceland and the small colony of Irish monks. By 1100 there were 50 000 people of Norse and Irish blood on the island. At the beginning of the tenth century, Gunnbjorn discovered islands to the west of Iceland, a discovery which led Eric the Red westward in 982 to discover the colonize Greenland. And, in about 1000 A.D., Eric's son Lief is thought to have explored the coast of North America.

The 1400s saw the beginning of the Golden Age of Discovery under the guidance and inspiration of Prince Henry the Navigator of Portugal. In 1419 João Gonçalves Zarco and Tristan Vaz Teixeria sailed southward to discover the islands of the Madeira group. Several subsequent voyages sailed as far south as Cape Bojador, then thought to be the beginning of the Sea of Darkness where the ocean ended in a vast swamp. In 1434 Gil Eanes broke the taboo and sailed south of Cape Bojador. By the time Henry died in 1460, his ships were trading and exploring as far south as Cape Palmes. And, in 1488, Bartholomeu Dias rounded the Cape of Good Hope, paving the way for Vasco da Gama's voyage to India in 1497. Thus, some 2200 years elapsed between the first and second recorded circumnavigation of Africa.

There is some evidence that in 1472 João Vaz Corte-Real, navigator for a joint Portuguese-Danish voyage, explored the coast of Newfoundland and reached the coast of North America. And, in 1492, João Fernandes explored Greenland, Newfoundland, and, perhaps, the coast of North America. However, history has awarded the credit for the discovery of the New World to Columbus for his four voyages (1492, 1493, 1498, and 1502) to the Caribbean area. Following up on the discoveries of Columbus, Pedro Alvares Cabral discovered Brazil in 1500. And, in 1501, a three-ship expedition under the command of André Gonçalves touched the coast of Brazil and then sailed 1600 mi (2574 km) south along the coast of South America from Cape São Roque to Rio Grande do Sul. The exploration of the east coast of South America was completed in 1519–1520 when Ferdinand Magellan sailed into the Pacific through the Strait of Magellan. This feat was duplicated in 1577 by Sir Francis Drake, in 1586 by Thomas Cavendish, and in 1594 by Richard Hawkins. However, it was not until 1615 that an expedition, under the command of Jacob le Maire and Jan Schouten of Holland, was able to "round the Horn" and sail through Drake Passage between the tip of South America and the ice-bound coast of Antarctica.

The early exploration of the Atlantic came to a close with the 1772 voyage of James Cook of England. Legend held that a great continent—called *Terra Australis* by the ancients—lay far to the south. Cook circled Antarctica without sighting land, but demonstrating that if a continent existed, it was much smaller than expected. The coast of Antarctica was first sighted in 1826 by John Briscoe. See COOK, JAMES.

Oceanographic exploration and research in the Atlantic began with the first deep soundings taken by J.C. Ross in 1839 and the biological studies made by Louis Agassiz and William Thomsen in 1860. Systematic research began with the efforts of Matthew Fontaine Maury who, in 1853, suggested the collection of hydrographic and meteorological data from the observations made by merchant ships. This effort provided the basis for the present-day pilot charts published each month. See AGASSIZ, LOUIS; MAURY, MATTHEW FONTAINE.

Bottom Topography There is no significant evidence that the Atlantic Ocean existed prior to about 100–120 million years ago. The widely accepted explanation of its creation is that within the past 200 million years North and South America began to drift away from Europe and Africa. (See CONTINENTAL DRIFT.) The pivotal point of this separation lay somewhere north of Iceland so that the continental margins opened like the blades of a pair of shears. The midline of this separation—a process that still continues—is the Mid-Atlantic Ridge. The Ridge is part of the greatest mountain chain on earth, the Mid-Ocean Ridge, which runs for 40 000 miles (64 360 km) and crosses all the ocean basins. The Ridge is typically about 931 mi (1500 km) in width and begins with the first significant change in elevation seaward of the ABYSSAL PLAIN. From this point the Ridge rises gradually to a rugged, highly fractured crest that lies about 6000 ft (1828 m) below the water surface and rises above the adjacent ocean floor a distance of 3000–9000 ft (1000–3000 m). Along the centerline of the Ridge is a rift valley which represents the zone of separation between the two sides of the Atlantic floor. The floor of the rift valley lies about 6000 ft (1829 m) below the flanking peaks and is from 15 to 30 mi (24 to 48 km) in width. The Ridge is made more rugged by transform faults which, at intervals, have acted to relieve slowly accumulating stresses by allowing the rock to slip along a line that is generally east-west. In places, such as the Romanche Trench between the shoulders of South America and Africa, the centerline of the Ridge has been offset by many miles.

Running up the centerline of the Atlantic Basin, the Mid-Atlantic Ridge passes through Iceland and into the Arctic Ocean. In the south, at about 55° S, the Ridge joins the southwest Indian Ocean Ridge that ultimately runs northward into the RED SEA. Periodically the volcanic peaks of the Ridge pierce the surface to form islands such as Iceland, St. Paul, Ascension, St. Helena, Tristan da Cunha, Gough, and Bouvet. The highest of these peaks is Pico in the Azores, which rises 7615 ft (2321 m) above the water surface.

Between the flanking edges of the Mid-Atlantic Ridge and the bordering continental shelves the floor of the Atlantic Basin is relatively uncomplicated when compared with that of the Pacific and Indian oceans. For instance, significantly fewer volcanic seamounts and guyots rise from the seafloor; of those that do, some account for such islands as the Faeroes, Madeira, Fernando Póo, Principe, São Tomé, Annobon, Fernando Noronha, Trinidad, and the South Sandwich Islands. (See CONTINENTAL SHELF; GUYOT; SEAMOUNT.) Also, the Azores, Canary Islands, Cape Verde Islands, and the Lesser Antilles are, predominantly, of volcanic origin. The island of Bermuda is a coral (see CORAL REEF) cap resting on an old volcanic cone. The other islands of the Atlantic, such as the British Isles, Greater Antilles, Falkland Islands, and South Georgia, are continental in character. The floor of the ocean in this zone is occupied by a series of basins slowly filling with sediment from the continental shelves and separated by ridges. An example is the Angola Basin, seaward of the mouth of the Congo River. The basin occupies an area of 54 040 mi^2 (140 000 km^2) and is fed with sediment in the north by the Congo SUBMARINE CANYON—the largest canyon in the Atlantic. The Angola Basin is separated from the Guinea Basin in the north by the Guinea Ridge and from the Cape Basin in the south by the Walvis Ridge. The deep floor of the Atlantic is populated with such arrangements from its northern to its southern boundary.

The sediments of the Atlantic basin, including the adjacent seas, are predominantly pelagic (origi-

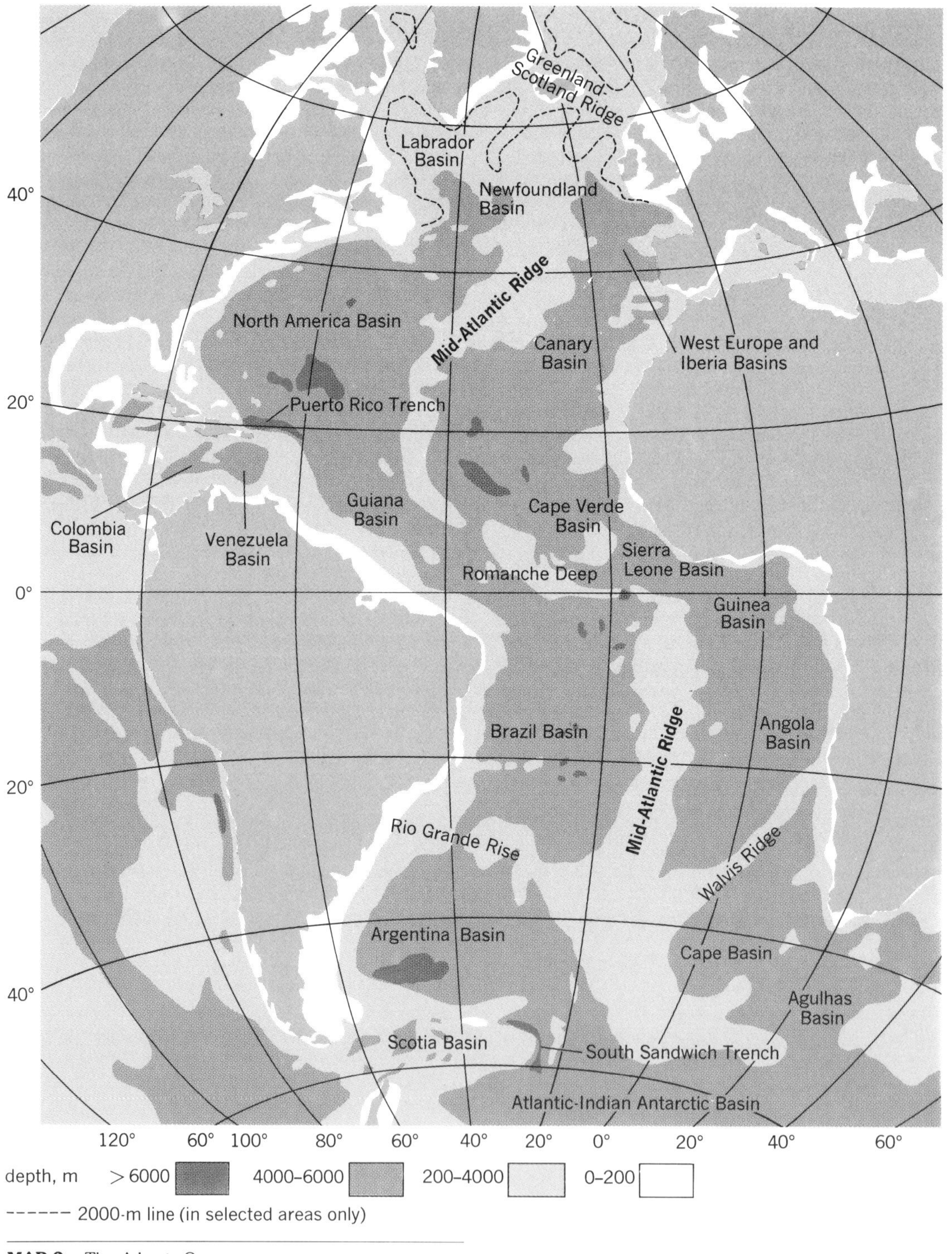

MAP 2. The Atlantic Ocean.

nating in the ocean itself) and make up about 75 percent of the total sediment. About 47 percent is calcareous globigerina ooze (see MARINE SEDIMENTS), 5 percent siliceous diatom ooze, and 18 percent red clay. (See DIATOMS.) The remainder of the sediment is hemipelagic (littoral sediment derived largely from the land) and covers the continental shelves, slopes, and rises. Through the work of the TURBIDITY CURRENT a great deal of this sediment has been carried out onto the ABYSSAL PLAIN.

The continental shelf bordering the Atlantic is typical of that which occurs along the east and west coasts of continents; i.e., the shelf is better developed off the east coast of North and South America than off the west coast of Europe and Africa. The exception is the extensive shelf along the coast of Norway, under the NORTH SEA, and around the British Isles. Otherwise the shelf is relatively narrow all the way to the Cape of Good Hope.

Currents Essentially, the surface circulation in the Atlantic consists of two large gyres located north and south of the equator, the major difference between the two being that the North Atlantic gyre revolves clockwise and the South Atlantic gyre revolves in a counterclockwise fashion. The North and South Equatorial currents (see EQUATORIAL CURRENT SYSTEM), drifting westward across the Atlantic in the equatorial region, pile up warm saline waters along the east coast of North and South America. Part of the South Equatorial Current turns northward to enter the CARIBBEAN SEA as the CARIBBEAN CURRENT, while the remainder turns south as the BRAZIL CURRENT. The North Equatorial Current turns northward as the ANTILLES CURRENT, which is joined in the vicinity of the Bahamas by the Caribbean Current (Florida Current), after it has traveled through the GULF OF MEXICO and out through the Florida Straits to form the GULF STREAM. Off Newfoundland this current, now called the NORTH ATLANTIC CURRENT, turns eastward to flow across the Atlantic. Part of it splits off to flow into the NORWEGIAN SEA as the NORWEGIAN CURRENT, while the remainder turns southward to flow along the coast of Europe and Africa as the CANARY CURRENT. During its west-east transit, the North Atlantic Current is influenced by the cold waters of the LABRADOR CURRENT and the EAST GREENLAND CURRENT flowing out of the Arctic Basin.

The gyre of the North Atlantic Current system is responsible for the formation of the SARGASSO SEA which occupies the center of the gyre. The Sargasso consists of a large body of relatively warm water that slowly revolves under the influence of the peripheral currents and the Coriolis force. The Sargasso Sea gets its name from the masses of seaweed (sargassum) that dot its surface.

In the South Atlantic a part of the South Equatorial Current turns south along the coast of Brazil as the BRAZIL CURRENT, which, in turn, gradually turns back to the east to join the WEST WIND DRIFT CURRENT that circles Antarctica. Near the coast of Africa a part of the West Wind Drift Current turns north to flow along the west coast of Africa as the cold BENGUELA CURRENT. This current, like the southward-flowing Canary Current, joins the westward-flowing North and South Equatorial currents to complete the gyres.

Surface currents do not, of course, represent the total movement of water in the Atlantic. There are slowly moving currents at several depths which, initially, result from the cooling of waters in certain locations, causing the waters to sink. Upon reaching a depth where their newly acquired density is equal to the surrounding waters, these water masses spread out and flow as a relatively sluggish current. Examples are the subarctic bottom water, chilled in the Arctic Basin, which flows out over the ridges between Greenland and Scotland. However, this cold water rarely reaches further south than 50° north LATITUDE. Antarctic bottom water, on the other hand, sinks off the coast of Antarctica and flows as far north as 40° N. It is found to the west of the Mid-Atlantic Ridge and flows into the eastern basin through the Romanche Trench. North Atlantic deep water sinks in the LABRADOR SEA and IRMINGER SEA and flows southward at a depth of 6500–9800 ft (2000–3000 m). On its way south this submerged river joins the cold, dense water flowing out of the Mediterranean Sea through the Strait of Gibraltar, and can be traced as far south as Antarctica, where it rises to the surface.

The sinking and rising of these great masses of water result in the Atlantic being better oxygenated than the Indian or Pacific oceans. This fact has a profound effect on the organic production of this vast area.

Tides The Atlantic is characterized by semidiurnal (twice daily) tides which vary considerably in amplitude from one location to another. An amphidromic point, or node, in the tide pattern exists at 55° S, between Brazil and the Gold Coast of Africa, between the Lesser Antilles and West Africa, and at 55° N. The mean tidal range in the open ocean is about 3 ft (1 m), but in the node off Rio Grande do Sul (Brazil) it is only 6 in (16 cm), and at Puerto Rico it is only 3.5 in (9 cm). Away from these nodes the tidal range increases: 31.9 ft (9.74 m) in the Bay of Bahia Grande in Patagonia, 34.7 ft (10.58 m) in the

Bay of St. Malo in the English Channel, and 37.6 ft (11.47 m) in the British Channel. The highest spring tide (see TIDES) in the world is in the Bay of Fundy at 46.4 ft (14.14 m).

Resources The Atlantic contains some of the most prolific fishing grounds in the world's oceans. However, these grounds are located, primarily, in the North Atlantic (off the coasts of Europe, North America, Greenland, and Iceland) and, to a lesser extent, in the waters off Antarctica. Among the more popular fishes taken by the fishing nations are HERRING, sprat, MACKEREL, SARDINES, pilchard, COD, HALIBUT, plaice, hake, and SOLE.

Since the first whaling expedition to antarctic waters in 1904, that region has been the most important hunting ground for whales in the Atlantic. However, whaling is a declining business as the numbers of whales decrease, as the market for whale products declines, and as the pressure from conservationists around the world increases. The Norwegians, Japanese, and Russians still take advantage of the December to March hunting season. The chief whaling areas are off the southwest coast of Africa, the waters off the Falkland Islands, and the ROSS SEA of the SOUTHERN OCEAN.

The mineral resources of the Atlantic Basin come primarily from the continental shelf and are of considerable importance. Sand and gravel for construction are dredged from shallow offshore areas of both North America and Europe. Placer diamonds are found off the mouth of the Orange River of Africa. The metal magnesium is extracted from SEAWATER by large plants located at Freeport, Texas, and at Hartlepool on the Tees ESTUARY in England. Bromine is also extracted at Freeport. Oil and gas, of course, are the most profitable resources extracted from the offshore areas. The drilling of offshore wells began off Texas in the Gulf of Mexico during the 1930s and has expanded rapidly in recent years. A recent strike took place in the North Sea, and parts of the shelf off the east coast of the United States have been leased to competing oil companies for exploratory drilling.

ATLANTIS See MYTHS AND LEGENDS.

ATOLL See CORAL REEF.

AZORES-BERMUDA HIGH See HURRICANE.

AZOV, SEA OF See INLAND SEA.

BACTERIA is the name for microscopic-sized organisms which have a primitive type of cell construction called prokaryotic; from one taxonomic standpoint, the phylum Schizophyta of the kingdom Procaryota contains the bacteria. In general, bacteria reproduce by a binary fission mode, in that each bacterial cell is formed by the division of a preexisting cell into two equal or nearly equal parts, each of which grows to parental size and form.

In the world's oceans, bacteria may be found at any LATITUDE and, as far as is presently known, at every depth. In coastal areas they are much more profuse than in the open ocean. The amounts of bacteria present vary proportionally with the distance from land, depth of water, presence of PLANKTON, etc. If one were to assume that there are 10^{10} bacteria per gram in the most productive sediment and only one bacterium or less per liter in open ocean water, the standing crop would be astronomical in amount, and this figure would double every 1 to 8 days depending on species. However, bacteria are eaten or they die or sink as fast as they reproduce. Along with ALGAE, bacteria constitute the most numerous organisms found in all the marine environments. However, relatively little is known about their dynamics or the scope or potential effect of their activity.

Some major bacterial species indigenous to the ocean are pathogenic (disease-carrying). Lobsters (and other shellfish), KELP, and SALMON are some forms of marine life that are especially susceptible to such bacterial diseases. Bacteria that are introduced into the oceans by human beings (e.g., coliform bacteria in the discharge of sewage) can infect marine life and can then be passed on to human beings if they bathe in the waters or eat the infected plant and/or animal. The majority of bacteria, however, are apparently useful. For example, there are 186 kinds of heterotrophic bacteria, that is, those bacteria that act upon dead organic matter (such as dead plants and fish) to decompose it into plant nutrients. They also convert dissolved organic matter into bacterial cell substances which can be assimilated by marine animals from PROTOZOA (e.g., sarcodinians such as FORAMINIFERANS and RADIOLARIA) to herbivorous fish and large invertebrates. Many bacteria are DETRITUS feeders in the oceanic food chains, and this PARTICULATE MATTER, or detritus, tends to concentrate dissolved organic materials from SEAWATER.

Changes in the composition of seawater by bacterial action are the decomposition of cellulose, lignin, chitin, urea, and petroleum hydrocarbons; the oxidation of ammonium to nitrite, nitrite to nitrate, and sulfide to sulfate; the use of molecular nitrogen, ammonium, urea, nitrate, nitrite, and phosphate in the bacterial cell; the release of ammonium and phosphate after death of the cell; and the formation of iron and manganese oxide concretions. Although the ubiquitous bacteria constantly change the nature of the water and sediments by such actions, only a few of their activities have been well documented.

For instance, it is believed that oceanic bacteria may be the marine atmosphere's principal source of particles needed to trigger precipitation. The bacteria that live in association with the tiny floating PLANT LIFE IN THE OCEANS (PLANKTON) are projected into the air by SEAFOAM bubbles that burst on the surface of the WAVES. Certain bacteria contained in these bursting bubbles find their way high into the atmosphere and may thus become the core or nucleus around which water droplets form. These droplets fall to the surface as rain or snow.

In addition, it has long been difficult for scientists to describe a true marine bacterium except in terms of its sodium chloride requirements. Sodium has been shown to be a requirement for some of these microorganisms. The requirements for other elements characteristically found in seawater may be equally important. Spectrographic analysis of marine bacteria shows that they concentrate most of the trace elements in varying amounts. Nickel and titanium, for example, are present in many marine bacteria. Calcium also enters bacterial cells as a chelate to the amino acid alanine. The alanine metabolizes, and the carbon dioxide is used to combine with the calcium to produce aragonite crystals.

In such bacterial activities as these, cations are both concentrated in or on bacterial cells and released from decomposing protoplasm. The rate of ion exchange between decomposing organisms in sediments and the water is as yet not well understood. However, it may be very possible that many of the deposits of minerals in ancient sediments might have their origin in microbial activity. During the activities of the bacteria, changes in pH and eH of the sediment's diagenesis occurs, with SILICA dissolving at high pH and carbonates at low pH. Several of the precipitated minerals also will be affected by pH change and especially by redox or eH changes. See REDOX POTENTIAL; PH.

It has also been recently shown that crystals of lodestone or magnetite (Fe_3O_4) appear to form within the cells of certain bacteria. These organisms are thus equipped with built-in compasses to sense the direction of the earth's magnetic field. Observations indicate that such bacteria, which are anaerobes (those that do not require air or free oxygen to maintain their life processes) use this ability to swim downward at various latitudes in the northern hemisphere to find the bottom conditions of the ocean most favorable to their growth.

At such bottom conditions bacteria serve as food for filter-feeding and burrowing marine organisms. In the deep basins, bacteria bring about oxidation in the absence of dissolved oxygen. However, not all bacteria in the SULFUR CYCLE produce hydrogen sulfide; some produce pure sulfur and transfer it back into this cycle by oxidation.

In summary, bacteria are clearly vital to the functioning of the marine ecosystem, but as yet the dynamics of bacterial life in the world's oceans are far from being known and understood in detail.

See also BASIN; ELEMENT; INVERTEBRATE; LOBSTER.

BALEEN is a horny substance which grows as fringed filter plates suspended from the upper jaws of baleen WHALES. This horny substance may be keratin or any of various albuminoids characteristic of epidermal derivatives, such as nails and feathers, which are insoluble in protein solvents, have a high sulfur content, and usually contain cystine and arginine as the predominant amino acids.

BALI SEA, located in the western PACIFIC OCEAN, is separated from the INDIAN OCEAN to the south by a line from the eastern tip of Java through Bali, Lombok, and Sumbawa; on the east it is bounded by a line from the west coast of Sumbawa to the Paternoster Island group and then westward through the Kangean Islands to the northeastern extreme of Java. The Bali Sea covers an area of 45 934 mi² (119 000 km²), occupies a volume of 11 756 mi³ (49 000 km³), and has a mean depth of 1348 ft (411 m).

The floor of the Bali Sea is dominated by the Bali Trough which has a maximum depth of 5216 ft (1590 m). The seafloor sediments are largely terrigenous materials, volcanic ash, and coral sands and muds. Access to the Indian Ocean is through the Bali and Lombok straits. The main inflow of water is from the Pacific through the BANDA SEA and the FLORES SEA.

Most oceanographers do not recognize the Bali Sea as a separate sea because its waters are not sufficiently different from those of the Java and Flores seas.

BALTIC SEA is an enclosed sea (intracontinental) in northern Europe and is bordered by Sweden, Finland, Russia, Poland, Germany, and Denmark. The Baltic communicates with the NORTH SEA through a series of winding channels between Sweden and Denmark—the Skagerrak, the Kattegat, and the Danish Sound. The latter is made up of the Little Belt (Lillebaelt), Great Belt (Storebaelt), and the Sound (Öresund). The Baltic is about 930 mi (1496 km) long and ranges from 50 to 425 mi (80 to 683 km) wide. Its coastline is about 5000 mi (8045 km) long. The Gulf of Bothnia is a long arm of the sea between Sweden and Finland, the Gulf of Finland extends between Finland and Russia, and the Gulf of Riga is a U-shaped indentation in the coast of Russia. Smaller indentations in the German coast bound the bays of Pomerania, Mecklenburg, and Kiel. The principal islands are Rugen, Bornholm, Öland, Gottland, Osel, Dagö, and the Åland and Danish groups. The Baltic covers an area of 148 996 mi² (386 000 km²), occupies a volume of 7917 mi³ (33 000 km³), and has a mean depth of 282 ft (86 m). The greatest depth is 1506 ft (459 m).

The Baltic Sea occupies a depression in the basic rock structure of northern Europe, and was a part of the ocean several times before the last Ice Age (11 000–8000 years ago). During the Ice Age the Baltic region was covered by a continental glacier

that extended into Europe from the north. As this glacier retreated northward, it left behind it the Baltic Ice Lake which was dammed by both the leading edge of the glacier and an uplift in the region of the Danish Sound and Belt region. For a time, after the glacier had retreated into northern Sweden, the Baltic Ice Lake drained into the WHITE SEA on the rim of the ARCTIC OCEAN. Then, about 6000–5000 B.C., the Danish Sound and Belt region subsided to allow drainage into the NORTH SEA. This transition to the present Baltic is known as the Littorina Sea.

BALTIC SEA. Northern Germany and Denmark are the highly visible land masses in this Skylab 3 photo, and the Baltic Sea stretches out to the right. Sweden lies just above it, swathed in cloud cover. (*NASA*)

The water around the cluster of islands at the entrance to the Gulf of Bothnia is referred to as the Åland Sea. The depth in this region reaches 984 ft (300 m). The maximum depth of the Gulf of Bothnia is 833 ft (254 m), the Gulf of Riga is 164 ft (50 m), and the Gulf of Finland is 328 ft (100 m). The sediment covering the floor of the Baltic is pre-

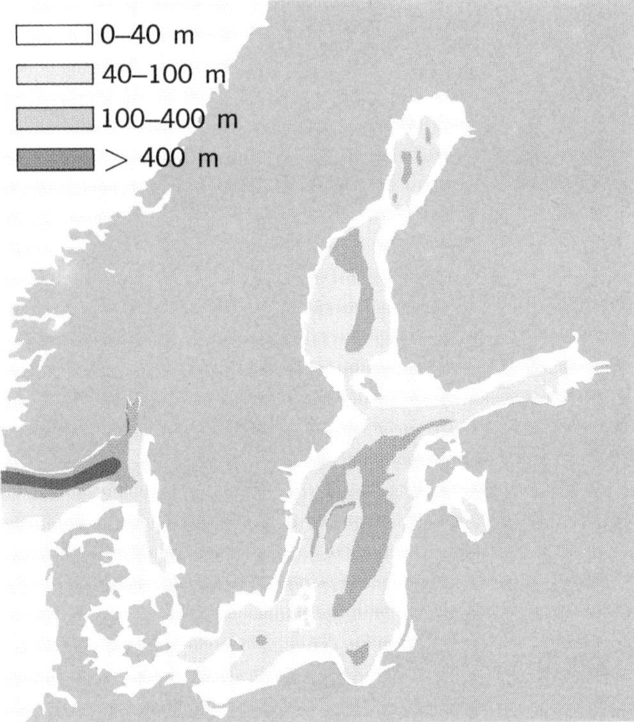

MAP 3. Contours of the Baltic Sea clearly show depth variation.

dominantly muds and clays, with some sand in the shallower coastal areas.

The water circulation in the Baltic Sea is strongly influenced by the inflow of fresh WATER from the surrounding landmass. The annual inflow of fresh water, plus precipitation at sea, minus evaporation, amounts to one-fortieth of the total volume of the sea. The low-salinity surface water (6–20 ppt) flows into the North Sea through the Sound and Belt region; while the higher-salinity North Sea water (20–30 ppt) flows into the Baltic beneath the lighter water. A fairly constant salinity is maintained in this fashion.

The annual range of temperatures is 32–64.4° F (0–18° C) for surface water and 35.6–41° F (2–5° C) for the deeper water. In very severe winters the entire Baltic freezes over with ice that ranges to 2½ ft (0.76 m) in thickness. In milder winters only the northern section of the Baltic becomes ice-covered. Navigational hazards due to fixed or drifting ice last from December to May. The ice disappears by June.

Tidal ranges in the Baltic Sea are very small, being on the order of 6 in (15 cm) off Copenhagen and less than 1 in (2.5 cm) off Stockholm. However, 10-ft (3-m) changes in sea level may be encountered in the Gulf of Bothnia as a result of storm surges.

The Baltic Sea is the largest body of brackish water in the world. The combination of brackish water, low nutrient level in the northern rivers, and restricted flow through the Danish Sound region makes for very poor FISH populations as compared with the more fertile North Sea. However, commercial catches of HERRING, TROUT, SALMON, FLOUNDER, COD, PLAICE, and EEL are taken.

BANDA SEA, located in the western PACIFIC OCEAN, is bounded by Sulawesi on the west, by the islands of Buru and Ceram on the north, by Ewab on the east, and by Tanimbar, Timor, and Flores on the south. The Banda Sea covers an area of 268 270 mi² (695 000 km²), occupies a volume of 510 768 mi³ (2 129 000 km³), and has a mean depth of 10 052 ft (3064 m). The maximum depth is 24 409 ft (7440 m).

Ridges with complicated topography divide the Banda Sea into the North and South Banda basins, Ambalau Basin, Manipa Basin, and the Batung and Weber troughs. Terrigenous and volcanic muds and globigerina ooze comprise the bottom sediments.

Pacific Ocean WATER flows into the Banda Sea through the MOLUCCA SEA. Surface currents flow generally eastward during the winter and westward during the summer in response to the prevailing monsoons. Surface temperatures range from 84° F (28.9° C) in summer (Southern Hemisphere) to 78° F (25.6° C) in winter.

BANK is the term for a relatively flat-topped elevation of the ocean floor over which the WATER is fairly shallow but still sufficient for surface navigation. Examples are the famous fishing grounds off the Newfoundland coast known as the Grand Banks, and the area off the southern coast of Iceland known as the Lousy Banks.

BAR See BEACH.

BARBER FISHES are a group of marine fishes that are distinguished by their behavior of cleaning other fishes of parasites, growths, and damaged or infected tissue. They are sometimes called doctor fishes and cleaner fishes. See WRASSE.

BARENTS SEA, a sea of the ARCTIC OCEAN, is located off the northwest coast of the USSR and the north coast of Norway. On the north, the Barents is separated from the Arctic Basin by the Spitsbergen

(Svalbard) Archipelago at the western corner and Franz Josef Land at the eastern corner. In the east the Barents is separated from the KARA SEA by the long island of Novaya Zemlya and a line north to Graham Bell Island. On the west the Barents merges with the waters of the NORWEGIAN SEA, the established boundary between the two being a line running from North Cape on the coast of Norway, through Bear Island, to the south point of West Spitsbergen. The boundary between the Barents and the WHITE SEA to the south is a line joining Svyatoi Nos on the Murman Coast to Cape Kanin on the Kanin Peninsula. Thus defined, the Barents has an area of 501 800 mi² (1 300 000 km²), a mean depth of 751 ft (229 m), and a volume of 77 251 mi³ (322 000 km³).

The waters of the Barents Sea as far north as Spitsbergen and as far east as Novaya Zemlya were familiar to Russian and Scandinavian sailors, seal hunters, and fishermen at least as far back as the twelfth century. The sea takes its name from Willem Barents, a Dutch ship captain, seal hunter, and navigator who crossed the Barents in an unsuccessful attempt to find the Northeast Passage to the Orient in 1596–1597. Nearly 300 years later (1878–1879) Nils Adolf Erik Nordenskjöld, a Swedish polar explorer, mineralogist, and map authority, entered the Barents through the Norwegian Sea to become not only the first to study it scientifically but also the first to negotiate the Northeast Passage by sailing on eastward through the Kara Sea, LAPTEV SEA, and EAST SIBERIAN SEA, and out into the PACIFIC OCEAN through the Bering Strait. The next, and more thorough, scientific study of the Barents was made in 1893 by Fridtjof Nansen, the Norwegian polar explorer, marine zoologist, and pioneer oceanographer, during his attempt to reach the North Pole aboard the *Fram*.

The geographic location of the Barents Sea and several of its physical features combine to produce characteristics that are quite distinctive among the Arctic seas. In addition to being totally confined to a CONTINENTAL SHELF, it rests on one of the widest continental shelves in the world at 750 mi (1206 km). At the northern limit of the shelf, the landmasses of Spitsbergen and Franz Josef Land shield the Barents from the permanent ice pack of the Arctic Ocean, while the long island of Novaya Zemlya acts as a barrier to the near-year-round ice of the Kara Sea to the east. On the west the Barents is open to the ATLANTIC OCEAN by way of the Norwegian Sea. As a result, the comparatively warm (4–12° C) and saline (over 35 ppt) waters of the North Cape Current, a branch of the GULF STREAM, flow in from the west between North Cape and Bear Island

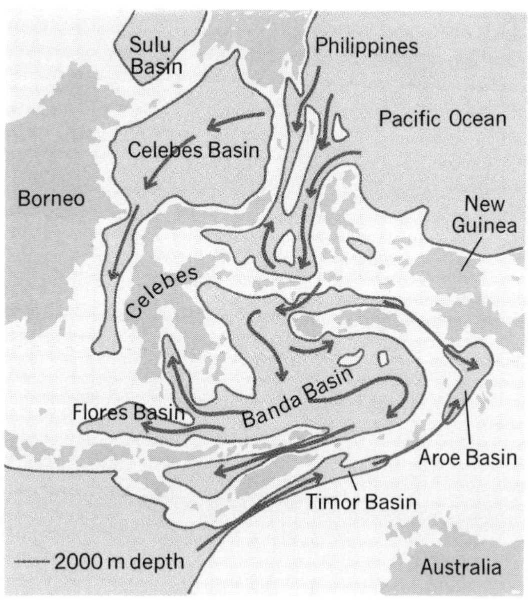

MAP 4. Current flow on the Banda Sea.

to break into many branches and mix with the colder and less saline waters of the Barents Sea.

The results of this combination of factors are quite remarkable. The Barents lies in the path of warm cyclones from the North Atlantic and cold anticyclones (see HURRICANE) from the Arctic, producing a very unstable climate and one of the most storm-ridden stretches of WATER in the world; storm waves reaching 12 ft (3.6 m) in height often pound the southern coast. On the other hand, higher air temperatures produce more temperate winters than are found in other Arctic seas, and precipitation is well above average for the region. The northern three-quarters of the Barents usually freezes during the winter, but the southern quarter, except for fjords along the Norwegian and Murman coasts, is open to water transportation all year. Except in the middle of the sea and along the coast of Novaya Zemlya, floe ice melts in the summer even during the colder years. The boundary between the warm Atlantic and cold Arctic waters on the west produces a phenomenon known as a "polar front," whereby rich nutrients, such as phosphorous and nitrogen, are mixed with more habitable waters to attract and sustain an abundance of plant and animal life.

Many small rivers flow into the Barents Sea, but the only major one is the Petchora. Small icebergs 39–49 ft (12–15 m) long are supplied by the glaciers on Novaya Zemlya and Franz Josef Land, but they usually melt in waters near their point of origin. This scarcity of sediment-carrying rivers and icebergs contributes to the exceptional clarity and transparency of the Barents Sea water. See FJORD; ICEBERG.

The bottom of the Barents Sea is complex and irregular. There is a gentle tilt from east to west with depths of 328–1145 ft (100–350 m) over most of the sea, dropping to 1968 ft (600 m) near the boundary with the Norwegian Sea. Locally, the bottom undulates with gentle rises and basins which complicate the distribution of water masses and sediment types.

Maximum tides in the Barents Sea range from 13 ft (4 m) at North Cape and 30 ft (7 m) at Strait Gorlo on the WHITE SEA to 5 ft (1.5 m) at Spitsbergen and 2.6 ft (0.8 m) in the vicinity of Novaya Zemlya.

BARNACLES, the common name for several species of CRUSTACEANS, which constitute the subclass Cirripedia. Although barnacles somewhat resemble CLAMS they are, in fact, cousins to lobsters. Unlike their cousins, barnacles are unable to move from a site (e.g., rock, piling, bottom of a ship) once they fix and attach themselves. Barnacles feed by projecting their legs out through an opening in the top of the shell. When the legs are out in the water they uncurl, and the feeding appendages or cirri located at the tips of each of the six legs are curved toward the mantle of the animal. These cirri on small hairlike projections hence act as a net to collect DIATOMS and other floating material in the water.

There are four orders of barnacles and over 800 recognized species. The order Thoracia contains such genera as *Lepas, Mitella,* and *Balanus,* and the term *barnacle* is usually associated with this group, which contains both the free-living and commensal forms, most of which are hermaphroditic.

BARRACUDA is the common name for some 20 species of marine fishes that belong to the genus *Sphyraena* (family Sphyraenidae) in the order Perciformes.

Barracuda are widely distributed in the tropical and subtropical waters of the world's oceans. The great or predatory barracuda, *Sphyraena barracuda,* is found off Brazil, in the Caribbean, off Florida, and in the Indo-Pacific area from the RED SEA to the Hawaiian Islands. Swift-swimming torpedo-shaped fishes, often attaining 6–8 ft (1.8–2.4 m) in length and possessing a long, jutting lower jaw with numerous long and sharp caninelike teeth, they are feared more than the sharks in some areas. While the reputation of these fishes as dangerous to swimmers is perhaps somewhat disputable, barracuda are unpredictable in behavior and are voracious predators that hunt by sight rather than by smell, as do the sharks. They often will attack any bright moving object indiscriminately, and they have been known to attack people in the water. See SHARK.

The northern barracuda species that inhabits the western north ATLANTIC OCEAN attains a length of about 18 in (46 cm), while the European barracuda, or spet, of the MEDITERRANEAN SEA and Eastern Atlantic is around 3 ft (0.9 m). Other species are the Indian barracuda and Commerson's barracuda, both of the INDIAN OCEAN, and the California barracuda.

The barracuda and the WRASSE fish have established a relationship, known as mutualism, which is beneficial to both. The wrasse swims fearlessly into the mouth of the barracuda and among its dagger-sharp teeth to help keep its host infection-free by feeding on the BACTERIA that grow in the host's mouth.

Although all the barracuda are considered to be good food fishes, at certain times the flesh of these fish becomes poisonous. This condition is caused by toxins that originate from toxic ALGAE and DIATOMS, which are consumed by plant-eating fishes or herbivores that, in turn, are eaten by predators such as the barracuda. The poison is known as ciguatera.

BARRIER BEACH See BEACH.

BARRIER REEF See CORAL REEF.

BASIN denotes a large, more or less circular or oval, depression on the ocean floor. The term *basin* is also often used to describe the entire depression that holds the world's oceans. Examples of smaller basins are the Natal Basin south of Madagascar and the Wharton Basin off the northwest coast of Australia.

BATHYTHERMOGRAPH is the name for a device used to obtain a record of water temperature as a function of depth (actually, HYDROSTATIC PRESSURE) in the ocean from a ship underway; it is abbreviated BT. A BT may be operated mechanically or electronically and is sometimes called a bathythermosphere.

See also INSTRUMENTATION; OCEANOGRAPHY.

BAY See OCEAN.

BAY OF BENGAL, an arm of the INDIAN OCEAN, is partially enclosed by Sri Lanka (Ceylon) and India on the west and north, and by Burma and the Andaman-Nicobar Ridge on the east. The southern boundary is a line joining Dondra Head on the southern tip of Sri Lanka to Poeloe Bras at the northwestern tip of Sumatra. The distance along this southern boundary is about 1000 mi (1609 km). The Andaman-Nicobar Ridge, which supports the Andaman Island group to the north and the Nicobar Island group to the south, separates the bay from the ANDAMAN SEA to the east. The great delta of the Ganges and Brahmaputra rivers lies at the head of the bay. Other rivers entering the bay are the Mahanadi, Codavari, and Kistna rivers of India. The Bay of Bengal (taken from the name of the Mogul state of India) covers an area of 838 392 mi^2 (2 172 000 km^2), occupies a total volume of 1 347 335 mi^3 (5 616 000 km^3), and has a mean depth of 8484 ft (2586 m).

The CONTINENTAL SHELF of the Bay of Bengal is up to 100 mi (161 km) in width—being wider on the north and east—and has an average depth of 600 ft (183 m) at its seaward edge. The shelf, composed mostly of sand grading seaward to clay and mud, is cut in several places by submarine canyons. Among these are the Ganges Canyon off the Ganges-Brahmaputra Delta with a depth of up to 2400 ft (732 m), and the Andhra, Krishna, and Mahadevan canyons along the western rim of the Bay. See SUBMARINE CANYON.

The deep basin of the Bay of Bengal is roughly U-shaped and lies at depths of up to 14 764 ft (4500 m). The basin floor is dominated by two features: the northern extremity of the remarkably straight, 3107-mi (5000-km) long Ninety East Ridge (see INDIAN OCEAN), and the Ganges Cone built by the flow of sediment into the deep basin from the shelves. The top of the Ninety East Ridge lies at a depth of around 7000 ft (2134 m), and its northern tip is being covered by sediment from the Ganges Cone. The cone shows a dendritic pattern of channels (fan valleys) down which sediment is moved long distances to the deep basin.

The seasonal monsoons strongly influence the surface circulation in the Bay of Bengal. During the spring and summer the wet southwest monsoon sets up a clockwise circulation in the Bay which reverses to a counterclockwise flow under the drag of the northeast monsoon of fall and winter. The shape of the Bay tends to focus the forces that act on it so that significant tidal ranges, seiches, and internal WAVES are experienced.

BEACH, in proper use, refers to that zone of unconsolidated material which extends landward from the low-WATER line to the point where there is a marked change in material or physiographic form, or to the line of permanent vegetation (usually the effective limit of erosion by storm waves). Beaches, by this definition, exist in a high-energy environment; therefore, their structure is complex, they are always in a state of transition, and their form is very sensitive to the size of beach material, the existence of offshore bars, the tidal range, frequency of storms, and the wavelength, height, and direction of waves which break upon them. The word *beach* brings to mind the white sands typical of those along the East Coast of the United States and the Gulf Coast of Florida which result from the weathering of granite into its two major components: quartz and feldspar. However, many Pacific islands have black beaches derived from the weathering of the volcanic rock out of which the island is created, and many European beaches are made up of flat rocks, called shingles, which range up to a foot in diameter and are created by sliding back and forth in the surf zone over many years.

The United States has a total of 84 249 mi (135 542 km) of shoreline, with Hawaii accounting for 930 mi (1496 km) and Alaska 47 300 mi (76 106 km). Some 34 520 mi (55 543 km) (41 percent) of the total is exposed shoreline, and 49 720 mi (80 000 km) (59 percent) is sheltered in bays, lagoons, and estuaries. Exclusive of Alaska, about 12 150 mi (19 550 km) (33 percent) of the U.S. shoreline have beaches, while the remaining 24 790 mi (39 887 km) do not. At present, about 20 500 mi (32 985 km) (24 percent) of the total shoreline is being eroded to some significant degree.

The anatomy of a slightly idealized, though fairly typical beach might begin with an offshore bar some distance seaward of the low-water line. The offshore bar, though not properly a part of the beach, is typical of all beaches which are exposed to significant wave action throughout the year, including those around the Great Lakes. The bar is built by material eroded seaward from the beach by extreme wave action such as that during storms. And, since more storms occur during winter, the bar is more pronounced during that season. Under the more gentle surf action of summer, the bar, along with the area just behind it (seaward), feeds some of its material back to the beach. The crest of a typical offshore bar lies at a depth roughly equal to the average height of the surf which breaks across it.

The offshore bar is an important feature in that it serves as a less-than-perfect barrier against beach erosion. The friction of the abruptly shoaling bottom causes incoming waves to break and lose much of their energy before they reach the beach. If the waves are large enough, they may form again to break against the beach, but with greatly reduced energy. Occasionally, breaks occur in the offshore bar; these then serve as a raceway for the return of water carried across the bar by waves; the result is the dangerous rip currents which can be troublesome to swimmers. However, though powerful, these currents are usually very narrow, and a swimmer can escape by swimming parallel to the beach.

Under certain conditions of long duration, offshore bars may grow into permanent barrier beaches which stand many feet above the high-tide line. Typical of barrier beaches are those which extend down much of the East Coast of the United States to form the well-known "outer banks" on which Cape Hatteras is located. Protection provided by the barrier beach is so good that the "beach" to which the tourists flock is now on the seaward edge of the barrier beach, while the former beach is a rather quiet area with little sand and with vegetation running down near to the high-tide line.

Between the offshore bar and the low-tide line is a trough which often runs to 6 ft (1.8 m) or more in depth. It is into this trough that the waves plunge as they break over the offshore bar, and it is along this trough that powerful currents, called longshore currents, transport large quantities of sand parallel to the beach. Longshore currents are particularly pronounced where waves strike the beach at an angle.

Lying between the low-tide and high-tide lines is a strip of sand called the low-tide terrace, the width of which depends on the tidal range and the steepness of the terrace. The steepness, in turn, depends to some degree on the coarseness of the material out of which it is made; the coarser the material, the steeper the terrace. It is along this terrace that bathers like to walk at low tide to search for seashells. Behind the low-tide terrace is a second terrace, called a *berm,* which is exposed only to those water levels which accompany storm waves. Some beaches may have more than one berm, and it is from these terraces that sand is depleted during the winter and renewed during the summer. It is also that area of the beach where most sunbathing takes place.

The cycle of beach erosion and replenishment has been repeating itself since the first continent stood above sea level. And the wearing away of the continent by wind, water, and chemical erosion has supplied durable sands such as quartz to the rivers which, in turn, have carried these sands to the sea to replace what has been lost to the beach cycle system. But now human beings are growing increasingly reluctant to have the continents eroded out from under them and are taking steps to stabilize farmlands, timberlands, and the sites of their cultural developments. Therefore, the supply of new sand to the sea has been reduced in some areas, though not alarmingly so.

In the face of some reduction in the rate that new sand is supplied to the sea, the rate of beach erosion and sand migration along the coast goes on with but little retardation. The steep, high-frequency waves of winter storms steal as much as 8 ft (2.4 m) of sand from some beaches, and the strong longshore currents, which have a single, predominant direction of flow off most beaches, move the eroded sand parallel to the beach until the long, low surf of summer uses it to renew some distant beach. Along some coasts the amount of sand moving past a given point may range from 100 000 to 2 million (2×10^6) yd^3 (76 500 to 1 530 000 m^3) in a year. Along deserted coasts this process is little noticed, but in resort areas, marinas, and commercial ports the process not only is of intense interest but can be disastrous if it becomes unbalanced. Therefore a number of structures and techniques have been devised to control instabilities in the natural system and to correct for those imposed by human beings.

Bulkheads, seawalls, and revetments made of steel, wood, concrete, or large rocks are used well back on the beach to serve as a last defense against the damage of severe storm waves. Care must be taken in the building of these structures because the waves strike with great force, and the downrush of water from a breaking wave may seriously undermine the sand at the base. A stone apron is often necessary to prevent scouring and undermining.

Breakwaters, usually oriented parallel to the beach and constructed by piling large stones on the bottom until the structure rises above the high-tide level, are placed offshore to shelter a beach area from wave action. However, such a structure often creates its own problems because the velocity of the longshore current flowing between the breakwater and the beach is reduced, causing the migrating sand the current is carrying to be stopped. This process not only causes the sheltered beach to grow seaward but also robs the beaches downdrift of the sand needed for renewal. The breakwater and beach at Santa Monica, California, comprise an excellent example of this problem. In such cases the stalled sand must be pumped across the sheltered area before it will continue on its way through natural forces.

Groins are low structures that begin in the backshore area and run a short way out to sea, usually perpendicular to the beach. Their purpose is to retard erosion on the beach and to build depleted

BEACH. (*Left*) The terrace rises smoothly to the undulating berm, and small dunes are visible in the background. The fences are part of an effort to catch the moving sands and build up dunes so that the beach will not be eroded by the wind and water. (*Right*) Westward-moving currents are transporting sand from the east side of Fire Island, a barrier beach, to the western end. Eventually, this process will in effect move the island. (*John Sanders*)

beaches seaward. Groins also retard the longshore drift of sand until the beach area has built out and stabilized. Often this stabilization is hastened by trucking sand in to build the beach back more quickly. Jetties are similar to groins, but are generally longer since their main purpose is to protect inlets and shipping channels from being blocked by the buildup of sand.

Since none of these structures works perfectly, a technique called artificial beach nourishment is used in some beach areas. This is simply the replenishment of beach sand from other sources such as sand dunes and from areas beyond the offshore bar.

BEAUFORT SEA, a sea of the ARCTIC OCEAN, is located off the northern coast of Alaska and western Canada. The Beaufort is bounded on the west by the CHUKCHI SEA in the vicinity of Point Barrow, Alaska; on the north by a line from Point Barrow to Cape Land's End on Prince Patrick Island; and on the east by a line from Cape Land's End through Banks Island and Cape Kellitt to Cape Bathurst on the Canadian mainland. The Beaufort Sea covers an area of 183 736 mi^2 (476 000 km^2), occupies a volume of 114 676 mi^3 (478 000 km^3), and has an average depth of 3294 ft (1004 m).

The Beaufort Sea takes its name from Admiral Sir Francis Beaufort, hydrographer of the British Navy, who, in 1806, devised the BEAUFORT WIND SCALE for reporting the estimated wind force at sea. Alexander Mackenzie was the first European to sight the Beaufort Sea when he followed the Mackenzie River to its mouth in 1789. Thirty years later, in 1819, Lieutenant (later Sir) William Edward Parry of the British Navy, commanding the *Hecla,* sighted the Beaufort Sea at the western end of

The Modern Beaufort Wind Scale

Beaufort number	Wind speed kn	Wind speed km/h	Nautical term	U.S. Weather Bureau term	Hydrographic Office: Term and height of waves, ft	Hydrographic Office: Code
0	under 1	under 1	Calm		Calm, 0	0
1	1–3	1–5	Light air	Light	Smooth, less than 1	1
2	4–6	6–11	Light breeze		Slight, 1–3	2
3	7–10	12–19	Gentle breeze	Gentle	Moderate, 3–5	3
4	11–16	20–28	Moderate breeze	Moderate		
5	17–21	29–38	Fresh breeze	Fresh	Rough, 5–8	4
6	22–27	39–49	Strong breeze			
7	28–33	50–61	Moderate gale	Strong		
8	34–40	62–74	Fresh gale		Very rough, 8–12	5
9	41–47	75–88	Strong gale	Gale	High, 12–20	6
10	48–55	89–102	Whole gale		Very high, 20–40	7
11	56–63	103–117	Storm	Whole gale	Mountainous, 40 and higher	8
12	64–71	118–133				
13	72–80	134–149				
14	81–89	150–166	Hurricane	Hurricane	Confused	9
15	90–99	167–186				
16	100–108	184–201				
17	109–118	202–220				

McClure Strait, but because of heavy ice was forced to turn back before actually entering the sea. In 1825–1826 Parry's friend John Franklin entered the Beaufort Sea via the Mackenzie River and explored as far west as Beechey Point on the north coast of Alaska. Eleven years later, in 1837, Thomas Simpson of the Hudson's Bay Company explored the coast between Beechey Point and Point Barrow on the edge of the Chukchi Sea to complete the exploration of the southern rim of the Beaufort Sea.

The final conquest of the Beaufort Sea, and much that remained unknown of the Canadian Arctic, was accomplished during the decade that began in 1848. This surge of exploration, consisting of some 40 expeditions, was largely due to the search for the crews of the *Erebus* and the *Terror* under the command of Sir John Franklin. Franklin departed England in 1845 in search of the Northwest Passage, only to be locked in the ice off King William Island. In April 1848, with supplies running dangerously low, what remained of both crews departed their imprisoned ships in a hopeless attempt to reach the Canadian mainland; none survived. Most significant to the exploration of the Beaufort Sea were the searches carried out by Robert McClure aboard the *Investigator* and Richard Collinson aboard the *Enterprise.* Both ships rounded Cape Horn in 1849 and proceeded, independently, to the Arctic through the Bering Strait. McClure sailed across the Beaufort Sea to Banks Island where he discovered Prince of Wales Strait, a key link in the elusive Northwest Passage. Collinson, the following year, also searched the Beaufort Sea as far east as Banks Island.

Although the term *Beaufort Sea* is firmly established in the geographical literature and recognized by the International Hydrographic Bureau, no physiographic or oceanographic justification exists for its separate identity. The characteristics of the WATER mass and its circulation make the Beaufort an integral part of the Arctic Ocean. The CONTINENTAL SHELF, rarely exceeding 93 mi (150 km) in width, is the narrowest of those on the rim of the Arctic.

The major rivers emptying their load of sediment and gravel into the Beaufort Sea are the Anderson, Colville, and Mackenzie. Annual ice scouring, and perhaps some glaciation during the Quaternary period, have transformed these riverborne gravels

International		Estimating wind speed	
Terms and height of waves, ft	Code	Effects observed at sea	Effects observed on land
Calm, glassy, 0	0	Sea like mirror.	Calm; smoke rises vertically.
		Ripples with appearance of scales; no foam crests.	Smoke drift indicates wind direction; vanes do not move.
Rippled, 0–1	1	Small wavelets; crests of glassy appearance; not breaking.	Wind felt on face; leaves rustle; vanes begin to move.
Smooth, 1–2	2	Large wavelets; crests begin to break; scattered whitecaps.	Leaves, small twigs in constant motion; light flags extended.
Slight, 2–4	3	Small waves, becoming longer; numerous whitecaps.	Dust, leaves, and loose paper raised up; small branches move.
Moderate, 4–8	4	Moderate waves, taking longer form; many whitecaps; some spray.	Small trees in leaf begin to sway.
Rough, 8–13	5	Larger waves forming; whitecaps everywhere; more spray.	Larger branches of trees in motion; whistling heard in wires.
		Sea heaps up; white foam from breaking waves begins to be blown in streaks.	Whole trees in motion; resistance felt in walking against wind.
Very rough, 13–20	6	Moderately high waves of greater length; edges of crests begin to break into spindrift; foam is blown in well-marked streaks.	Twigs and small branches broken off trees; progress generally impeded.
		High waves; sea begins to roll; dense streaks of foam; spray may reduce visibility.	Slight structural damage occurs; slate blown from roofs.
High, 20–30	7	Very high waves with overhanging crests; sea takes white appearance as foam is blown in very dense streaks; rolling is heavy and visibility reduced.	Seldom experienced on land; trees broken or uprooted; considerable structural damage occurs.
Very high, 30–45	8	Exceptionally high waves; sea covered with white foam patches; visibility still more reduced.	
Phenomenal, over 45	9	Air filled with foam; sea completely white with driving spray; visibility greatly reduced.	Very rarely experienced on land; usually accompanied by widespread damage.

into numerous low islands near the coast. The largest of these are Barter and Herschel islands with an area of 5.4 and 7.3 mi² (14 and 19 km²), respectively. Further from the coast the continental shelf is incised by three submarine canyons.

The water column in the Beaufort Sea consists of four distinct layers: a layer of Arctic water at the surface which is about 328 ft (100 m) thick and is the coldest of the four layers [28.9° F (−1.7° C) by the end of winter]; a layer of warmer Pacific water which enters through the Bering Strait and is about 328 ft (100 m) thick; a 2296-ft (700-m) thick layer of Atlantic water, the warmest of the four layers [32° F (0° C)]; and a bottom water mass which lies between the bottom and a depth of about 2953 ft (900 m). The predominant circulation in the sea is clockwise and serves to pile ice along the Canadian and Alaskan coasts for much of the year. Assured navigation can only be expected from mid-August through September.

BEAUFORT WIND SCALE is a system of estimating and reporting wind speed. It was introduced in 1806 by Sir Francis Beaufort, a famous British hydrographer. Beaufort devised the scale as an aid in determining the amount of canvas that a full-rigged frigate of the period could effectively carry under different wind speeds. With the advent of steam, the scale was modified to accommodate the more modern requirements. In its present form, the scale equates Beaufort force (or Beaufort numbers) to wind speed, includes descriptive terminology, and describes the visible effects of the wind on land objects and the sea surface. The modern Beaufort scale is given in the table on pp. 34–35.

BELLINGSHAUSEN SEA is located off the coast of Antarctica and is a marginal sea of the South PACIFIC OCEAN. The sea lies between the Thurston Peninsula on the west and the Antarctic Peninsula on the east. The ANTARCTIC CIRCLE serves as its northern boundary.

The Bellingshausen Sea is named for Baron Fabian Gottlieb von Bellingshausen, a Russian admiral, who, in 1819, led an expedition to Antarctic waters at the direction of Alexander I. Bellingshausen is generally considered to be the first great antarctic explorer after James Cook of England (see COOK, JAMES). Commanding the *Vostok* and *Mirny*, Bellingshausen circumnavigated the continent, as had Cook, and sighted land on several occasions. Though he did not claim credit for it, he is considered the first to actually discover the continent that others had missed because of ice and low visibility.

The Bellingshausen Sea includes Ronne Bay and Marguerite Bay. Islands in the sea are Alexander I Island and Charcot Island on the eastern side, Fletcher Island on the west, and Peter I Island between the coast and the Antarctic Circle. WATER depths drop off to over 12 000 ft (3658 m) in the vicinity of the Antarctic Circle. The bottom sediments show the strong influence of glaciation. The water behind the islands is usually icebound all year, and ice extends out to sea during the Southern Hemisphere winter.

BENGUELA CURRENT flows northward along the west coast of Africa and is distinguishable between about 15 and 35° south LATITUDE. The Benguela is less swift than the AGULHAS CURRENT, part of which at certain times of the year rounds the Cape of Good Hope to add its warmer waters to those of the cold Benguela. The name Benguela is reserved for a narrow area of UPWELLING within about 100 mi (160 km) of the coast. The main northward flow of water further west is called the SOUTHEAST TRADE-WIND DRIFT. The temperature of the Benguela WATER ranges from a low of around 59° F (15° C), in the vicinity of the Cape of Good Hope, to 71.6° F (22° C) in the LATITUDE of Cape Frio. The current velocity rarely exceeds 10 in/s (25 cm/s).

The cold, nutrient-laden water which rises to the surface to form the Benguela Current supports abundant marine life and a sizeable fishing industry. However, if for any reason the forces (wind direction and velocity) which sustain the upwelling conditions are interrupted for an extended period of time, the oxygen supply to the shallow areas over the CONTINENTAL SHELF is reduced. In the northern part of the Benguela Current this depletion of oxygen causes an explosion of bacterial action in the bottom sediments. These sulfate-reducing bacteria produce hydrogen sulfide (H_2S) in such quantities that large fish kills may result, and the stench of hydrogen sulfide may be detectable up to 40 mi (65 km) inland.

BERING SEA lies on the northern rim of the PACIFIC OCEAN and is bounded by Alaska, Siberia, and the Aleutian Islands. Officially the northern limit of the Bering is taken to be the southern limit of the CHUKCHI SEA, which in turn, is the ARCTIC CIRCLE. A more practical northern boundary, however, is the narrow point of the Bering Strait. The southern boundary is provided by a line running from Kabuch Point on the Alaska Peninsula, through the Aleutian Islands, to the southern extreme of the Komandorski Islands, and on to Cape Kamchatka in such a way that all the narrow WATER between Alaska and Kamchatka is included. The Bering Sea covers an area of 889 344 mi² (2 304 000 km²), has a mean depth of 5243 ft (1598 m), and occupies a volume of 883 359 mi³ (3 683 000 km³). Its greatest depth is 14 501 ft (4420 m).

Evidence indicates that the Russian Cossack Simon Dezhnev was the first to explore the Bering Sea. Dezhnev and a small party are reported to have sailed eastward from the mouth of the Kolyma River on the EAST SIBERIAN SEA in 1648. Rounding East Cape (Cape Dezhnev), they sailed through Bering Strait, into the Bering Sea, and westward to the mouth of the Anadyr River on the southern edge of the Chukchi Peninsula. The Sea, of course, was named for Vitus Bering, the Danish navigator who served with the Russian navy during the Great Northern Expedition which lasted from 1724 to 1749. In 1728 Bering departed from his staging area in the Sea of Okhotsk and sailed through the Bering Sea and northward through the Bering Strait to enter the southern Chukchi Sea. See OKHOTSK, SEA OF. And, in 1778, Captain James Cook, aboard the *Resolution*, sailed northward through the Bering Strait to become the first man to cross both the ARCTIC CIRCLE and the ANTARCTIC CIRCLE. See COOK, JAMES.

The Bering Sea floor is divided into two zones

BERING SEA. Fairway Rock is a mile-long bare islet in the Bering Strait near the USSR territorial line. This photograph was taken through the periscope of the nuclear submarine *U.S.S. Nautilus* as it prepared for the historic submerged run to the North Pole in 1957. *(Oceanographer of the Navy)*

which are approximately equal in area. The northeastern half of the sea is underlain entirely by one of the largest continental shelves in the world, extending as much as 400 mi (643 km) offshore. Extending through the Bering Strait and into the Chukchi Sea, this shallow CONTINENTAL SHELF of less than 656 ft (200 m) restricts the flow of all but surface water into the Arctic Basin. The second zone occupies the southwestern half of the Sea and consists of a deep basin, the maximum depth of which is 14 501 ft (4420 m). This remarkably smooth basin floor, which generally lies at a depth of 12 467–12 795 ft (3800–3900 m), is broken by two ridges: Olyutorski Ridge, which penetrates the basin from the north, and the unusual Rat Island Ridge which curls into the basin in a counterclockwise fashion from its beginning on the Aleutian Island arc. The two ridges divide the deep zone into an eastern and western basin. The deep zone also represents a rapidly filling sediment basin in which some 1.2–2.5 mi (2–4 km) of sediment now rests on the basaltic basement rock.

Several islands rise from the smooth floor of the Bering shelf: notably St. Lawrence Island, Nunivak Island, and the Pribilof Islands. At its edge the shelf plunges steeply to the deep basin with a slope of 4–5°. In the southeastern corner near the Aleutian chain the shelf is deeply cut by Bering Canyon, perhaps the largest SUBMARINE CANYON in the world. The canyon is over 100 mi (161 km) long, has in excess of 50 tributaries, and is over 20 mi (32 km) across where it is most deeply incised. In places the canyon walls stand 6000 ft (1829 m) above the gently sloping (½°) floor. The sediments of the Bering shelf are sands and gravels grading to silts at the base of the slope. The deep basin, by contrast, is covered by diatomaceous ooze.

Air temperature over the Bering Sea ranges from −13° F (−25° C) during winter to 50° F (10° C) in summer. During the winter as much as 90 percent of the Sea becomes ice-covered, but during the summer the Sea is entirely free of ice.

Water currents in the Bering Sea are produced by wind action, the inflow of water from the Pacific

BIOLOGICAL OCEANOGRAPHY. (*Top*) The research vessel *Albatross IV,* specifically designed for the study of marine biology, is equipped with a large opening at the stern for trawling. (*Arnold Gordon*) (*Right*) A diver examines coral during a saturation dive off Grand Bahama Island. The base of operations for the four participating divers was a cylindrical 5 × 2.5 m (16 × 8 ft) Perry Hydro-Lab habitat submerged to a depth of about 15 m (50 ft). (*Dr. George Lewbel*)

Ocean through the Aleutian Island chain, tides, and the inflow of fresh water from rivers. The primary current pattern is a cyclonic gyre over the deep basin. Part of this current flows northward through the Bering Strait and part flows back into the Pacific. Currents over the shelf are primarily due to tidal action, except near the Alaskan Coast. Here fresh water from numerous rivers flows north, with most of it passing through the Bering Strait into the Chukchi Sea.

The Bering Sea is exceedingly fertile in marine life. Two PLANKTON blooms—one in the spring and the other in the fall—consist primarily of DIATOMS and provide the basic link in the food chain. King crab, SHRIMP, and some 315 species of FISH, 25 of which are of commercial value, make the Bering a valuable fishing ground. WHALES such as the killer, beluga, beaked, sei, finback, right, humpback, and sperm are plentiful. The Pribilof and Komandorski islands are nurseries for fur SEALS, and SEA OTTER, SEA LIONS, and WALRUS are numerous.

In the early 1800s the United States tried to exercise control over the entire Bering Sea in order to protect the seals from being reduced to extinction by hunters from Canada, Mexico, Japan, and Russia. Great Britain protested this act, and an international tribunal in 1893 ruled that the United States could not control these waters. However, the tribunal did place some limitation on the killing of seals in the area. Subsequent agreements signed in 1911 and 1957 further protect the seals in this great breeding ground.

BERM See BEACH.

BERMUDA TRIANGLE See MYTHS AND LEGENDS.

BIGHT See OCEAN.

BILLFISH is a common name applied to any of various ocean fishes having bill-shaped jaws, such as a SPEARFISH.

The billfishes, such as marlin, swordfish, SAILFISH, and spearfishes, have been called the world's most prized marine game fish, especially the black marlin and the swordfish. Black marlins 15 ft (4.57 m) long and weighing in excess of 1500 lb (680 kg) have been caught on rod and reel by sportsfishermen.

The billfishes eat other fishes—primarily MACKEREL, TUNA, and flying fishes—as well as SQUID. They prey on schools of fish by pursuing them, often at speeds of 40 mph (64 km/h), and enter the school striking from left to right with their beaks. They then feed on the dead and wounded victims at their leisure.

The sharks, chiefly large sharks such as the tiger SHARK, are practically the only enemies of the large and speedy billfishes.

Speaking of enemies of these fishes, in 1967, one irate 200-lb (90.7-kg) swordfish classified the oceanographic research submersible *Alvin* as a foe. The *Alvin* was on a routine geology dive on the Blake Plateau (east-southeast of Charlestown, South Carolina), when, shortly after the submersible reached the bottom at 1800 ft (545 m), the swordfish made a deliberate and fatal (for the swordfish) attack on her. The fish became wedged into the structure and caused a small leak of SEAWATER into the submersible, which then aborted its dive and returned to the surface with the swordfish flailing helplessly all during that time. When the fish was gutted for examination of its stomach, it was found to contain squid almost exclusively.

BIOLOGICAL NOISE See UNDERWATER SOUND.

BIOLOGICAL OCEANOGRAPHY, sometimes referred to as marine biology, is the study of temporal and spatial distributions of populations of marine organisms and their interactions with each other and the environment.

Biological oceanography is, as can be deduced from the definition given, concerned primarily with systematics and ECOLOGY. However, the field is vast, and the interests of the marine biologist are many-faceted.

OCEANOGRAPHY evolved as a science when biologists interested in the world's oceans began to answer such questions as: What types of life exist in the oceans? How do these organisms live and survive? Where do they live and migrate? What controls their distribution patterns?

Such difficult questions are still being pondered by the biological oceanographer whose studies extend from experiments under controlled conditions in the laboratory to investigations of life forms at the various depths of the ocean.

The scientific disciplines which contribute to such studies, aimed at a better and more organized understanding of life in the oceans, range from molecular physics and biology to the physiology of parts of the organisms, the ecological interrelationships of the organisms to one another and to their habitat, and the natural life cycles of organisms and populations. In order to be meaningful, all such information, from molecular biology to life-cycle histories, must be based upon the fundamental knowledge of species, as exemplified by TAXONOMY and systematic biology. Such systematic studies are less than 200 years old, while the development of marine life is as old as life on this planet.

BIOLUMINESCENCE (OCEAN) is the emission of visible light produced by living marine organisms, both plant and animal. Thousands of species of marine organisms, including PLANKTON, the lantern fishes, and types of SHRIMP (EUPHAUSIIDS), possess luminous organs and thus emit visible light. Oxygen appears necessary to the process, which is a transformation of chemical energy into light energy. This production of light, often called *cold light,* is brought about by the interaction of two chemical substances, luciferin and luciferase. The ability to produce such light is limited almost entirely to marine life. Very few land forms and almost no freshwater organisms have this property.

The Romans were among the first to record a "phosphorescence" of the sea. Columbus observed this also when he reported seeing mysterious lights in the water before he landed on San Salvador. Darwin wondered about such lights as he witnessed "livid flashes" from the *Beagle* along the South American coast. (See DARWIN, CHARLES.)

This phenomenon of light production is caused by thousands of species of marine organisms (mostly animals), and it takes place in various parts and depths of the temperate ocean world, particularly during the warm season.

In disturbed warm surface waters, some of the single-celled animals and MICRO-ALGAE, or PLANKTON, often make flashing displays of bioluminescence. This phenomenon has resulted in rendering surface vessels, submarines, and mine fields detectable from the air. During World War II, a number of ships, including torpedo boats, were detected and attacked because of the bioluminescent wakes resulting from mechanical stimulation of light-emitting organisms. Chiefly, such luminescence is caused by the DINOFLAGELLATES, which comprise a large portion of the ocean's plankton and possess both animallike attributes, for they move like animals and ingest food, and plantlike attributes, for they are also capable of PHOTOSYNTHESIS. Some others are holozoic and obtain food in the manner of most animals by ingesting complex organic matter. When stimulated, these dinoflagellates produce a light caused by a biochemical reaction catalyzed by an enzyme in the presence of water and oxygen. Light-emitting luciferin is oxidized by the enzyme luciferase, and some luciferin molecules absorb energy and thereby reach an excited state; each molecule of luciferin releases one photon (unit) of light. If SEAWATER is stirred, the luminous discharges of individual microorganisms look like sparkling crystals. If the water is agitated rapidly, the points of light emitted by dinoflagellates fuse into a bright glow. Turbulence resulting from the swimming motion of fishes provides the mechanical stimulation that outlines their bodies with light and leaves behind a luminous trail.

Examples of luminescent dinoflagellates are the *Noctiluca militaris, Pyrodinium bahamense,* and *Gonyaulax polyedra.*

Luminescent bacteria in seawater also make decaying fish shine in the dark. Still other, higher-order, organisms produce their own light, and these organisms range from JELLYFISH, SHRIMP, and SQUID to certain fishes and sharks.

Many groups of DEEP-SEA FISH are luminiferous and are equipped with unusual forms of light-producing antennas (as in the case of the angler fish) as well as with luminous silvery plates and light organs patterned in various parts of their bodies. It is believed that in this case the property of blue or blue-green type of luminescence is used as a defense against predators, as a means (and bait) for hunting food, and as an aid for individuals of the same species, but of different sex, to identify each other in the darkness of their environment.

Since the major concentrations of luminescing organisms inhabit the upper 330 ft (100 m), usually in waters where most pelagic fishes abound, the phenomenon of bioluminescence has been used by commercial fishermen to detect and capture various species of fish (e.g., MACKEREL, *Scomber japonicus;* bluefin TUNA, *Thunnus thynnus;* MENHADEN, *Brevoortia patronus;* and SARDINES, *Sardinella aurita*).

BIRDS See MARINE BIRDS.

BISMARCK SEA is located on the northwest corner of the South PACIFIC OCEAN. It is bounded by New Guinea on the southwest, and by the Bismarck Archipelago on the north, east, and south. Beginning on the north coast of New Guinea, the boundary is defined by a line beginning at Baudissin Point and running clockwise through Wuvulu Island, the Ninigo Group, Hermit Island, the Admiralty Islands, New Hanover, New Ireland, and New Britain, then across Umboi Island to Teliata Point on New Guinea. The Bismarck Sea covers an area of 15 440 mi² (40 000 km²) and occupies a volume of 14 395 m³ (60 000 km³).

The Bismarck Archipelago, consisting of over 300 islands, was discovered in 1700 by the English navigator William Dampier, but it remained unclaimed until Germany took possession in 1884. At the end of World War I the islands became a mandate under the League of Nations and were administered by Australia. After World War II they became a part of the United Nations Trust Territory of New Guinea and are still administered by Australia. In 1511, Antonio de Abrea, a Portuguese navigator, first sighted the island of New Guinea (second largest in the world after Greenland), and the Spanish explorer Ortiz de Reyes became the first Caucasian to touch foot on the island in 1542.

As compared with the SOLOMON SEA to the south, the floor of the Bismarck Sea is relatively uncomplicated. The near-oval basin consists of two major depressions: the New Ireland Basin on the east and the New Guinea Basin on the west. The two basins, which exceed 6562 ft (2000 m) in depth, are separated by a rise which strikes northwest-southeast. Numerous seamounts dot the floor of the sea, and the basin is surrounded by largely dormant volcanic cones. See SEAMOUNT. The deep basin is covered with globigerina ooze. Volcanic ash and muds, along with carbonate reef debris, are found in localized areas. Circulation of the water is provided by the South Equatorial Current (see EQUATORIAL CURRENT SYSTEM) for much of the year.

BIVALVE is the common name for a wide number of diverse animals, having soft bodies enclosed in calcareous two-part SHELLS. Such animals include

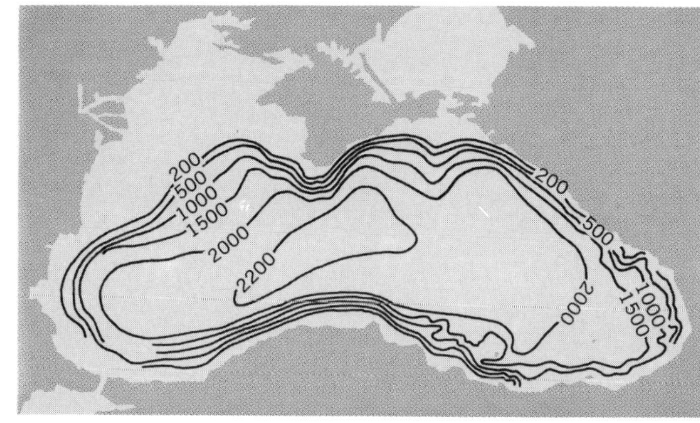

MAP 5. The Black Sea, showing depth variation.

the mollusks, some CRUSTACEANS (the ostracods), and BRACHIOPODS. See BIVALVIA; CLAMS; MOLLUSK.

BIVALVIA, a large class of the phylum Mollusca, contains the CLAMS, oysters, and other bivalves. See MOLLUSK; OYSTER.

BLACK SEA is an intracontinental sea that is surrounded by the Soviet Union on the north and east, Turkey on the south and southwest, and Bulgaria and Rumania on the west. The Black Sea communicates with the MEDITERRANEAN SEA through the Bosporus and Dardanelles which lie, respectively, on the eastern and western edges of the Sea of Marmara. The Sea of Azov lies on the northern rim of the Black Sea with access through the Kerch Strait. The Sea of Azov, in turn, is connected to the Caspian Sea on the east by the Don and Volga rivers and the Don-Volga Canal. The Black Sea and the Sea of Azov cover an area of 117 946 mi² (461 000 km²), occupy a volume of 128 832 mi³ (537 000 km³), and have a mean depth of 3825 ft (1166 m). The Sea of Azov has a maximum depth of only 44 ft (13.5 m), and an area of 14 668 mi² (38 000 km²). The maximum depth of the Black Sea is 7365 ft (2245 m), and it measures 750 mi (1207 km) long by 380 mi (611 km) wide. Heavy winter fogs give a dark cast to the water—hence the name.

Several times during their geologic history the Sea of Marmara, the Sea of Azov, and the Black and Caspian seas have been connected to the Mediterranean Sea and both the ATLANTIC OCEAN and the INDIAN OCEAN. On the other hand, prolonged lowering of sea level has periodically severed the connection between each of these basins because of

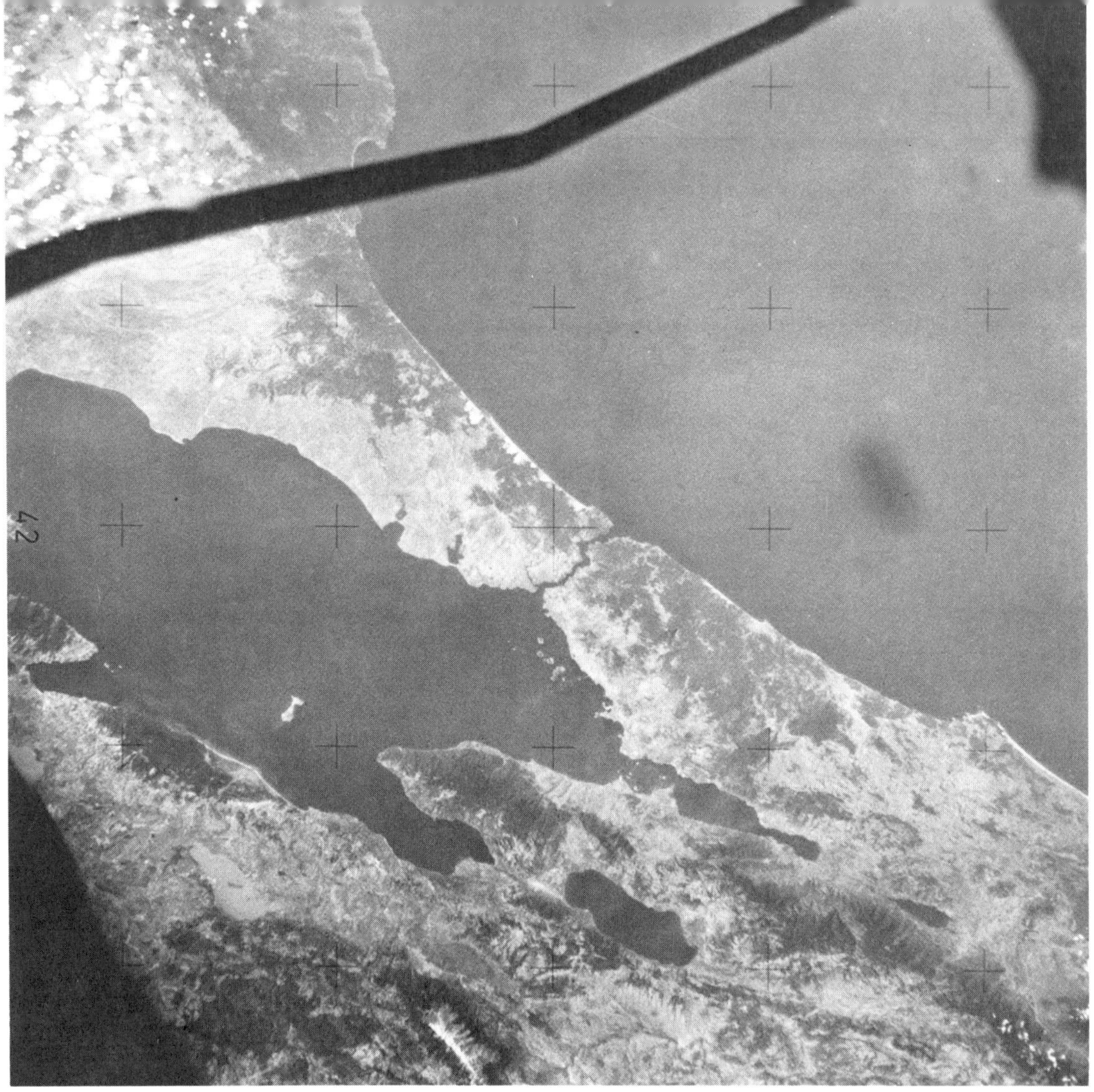

BLACK SEA. The Bosporus cuts across the northwest edge of Turkey, connecting the Sea of Marmara and the Black Sea. The continental shelf is relatively wide here and northward to the Crimean Peninsula, but the rest of the Black Sea's coastline plunges quickly to great depths. (*NASA*)

the shallowness of the connecting channels. [The Bosporus Strait is only 90 ft (27 m) deep.] The present connection between the Mediterranean and Black seas was established about 6000 years ago. At that time the level of the Black Sea was well below that of the Mediterranean. A state of equilibrium now exists.

The basin of the Black Sea is roughly elliptical in shape and has a very narrow CONTINENTAL SHELF except in the northwest corner between the Bosporus and the Crimean Peninsula. The shelf is cut in many places by submarine canyons. See SUBMARINE CANYON. The floor of the deep basin, to which the shelf plunges rather abruptly, is relatively flat. The sediment cover grades from sandy terrigenous materials in the coastal areas to calcareous muds on the floor of the deep basin. Various forms of ferrous sulfide are incorporated in the sediments of the deep basin.

The Black Sea receives the drainage [96 mi^3 (400 km^3) per year] of a large part of Central Europe through the Dnieper, Dniester, Bug, and Danube rivers; of the Soviet Union through the Don, Kuban, and a number of smaller rivers; and of northern Asia Minor through the Tehoruk, Yeshil Irmak, Kizil Irmak, and Sakaria. This runoff plus precipitation at sea exceeds evaporation and results in a net outflow of WATER through the Bosporus [84 mi³ (350 km³)]. This surface discharge is partially compensated [48 mi³ (200 km³)] by an inflow of Mediterranean water beneath the surface outflow. Because of the heavy discharge of river water the surface layer of the Black Sea has a relatively low salinity (16–18 ppt). The salinity at depths below 650 ft (200 m) reaches only 21–22 ppt. But the rapid transition from the lower to the higher salinity establishes a sharp density gradient that strongly inhibits vertical mixing between the surface and deeper layers. This, combined with the shallowness of the connecting channels with the Mediterranean, makes the Black a stagnant sea. As a result of this stagnation, the dissolved oxygen disappears at a depth of about 650 ft (200 m), and hydrogen sulfide (H_2S) is present at all greater depths. This means that no life, except anaerobic bacteria, exists below this level. This condition permits the entrapment and decomposition of large amounts of organic matter at depth, with the resulting production of phosphate ions, carbon dioxide, ammonia, and methane.

The surface circulation in the Black Sea is influenced both by the wind and by river discharge. The result is two counterclockwise gyres—one in the eastern and one in the western sector. Smaller gyres exist at the boundary between these two circulatory systems. Surface temperatures range from 32° F (0° C) in winter to 71.6° F (22° C) in summer. At depths below 650 ft (200 m) the temperature is fairly constant at 47.3° F (8.5° C). During winter the heavy inflow of fresh water allows ice to form, sometimes reaching as much as 70 mi (113 km) offshore.

The animal population of the Black Sea consists of about 1500 species, derived mainly from the Mediterranean. All, of course, live in the top 650 ft (200 m) of the sea. There are some 170 FISH species, 20 percent of which are of commercial value. Among these are MACKEREL, bonito, anchovies, HERRING, mullet, carp, and sturgeon. Fishing is highly developed but is largely confined to the inshore area. Of the 15 percent which is carried out in the open sea, some 90 percent is devoted to catching PORPOISE. SEALS are also quite common in the Black Sea. Some 285 species of red, brown, and green ALGAE exist in the sea, and some of these species form vast beds in shallow coastal waters.

BLOWFISH See PUFFERFISH.

BLOWHOLE is the nostril on top of the head of CETACEANS. (See WHALES; PORPOISE.) A blowhole is also a longitudinal passageway or tunnel in a sea cliff on the upland side away from shore. During storms, columns of sea spray usually are thrown up through the opening.

BLOWHOLE. Spray shoots high in a blowhole on the rocky California coast. *(Scripps Institution of Oceanography)*

BLUEFIN See TUNA.

BONY FISH is the name applied to fishlike vertebrates that are members of the class Osteichthyes, which includes two subclasses, 32 orders, about 357 families, some 3570 genera, and around 18 000 species. The bony fishes are distinguished by having a bony skeleton, a SWIM BLADDER, a true gill cover, and mesodermal ganoid, cycloid, or ctenoid scales. This class of vertebrates includes most of the familiar fishes. See also TELEOSTS.

BOWDITCH, NATHANIEL (1773–1838), an American mathematician, astronomer, navigator, and insurance executive, is best known for his book *The Practical Navigator,* published in 1799. Actually, the book began as a revised and edited American version of J.H. Moor's book by the same title. Bowditch corrected some 11 000 errors in the original

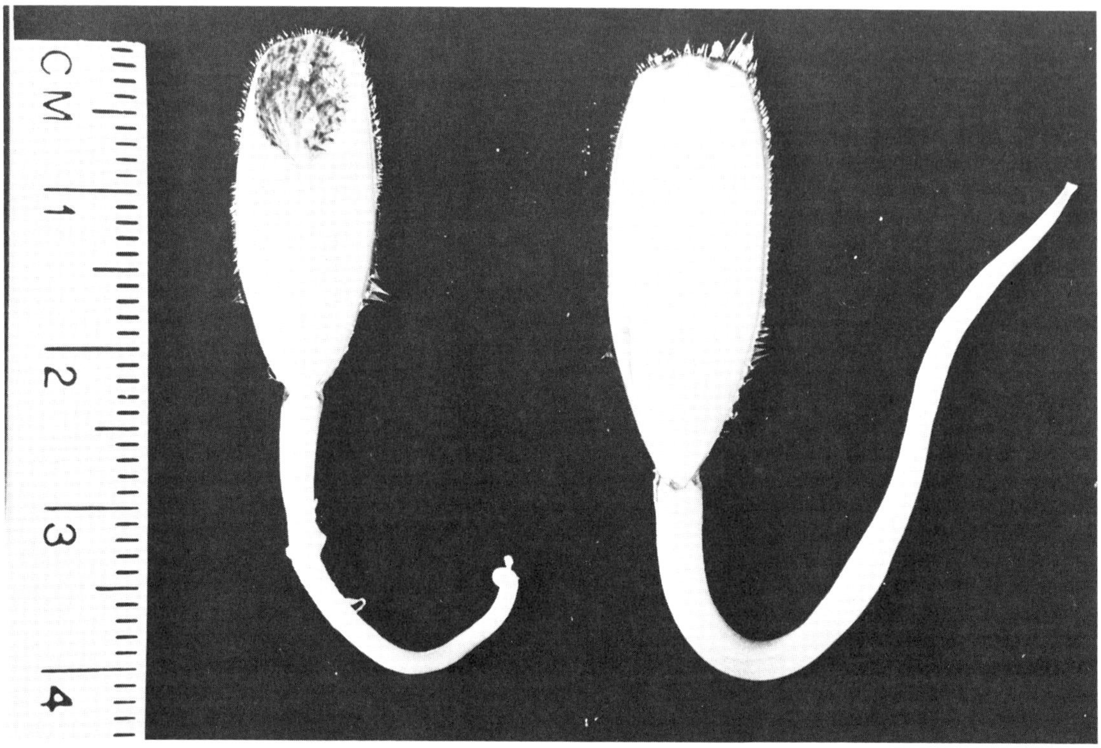

BRACHIOPODS. Two brachiopods lie next to a metric ruler to give an indication of size. The ventral cavity, which holds the feeding organ, gapes open between the valves in the animal on the left. (*NOAA*)

and so expanded it that the title was changed to *The New Practical Navigator* in its third edition, published in 1802. The book became and remains the standard navigational aid. To date the book has gone through 65 reprints or editions.

Bowditch was born March 26, 1773, in Salem, Massachusetts. His father was a shipmaster and cooper. Bowditch's formal schooling came to an end in his tenth year when the state of family finances forced him to go to work. At 12 he became a clerk in a ship chandlery and began a lifelong process of self-education; his favorite subjects were mathematics, astronomy, and foreign languages. In 1795 Bowditch went to sea for a total of five voyages, beginning as clerk and finishing as master of the ship. It was during this period that he became interested in navigation and published the book that earned him an international reputation.

Bowditch married Elizabeth Boardman in March 1798; she died the same year. In October 1800 he married his cousin Mary Ingersoll by whom he had six sons and two daughters.

In 1803 Bowditch retired from the sea and accepted an appointment as president of the Essex Fire and Marine Insurance Company. This position allowed him sufficient income and leisure time to pursue his scientific interests. In 1823 he became an actuary for the Massachusetts Hospital Life Insurance Company. During this period he conducted a survey of Salem harbor and published some 23 papers. He also produced a critical and annotated translation of Pierre Simon de Laplace's *Mécanique céleste,* the publication costs of which were paid out of his own pocket.

Bowditch was not noted for creativity and original thought, but he was a meticulous and exhaustive critic with exceptional mathematical skills. He died March 17, 1838.

BRACHIOPODS are a phylum of exclusively marine, bivalved enterocoelomates having symmetrically placed but dissimilar valves located about a median longitudinal plane. These invertebrates are small and typically attached to some substrate by a posteriorly located fleshy stalk or pedicle. Anteriorly, a relatively large mantle

cavity is always developed between the valves, and the filamentous feeding organ, or lophophore, is suspended in it, projecting forward from the anterior body wall. The whole shell or encasement of the animal resembles an ancient Roman lamp and hence these animals are often referred to as lampshells.

The brachiopods were very prevalent on the Paleozoic ocean floor some 300 to 500 million years ago and many of them are found in fossil form today. However, a few hundred living species of brachiopods such as *Pelagodescus atlanticus, Crania boetteri* and *C. eucalathis* now remain. These animals resemble CLAMS although they have a simpler internal anatomy, and their hinged shells form the top and bottom of the animal rather than the left and right halves. The shells open at one edge, revealing a filtration apparatus within. The water enters on each side of this filter and leaves the filter from the center of the opening. Brachiopods occur most commonly beneath the waters of the continental shelves. *Pelagodescus atlanticus* is the most numerous and most widespread species found in the world's oceans. Most species are DETRITUS and PLANKTON feeders.

BRAZIL CURRENT. Warm (dark) water flows south with the Brazil Current and extends into the cooler (light) water moving north with the Falkland Current in this infrared satellite photo of the Argentinian coast and adjacent waters. *(NOAA)*

BRAZIL CURRENT is an extension of a part of the warm, westward-flowing South Equatorial Current (see EQUATORIAL CURRENT SYSTEM) which turns south to flow along the Brazilian coast. At about 35° south LATITUDE it meets the northward-flowing FALKLAND CURRENT, and the two turn eastward to become the South Atlantic Current.

The Brazil Current is a relatively weak current, flowing at the rate of about 2 kn and forms a shallow surface current of only about 656 ft (200 m) thickness. Because of its long journey across the Atlantic in the vicinity of the equator, and the resulting evaporation, the water in this current is very saline (36–37 ppt).

BREAKWATER See BEACH.

BULKHEAD See BEACH.

BULLARD, SIR EDWARD CRISP (1907–), British geophysicist, made important contributions to the development, both theoretical and experimental, of the physics of the earth. Much of his work has been devoted to the origin of the earth's magnetic field. However, he also made significant contributions to seismic activity at sea, heat flow on both land and sea, and to the expanding fields of PLATE TECTONICS and CONTINENTAL DRIFT.

Bullard, the son of an English brewer, was born September 21, 1907, in Norwich, England. He was educated at Cambridge University, receiving his B.A. in 1929 and his Ph.D. in 1932, for work on the scattering of slow electrons and the measurement of gravity. He remained at Cambridge until the outbreak of the war in 1939. During the war years he worked on the protection of ships against magnetic influence mines, and on operations analysis. After the war he returned to Cambridge. He served as professor of physics at the University of Toronto during the period 1948–1949. For the next 6 years he served as director of the National Physical Laboratory of England before again returning to Cambridge where he was appointed professor of geophysics and head of the department in 1964. In 1963 Bullard accepted a concurrent appointment as professor of geophysics at the University of California, San Diego. He was married to Margaret Ellen Thomas in 1931 and to Ursula Margery Curnow in 1975. He had four daughters by his first marriage.

Bullard's work on terrestrial magnetism was influenced by a suggestion made in 1919 by Sir Joseph Larmor that the sun's magnetic field might be due to motion of material in the sun acting as a dynamo. To Bullard, variations in the earth's magnetic field might be influenced by motion in the earth's electrically conducting core. By using one of the early computers, Bullard was able to show that the process is, at least, possible. Today, the hypothesis that the earth's magnetic field and its variations result from the core acting as a dynamo is accepted by most scientists.

BURMA SEA See ANDAMAN SEA.

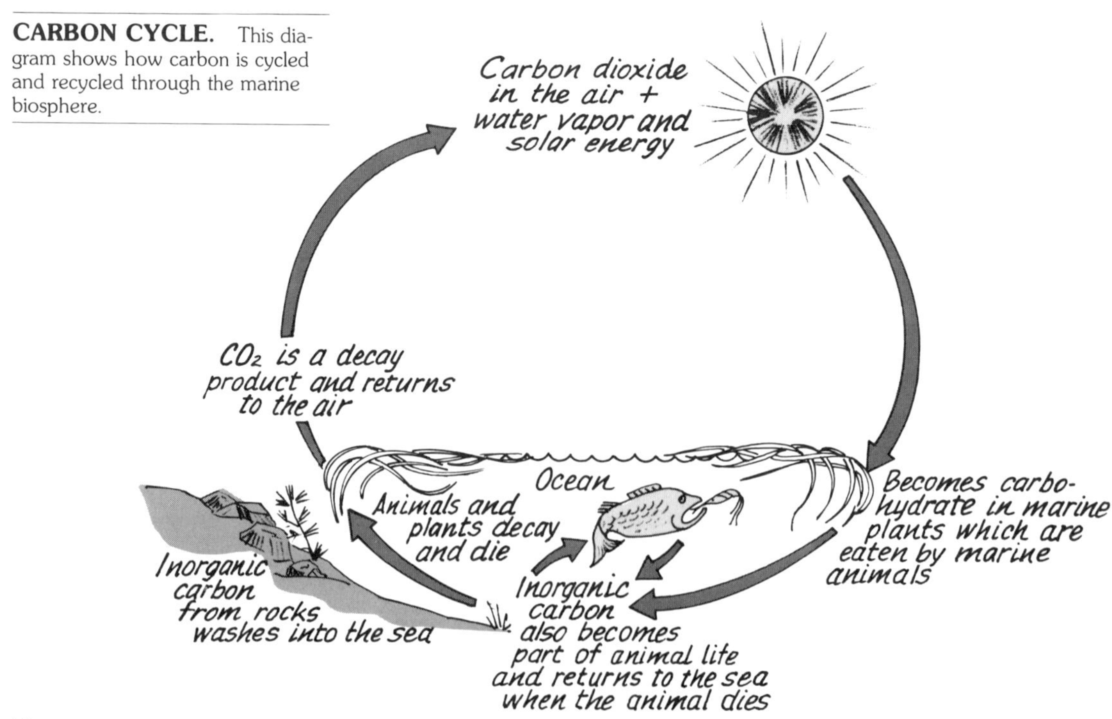

CARBON CYCLE. This diagram shows how carbon is cycled and recycled through the marine biosphere.

CALCAREA See SPONGES.

CALIFORNIA CURRENT is the eastern limb of the wind-driven current system flowing clockwise around the North Pacific Basin which connects the North Pacific Current in the north with the North Equatorial Current (see EQUATORIAL CURRENT SYSTEM) in the south. It is identifiable along the coast of North America from Washington state to Baja California and occupies a broad band from the coast to about 621 mi (1000 km) offshore. The net transport of WATER to the south by the California Current is approximately 388 million (388 × 10^6) ft^3/s (11 million m^3/s). A countercurrent setting to the northwest exists within a zone from the coast to about 62 mi (100 km) offshore. The countercurrent is most obvious below about 492 ft (150 m) of depth, and flows at a rate of around 9.8 in/s (25 cm/s). Surface currents in this nearshore zone vary in direction with the season. During the winter they generally flow to the northwest, and during the summer to the southeast.

Since the waters of the California Current have just completed a trip across the northern rim of the PACIFIC OCEAN, they are, understandably, cool. Temperatures of 48.2° F (9° C) in the north and 78.8° F (26° C) in the south are not uncommon. However, the cold coastal waters which keep surfers in wet suits even on the hottest summer days is due to deep water rising to the surface in a process known as UPWELLING. The prevailing wind direction throughout the year is from the northwest. The influence of the wind combined with that of the CORIOLIS EFFECT causes the surface water to move offshore to the southwest. Cold, deep water rises to replace the surface water. In the summer the warm moisture-laden air moving in from the northwest is chilled by nearshore surface water—the result is the famous California fog.

CALLAO PAINTER See EL NIÑO.

CANARY CURRENT is part of the circulation about the North ATLANTIC OCEAN. It is an extension of the North Atlantic Current which flows south along the coast of Europe and Africa. It is best identified between the Canary Islands and the Cape Verde Islands. In the vicinity of the latter, it divides, one branch flowing westward as the North Equatorial Current (see EQUATORIAL CURRENT SYSTEM), and the other flowing southeast to join the GUINEA CURRENT. The Canary flows at a speed of about 0.2 kn and transports about 565 million (565 × 10^{6-}) ft^3/s (16 million m^3/s) of WATER. This relatively cool current plays a significant role in modifying the weather along the coastal areas of Europe and Africa. In a manner similar to that of the CALIFORNIA CURRENT it produces sea fog off the coasts of Spain and Portugal. See CORIOLIS EFFECT

CAPILLARY WAVE See WAVES.

CARBON CYCLE is the complex equilibrium system of nature in which carbon dioxide (CO_2), converted from carbonates, is transmitted to living matter, and then via metabolic processes is regenerated back to carbon dioxide. (See diagram opposite.)

The carbon cycle impinges fundamentally upon every facet of the earth sciences—meteorology, OCEANOGRAPHY, geology—as well as ECOLOGY.

The biosphere (or that part of the earth in which life exists) contains many carbon compounds (e.g., coal, petroleum, natural gas, limestone, and other carbonate rocks such as shells, coral rocks, peat, etc.). These compounds are continuously being formed, transformed, and decomposed. The dynamics of this relationship is essentially maintained by various mechanisms: (1) the ability of SEAWATER to dissolve large quantities of carbon dioxide (CO_2);

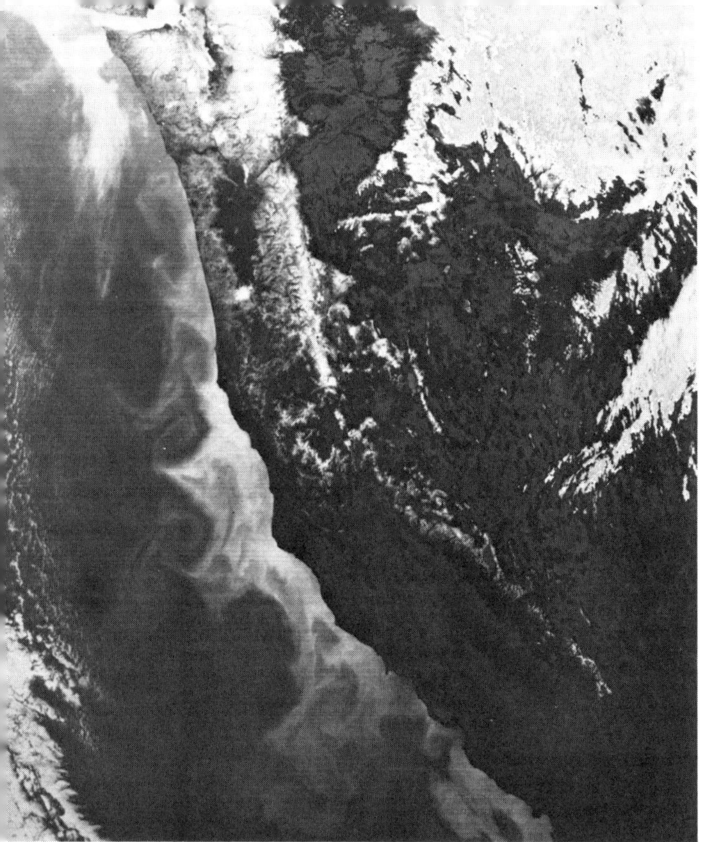

CALIFORNIA CURRENT. The infrared satellite image shows irregular mixing of the coastal water and the offshore California Current. The light-toned coastal waters are near 10° C, while the darker current is about 20° C. *(NOAA)*

(2) the physical and chemical processes such as leaching, combustion, or volcanic actions; (3) the ability of PLANKTON in the ocean and land plants to trap sunlight energy and carbon dioxide and utilize it in PHOTOSYNTHESIS (and nonphotosynthetic processes) to transform CO_2 into organic molecules; and (4) the complex interaction of plants and animals to degrade carbon dioxide compounds, thereby releasing CO_2 into the atmosphere. Although CO_2 comprises only about 0.044 percent of the atmosphere, its significance for life is of tremendous importance since it is the source of the carbon atoms used by living matter to construct the wide variety of organic compounds of which their bodies are composed. In this complex chain, simple animals of the sea live on plants or organic DETRITUS and, in turn, form food for other animals.

The process which releases carbon dioxide is called respiration, which may be thought of as the reverse of photosynthesis, with oxygen being consumed and CO_2 released.

In the oceans, dissolved CO_2, bicarbonate and carbonate ions are all present, and these can effectively buffer the solution at a pH close to 8. Also involved in the system is equilibration across the air-sea interface, precipitation of shells of animals, the dissolution of carbonates from the sediments, as well as some localized inorganic precipitation of calcium carbonate. The fact that the pH of seawater is controlled directly affects many of the chemical equilibria of the seas and is one of the factors that led both to the genesis and maintenance of life in this environment. See ION.

In the world's oceans the carbon cycle is more or less self-contained; the phytoplankton take up the dissolved CO_2 and give up oxygen which is dissolved in the seawater. Zooplankton and FISH consume the carbon fixed by the phytoplankton and utilize the dissolved oxygen for respiration. The decomposition of organic matter in the ocean replaces the CO_2 assimilated by the phytoplankton.

However, one of the most serious questions relative to the carbon cycle concerns the rates of exchange between CO_2 in the atmosphere and the total carbonic concentration of the ocean and sediments. Atmospheric CO_2 increases (approximately from 290 parts per million (ppm) to 330 ppm in the last 100 years) have been caused by increased burning of fossil fuels, changed agricultural practices, deforestation, and industrial production. The effects of higher CO_2 concentrations on climate are controversial. Atmospheric warming due to an increased "greenhouse effect" may take place, but definite predictions cannot yet be made in the absence of data on what effect atmospheric moisture and closed coverage, variations in ARCTIC OCEAN ice, and the amount of atmospheric PARTICULATE MATTER will have on climatic conditions. However, if increased atmospheric CO_2 causes changes in sea TEMPERATURE, the possibility exists that the surface layer of the seas will not be able to assimilate the increased CO_2 and will become generally undersaturated with calcium carbonate. The solubility of CO_2 while not known with a high degree of accuracy, decreases with increasing temperature. A condition of undersaturation of the surface layers of the world's oceans, especially the shallow seas, caused by a disruption of the major CO_2 cycle or the circulation of carbon in nature, could have very serious consequences on world climate and sea life as well as life in general.

Modern studies of CHEMICAL OCEANOGRAPHY play an important role in providing a better understanding of the CO_2 cycle and of the transfer of heat and moisture at the ocean surface–atmosphere interface, both of which play a vital role in the distribution of rainfall and temperature.

CARBON-NITROGEN-PHOSPHORUS RATIO is a nearly constant relationship between the concentrations of the elements carbon, nitrogen, and phosphorus in PLANKTON and the carbon, nitrogen, and phosphorus in SEAWATER. This phenomenon is due to the removal of the elements by the organisms in the same proportions in which the elements occur and the return of these elements upon decomposition of the dead organisms.

By the analysis of marine organisms, it has been shown that in their protoplasm the elements carbon, nitrogen, and phosphorus are found in the ratio of 100:15:1. These elements are taken from the water in these ratios for the production of organic matter. During decomposition of this matter, the elements are returned to the water in the same proportions. In coastal waters, and especially in the confined seas (due to the fact that decomposition of organic material is not an instantaneous process which releases all elements at the same time), it is not uncommon to encounter different ratios of elemental composition. Particularly, the nitrogen-to-phosphorus ratio may vary widely from the 15:1 composition.

CARIBBEAN CURRENT is a warm, swift current, flowing northwestward through the CARIBBEAN SEA. Originating in the North Equatorial Current (see EQUATORIAL CURRENT SYSTEM) and Guiana Current to the south, the current flows into the Caribbean through the straits in the vicinity of St. Lucia. The current passes into the GULF OF MEXICO through the Yucatan Channel. From here it flows eastward to form the FLORIDA CURRENT. The current flows at a speed of about 0.8 kn and transports a volume of about 1.06 billion (1.06×10^9) ft³/s [30 million (30×10^6) m³/s].

CARIBBEAN SEA, a marginal sea of the western ATLANTIC OCEAN, is bounded by Central and South America on the west and south, and the islands that make up the Greater and Lesser Antilles on the north and east. Officially, the boundary is a line running clockwise from Cape Catoche on the Yucatan Peninsula, across the Yucatan Channel to Cuba, and on to Hispaniola (Haiti, Dominican Republic), Puerto Rico, across the Anegada Passage to the Lesser Antilles, and along the outer edge of these islands to Baja Point in Venezuela. The sea is separated from the GULF OF MEXICO by a line through the Yucatan Channel. The Caribbean is roughly 1700 mi (2735 km) long (east-west) and 500–800 mi (805–1287 km) wide (north-south). It covers an area of 1 063 044 mi² (2 754 000 km²) occupies a volume of 1 654 783 mi³ (6 860 000 km³), and has a mean depth of 8172 ft (2491 m). The maximum depth of 23 294 ft (7100 m) is found in the Cayman Trench.

The coastline of Central and South America is indented by several important bodies of water: the Gulf of Honduras off the coast of Guatemala and Honduras; Mosquito Gulf off the coast of Panama; the Panama Canal at Colon, Panama; the Gulf of Darien bordered by Panama and Colombia; the Gulf of Venezuela off the entrance to Lake Maracaibo in northern Venezuela; and the Gulf of Paria bordered by Venezuela and the island of Port of Spain. Most of the rivers of Central America drain into the Caribbean, but those of the South American sector are, for the most part, shunted into the Orinoco River which empties into the Atlantic just south of Port of Spain. The major passages into and out of the Caribbean are the Yucatan Channel between Yucatan and Cuba, the Windward Passage between Cuba and Hispaniola, Mona Passage between Hispaniola and Puerto Rico, Anegada Passage between the Virgin Islands and St. Martin, and Dominica Passage north of the island of Dominica. Each of these passages is over 3281 ft (1000 m) deep.

The basin of the Caribbean Sea is a complex of basins and trenches separated by ridges. The northernmost of these is the Yucatan Basin with a depth of around 16 404 ft (5000 m). The basin is bordered on the north by the 137-mile (220-km) wide Yucatan Passage and on the south by the Cayman Ridge which separates it from the Cayman Trench. The ridge runs from Cuba to the vicinity of the Central American coast and breaks the surface in the east as the Cayman Islands. The relatively narrow Cayman Trench contains the greatest depth in the Caribbean at 23 294 ft (7100 m), and is separated from the Colombia Basin further south by the broad, wedge-shaped Nicaraguan Rise which supports the island of Jamaica. The Colombia Basin lies at depths as great as 12 026 ft (3666 m) and is connected to the Venezuelan Basin further to the east by the North Venezuelan Trough. Otherwise, it is separated from the Venezuelan Basin by the Beta Ridge which projects westward from Hispaniola. The Venezuelan Basin has depths to 16 594 ft (5058 m) and is bordered by the narrow, arcuate Aves Swell. Between the Aves Swell and the Lesser Antilles loop lies the Granada Trough.

The origin of this complex structure has been long debated without clear resolution. One line of

CARIBBEAN SEA. Dense vegetation crowds a narrow beach on a typical Caribbean island, St. Thomas in the Virgin Islands. *(U.S. Navy)*

reasoning holds that the Caribbean Basin is oceanic, but that the combination of uplift, volcanic activity, and sediment inflow is gradually converting it into a continent that will eventually be annexed by the surrounding landmasses.

The circulation within the Caribbean Sea begins with the northward-flowing Guiana Current, part of which flows into the sea through passages between islands of the Lesser Antilles. The resulting surface current winds its way through the sea and exits into the Gulf of Mexico through the Yucatan Passage. Deep circulation is restricted by the ridge and basin structure and the shallowness of most passages through the island chain. The temperature of the surface water in the Caribbean ranges from around 82.4° F (28° C) in summer to about 77° F (25° C) in winter. Due to evaporation, the surface salinity runs around 36 ppt.

CASPIAN SEA See INLAND SEA.

CATHODIC PROTECTION See CORROSION.

CEPHALOPOD is a common name for a group of exclusively marine mollusks. The cephalopods (class Cephalopoda) number about 650 species and vary widely in size, form, and structure. They differ greatly from the rest of the mollusks. Some examples of these animals are the OCTOPUS, SQUID, and NAUTILUS. See MOLLUSK.

CERAM SEA is located in the western PACIFIC OCEAN between the islands of Sulawesi (Celebes) and New Guinea. It is bounded on the north and northeast by a line running from the Soela Islands through Obi Major, Tobali, Kekek, Pisang, and the Kofiau Islands to the western point of New Guinea and down the coast to Tanjong Namaripi; on the southeast, by continuing the line down the Aroe Islands and over to the Tanimbar Islands; on the south, by a line running through Ceram to the northwestern point of Buru; and on the west, by a line from Buru through Sanana Island to Mangole Island. The Ceram Sea covers an area of 72 182 mi² (187 000 km²), occupies a volume of 54 459 mi³ (227 000 km³), and has a mean depth of 3966 ft (1209 m).

Some maps show a Buru Sea in the western part of the Ceram Sea, but the Buru is not recognized as a separate sea by most oceanographers. The Buru

Basin, with a depth of over 9842 ft (3000 m), underlies this western portion. The topography of the basin is complicated by several east-west ridges. Pacific WATER enters this basin through the Lifamatola Strait and continues southward through the BANDA SEA and into the INDIAN OCEAN.

CETACEANS are members of the order Cetacea, which comprises the WHALES, porpoises, and dolphins. (See DOLPHIN; PORPOISE.) The cetaceans are the mammals most completely adapted to an aquatic existence. See MARINE MAMMALS.

***CHALLENGER* EXPEDITION** (of 1872–1876) was the 69 000-nautical-mile world voyage of the British ship *H.M.S. Challenger* whose findings marked the start of scientific OCEANOGRAPHY. It is generally agreed that every branch of oceanography can trace its foundations to methods developed and oceanic observations made during this nineteenth century expedition.

The *H.M.S. Challenger* was a 225-ft (68-m) long wooden corvette (combination sail and coal-burning steam-engine propulsion) of 2306 tons (2092 metric tons) assigned for the use of Wyville Thomson, a professor of natural history at the University of Edinburgh. It was a naval ship refitted with the best equipment of the time for carrying out the task of investigating "the conditions of the Deep Sea through the Great Oceanic Basins"; the vessel spent 3½ years at sea. During this time, in which various types of ocean samples and specimens were taken at over 350 stations in all the world's oceans except the ARCTIC OCEAN, a vast amount of new information about living things of the oceans, the ocean waters, and the seafloor itself was collected. Scientists of the expedition were headed by John Murray. These men, using rather primitive techniques and devices according to present-day standards, nevertheless mapped sections of the ocean bottom, classified deep-sea MARINE SEDIMENTS, and collected an immense number of SEAWATER samples and biological specimens.

The results of the expedition, published in 50 folio volumes over the 20 years subsequent to 1876, aroused tremendous interest among scientists and the public at large. However, although the pioneering work of the *Challenger* expedition was perhaps relatively slow in stimulating other scientific work at sea, the type of patterns of ocean investigations set by it provided the modern-day beginnings of the science of oceanography.

CHANNEL See OCEAN.

***CHALLENGER* EXPEDITION.** *H.M.S. Challenger,* whose 1872–1876 voyage became the first scientific oceanographic expedition and laid the foundations for all of oceanography, battles waves in the Indian Ocean just off the subantarctic island Kerguelen.

CHEMICAL OCEANOGRAPHY, or, as it also called, *marine chemistry,* applies the principles of the science of chemistry and uses chemical techniques for the study of the properties and chemical interactions of substances present in the ocean world.

Chemical oceanography is a vast and complex field of endeavor. Its main parts may be said to be those concerned with the chemistry of (1) SEAWATER, (2) marine sediments, (3) living ocean organisms, and (4) ocean boundary physicochemical interactions.

In order to carry out such work effectively, the chemical oceanographer must not only be a chemist but must also have a good working knowledge of the other interrelated ocean science disciplines, e.g., PHYSICAL OCEANOGRAPHY, GEOLOGICAL OCEANOGRAPHY, and BIOLOGICAL OCEANOGRAPHY.

The ocean chemist must deal with the challenge of understanding the multitude of chemical interactions that are taking place in a huge reaction vessel. These, for example, include virtually all the biochemical changes associated with countless creatures from microscopic organisms to WHALES; compositional changes in a tremendous number of complex organic and inorganic substances with such changes occurring as the result of ocean circulation and mixing phenomena as well as from internal processes; and where all of the known naturally occurring elements are present, but with most of them in extremely small amounts: less than one part per million (ppm) parts of water. (See ELEMENT.)

Because of such conditions, it can be said that OCEANOGRAPHY is dependent upon chemical information and the techniques of chemistry.

CHIMERA is a deep-water chondrichthyan fish of the subclass Holocephali, comprising the ratfishes. Chimeras (sometimes spelled chimaeras) are found in both the ATLANTIC OCEAN and the PACIFIC OCEAN. The Atlantic (*Chimaera monstrosa*) and the Pacific (*Hydrolagus collieri*) species both possess poisonous spikes or stings located along the leading edge of the first dorsal fin. (See FINS.)

CHITON is the common name for some 600 living species of marine mollusks which are members of the order Polyplacophora in the class Amphineura. See MOLLUSK. These animals chiefly inhabit the shallow coastal waters of the oceans. Highly specialized for life on hard and uneven substrates, chitons occur on all rocky shores which are free of ice. However, a few species are abyssal. See HABITABLE ZONES.

Most species vary from 1 to 3 in (2.5 to 7.5$^+$ cm) in length and resemble elongated limpets. The body is elliptical with a dorsal shell comprising eight calcareous articulating plates overlapping posteriorly. This allows the animal to roll into a ball. See LIMPET.

Their diet consists of ALGAE and microorganisms. Some of the chitons themselves are considered edible by humans.

CHLORINITY is the measured chemical value (see MOHR TITRATION) of the approximate amount of chloride in a sample of SEAWATER. Expressed in grams per kilogram of a seawater sample, chlorinity has been defined as the mass of pure silver required to precipitate the HALOGENS in 0.3285233 kg of seawater. Since these halogens consist of fluorine, bromine, iodine, and astatine in addition to chlorine, the chloride content is about 1,00045 times the chlorinity as determined by classical standard titration practices. The SALINITY or total salt content of seawater (this content can be variable), can be related for many practical and theoretical oceanographic purposes to the measured chlorinity (Cl). This relationship is usually expressed by the equation

Salinity (*S*, in ppt, or PARTS PER THOUSAND, the units also being designated by the symbol ‰)

$$= 1.80655 \times \mathrm{Cl(ppt)}$$

Thus, if the chlorinity is found to be 19.378 ppt (the value of NORMAL WATER or standard seawater), the salinity is equal to 35.007 ‰ (= 19.378 × 1.80655).

Laboratory salinometers (which measure conductivity) can now be used to measure salinity with a greater degree of accuracy than is obtained with the more classical techniques of chemical quantitative analysis. Physical oceanographers are primarily interested in the salinity values since these are used as a means of calculating the DENSITY of seawater—a value that is relatively difficult to accurately obtain otherwise, either *in situ* or in the laboratory. The electrical conductivity of a sample can be related to salinity and density, and, after TEMPERATURE, is one of the most frequently measured physical-chemical properties of seawater.

See also INSTRUMENTATION.

CHLOROSITY is a measure of the chloride and bromide content of 1.057 qt (1 L) of SEAWATER. It equals the chlorinity of the sample times its density at 68°F (20° C).

Chlorosity is a less commonly used term in work involving the chemical analysis of seawater. It is the property corresponding to the CHLORINITY expressed as grams per liter (20° C).

CHUKCHI SEA, a sea of the ARCTIC OCEAN, is the easternmost sea off the Siberian Coast and the most western of those off the Arctic rim of North America. The northern boundary of the Chukchi Sea is the edge of the CONTINENTAL SHELF, running from Point Barrow to 75° N in the longitude of Wrangel Island. The western boundary is provided by the EAST SIBERIAN SEA, or a line from the edge of the continental shelf through Wrangel Island to

Cape Yakan on the mainland of the USSR. The recognized southern boundary is the ARCTIC CIRCLE, but the narrow point of the Bering Strait further south is considered by some to be a more practical boundary. The Alaskan Coast, from Point Barrow to the Seward Peninsula, serves as the eastern boundary. The Chukchi Sea, which covers an area of 224 652 mi² (582 000 km²), has a mean depth of 289 ft (88 m), and a volume of 11 995 mi³ (51 000 km³).

As a result of a suggestion made by H.U. Sverdrup, the Chukchi Sea takes its name from the hardy Chukchi Indians who inhabit the rugged Chukchi Peninsula in eastern Siberia. (See SVERDRUP, HARALD ULRICK.) The first to explore the Chukchi, undoubtedly, were the ancestors of the Eskimo and North American Indians who migrated across its bottom before it ceased to serve as a land bridge between Siberia and Alaska some 8000–11 000 years ago. During recorded history evidence exists that the Russian Cossack Simon Dezhnev was the first man to explore the Chukchi. If true, Dezhnev's voyage is one of the most remarkable in the entire history of Arctic exploration. According to the record, Dezhnev's party, using flat-bottomed boats about the size of a sailing dinghy, sailed eastward from the mouth of the Kolyma River on the East Siberian Sea in 1648. Before the voyage had ended, Dezhnev had sailed over 2000 mi (3218 km), around the East Cape (Cape Dezhnev), through the Bering Strait, and to the mouth of the Anadyr River in the Gulf of Anadyr.

The southern edges of the Chukchi Sea were explored during the Great Northern Expedition, begun by Peter the Great of Russia, which lasted from 1724 to 1749. In 1728, Vitus Bering, a Danish navigator serving with the Russian navy, sailed northward through the Bering Strait from his staging area on the Sea of Okhotsk. (See OKHOTSK, SEA OF.) On the basis of this voyage Bering was placed in command of the Great Northern Expedition. He made another voyage to the southern Chukchi in 1741 but died on the return trip. In 1778, Captain James Cook, British navigator and explorer, sailing aboard the *Resolution,* reached the vicinity of Point Barrow, thus becoming the first explorer to cross both the ANTARCTIC CIRCLE and the ARCTIC CIRCLE. (See COOK, JAMES.) On his way southward through the Bering Strait Cook realized an opportunity which escaped Bering on both his voyages: "The weather clear, we had the opportunity of seeing, at the same moment, the remarkable peaked hill, near Cape Prince of Wales, on the coast of America, and the East Cape of Asia with the two connecting islands of Saint Diomede between them." During the period 1820–1824, Ferdinand von Wrangel, a Russian naval officer, polar explorer, and (later) governor of Alaska, led an overland expedition which reached North Cape on the Bering Strait. Wrangel, using dog sledges, tried unsuccessfully to reach the island on the western boundary of the Chukchi which was later named in his honor. Thomas Long, captain of an American whaling vessel, discovered Wrangel Island in 1867 and named it for the man who had sought it in vain. (Some authorities credit Robert Parry of the U.S. Navy with the discovery of this island.) In 1879, Nils Adolf Erik Nordenskjöld, the Swedish polar explorer, studied the Chukchi Sea as he passed through on his historic west-to-east transit of the Northwest Passage. Extensive scientific research was carried out in the Chukchi during the Norwegian North Polar Expedition aboard the *Maud* in 1918–1925. The Chukchi continues to be of great interest to oceanographers because it is an area of exchange between the waters and sea life of the Arctic Ocean and the PACIFIC OCEAN.

The climate of the area occupied by the Chukchi Sea is Arctic, with average air temperatures during the coldest month (February) ranging from −5.8° F (−21° C) in the south to −16.6° F (−27° C) in the north, with extremes as low as −50° F (−46° C). In July average temperatures range from 42.8° F (6° C) in the south to 35.6° F (2° C) in the north. Under the influence of this harsh environment the Chukchi is completely ice-covered except for 2–3 months during the summer. It is only in the extreme southern reaches, particularly along the Alaskan coast, that the ice ever completely disappears during the summer.

The tidal range in the Chukchi Sea is exceedingly small, being on the order of $\frac{1}{2}$ ft (0.15 m) in the vicinity of Point Barrow, Alaska. The current system in the sea is dominated for most of the year by a flow of warmer (4–12° C) WATER setting northward through the Bering Strait and flowing northeastward along the coast of Alaska, and a cold current which enters from the East Siberian Sea. Very little of the Chukchi water is returned southward through the Bering Strait.

The bottom of the Chukchi Sea is covered with muds, sands, and gravel over most of its area. Bottom depths vary between 131 and 197 ft (40 and 60 m) over much of the sea. Exceptions are the two submarine canyons which cross its floor. (See SUBMARINE CANYON.) Herald Canyon begins at about 70° N and runs northward to the continental margin, with canyon depths ranging up to 295 ft (90 m). Barrow Canyon begins about 93 mi (150 km) west of Point Barrow and runs northeastward into the

BEAUFORT SEA, with canyon depths ranging from 164 to 328 ft (50–100 m). Barrow Canyon was used by the U.S. nuclear submarine *Nautilus* to gain entrance to the Arctic Basin during its historic submerged transit of the Arctic Ocean in 1957.

Being strongly influenced by the warm current setting northward through the Bering Strait, the Chukchi Sea has a richer and more varied sea life than most Arctic seas. The Arctic char and COD are the most common FISH, and WALRUS, SEALS, POLAR BEARS, and WHALES are abundant.

CIGUATERA See BARRACUDA; SNAPPER; VENOMOUS MARINE LIFE.

CLAMS (AND OTHER SHELLFISH). Clams is a name generally applied to a number of worldwide aquatic, INVERTEBRATE animals that belong to the class BIVALVIA (also called Pelecypoda) of the phylum Mollusca, which contains the oysters, MUSSELS, scallops, and other bivalves as well as the clams. See OYSTER; SCALLOP. The bivalves which are popularly called clams are distinguished, in general, by their two symmetrical laterally compressed, calcareous, hinged valves, which form a layered shell structure enclosing and protecting the soft body parts of this bottom-dwelling MOLLUSK.

There are a total of about 10 000 species of bivalve mollusks and it is estimated that bivalves, as a class, comprise two-thirds of all invertebrates taken in by commercial fishing activities throughout the world. See COMMERCIAL OCEAN FISHING. Of these, the clams and oysters are of greatest economic importance because of their food value both to man and to certain demersal fish such as the FLATFISH and COD, which also provide a valuable source of food for humans.

As a class, the bivalves are found in all the world's oceans from shallow inshore waters to the greatest depths, as well as in most lakes, rivers, ponds, and other freshwater bodies. In the ocean environment, they are vulnerable to many predators, most especially the boring snails and CRABS (e.g., the green crab, *Carcinus maenas*). In addition, certain DINOFLAGELLATES, such as *Gymnodinium brevis,* can cause their death by toxic poisoning when they are swarming in red tides. See RED TIDE.

Clams In general structure, clams are quite similar, although their shape and size may vary considerably. All have an acid- and salt-resistant shell composed of calcium carbonate, which is secreted within the matrix of conchyolin, an organic, nitrogenous substance closely allied to chitin. The mantle is composed of right and left parts or valves and these form a very loose covering over the animal's soft body. The two valves are primarily connected dorsally by an elastic ligament which acts as a spring and tends to pull the dorsal edges of the valves together. Such an action is opposed by two strong transverse abductor muscles connecting the two valves together. These muscles are also attached to the body which is a mass much smaller than the mantle cavity. Feeding is carried out by the gills located on each side of the small mouth at the anterior end of the body. Fine particles of food are trapped in these gills and are carried to the mouth and enlarged stomach. Slitlike openings or ciliated siphons, which are located between the sides of the posterior end of the mantle, circulate water currents through the mantle cavity, bringing water in at the ventral opening and out at the dorsal. A large muscular lobe called the *foot,* located between the gills, extends ventrally and anteriorly from the body. This foot is used for burrowing or locomotion. Relaxation of the abductor muscles allows the valves to separate. Then the foot is extended, the tip is anchored to a surface, and the rest of the clam is drawn up to that surface. This is repeated in the animal's method of movement. In some species, this foot action may be assisted by "byssal" threads or horny growths on the foot.

The nervous system consists of three pairs of ganglia and these have connecting fibers; the heart is dorsal and as in most bivalves the intestine passes through it.

In the marine clams, reproduction is carried out by the discharge of sperm and eggs into the water by the separate sexes, although it is not uncommon for a clam to change sex from male to female. The fertilized eggs develop into larvae and these eventually become mature clams. See LARVA. Concentric rings are laid down on the shells as the clam grows. Some species have ridges which radiate from the hinge to the edge of the shell.

The largest clams ever evolved are of the Indo-Pacific family Tridacnidae. Actually, the family Tridacnidae comprises two genera, *Tridacna bruguiére* with five species (*T. gigas, T. derasa, T. squamosa, T. maxima,* and *T. crocea*) and the monotypic genus *Hippopus* represented by the species, *Hippopus hippopus.* Of these six species, *T. gigas* is the largest, reaching a valve length in excess of 4¼ ft (130 cm) and a valve weight of over 331 lb (150 kg). This species of giant clam has a pronounced scalloped margin on the mantle. The other species are intermediate in size with *T. crocea* being the smallest, having a maximum valve length of about 7.9 in (20 cm). All these clams, except possibly *T. crocea,* are of value both for their meat, which is consumed locally, and for their shells, which are of value to collectors. *Tridacna gigas, T. derasa,* and *H. hippopus,* the larger species, are particularly sought for their shells.

The American soft-shell or long-necked clam, *Mya arenaria,* is abundant in ATLANTIC OCEAN waters off the east coast of the United States, north of Cape Cod, Massachusetts. It is also common around the waters of Great Britain.

Another typical Eulamellibranch clam is the quahaug, or littleneck, *Mercenaria mercenaria* (or *Venous mercenaria*). This species, the only known surviving species of the family Arctididae, has been especially plentiful on the North Carolina and Florida shores of the United States where they have been dredged out in huge quantities by commercial fishermen. The quahaug is also native to large parts of the CONTINENTAL SHELF along the Atlantic Coast from Cape Hatteras, North Carolina, to the ARCTIC OCEAN at depths of 36 to 5400 ft (10.9 to 164 m). Once thought inexhaustible, resources of hard clams (*Mercenaria mercenaria*), sea clams (*Spisula solidissima*), and soft-shell clams (*Mya arenaria*) are dwindling rapidly.

U.S. Pacific coast clams include the razor clams (*Siliqua patula*), butter clams (*Saxidomus nuttalli* and *S. giganteus*), littleneck clams (*Protothaca staminea*), the Atlantic soft-shell clams (*Mya arenaria*) which have been transplanted, and the geoducks (*Panope generosa*).

Oysters Among the 100 or more species of edible oysters (family Ostreidae) are the European flat oysters (*Ostrea edulis*) and the elongate, cupped Japanese, Portuguese, and American oysters (*Crassostera gigas, C. angulata,* and *C. virginica,* respectively). True oysters are distinguished by having dissimilar lower and upper shells, and these shells or valves are hinged together by a complex elastic ligament. The upper valve of the shell is normally flat, while the lower is concave, providing space for the body of the oyster. The two valves fit together to form a watertight seal when the oyster closes, provided that the shell has not been damaged. Near the center of the oyster's body is an abductor muscle, attached to both valves, which controls the opening and closing of the shell.

Aristotle wrote that oysters "have no sensations or sex, but arise spontaneously from the foam around stationary ships." However, the fact of the matter is that the British oyster is bisexual and reproduces by a complex and time-consuming process of being male, and then female, and then male again. Portuguese oysters are either male or female and breed by a process not too different from mammalian reproduction.

The tidal estuarine waters that form the principal habitat of most commercial mollusks, such as clams and oysters, comprise one of the most complex environments in nature. For example, while the pH of the open ocean usually ranges from 7.5 to 8.5, the pH in tide pools, bays, and estuaries may decrease to 7.0 or lower due to dilution, H_2S production, and OCEAN POLLUTION. (See BAY; ESTUARY.) Since clam and oyster larvae must, at times, encounter a wide range of pH in their natural habitat, the success or failure of recruitment of these mollusks to some areas may be determined by variations in pH. In the case of clams, the pH must not be below 7.0, and for oysters, 6.75. Neither reproduce in estuarine waters having a pH of 9.0 or greater. Oysters, can, however, adapt to living in waters with considerable changes in salinity and TEMPERATURE, but their growth is more rapid in warm waters.

Marine oysters, in general, are mostly distributed on the coastlines between the 45th parallels north and south of the equator, but because of the nature of the GULF STREAM, they are also found as far north as the 62d parallel on the European side of the Atlantic (although some species live in very deep waters). In North America, Europe, and parts of Asia, they are very valuable as food; and, in addition to this food value, a few produce (via parasites and secretions) mother-of-pearl shells and pearls.

Mussels The mussel is also a mollusk with two equal shells. Most widely known and distributed of the marine mussels is the species *Mytilus edulis.* These thin-shelled shallow-water bivalves grow securely attached to the ocean bottom and occur in large-size beds. The mussel's powers of propagation are tremendous, and a medium-sized animal of this species produces 7–12 million ($7–12 \times 10^6$) eggs per year. This egg-laying capacity is surpassed by the American oyster's 100 to 500 million per year capacity. The inside lining of the mussel shell is pearly and iridescent. The animal is very well adapted to its tidal environment and is the most abundant bivalve along the northern coasts. One reason for this success is its ability to attach itself to almost anything by means of a mass of horny threads, the byssus, which is formed by a gland in

CLAMS AND OTHER SHELLFISH

A clammer watches as his dredge brings up a haul of clams in the Chesapeake Bay. *(NOAA)*

A ten-centimeter ruler (above) gives the scale for a butter clam (left) and a cockle taken from Alaskan waters. A clam extends its foot for burrowing or for locomotion by relaxing its abductor muscles so that the valves may separate. *(NOAA)*

Here are four different types of oyster: (1) the Japanese, or Pacific, oyster, now the principal species grown commercially on the American west coast; (2) the European oyster, the common oyster on the continent, which has been introduced into Maine, and is experimentally grown on the west coast, but is not yet commercially important in the United States; (3), the Eastern oyster, also known as the Chesapeake Bay oyster, Virginia oyster, and "elephant's foot," which is the principal commercial oyster of the U.S. east coast; and (4) the Olympia oyster, once the principal species on the west coast, but whose population has decreased. *(NOAA)*

Where spawning conditions are good, oysters in their natural state will attach themselves one generation on top of another, until great ridges of them exist up to high tide levels. These oysters are of little commercial value as they are of poor quality. They can be improved greatly by spreading them out so that they have room to grow into desirable market sizes. *(Fish and Wildlife Service)*

Although the mussel's powers of propagation are great (often 7 to 12 million eggs produced per year), this species, *Pleurobema clava,* is proposed for listing as a threatened species. *(Hans Stuart)*

A viscous mass of muscle and other tissues occupies the lovely shell of the giant sea scallop. Complex eyes, which appear as tiny dark dots, line the edge of the mantle. *(NOAA)*

the foot. This attachment is so firmly moored that it can withstand the heaviest storm action. The mussel, like other bivalves, is a filter feeder. Its protective shells are tightly closed if exposed at low tide, but when covered with water they are held slightly open to allow the water to be drawn through the animal. Water flows through the gills for respiration, and the food in the form of PLANKTON (animal and vegetable organisms) and DETRITUS (fine inorganic and organic debris) is filtered into the mouth. Large quantities of water pass through every mussel. An adult about 3 in (7.62 cm) long will pass through itself as much as 10 to 15 gallons (37.9 to 56.8 L) in a 24-hour period. Plankton organisms and detritus provide the primary food, and this is rapidly and efficiently converted by the mussels to excellent flesh for human consumption. However, at times and in certain locations mussels should not be eaten since they and other bivalves can accumulate toxic substances when they are present in the water around the mollusks.

Scallops and Cockles Many species of scallops and cockles are widely distributed throughout the world from the polar regions to the tropics. Throughout the centuries, many historical events have utilized the scallop shell as a symbol. Buildings in ancient Pompeii were ornamented with scallop shell designs. During the Crusades scallop shells were the symbol of the holy pilgrimages. Artists have so admired their symmetry that they were often used in paintings.

Scallops, like clams and oysters, are mollusks having two shells. They differ, however, from those shellfish in that they are active swimmers. Scallops swim freely through the water by snapping their shells together. They rarely travel any distance from their beds, which are mostly in shallow waters. For instance, the bay scallop, *Argopecten irradians,* an important commercial organism in eight Atlantic coast states of the United States, is most often found associated with EELGRASS and other SEAGRASSES. However, some species do live offshore at considerably greater depths [viz., 2800 ft (853 m)]. Scallops are equipped with complex eyes, arranged in a row on the edge of the mantle. These remarkable organs, which are very similar in basic structure to human eyes, have been the object of much scientific study.

Summary Many factors, some beyond human control, such as FUNGUS infestation in the oysters especially, affect the abundance of these animals. However, it is well known that pesticides; pollutants such as fuel oil, gasoline, and other liquid petroleum products; and the dumping of industrial wastes destroy or otherwise damage this valuable resource.

CLAPOTIS See WAVES.

CLIMATE gives the average weather conditions over a period of years. Microclimate refers to such an average for a small area, such as in the case of the living area occupied by a particular species of organism, or colony; macroclimate is that climate typical of a large area.

The world's oceans play a vital role in determining the weather and climate in the world. These oceans, which cover an area of the earth's surface of over 139 million (139×10^6) mi² (361 million km²)—more than double the surface of Mars or 9 times the area of the moon—can be referred to as a magnificent, albeit at times unpredictable, systems design. It is made of several separate and interacting subsystems of a physical, chemical, biological, geological, and meteorological nature, as well as others operating at the interface.

This overall system—a vast heat engine—is straightforwardly designed by nature to serve human beings well, since over 35 000 times the existing annual ENERGY consumption of the world is delivered to our planet as solar radiation, and a major portion of this is absorbed by the sea. Indeed, the top 30-ft (10-m) layer of the oceans has as much heat-absorption capacity as our entire atmosphere.

All atmospheric processes and weather conditions are ultimately caused by solar energy reaching the earth. For example, in the tropics there is intense solar radiation and hence a net surplus of radiation energy while the converse is true of the polar regions. These variations result in the differential heating and cooling of the atmosphere which generates a circulation of large amounts of heated and moisture-laden air to areas of deficiency—usually into the higher latitudes. Such global air currents together with ocean currents play a very significant role in modifying world climate. Moreover, these air currents are typically unsteady, and because of this various weather changes take place.

The circulation of the atmosphere is substantially influenced by various actions taking place at the air-ocean interface. (See WATER CYCLE.) In short, water evaporated from the oceans is made available to the atmosphere and transported by maritime air masses to the land where it is partially precipitated as rain, snow, hail, frost, or dew. Also, since SEAWATER has a very high capacity for storing heat, the world's oceans serve as huge heat reservoirs moderating the high temperatures of summer and the cold of winter. Thus, the ocean—by the exchange of energy and water between it and the atmosphere—is a giant flywheel that significantly influences the climate and weather of the world.

See also OCEANOGRAPHY.

COCKLE is the common name for several species of marine mollusks in the class BIVALVIA. It is characterized by a shell having convex radial ribs. See CLAMS; MOLLUSK.

COD or **CODFISH** is the name for various fishes of the family Galidae in the order Gadiformes. This order is composed of 8 families, about 185 genera, and some 750 species; the family Galidae is the best-known family because it comprises many commercially valuable species. The fish species within the other families of Gadiform fishes are mostly deep-water inhabitants.

The cod family contains 150 species which include, among others, the haddock (*Melanogrammus aeglefinus*) and pollack (*Pollachius pollachius*) that are valuable food fishes. However, one of the best-known and most fished of all the commercial species is the Atlantic cod (*Gadus morhua*). The Atlantic cod frequently reaches a weight of 50–200 lb (22.7–90.7 kg) and a length of about 6 ft (1.8 m). Its color is olive green to brown, the back and sides marked with spots, and the belly silvery. The fish is covered with small cycloid scales, has pelvic fins on the throat, and two anal and three divided dorsal fins. The mouth is large, containing strong sharply-pointed teeth, and the upper jaw projects slightly beyond the lower, which contains a barbel under the chin. It is extensively fished off the Newfoundland banks, while the Pacific cod (*G. macrocephalus*) occurs in the North PACIFIC OCEAN, and a circumpolar species, the Arctic cod (*Boreogadus saida*), is found around the ice packs in the summer.

Cod are cold-water fish that live on the CONTINENTAL SHELF, usually swimming in schools near the bottom where they voraciously feed on any animals of appropriate size that they encounter—FISH, SHRIMP, CRAB, SQUID, and other MOLLUSKS.

They are prolific breeders, with the 4–5 year old sexually mature female producing as many as 4–10 million eggs of which large numbers are eaten by other fishes as well as other predators.

The annual catch of cod is well over a billion pounds, and it is sold as flakes, shredded, pickled, green or smoked, in salted slabs, whole, and in fresh and frozen filets. The livers are processed for cod liver oil which is rich in vitamins.

COELACANTH is a name for a very rare species, *Latimeria chalumnae,* of "four-legged" ocean fish in the superorder Crossopterygii of the class Osteichthyes (the bony fishes). The Crossopterygians are the oldest members of the vertebrates and survivors of the earth's ancient history. However, most of its members are extinct.

Coelacanths disappeared from the fossil record over 60 million (60×10^6) years ago, and they were thought to be extinct until the first documented living specimen was caught off the east coast of Africa in 1938. Since then, 10 more of these living fossils have been recovered and studied by marine zoologists. A 30-lb (13.6-kg) specimen is at the SCRIPPS INSTITUTION OF OCEANOGRAPHY (San Diego, California). An adult coelacanth may weigh as much as 200 lb (90.7 kg) and exceed 5 ft (1.5 m) in length.

Thought to have once been an inhabitant of fresh water, the coelacanth, amazingly, has survived almost unchanged since long before the dinosaur age. The coelacanths have paired leglike pectoral and paddlelike pelvic fins protruding from their body, and these muscular, scaled stalks operate much more like the limbs of land vertebrates than the fins of any ordinary fish. The discovery of this "missing link" has reinforced the biologists' long-held belief that the ocean fishes had developed appendages of this type before the first amphibians left the sea to invade the land at the end of the Devonian period some 300 million years ago.

COELENTERATES or **CNIDARIANS** are names given to chiefly marine animals of the phylum Cnidaria or Coelenterata whose members typically have a simple structure, usually consisting of a circle of tentacles at the mouth, a gastrovascular cavity, and the presence of stinging cells called *nematocysts.* While the terms, colenterate and cnidarian are synonymous, cnidarian is now preferred by most authorities.

As a phylum, the cnidarians consist of three classes of predominantly marine animals. These are

1. The class Hydrozoa, which is comprised of five orders that include more than 2700 species of marine HYDROIDS, fresh-water hydras, the smaller JELLYFISH, special CORAL, and siphonophores. The PORTUGUESE MAN-OF-WAR (*Physalia physalia*) is a familiar siphonophore, and the Fire Coral (*Millepora alcicornis*) is representative of the special corals.

2. The class Scyphozoa, which are mostly large jellyfish, such as typified by the common jellyfish (*Aurelia aurita*).

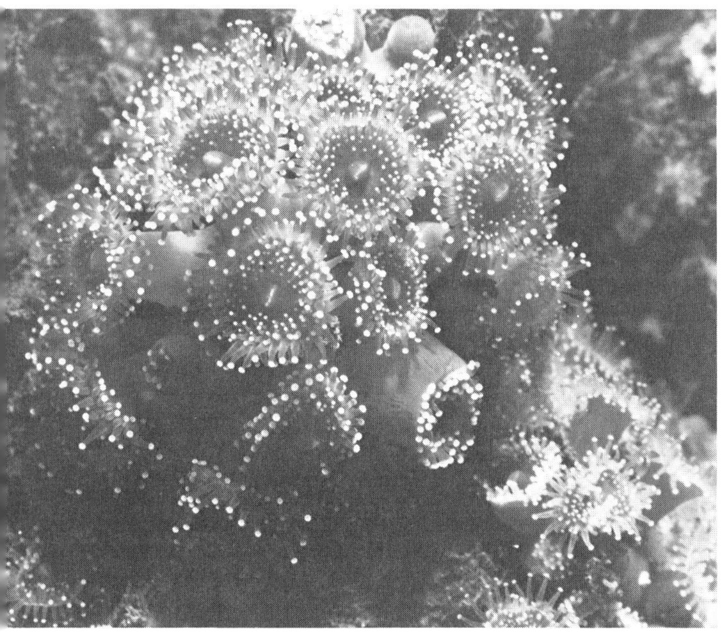

COELENTERATES. (*Above*) The ends of these polyps' tentacles appear phosphorescent. This colony was at a depth of 100 ft in the Point Loma Kelp Beds near San Diego, California. (*Oceanographer of the Navy*) (*Right*) Coelenterates of the Southern Ocean wave their tentacles equipped with stinging cells, searching for food. (*Oceanographer of the Navy*)

3. The class Anthozoa to which the SEA ANEMONE and most corals belong. This class comprises over 6000 marine species.

There are two principal body forms of cnidarians, the polyp and the medusa, and these have many common features. The most obvious external character in both is radial symmetry, and a plane taken through any lengthwise direction yields identical halves. The polyp, or hydroid, in general, resembles a cylinder open at one end. The mouth and tentacles are located at this open anterior end, and the closed and or base is usually attached directly or by a rootlike growth to some substrate on the ocean floor. The polyp is covered with an outer epidermis and lined by a second tissue layer, the gastrodermis, and the two layers are joined by an essentially noncellular layer, the mesoglea. The free-swimming medusa resembles a bell or umbrella-shaped gelatinous undulating mass, with the convex surface uppermost and the mouth suspended in the center of the underside of the bell on a short projection or manubrium. The mouth leads to the central stomach which in turn gives rise to a network of canals in the animal. The tentacles are located on the margin of the bell.

The cnidarians are carnivorous and feed chiefly on small fish and their eggs, small CRUSTACEANS, mollusks, COPEPODS, and other cnidarians. (See MOLLUSK.) They capture these and convey them to their mouth by means of their tentacles which possess nematocysts, stinging cells having paralyzing and adhesive powers. Reproduction can be asexual (e.g., by budding, or the splitting off of cells from one, not two, living entities to form a new individual) or sexual (by the union of two distinct entities such as gametes being shed into the water and fertilized).

COELOMATE, the equivalent name for Eucoelomate, is a member of a large sector of the animal kingdom which is characterized by having a true coelom or body cavity. See BRACHIOPODS; CRUSTACEANS.

COLLOID is a system consisting of an intimate mixture of two substances, one of which (the colloid) is very finely divided and dispersed uniformly throughout the second substance (the medium) which can be a gas, liquid, or solid. Colloids are extremely small, submicroscopic in size, and are best-detected and measured by an electron microscope.

The colloids present in SEAWATER usually range in size from 10^{-7} to 10^{-5} cm. They are produced largely by dispersion processes, whereby a coarser material is broken up into finer particles, or by condensation, in which atoms or molecules are induced to aggregate from solution. Classified as being either hydrophilic or hydrophobic, depending upon whether or not they are attracted by and associated with water, colloids in seawater consist of mineral substances, large molecules, biopolymers, DETRITUS, and the products of hydrolysis and precipitation processes. Colloids have unique properties, including stability which is principally due to the large surface area of the dispersed phase. ION adsorption takes place because of this, and these ions impart an electric charge to the colloidal particles so that they repel each other, preventing their further aggregation and subsequent precipitation. (See ADSORPTION.)

Very little is presently known about colloidal suspensions in seawater and especially the importance of roles that they play in the oceans (see PARTICULATE MATTER). However, the colloidal precipitates of calcium, magnesium, iron, and aluminum, as well as the hydroxides and carbonates, apparently concentrate organic matter to a very high degree. The significance of this has yet to be firmly assessed.

COMMENSALISM See MUTALISM; SHRIMP; SYMBIOSIS.

COMMERCIAL OCEAN FISHING refers to the organized method(s) of taking FISH from the world's oceans on a commercial or profit-making basis. On the average, about 76 million (76×10^6) tons (69 million metric tons) of marine fish have been taken annually from the world's oceans in recent years. This amounts to approximately 3 percent of the world's total food supply.

Commercial fishing operations which predominate in the Northern Hemisphere are involved primarily in the removal of the bony fishes of the world's oceans. Other forms of marine life captured, for instance, WHALES, SEALS, SHRIMP, LOBSTER, CLAMS, etc., are not discussed here but are covered in some cases under their headings.

In recent times, ocean finfish have been caught chiefly by five methods: (1) the drift net, (2) the purse seine, (3) the trawl net, (4) longline gear, and (5) pole and line.

In the drift net, which is employed principally to catch MACKEREL and HERRING, a net is played out from the fishing vessel across an incoming or outgoing tide. (See TIDES.) This net hangs wall-like in the water, supported by floats at the top and weighted down at the bottom. When schools of fish attempt to pass through this wall, their gill covers become entangled in the mesh entrapping them.

In the purse seine, a net is lowered into the water from two boats accompanied by a larger ship. The two smaller boats launch the net so as to encircle a large school of fish such as TUNA and MENHADEN. In this method, the net is equally divided between the two smaller vessels which are lashed together as soon as they are launched. The purse vessels move to the school of fish, separate, and swing into a big circle, surrounding the fish. The top of the net is equipped with floats, while the bottom is weighted with lead and brass rings. When the purse boats meet the far side of the school, the ends of the seine are made fast and the bottom is drawn together by means of a rope passed through the brass rings. This pursing effect traps the fish in the bowl-shaped net, and the fish are off-loaded in the large ship. These expensive nets sometimes are damaged when a SHARK is caught with the menhaden and cuts its way loose. In addition to the unwanted sharks, large numbers of porpoises, unfortunately, are also trapped and killed, especially in yellowfin tuna fishing with these nets. (see PORPOISE.)

The trawl net is the most prevalent method used in commercial fishing. In its employment, an open-ended bag-shaped net is drawn along the ocean bottom (usually at depths of no greater than 2400 ft (732 m), and when it is thought that the net has been down sufficiently long, it is power-hauled aboard the ship. This method is especially effective in catching FLATFISH.

In pole and line gear fishing, which is used for catching skipjack tuna primarily, the method is as the name implies. Live bait is used as chum, and the fish are hooked on lines and brought aboard manually by poles to which the hook and line are attached.

In longline fishing, also used for tuna, long fishing lines equipped with a series of baited hooked lines are suspended vertically from one main line. These lines are brought in with the hooked fish.

COMMERCIAL OCEAN FISHING

Chesapeake Bay fishermen dip a load of fish from a pound net. The U.S. Fish and Wildlife Service notes that pound nets are among the most important commercial gear in this vicinity, taking large catches of croakers, sea trout, shad, and herring. *(NOAA)*

Men on a Japanese factory ship fish for pollock in the Bering Sea. *(NOAA)*

COMMERCIAL OCEAN FISHING

(*Right*) Alaskan fishermen inspect a young salmon tow net. (*NOAA*) (*Below*) The diagram explains the workings of a floating salmon trap in Alaska. Many salmon are trapped as they approach rivers to spawn. (*NOAA*)

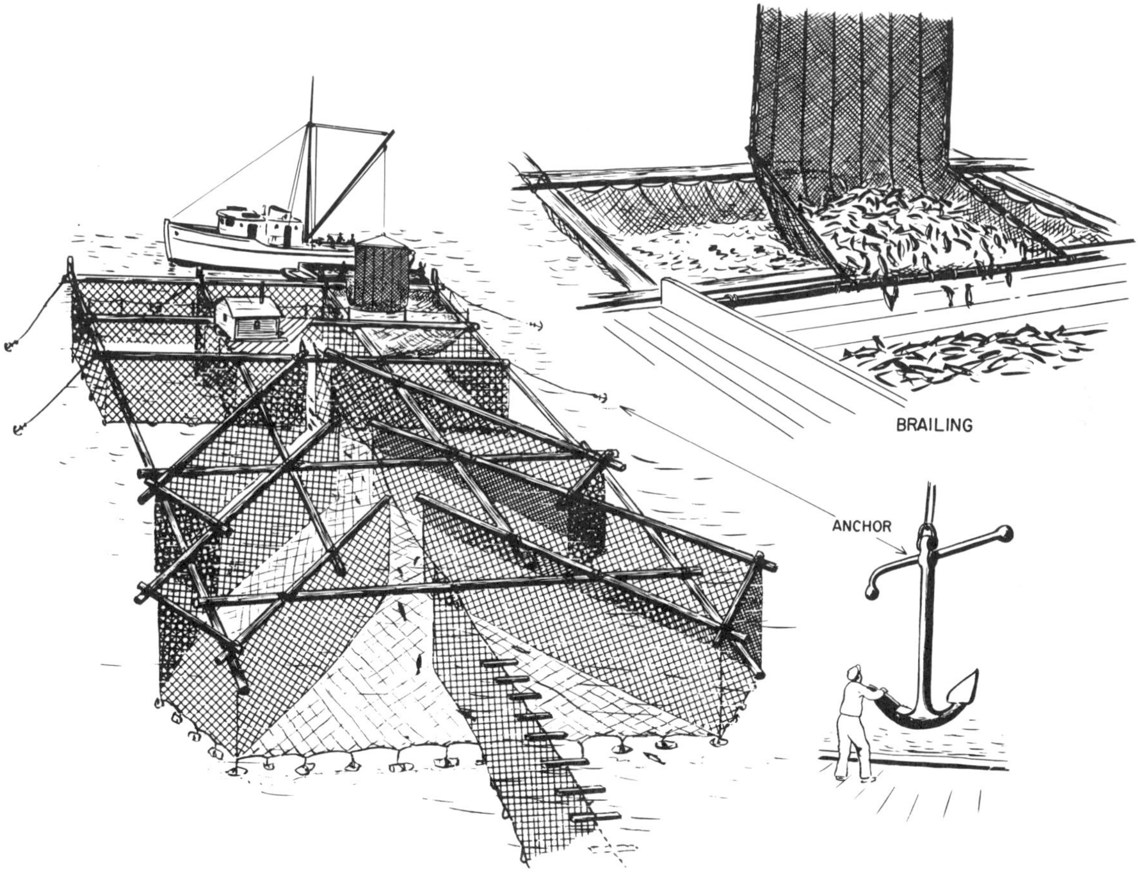

BRAILING

ANCHOR

The large fishing tonnage has grown to its present size by more than tripling in the past 25 years. This increase was the result of the above catch methods plus the introduction of new technologies that have made it possible to catch many more of the available stocks than was possible in the past. Moreover, the catches are now being made with larger and more efficient vessels, at greater depths, and at greater distances and times away from home port. Such ships use a stern (net) trawling method which makes possible the employment of larger nets than formerly used. Many of these long-distance stern trawlers are really floating fish factories that can catch, package, and process the fish at sea.

Of the some 25 000 species of fish in the world's oceans, only a few dozen (e.g., COD, tuna, mackerel, HADDOCK, FLOUNDER, SALMON, etc.) have been exploited in large numbers because of consumer reluctance to try new species. These numbers, as it turns out, have been excessively large, and that particular amount of available life in the oceans appears to be dwindling at an alarming rate. This lack of availability is being reflected now in a leveling off of the catch, in spite of more fishing ships and better equipment.

In view of this situation, there has been a growing sensitivity on the part of the major fishing nations (U.S.S.R., Japan, and the United States) of the world to promote consumer interest in new fish stocks. For instance, the Soviets have tried to market KRILL (the small, protein-rich, shrimplike CRUSTACEANS that furnish the diet for whales, various seals, some MARINE BIRDS, and SQUID). The success of such a venture on a large scale has yet to be proven. This question is true for other postulated marketable animals such as the deep-sea lantern fish, squid, the red crab, jack mackerel, and others formerly characterized as "trash-fish."

While perhaps much of the increase in any future catches will come from these sources, there is also some realization among the fishing nations that the ocean food resource must now be protected and managed wisely. Not surprisingly, this concept raises conflicting issues within the fishing industry.

For the domestic industry, a quota system designed to limit the catch to a desired level is politically difficult to execute because of the many intrinsic problems involved in apportioning catches fairly among an increasing number of claimants while still protecting the stocks from depletion. At the international level, there has long been much ill feeling among fishing nations as to what constitutes their particular resource zone. The LAW-OF-THE-SEA CONFERENCES have failed to define, on any accepted treaty basis, such a zone. As a result, several countries, including the United States, have set a 200-nautical mile (370 km) limit, measured from their shores to an open sea, as the offshore region wherein the resources belong exclusively to them. As the protracted Law of the Sea negotiations continued in 1979, the conduct of marine research at sea was subjected to increased control by coastal states with the adoption of policies regarding such research to be exercised within the 12-mile territorial sea, on the continental shelf and to a distance of 200 miles offshore in the Exclusive Economic Zone. In many ways, these developments provide no solace to those concerned with better international cooperation, with marine research or to those concerned with preservation of fish stocks which may be damaged before more corrective measures can be taken.

CONCH is the common name for any of the large GASTROPODS in the "stromb" (strombidae) group. Over 60 species are recognized with the best known being the West Indian conch (*S. gigas*). See SEA-SHELLS.

CONGER. See EEL.

CONTINENTAL DRIFT is the name of a theory generally attributed to the German astronomer and meteorologist, Alfred Wegener, which holds that as late as 200 million (200×10^6) years ago all the land masses of Earth were nestled together in one giant continent known as Pangaea (from the Greek, meaning "all land"). And then, for reasons still under debate by earth scientists, Pangaea began to break up and the resulting continents began to drift apart at an average rate of about one inch (2.54 cm) per year. The present placement of the continents is thought to be due to this continuing process.

According to the theory, Pangaea (some scientists prefer to think of two giant continents—Laurasia to the north and Gondwana to the south) lay roughly in the region now occupied by the ATLANTIC OCEAN. An indentation in the supercontinent between Africa and Eurasia was occupied by the Tethys Sea—the ancestral MEDITERRANEAN SEA. The remainder of Earth's surface was covered by Panthalassa—the ancestral PACIFIC OCEAN.

The breakup of Pangaea did not occur as a single cataclysmic event, but occupied many millions of years. The separation of North America from Eurasia appears to have been the first event, occurring almost 225 million years ago. This was followed by the separation of Africa and South America about 150 million years ago, and Africa from India, Australia, New Zealand, and Antarctica about 110 million years ago. The growing separation between North and South America from Africa and Eurasia formed the Atlantic Ocean and enclosed the ARCTIC OCEAN. To the east of Africa, India broke free from Antarctica and Australia and "sailed" northward until it collided with Asia. The present surface expression of this great collision is the Himalayan Mountains. Finally, Australia broke free of Antarctica and drifted northeastward and then northward to form the eastern boundary of the INDIAN OCEAN.

Seafloor spreading and plate tectonics are the physical processes used to explain the ability of a continent to "move through" the solid crust of Earth. Earth's crust is divided into a number of crustal plates having definite, though not always well-defined, boundaries. Some plates are totally oceanic, while others are both oceanic and continental. The North American Plate, for instance, extends to the Mid-Atlantic Ridge in the east and to California's San Andreas Fault in the west.

As one plate moves away from another, new material is added to the trailing edge of both plates by lava welling up and solidifying along the axis of separation. This axis is the Mid-Ocean Ridge—the greatest mountain range in the world—which runs for some 40 000 miles (64 360 km) through all the world's oceans. At their leading edge, plates are either forced under the opposing plate and consumed in the earth's interior, or they override the opposing plate. Fortunately, continental plates are lighter than oceanic plates and tend to ride over rather than plunge beneath the latter. The deep ocean TRENCHES, such as that running along the western edge of the South American continent, represent areas in which an oceanic plate is plunging beneath a continental plate.

The Theory and its Development Almost as soon as the continents bordering the Atlantic appeared on a single map, scientists and explorers began drawing attention to the similarity of certain coastlines on either side of the ocean. Particularly noticeable were the coastlines of eastern South America and western Africa which appeared to fit together like two pieces of a jigsaw puzzle. Francis Bacon called attention to these similarities in 1620, but did not go so far as to say that the continents might once have been joined. In 1858, Antonio Snider-Pellegrini did suggest that the continents were once joined. He based his speculation on the similarity between certain European and American fossil plants of the Carboniferous period (about 300 million years ago) and suggested that the biblical Great Flood had caused the continents to separate. Around the turn of the century, the Australian geologist Edward Suess, noting the similarity of certain geologic formations in Africa and South America, fitted the continents together into a single supercontinent he called Gondwanaland, (taken from Gondwana, a key geological province in East Central India). And, in 1908, F.B. Taylor of the United States outlined ways in which the continents might have moved laterally to their present positions. But it was the carefully considered theory of Alfred Wegener of Germany, presented in 1912, that became the center of the controversy that raged throughout the scientific world until its general acceptance in the 1960s.

The all but total acceptance of the main elements of continental drift by geologists, geophysicists, and oceanographers has resulted in both a revolution within the earth sciences and a significant advancement in human understanding of the planet on which we live.

From the standpoint of educational background, Wegener looms as an unlikely scientist to have contributed so significantly to geology and geophysics. He was born in Berlin in 1880 and studied at the universities of Berlin, Heidelberg and Innsbruck. He took his doctorate in astronomy, but his interest was almost immediately captured by the comparatively new field of meteorology, and by a long-held ambition to explore Greenland. (He completed two expeditions and died while undertaking a third in 1930.) Ultimately, Wegener rose to the post of director of the Meteorological Research Department of the Marine Observatory at Hamburg and, in 1924, occupied a chair of meteorology and geophysics at the University of Graz in Austria.

It is somewhat uncertain as to when the idea that the continents might have moved laterally upon the surface of the earth first occurred to Wegener. Colleagues recall that he had been interested in the subject as early as 1903, but he said that he did not begin seriously to ponder the problem until 1910. His views on the subject were first presented to the scientific community in a lecture before the German Geological Association in Frankfort-am-Main in 1912, and published in two German journals later that year. A more comprehensive treatment was published in 1915 in a book titled *Die Entstehung der Kontinente und Ozeane* (*The Origin of Continents and Oceans*). But it was not until the publication of its third edition in 1922 that the book was translated into several foreign languages, including English.

During the nineteenth century geologists had developed a comfortable picture of the earth which

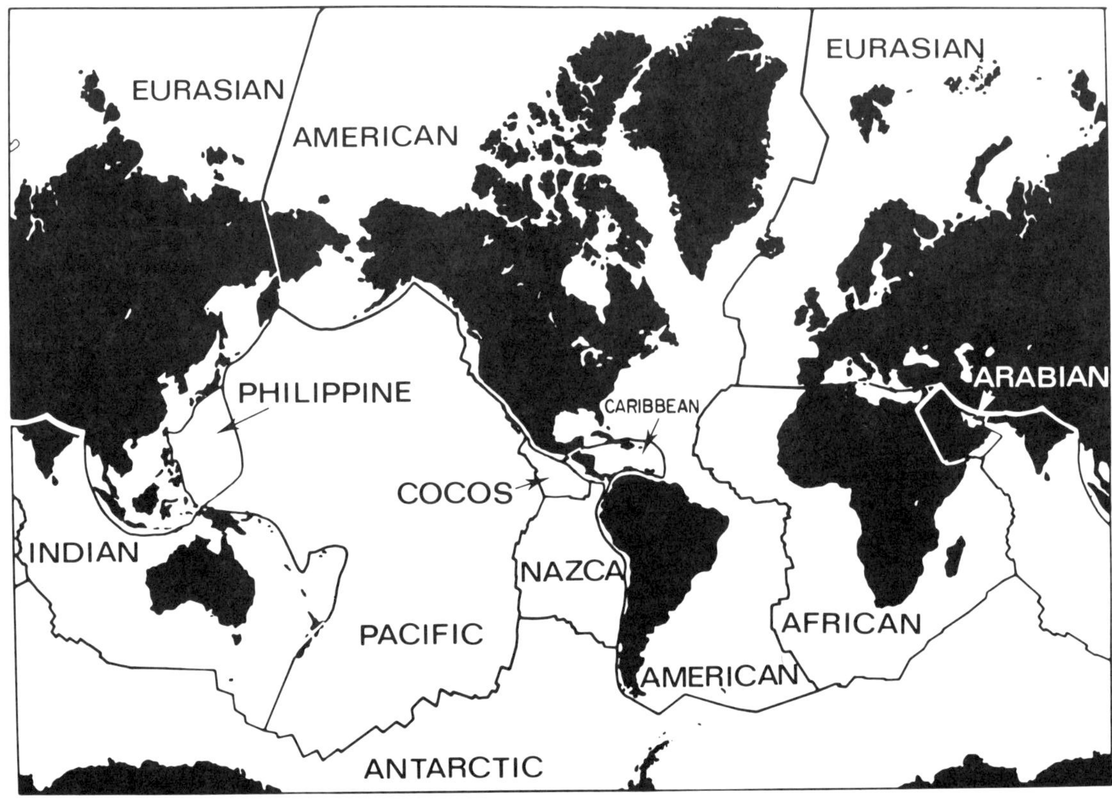

appeared to explain much of what they observed. Earth was thought to have solidified from a molten state, with the heavier elements such as iron settling to the center, and the lighter ones such as silicon and aluminum rising to the surface to form the crust. Since the cooling and solidification process was thought to be continuing, the crust had to contract in order to accommodate the reduction in the earth's size. Thus, mountain ranges were easily seen as vertical accommodation to compression, and ocean basins were explained as areas where great arches of the earth's crust had collapsed under their own weight. Land bridges between land masses, which had since collapsed, were postulated to explain what appeared to be the migration of plants and animals from one continent to another.

About the time that Wegener became interested in geological processes, this picture was made even more acceptable by the introduction of the theory of isostasy which held that the crustal features of the earth are in hydrostatic equilibrium. That is, the less dense continents "float" on the denser layers beneath them and, being lighter, stand higher than the heavier layers that make up the ocean basins. Under the principles of isostasy, continents could move vertically, but lateral movement on anything but a local scale was considered impossible.

To Wegener, the accepted picture of crustal processes was simply inadequate to explain many observed features, and there were too many contradictions. First, there was the intriguing fit between the continents of South America and Africa, a fit that was improved if the CONTINENTAL SLOPE rather than the coastline was used. Also, if mountains were caused by a fairly uniform contraction of Earth's crust, the distribution of those mountains should also be uniform. In contrast, however, mountains were seen to be restricted to curvilinear belts (e.g.,

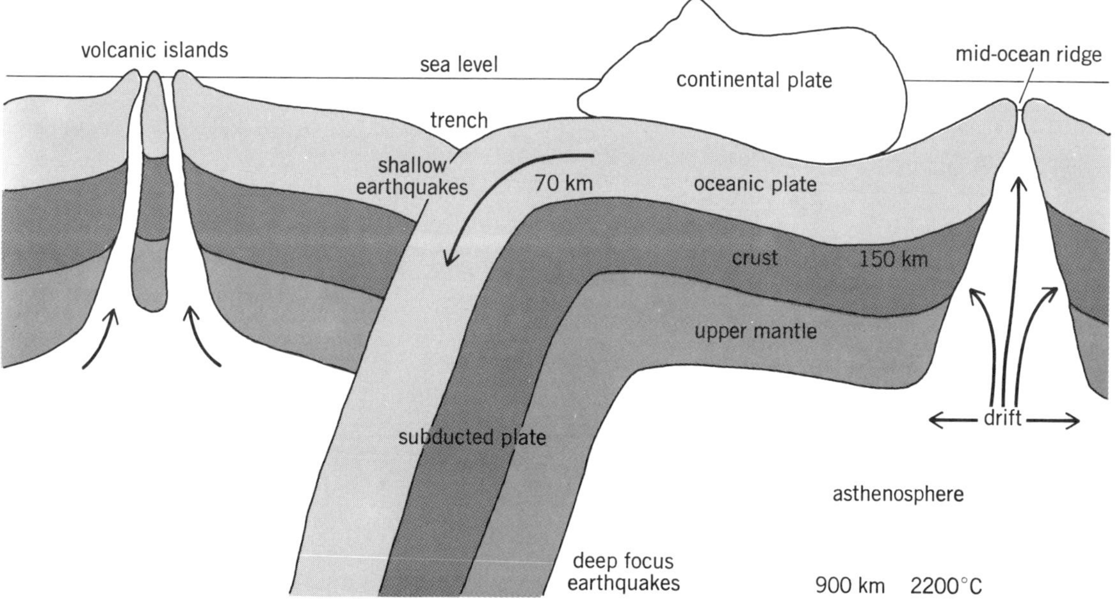

CONTINENTAL DRIFT. (*Left*) A simplified map of the major tectonic plates. The theory of plate tectonics holds that the earth's crust is composed of a number of independent plates, measuring some 60 mi (100 km) in thickness, that are in motion relative to each other at a rate of 1 to 10 cm/yr. Plate material is added at the spreading centers (Mid-Ocean Ridges) and destroyed at the plate's leading edge by mountain building or subduction. Most earthquakes and nearly all volcanoes are confined to plate boundaries. (*Woods Hole Oceanographic Institution*) (*Above*) Theoretically, continents ride as passengers on oceanic plates which are continually being renewed by lava welling up at the Mid-Ocean Ridges, and are destroyed when they plunge into the high-temperature asthenosphere. The subduction process creates a trench—sometimes as deep as 10 000 m—at the junction of the plunging and overriding plates. Molten material from the plunging plate frequently escapes through fractures in the overriding plate to form volcanic islands or island arcs. In this figure, the continent, since it is lighter and cannot be subducted, will eventually incorporate the island arc. In this and related ways the continents of the earth continue to grow.

the Andes of South America and the Rockies and Appalachians of North America). Wegener also pointed to ancient rock formations found in both Africa and South America, and suggested that the outlines of these formations were completed if the continents were properly fitted together. As he put it, "It is just as if we were to refit the torn pieces of a newspaper by matching their edges, then check whether or not the lines of print run smoothly across. If they do, there is nothing left but to conclude that the pieces were in fact joined this way. If only one line was available for the test, we would still have found a high probability for the accuracy of the fit, but if we have *n* lines, this probability is raised to the *n*th power."

Wegener also used the fossil record to support his argument. The reptiles, in particular, provided excellent examples. For instance, fossils of *Mesosaurus*, a small reptile about 18 in (46 cm) long that lived around 270 million years ago, is found only in certain areas of Africa and Brazil. Wegener claimed that such animals could not have crossed from one continent to another on land bridges that have since collapsed. He argued that for a light land bridge, composed of the same material as the continents, to sink into an ocean floor of greater density would violate the principles of isostasy.

In later editions of his book, Wegener incorporated data and arguments derived from his study of ancient climates. Particularly interesting were tillites (sediments left by the movement of glaciers) in South Africa and southern South America. These sediments were about 300 million years old and suggested to Wegener that the southern extremities of both continents were then located very close to the South Pole. Ninety degrees north of the tillite deposits, Wegener found deposits of coal that stretched from the eastern United States to China.

Fossil plants in these coal beds were of the tropical variety, which indicated that they were laid down near the Equator around 240 million years ago. Higher (younger) in the same sedimentary column, the coal deposits were replaced by salt deposits and desert sands similar to those now found in the trade wind belts on either side of the Equator. This gradation would indicate that either the Equator had moved or the continents had moved with respect to the Equator.

Having amassed a wealth of such observations and arguments, Wegener arrived at the conclusion that was to revolutionize the earth sciences: All the world's land masses were once part of a single giant continent he called *Pangaea.* About 200 million years ago Pangaea began to break up with the resulting pieces gradually drifting to their present positions.

Initial reaction to Wegener's theory was mixed, but with time it rose to nearly total rejection. In the mid-1920s organized resistance to the idea of continental drift solidified around the attacks of Harold Jeffreys of Cambridge. Jeffreys readily disproved Wegener's speculation that the force responsible for the movement of the continents was the tidal effect of sun and moon on the earth's crust. Jeffreys demonstrated by simple calculations that if the earth's crust was weak enough to be deformed by such forces, mountains would collapse under their own weight. After Jeffreys' attack, resistance to the theory rose to such a point that as late as the early 1960s it was considered professional suicide for an earth scientist to be identified as a proponent of continental drift.

Confirmation of the Theory It is one of the strengths of the scientific process that any new theory or hypothesis must successfully meet the challenge of at least the majority of its critics before it is even tentatively accepted. Such was the case with continental drift. Although acceptance came with surprising suddenness during the 1960s, that acceptance was based on decades of accumulated evidence and discoveries, all of which supported the original claim that continents not only could, but had migrated laterally "through" a solid crust.

The dawn of the Atomic Age brought new tools to the study of Earth. It was found that the age of Earth's crust could be reliably determined by first measuring the rate at which the unstable isotopes of certain elements decay, and then finding the ratio of daughter- to parent-isotopes in rocks. By this technique, the age of the earth's solid envelope was found to be about 4.5 billion (4.5×10^9) years: much older than previously thought and providing ample time for such gradual processes as continental drift.

Radioactive dating also demonstrated that, contrary to earlier belief, the continents have not remained relatively constant in size and general shape but have grown through accretion—progressively younger material being added to ancient "core" rocks called *cratons.* Also, it was found that the radioactive decay of uranium, thorium, and an isotope of potassium, all distributed throughout the earth, generated considerable heat. It was realized that this unexpected source of heat was sufficient to at least retard the rate at which the earth was cooling and contracting, thus seriously weakening the long-held idea that many of the earth's features and processes were due to a general and uniform contraction of the crust.

During the famous CHALLENGER EXPEDITION of 1873, soundings—laboriously made by paying out a weighted hemp line—revealed the existence of a remarkable ridge in the middle of the Atlantic Ocean. During the period 1925–1927, the German research ship *Meteor,* using echo soundings (see UNDERWATER SOUND), mapped parts of the ridge in greater detail. In 1929, a similar elevation—the East Pacific Rise—was found in the Pacific Ocean. Over the next 30 years fathometer traces were accumulated from thousands of navy, merchant, and research ships sailing throughout the world's oceans. By 1959, Bruce Heezen, Marie Tharp, and Maurice Ewing of Columbia University were able to identify and trace a major ridge system of some 40 000 mi (64 360 km) in length that ran through every ocean and the NORWEGIAN SEA. (See HEEZEN, BRUCE CHARLES; EWING, WILLIAM MAURICE.)

To demonstrate how the mind leaps to anticipate a conclusion based on a little data, it should be noted that the world-circling Mid-Ocean Ridge was sketched out by L. Kober of Germany in 1928.

The Mid-Ocean Ridge system is best developed in the Atlantic where it occupies the center one-third of that ocean. The ridge rises an average of 0.6–1.8 mi (1–3 km) above the adjacent ocean floor and has an average width of 900 mi (1500 km). The crest of the ridge, in the Atlantic, is marked by a rift valley 15–30 mi (24–48 km) in width and about 6000 ft (1800 m) in depth below the valley rim. The rim itself lies an average of 6000 ft (1800 m) below the surface of the water, although the ridge rises above the surface in places (e.g., Azores, Iceland, etc.). Away from the valley, in both directions, the rugged, highly fractured surface of the ridge gradually drops away until it is obscured by the sediments derived from the continents (see ABYSSAL PLAINS). In places, particularly the equatorial Atlantic, the ridge is offset by great transform faults that run at right angles to its axis and can be traced over distances as great as 2000 mi (3200 km). The ridge displays similar characteristics in the Indian Ocean, but the ruggedness and the central rift valley is missing in the Pacific. Throughout its entire length the Mid-Ocean Ridge is the site of shallow earthquakes.

Over 40 years ago the Dutch geophysicist Felix A. Vening Meinesz demonstrated that a submerged submarine could be used to take sensitive gravity measurements at sea. In the course of his work, he discovered that some of the largest gravity deficiencies on earth are found along the deep sea trenches that exist along the island arcs of the Pacific and along the western edge of South America. This deficiency in mass was interpreted as areas in which the crust (ocean floor) was being pulled down into the interior of the earth, creating trenches.

The gravity anomalies discovered by Vening Meinesz stimulated Arthur Holmes of the University of Edinburgh and David T. Griggs of the University of California at Los Angeles to revive an old concept of geophysics; namely, that the interior of the earth is in a sluggish state of thermal convection similar to that of water in a pan that slowly comes to a boil. Measurements of heat flowing upward through the sea floor indicated that along the Mid-Ocean Ridge system the heat flow was 2 to 8 times the average flow of a millionth of a calorie per square centimeter per second (1×10^{-6} cal $cm^{-2}s^{-1}$) found on land and elsewhere on the ocean floor. Further, it was found that in the trenches the heat flow falls to a tenth of the average value. These discoveries and observations led to the concept of thermal cells, in which material sluggishly rose toward the surface in the areas occupied by the Mid-Ocean Ridge System and descended again in the areas of the trenches. Many earth scientists believe that these thermal cells provide the driving force that causes continents to drift apart. This possibility is supported by the discovery that the ridges are areas of tension (pulling apart) rather than of compression.

In 1906, 6 years before Wegener published his theory, a fundamental observation was made that would ultimately break the back of the resistance to continental drift. In that year a French physicist, Bernard Brunhes, found some volcanic rocks which were magnetized exactly opposite to the earth's present magnetic field. Brunhes concluded that the earth's magnetic field had reversed since the volcanic rocks had solidified from a molten state. Unfortunately this work attracted little attention within the scientific community until the general concept of paleomagnetism began to receive attention during the 1950s.

When a rock bearing oxides of iron or titanium solidifies, either by cooling from a molten state or by precipitating from an aqueous solution, it is slightly magnetized in the direction of the earth's magnetic field. The rock then becomes a "frozen" compass, pointing to the position of the earth's magnetic poles at the time the rock solidified—unless, of course, the rock has moved in the interim. In the early 1950s, a number of scientists, including S.K. Runcorn, P.M.S. Blackett of the University of London, and Emil Thellier of the University of Paris, began measuring the polarity of rocks from several continents. They found that as the rocks became progressively older the direction of the magnetic poles shifted, but differently for each continent. Since it was unlikely that the position of the magnetic poles had changed much in their relationship to the geographic poles, the continents themselves must have moved independently of each other.

During the late 1950s and early 1960s, the pace of discovery began to quicken. Allan Cox, G. Brent Dalrymple, and Richard R. Doell of the U.S. Geological Survey, in a study of basalt rocks carefully dated by radioactive techniques (measurement of the amount of argon 40 formed by the decay of radioactive potassium 40), were able to show that over the past 3.6 million years the earth's magnetic field had not only reversed as Brunhes suggested in 1906, but had done so several times. Neil Opdyke and James D. Hays of Columbia University added confirmation of these measurements by finding similar reversals in marine sediments. This was followed by an unusual discovery by Ronald G. Mason and Arthur D. Raff of the Scripps Institution of Oceanography: Using a sensitive magnetometer towed behind a ship, they discovered that huge areas of the ocean floor were magnetized in long, parallel bands, and that the polarity reversed from one band to the next.

The stage was now set for all the pieces of research to come together in one grand test. Around 1960, Harry H. Hess of Princeton University had suggested the concept of seafloor spreading. He pictured the mid-ocean ridges as spreading axes. As the forces responsible for continental drift—perhaps convection currents—caused the crustal plates on either side of the ridge to drift apart, lava flowed upward at the center of the ridge and solidified at the same rate at which the plates drifted apart. Thus, new plate was continually being added at the ridges and continually being destroyed as it was subducted at the trenches. F.J. Vine and D.H. Matthews of the University of Cambridge now suggested that if seafloor spreading was taking place, then lava welling up in the ridges would be magentized in the direction of Earth's magnetic field as it solidified. Further, since the lava was added in about equal amounts to each of the plates moving in opposite directions from the ridge, an identical magnetized band should exist on the seafloor on either side of the ridge. When Earth's magnetic field reversed, of course, the band immediately following would have reverse polarity and the width of each band would be a measure of the time between reversals. Thus, if seafloor spreading existed, a series of magentized bands, running parallel to the Mid-Ocean Ridge, should cover the ocean basin on either side of the Ridge and, on the basis of width and polarity, both series should match. They suggested that this line of reasoning be put to the test.

Confirmation came swiftly. F.J. Vine and J. Tuzo Wilson demonstrated the symmetry of bands on both sides of a ridge near Vancouver Island. And, James R. Heirtzler, W.C. Pitman, G.O. Dickson, and Xavier LePichon of Columbia University demonstrated that the same symmetry of bands exists in the Pacific, Atlantic, and Indian Oceans. Eventually, 171 reversals spanning 76 million years were identified. While the same number of bands have been identified in each ocean, the widths of the bands are different because the spreading rate has been different for each ocean. Studies of these bands indicate that the longest period between reversals of the earth's magnetic field has been about 3 million years and the shortest about 50 000 years. The average interval between reversals is between 420 000 and 480 000 years. We have been in the present orientation of the magnetic field for 700 000 years. Perhaps we are due for a change. The cause of these reversals is not known. Scientists have suggested meteorite strikes, mountain building, and irregularities in the earth's orbit.

Much of the above research was reported at the annual meeting of the Geological Society of America in San Francisco, California, in 1966. When the meeting was over, so was the battle for acceptance of Wegener's theory of continental drift. (See GLOMAR CHALLENGER.)

Plate Tectonics The theory of plate tectonics serves to unify the older theory of continental drift and the more recent concept of seafloor spreading. According to the theory, Earth's crust is composed of a number of rigid plates ranging in size from the plate that is essentially coincident with the borders of Turkey, to the Pacific plate which covers most of the Pacific basin. Each plate is thought to be approximately 60 mi (100 km) in thickness. The continents actually sit on a plate and are rafted along as the plate moves. Thus, the difficulty of explaining how a continent can plow its way through an oceanic plate is avoided; it merely rides along as a passenger on an oceanic plate. Therefore, "oceanic plate drift" is, perhaps, a more descriptive phrase then "continental drift."

Plates are in constant relative motion in one of three ways: They can pull away from each other as on either side of a spreading axis (Mid-Ocean Ridge); they can slide past one another; or they can converge. In the case of convergence, an oceanic plate, being of greater density, will plunge beneath a continental plate; whereas two continental plates will buckle at their leading edges to form a mountain range.

The thickness of plates has been postulated on the basis of seismic studies. The velocity of seismic waves is high in rigid, dense rocks, and low in less rigid, lighter rocks. Also, the velocity increases with confining pressure (depth), and decreases with increasing temperature. Seismic studies reveal that the velocity of shear waves suddenly decreases at a depth of about 42 mi (70 km) below the ocean floor, and ~90 mi (150 km) below the continents. Shear-wave velocities then increase with depth, particularly between 210 and 270 mi (350 and 450 km), and just above 420 mi (700 km).

The interpretation of these data is that the outer layer of rock (the lithosphere) rests on a weaker and hotter layer (the asthenosphere) that becomes increasingly viscous with depth. It is this hotter, more viscous layer that is thought to permit the overlying, rigid plate to move horizontally. In other words, the more viscous layer, though still a solid, deforms plastically to allow the overlying plate to move across it.

Should one visualize the earth's crust as composed of a number of curved plates fitting snugly together, then it becomes obvious that, if the earth is to remain essentially the same size, plate material must be destroyed at the leading edge at the same rate as new material is added to the trailing edge. In the case of two continental plates, the destruction is in the form of vertical deformation of the material to produce a range of mountains. Otherwise, the plate of greater density is forced beneath the adjoining plate in a process called *subduction,* and ultimately melts in the asthenosphere.

During the late 1950s a plan emerged to drill a hole through the ocean floor and down to the Mohorovicic Discontinuity which lies at a depth of about 25 mi (40 km) beneath the continents and about 4.5 mi (7 km) beneath the ocean bottom. The discontinuity is thought to represent a phase change in which the rocks below the boundary, as a result of heat and pressure, have assumed a greater density than those above. The plan, known as the Moho

Project, was eventually cancelled, but one of its selling points had been that drilling through the sediment that covered the basaltic floor of the ocean would be like leafing backward through the pages of earth history all the way to the beginning of the oceans—some 600 million to 4.5 billion years ago (see AGE OF THE OCEAN). At that time (1950s) the ocean basins, like the continents, were thought to be static features that had steadily accumulated sediment, layer after layer, since their creation.

Subsequent investigations, of course, have indicated that the ocean basins contain no sediment older than the breakup of Pangaea some 200 million years ago. These "pages of history" of greater age have simply been destroyed by seafloor spreading and the subduction of crustal plates. It has been estimated that some 20 billion cubic kilometers of crustal material have been destroyed during this cycle of continental drift and that, at the present rate of plate movement, an area equal to the entire surface of the earth will be destroyed during the next 160 million years.

In the subduction process the subducted plate is gradually forced downward beneath the overriding plate until it stabilizes at an angle of 30–45°. This process is responsible for the formation of the deep sea trenches such as that along the Pacific rim of South America, the Aleutians, and Japan. The subducted plate, being of greater density than the surrounding material, continues its downward plunge at a rate equal to the drift rate of the plate: about 1–10 cm/yr. Since the temperature increases significantly with depth, the plunging plate eventually melts and is absorbed by the asthenosphere. The depth at which a plate melts and loses its identity depends upon the rate at which it is sinking: the more rapid the sinking rate the greater the depth before extinction. However, no plate has been identified below 420 mi (700 km) where temperatures commonly reach 4000° F (2200° C).

Forcing a slab of cold, dense basalt 60 mi (100 km) thick into the earth's interior at a slow but steady rate while gradually raising its temperature until it finally melts completely is, to say the least, a rugged and stressful process. The periodic release of these stresses make itself felt as earthquakes and volcanoes. It is no accident that the earthquake zones of the world are to be found along the mid-ocean ridges and along plate boundaries or that approximately 99 percent of all volcanoes are similarly located. The volcanoes appear to result either from a sudden release of lava at a spreading axis (Iceland), or several tens of miles behind a subduction zone. When volcanoes erupt behind a subduction zone the result is a feature known as an Island Arc, such as the Aleutian Island chain. Such eruptions probably result from the escape of molten material from the subducted plate through fractures in the overriding plate caused by the stresses of subduction. It is significant that no deep-focus earthquakes are known to occur below a depth of 420 mi (700 km)—the depth at which even the fastest-sinking plate loses its identity.

The centers of drifting plates are relatively quiescent; earthquakes and volcanoes are rare. However, there are a few areas well removed from plate boundaries—notably the Hawaiian Islands—where volcanic activity occurs. Such areas, known as *hot spots,* apparently represent a relatively stationary source of magma which periodically erupts through weak areas in the overlying plate to form seamounts or volcanic islands (see SEAMOUNT). Since the hot spot is stationary and the plate is moving, a line of such seamounts and islands should result. This is precisely what seems to have happened in the case of the Hawaiian chain. From the still active volcanos Kilauea and Mauna Loa on the island of Hawaii, the chain runs northwestward to the vicinity of Milwaukee Seamount; here it turns northward as the Emperor Seamount chain which terminates near the subduction zone at the western extremity of the Aleutian Islands. These islands and seamounts become progressively older the further they are removed from Hawaii, reaching an age of 70 million years near the end of the chain. Similar examples in the Pacific are the Tuamotu group beginning with the relatively recent Pitcairn Island, and the Austral group beginning with McDonald Seamount. Both groups show the same dogleg pattern as the Hawaiian chain. Obviously, the drift direction of the Pacific plate changed about 40 million years ago.

In a process often called "geostill," plate tectonics also appears to play a central role in the origin of primary deposits of such metals as copper, silver, iron, cobalt, etc. As lava wells up along the spreading axis of the mid-ocean ridges and cools, it forms cracks and fissures which permit seawater to circulate to considerable depths below the ocean floor. As the water is increasingly heated with depth it tends to rise back to the surface while cooler water takes its place. The hot, circulating brine leaches minerals from the cooling basalt. The metal-rich solution is vented at the surface, where it scavenges additional minerals from the overlying seawater before it precipitates to form a blanket on the surrounding seafloor. When the plate is eventually subducted, its surface, including the metal-rich blanket, is the first to melt. The melt is often transported upward into the overlying plate where it slowly cools, thus allowing the metals to crystallize out as economic deposits. The many primary metal deposits of the Andes Mountains are thought to have been formed in this way.

It is also interesting to speculate about the fate of the organic material that is subducted along with the sediment layer carried by the plunging plate. These organics reach retort temperatures long before the surrounding temperature is sufficient to melt the sediment of which it is a part. Whether the resulting gas and hydrocarbons contribute to the formation of economically exploitable oil and gas deposits is not known.

The Past There is convincing evidence that the present cycle of continental drift is not the first. That evidence exists primarily in the form of mountain belts older than the breakup of Pangaea. For instance, the Appalachian-Caledonian belt of North America and the Ural Mountains of Russia appear to have been caused by the collision of two continental plates in much the same way the Himalayas were formed by the collision of the Indian and Asian plates during the present cycle of drift. Each of the two older mountain systems contain a thin zone of material known as *ophiolites* which have a characteristic sequence of rocks typical of an oceanic plate. Therefore, it appears that the Appalachians were created when North America and Africa collided some 400 million years ago.

The absence of well-defined mountain belts older than two billion years suggests that some mechanism other than plate tectonics operated prior to that time. It may be that no crust thick enough to sustain the stresses of continental drift could form prior to two billion years ago.

The Future Continental drift proceeds at such a slow pace that it is perceptible to the observer only by the earthquakes and volcanoes it creates. Over the next 50 million years, however, the process will bring about monumental change. The Atlantic and Indian Oceans will continue to grow at the expense of the Pacific. Antarctica will remain essentially fixed, but Australia will move northward until it begins to rub against the Eurasian plate. Africa will move slightly northward, and continue to turn counterclockwise. The northward motion will close the Bay of Biscayne and nearly collapse the Mediterranean Sea. A portion of Africa east of the Rift Valley will break away and move northeastward. North and South America will continue to move westward, and new land will be created by the compression of the CARIBBEAN SEA area. And Baja California and that portion of California west of the San Andreas Fault will break away from the mainland and move northward. In 10 million years Los Angeles will be a suburb of San Francisco, and in 60 million years it will begin to move into the Aleutian Trench.

CONTINENTAL RISE See CONTINENTAL SHELF.

CONTINENTAL SHELF is the submerged border of landmasses which occupies the zone extending seaward from the low-tide line to a point where the ocean bottom abruptly slopes more steeply toward greater depth. The depth of the "breakpoint" ranges from 66 to 2094 ft (21 to 621 m), with a worldwide average of 506 ft (133 m). The 100-fathom [600 ft (183 m)] curve has long been used as the outer limit of the continental shelf, but this came into use largely because early sailing charts had only three depth contours—10, 100, and 1000 fathoms.

Continental shelves range in width from essentially zero for precipitous coastlines such as along the west coast of Panama, to 750 mi (1206 km) for the shelf underlying the BARENTS SEA; the worldwide average being 48 mi (78 km). The average seaward slope of the shelf is 0°07′, being a bit steeper on the inner than the outer half. The average depth of the flattest portion of the shelf is 230 ft (70 m). Continental shelves underlie only 7.5 percent of the total area of the world's oceans but represent an area equal to 18 percent of the total land area of the earth—roughly equal to that of Europe and South America combined.

Continental shelves may be divided into two basic categories: those underlain by igneous material (i.e., material solidified from a molten state), and those made up of sedimentary materials (gravel, sand, silt, and clays). The latter category appears to be produced by land-derived sediment accumulating between the shore and some form of dam lying submerged at the outer limit of the shelf. Such dams may be of tectonic origin, such as the upthrust blocks of igneous material that rim much of the Pacific Basin; algal reefs, such as those around Florida, the east coast of Central America, and the northeast coast of Australia; diapir dams (salt domes forced upward into the overlying sediment by the weight of that sediment on a thick layer of salt at great depth) such as those rimming much of the GULF OF MEXICO; and areas such as that along much of the east coast of North and South America where the dam has long since been covered and overrun by sediments and the outer slope is maintained now by the angle of repose of the sediment itself. The estimated average thickness of the sediment accumulated behind these dams is about 1.2 mi (2 km), resulting in a total volume of 12 million (12×10^6) mi^3

(50 million km³) tied up in the continental shelves of the world.

Geologists once held the rather simplistic view that continental shelves represented a relatively featureless, gently sloping plain on which the sediment graded from coarse sands and gravels near shore to fine-grained silts and muds near the shelf edge. However, detailed studies beginning during World War II and continuing until today have shown the contrary to be the case. The surfaces of the continental shelves show many irregularities which result from the long-term conditions under which they were formed. For instance, shelves that have been subjected to glaciation may have troughs, such as those off some of the Norwegian fjords, which range from 1000 to 5000 ft (305 to 1524 m) in depth. Such glacial troughs become progressively shallow as they extend out on the shelf. In contrast, other shelves, such as that off the coast of New York, may have a submarine valley and SUBMARINE CANYON (Hudson Canyon) which cuts through the shelf in ever-increasing depth until it penetrates the continental slope (see below). Other shelves, such as that along the east coast of North America, show ridges of 20–30 ft (6–9 m) in height which run roughly parallel to the present coastline. It is believed that these represent submerged barrier islands (see BEACH) which were formed at lower stands of SEA LEVEL. And, since most of the continental shelves of the world were exposed when sea level dropped during the last Ice Age, coarse sands and gravels were carried far out into the shelves by stream action so that such materials are now found even as far out as the breakpoint of the shelf.

Between the seaward break of the continental shelf and the ABYSSAL PLAIN of the deep ocean floor lie the continental slope and the continental rise. Both these features are intimately connected to the shelf in both origin and sediment characteristics.

Continental Slope In its most abbreviated form the continental slope is defined as "the declivity from the outer edge of the continental shelf into great depths," or "that relatively steep portion of the seafloor which lies at the seaward border of the continental shelf." The slope of the seafloor thus defined ranges from 3 to 6°, with an average of 4°, and the depth ranges from 328–984 to 4592–10 496 ft (100–300 to 1400–3200 m). The width ranges from 12 to 62 mi (20 to 100 km). If a depth of 6560 ft (2000 m) is taken as the average seaward border of the continental slope, then it occupies 8.5 percent of the ocean floor. The average height of the continental slope is 12 000 ft (3658 m), with some heights ranging to 30 000 ft (9144 m). Thus, continental slopes are one of the greatest relief features on the face of the earth.

Continental slopes appear to be the result of sediment derived from the continental shelf by current action and slumping and directly from the land by river transport. Mud makes up about 60 percent of the slope sediment, with sand (25 percent), rock and gravel (10 percent), and shell fragments and oozes (5 percent) making up the remainder.

The continental slope is usually gently curving in outline, but its surface is periodically cut be ravines and submarine canyons similar to those found on mountain slopes. Hills also exist which may be the result of slumping.

The continental shelf and the continental slope, taken together, are often referred to as the continental terrace or continental margin. In some areas the continental terrace does not fit the above description but is composed of many basins, rises, and plateaus. Areas off the coasts of southern California, New England, and the northern shelf of Venezuela are examples. The term *continental borderland* has been proposed for such topography.

Continental Rise The continental slope terminates at a depth of 4590–10 500 ft (1400–3200 m). Between this point and the abyssal plain is a gently sloping prism of sediment known as the continental rise which falls away at the rate of about 1 ft(0.3 m) in 2000 ft (610 m), although steeper slopes are fairly common. The rise has a width that ranges from zero where TRENCHES exist to 373 mi (600 km). This is the zone where the sediments, slumping from the continental shelf and slope and carried by some TURBIDITY CURRENT, comes to rest unless it is carried out onto the abyssal plain. It is also the area where the great fans spread out from the mouths of submarine canyons.

Shelf Resources It has been stated that the wealth to be won from the ocean is inversely proportional to the depth at which it is sought. To date this has proved to be the case. The shallow continental shelves have proved to be a true cornucopia of resources for human consumption. Approximately 90 percent of the world's marine food supply is harvested from the continental shelf and adjacent bays. As early as 1970 some 17 percent of all petroleum and 6 percent of all gas was being produced from wells on the continental shelf, and by 1980 this amount is expected to quadruple. Today some 100 mines, with the shaft entry on land, mine coal, iron ore, nickel-copper ores, tin, and limestone from the continental shelf. For instance, 30 percent of Japan's coal is mined from depths as great as 8000 ft (2400 m) below sea level, and 10 percent of British coal is taken from beneath the sea. Around the rim of the Gulf of Mexico sulfur is extracted from the caps of salt domes intruded into the continental shelf by forcing superheated water into the solidified sulfur and forcing the solution back to the surface with compressed air. Other minerals currently being mined include gold, diamonds, tin, phosphorites, lime in the form of shells and oolites, and sand and gravel.

CONTINENTAL SLOPE See CONTINENTAL SHELF.

COOK, JAMES (1728–1779), an English explorer, navigator, and cartographer, is famous for his three voyages of discovery by which he added significantly to our knowledge of the geography of the oceans—particularly the PACIFIC OCEAN. He is also credited with the discovery that scurvy could be prevented on long voyages by adjusting the crew's diet to include fruits and vegetables.

Cook was born in Yorkshire, to poor parents, on October 27, 1728. At 18 years of age he took a job with a shipowner and made several voyages to the BALTIC SEA. When the Anglo-French war broke out, he enlisted in the Royal Navy as an able-bodied seaman. Within a month he was promoted to master's mate, and to master 4 years later. In 1759 he was given command of a ship and took part in operations in the St. Lawrence River. After the war ended in 1763, Cook, as captain of the schooner *Grenville* was charged with surveying the coasts of Newfoundland, Labrador, and Nova Scotia. His findings over a 4-year-period were of such importance that they were published by the English government.

On August 26, 1768, Cook set sail aboard the *Endeavour* to observe the transit of Venus in the Pacific Ocean and to explore new lands in the area. He was accompanied by an astronomer, two botanists, a landscape artist, and a painter of natural history. Sailing south and west he rounded Cape Horn and reached Tahiti on April 13, 1769. After observing the transit of Venus on June 3, he sailed to New Zealand where he spent 6 months charting the two islands. He then sailed along the east coast of Australia, which he named New South Wales and claimed for England. After sailing to Java through the strait that separates Australia from New Guinea, he returned to England by way of the INDIAN OCEAN and the Cape of Good Hope, reaching England on June 12, 1771.

On July 13, 1772, Cook again set sail from England, this time to verify the reported existence of a great southern continent. Accompanied by the *Adventure,* he sailed south along the African Coast and around the Cape of Good Hope. He crossed the ANTARCTIC CIRCLE in January 1773. Finding no southern continent, he sailed to New Zealand from which he proceeded to explore the New Hebrides, chart Easter Island and the Marquesas, visit Tahiti and Tonga, and discover New Caledonia and the islands of Palmerston, Norfolk, and Niue. He returned to England on July 29, 1775.

Cook's third and final voyage departed England on July 12, 1776. The objective was to explore the northern Pacific and to search for a passage around North America to the ATLANTIC OCEAN. Rounding the Cape of Good Hope, Cook crossed the Indian Ocean to New Zealand, then to Tahiti, and then to an island sighted on Christmas Eve which he named for the occasion. Sailing further north, he discovered the Hawaiian Islands. In February 1778 he sighted the coast of present-day Oregon and turned northward through the BERING SEA and Bering Strait into the Arctic Ocean. Finding no passage to the east, he returned to Hawaii where he was killed by natives on February 14, 1779.

COPENHAGEN WATER See NORMAL WATER.

COPEPODS are small marine CRUSTACEANS that make up the largest group of permanent zooplankton, or the tiny [approx 0.197 in, (5 mm) on the average] ocean animals that drift from place to place, especially in the top layers of the world's oceans. (See PLANKTON.)

SHRIMP-like in form, copepods (with the possible exception of the nematodes or unsegmented worms) doubtlessly outnumber all other animals on earth; some 10 000 different kinds of copepods exist in the ocean, and it has been estimated that a single species may number up to 28 000 individuals per 35 ft^3 (1 m^3) in surface coastal waters. These animals tend to migrate vertically in SEAWATER; that is, they change their position upward and downward in time with the rising and setting of the sun.

Some copepods are herbivores and feed only on plants (phytoplankton), while others are carnivores and exist by feeding on other tiny marine animals.

The copepod *Calanus finmarchicus* is one of the commonest copepods found in the oceans. It consumes large numbers of DIATOMS but avoids the *Ceratium* DINOFLAGELLATES.

It is very difficult to give a general physical description of copepods because they exhibit a wide variety of forms and many can only be identified as copepods by a study of their life history. They lack a carapace (a chitinous case), and the number of body segments is generally nine. They have a varying number of pairs of trunk limbs, not exceeding six. Several species, such as *Cecrops latreillii, Sphyrion lumpi,* and *Lepeophtheirus salmonis,* are parasites and live on the OCEAN SUNFISH (*Mola mola*), the ocean perch (*Sebastes marinus*), and the SALMON, respectively.

Copepods, often called "insects of the sea," consume microscopic ALGAE and change the polyunsaturated fat in the algae into polyunsaturated oils which they then store in sacs for use as food supply during periods of famine.

Copepods (especially the common copepod) constitute the principal food for the common HERRING, one of the most abundant fish in the world's oceans. The COD feeds upon the herring, and the SEALS eat the cod. In this particular OCEAN FOOD CHAIN, the seals are eaten by the KILLER WHALES.

In some areas of the ocean, baby salmon, as they swim from their river birthplaces to begin life in the open ocean, eat only these oil-rich copepods. Since the young salmon eat only copepods, their whole metabolic system is geared to digest large amounts of fats which are turned into a fatty alcohol; they then convert the fatty alcohol into fatty acids. Fatty acids are the common fats that human beings eat.

However, to perform this chemical conversion, fatty aldehyde must be involved in an intermediate stage. So far, no trace has been found in salmon.

Fatty aldehydes also are a source of mystery in human metaoblism. They occur as major components of human heart muscle and the brain, but their existence has never been explained. Analyses of the various copepod oils and a deeper understanding of the metabolism of the salmon may provide a clue to the mysteries of how the fatty aldehydes got there or what they do.

CORAL is the name for certain solitary or colonial anthozoan COELENTERATES which form a hard external-skeleton covering of calcium compounds or other materials. Corals which form large reefs are limited to warm, shallow waters, while those forming solitary growths are found in colder waters to great depths.

The corals are invertebrates belonging to the phylum Coelenterata, which is Greek for "hollow-gutted." They are akin to the SEA ANEMONE, and like the sea anemone they are characterized by their polypoid feature. The coral LARVA, which is globular in shape, soon evolves into polyps (a word derived from the Greek word *polyporus,* meaning "many-faceted"). The larva requires a support of some type to form the polyp and to grow. As soon as this is found on the ocean bottom or some rocky surface, soft tubular bodies with oral openings surrounded by hollow tentacles at the upper ends develop from the anchored pedestal. The animal feeds by snaring PLANKTON and other tiny organisms around it with these venomous tentacles.

Most corals multiply and branch into large and complex forms—fans, staghorn, or branched and boulderlike or brain-shaped. They also occur in a variety of colors, with yellows, reds, oranges, and purples being the most common. The polyp may remain solitary in which case it is called ahermatypic, or the individuals may combine to form a colony (hermatypic corals). In this regard, corals not only reproduce sexually, in which case an egg is produced and a larva formed, but they also reproduce by asexual budding. The polyp as it grows become longer and branchlike. On this branch a bud appears, and this polyp begins to form its own formation. It then divides and redivides, thereby building a coral mass.

In the colonies, the individuals are bound together by their calcium carbonate (limestone) secretions. Colonial polyps also share a common digestive tract.

See also CORAL REEF.

CORAL REEF is a limestone structure in relatively shallow WATER that results from the complex association of many calcium-secreting plants and animals. Today, the blue-green, green, and red ALGAE are the most important plant contributors to the reef structure, and the CORAL is the most important reef-building animal. To a large degree, however, both the algae and the coral provide a matrix which serves to hold and slowly cement into a stony mass the great volume of calcareous material contributed by a broad range of reef-dwelling plants and animals. A coral reef, then, is also one of the most complex ecosystems to be found in the world's oceans and one that occupies a relatively narrow niche in the overall marine environment.

Today, living coral reefs are to be found only in the zone between 35°N and 32°S LATITUDE. For vigorous growth a coral reef must begin on a relatively firm bottom, at a depth not to exceed about 65 ft (20 m) in order to have sunlight available for photosynthesis. The water must be clear, its temperature should ideally fall between 82.5 and 86° F (25–30° C) and never go below 64.4° F (18° C), and the salinity should lie in the range of 27–38 ppt. Since most of the reef-building plants and animals are stationary (sessile), good water circulation is necessary to provide a continuous supply of food. The water level with respect to the reef should be such that the top of the reef is only exposed

CORAL. (*Top left*) Coral polyps form a colony near the Bahama Islands. (*U.S. Navy*) (*Bottom left*) Fish take a leisurely swim through the elkhorn coral of a reef near St. Croix in the Caribbean Sea. (*NOAA*) (*Top right*) Branching coral reaches upward from the ocean floor. (*Bottom right*) A coral polyp spreads feathery tentacles near High Cay in the Bahamas. (*U.S. Navy*)

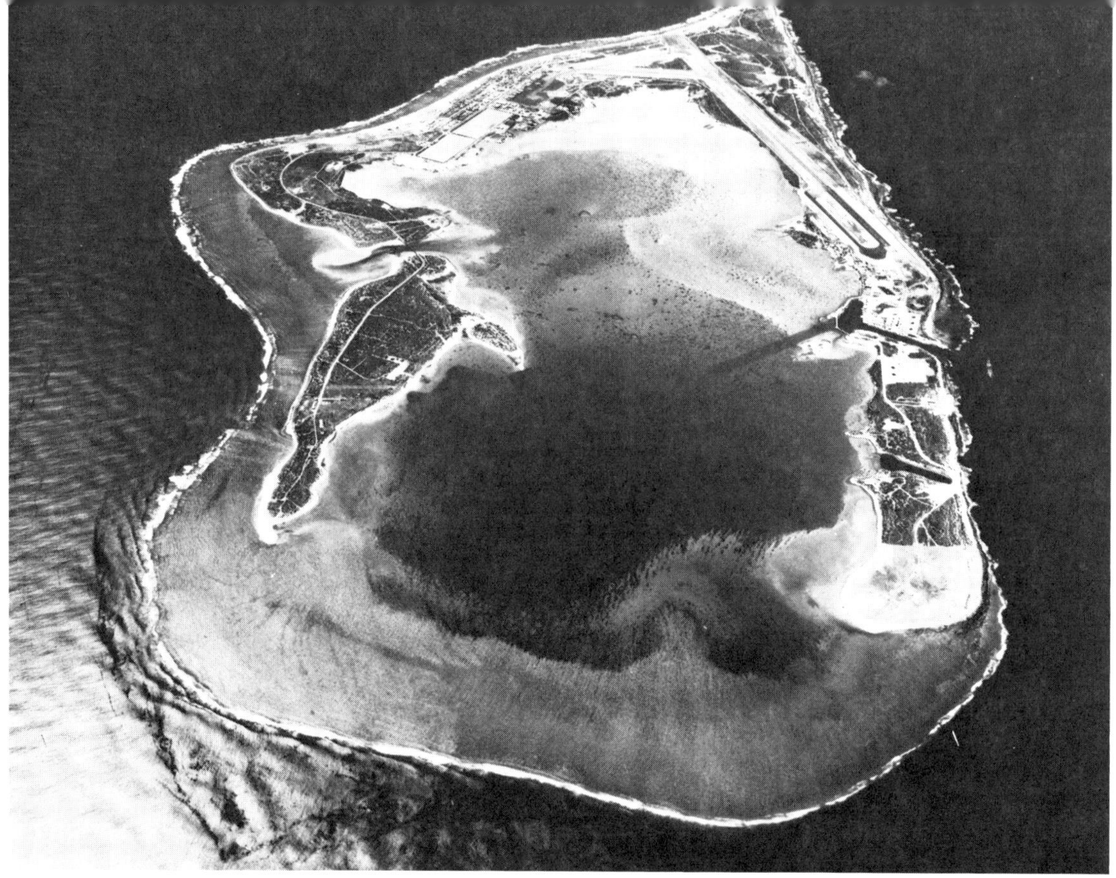

CORAL REEF. (*Above, left*) Wake Island, a center of U.S. naval air facilities, is actually an atoll, or circular reef around a lagoon. (*U.S. Navy*) (*Above, right*) Clouds float over Kwajalein Atoll in the Marshall Islands. This is the largest atoll in the world, covering 840 mi² (2175 km²). (*U.S. Navy*)

at low tide, if at all. Under proper conditions a coral reef can be expected to grow at the rate of about 1 in ($2\frac{1}{2}$ cm) per year. However, due to the destructive action of storm waves, boring marine worms, and certain species of predators (i.e., the crown-of-thorns starfish), coral reefs do not add to their mass at this rate over long periods of time.

Geologic History The first reefs began to form in the world's oceans with the advent of filamentous algae about 2 billion (2×10^9) years ago. These reefs, many of which are still visible in the fossil record, are quite distinctive in that they consist of laminated limestone flanked by reef debris. The name stromatolite (from the Greek, meaning "flat" and "stone") has been given to this reef form. The laminated structure results from layer after layer of fine calcium carbonate grains being trapped by the hairlike upper surface of the algae clump. As each layer builds up, the algae are forced to grow upward to reach sunlight, and a new trapping surface is formed.

The beginning of the Cambrian period some 600 million (600×10^6) years ago saw the introduction of the first animal into the reef-building community. This was a stony, spongelike animal called archaeocyathid (Greek for "ancient cup"). These animals associated themselves with the stromatolite clumps and contributed to the accumulation of reef material. And then, about 540 million years ago, the archaeocyathids simply disappeared without any apparent cause. For the next 60 million years the reef-building process was devoid of any animal, and all reefs of this interval were built by blue-green algae alone.

In middle Ordovician time, some 480 million years ago, animals rejoined the reef-building process. These animals included the stony sponges called stromatoporoid and the stony coelenterates—the first of the corals. Also joining the reef community was another plant, the coralline red algae—solenopora. For 130 million years the profitable association between algae, coral, and sponges thrived.

Then, around 350 million years ago, the reef community was again devastated by unknown factors or environmental changes. For the next 13 million years only scattered and poorly nourished stromatolite reefs survived.

With the beginning of the Carboniferous period the reef community again began to flourish. The revitalized community contained the rugged stromatolites, along with bryozoans, brachiopods, and corals. Joining the community were two new groups of calcareous green algae—dasycladateans and codiaceans, the chambered sphinctozoan sponge, and the crinoids, or sea lilies. For 115 million years this new community flourished—radiating into thousands of new species that accumulated giant reefs still preserved in the fossil record. And then, as had happened twice before, disaster struck. This time (225 million years ago) the destruction was total, for no reefs were known anywhere in the world for the next 10 million years.

About 215 million years ago another reef community began to establish itself slowly. As before, the algae were the most important members of the community, but several new families of coral—the scleractinians—became prominent. The stromatoporoid sponges reestablished themselves. A new coralline red algae, the lithothamnion, became a prominent member of the community. But the rugged and tenacious stromatolites were conspicuous by their absence. At this point there was another interval—20 million years in this case—in which reefs were unknown anywhere in the world. This period was followed by an impressive reef expansion in which a hitherto obscure bivalve mollusk known as a rudist played the most prominent role. For 60 million years the rudists rivaled both the algae and coral as a reef builder. Then came the great extinction that occurred at the close of the Cretaceous period, some 62 million years ago. Nearly a third of all animals known at that time were not to survive. Of the 115 genera of dinosaurs, none survived. In the reef community, the rudists, along with two-thirds of the coral genera, perished.

Again, a period of 10 million years was to pass before the reef community again began to flourish. Since that time the corals have become the most important single member of the community. There have been several periods of decline in reef building since that time, but no great extinctions. Oddly enough, the last Ice Age, with its great swings in sea level, had little effect on the reefs. Today, reef communities are worldwide but are restricted to a relatively narrow sanctuary on either side of the equator where conditions still favor their growth.

Reef Types There are a number of different types of reefs, with each type reflecting the topography and bathymetry of the bottom upon which it rests. The fringing reef, for instance, grows directly against a rocky coast and ranges up to several hundred feet in width. The barrier reef is the largest and best known of all reef types. These reefs are separated from the coast by a lagoon that may range from 65 to 325 ft (20 to 100 m) in depth. The reef itself is on the order of 1650 ft (500 m) in width, although some may reach several miles in breadth. The reef is usually interrupted in places by channels that lend access to the lagoon from the sea. In many cases the lagoon contains isolated pinnacles of coral that can prove hazardous to navigation. The seaward face of the reef is steep and usually conforms to the steep slope of the pedestal on which it sits. The Great Barrier Reef of Australia, the largest coral formation in the world, extends in broken chains along the northeast coast of Australia for 1250 mi (2011 km) and lies from 10 to 150 mi (16 to 241 km) offshore.

Atolls are similar to barrier reefs except that they are annular rings which enclose a lagoon. Some atolls are found on the CONTINENTAL SHELF, but most rest on isolated structures that rise from the deep seafloor. Atolls range in diameter from less than a mile to 20 mi (32 km). Small islands may populate the lagoon inside the larger atolls. The reef portion of the atoll is a low structure, often exposed at low tide, which ranges in width from a narrow ribbon to over a mile. The largest atoll in the world's oceans, Kwajalein Atoll in the Marshall Islands, covers and area of 840 mi² (2175 km²).

Reef Origin The origin of coral reefs has been debated by oceanographers for over a century. Since coral does not grow below about 65 ft (26 m) and can survive only brief periods above water, the question arises as to how coral reef formations several hundred feet in thickness could have formed. Charles Darwin, as a result of his voyage aboard the *Beagle* in 1831, suggested that such reef growth was made possible by the gradual subsistence of the pedestal upon which the reef first began to grow. In other words, the reef organisms grew upward to compensate for the gradual submergence of their platform. More recently, the gradual rising of the sea level in response to the melting of glaciers at the end of the last Ice Age has been added to the subsistence mechanism as a possible explanation.

Recently, scientists have developed a method of determining coral growth rates based on the assimilation of radium isotopes by the living coral. As the coral deposits its skeleton of calcium carbonate, the radium isotopes decay at known rates; thus, samples collected through a massive coral growth (head) may be dated to determine how fast the coral is growing. X-ray pictures of slices through coral heads show alternating light and dark banding, suggesting periodic growth increments. By x-raying corals analyzed by the radium isotope method, it has been confirmed that this banding is annual.

Through these studies, coral growth rates are now precisely measurable. The banding and radium methods may be used to review changes in growth rate as the coral adapted to changes in the environment in which it was growing. Data from corals also may be used to reconstruct the patterns of radioactive fallout and to monitor changes in the local water chemistry.

CORAL SEA is located on the northwestern corner of the South PACIFIC OCEAN and lies off the east coast of Australia. The western boundary begins at 30°S latitude in the vicinity of Grafton, runs north along the Australian coast, and crosses the Torres Strait to the Territory of Papua, New Guinea. The boundary line then runs eastward along the reef line to Rennell and San Cristobal in the Solomon Island chain, and then southward along the eastern side of the New Hebrides Islands to Aneityum. It then angles southwestward through New Caledonia to the parallel of 30°S and along this parallel to the coast of Australia. The waters of the Coral Sea mingle with those of the SOLOMON SEA to the north and the TASMAN SEA to the south. Connection to the INDIAN OCEAN is through the Torres Strait. The sea covers an area of 1 849 326 mi² (4 791 000 km²), occupies a volume of 2 751 768 mi³ (11 470 000 km³), and has a mean depth of 7854 ft (2394 m).

The Coral Sea area was opened up to the outside world by English, Dutch, Portuguese, and Spanish explorers during the sixteenth and seventeenth centuries. In 1511, Antonio de Abrea, a Portuguese navigator, first sighted the island of New Guinea, and the Spanish explorer Ortiz de Reyes went ashore on the northern coast in 1542. In 1567 the Spanish explorer Álvaro nendaña de Neyra discovered an island chain to the east of New Guinea to which he gave the name *Solomon* because he thought that he had found the source of the gold in King Solomon's temple. In 1606 Pedro de Queirós of Portugal discovered the New Hebrides Islands to the south, and James Cook of England explored them more completely in 1774. (See COOK,

JAMES.) During the same cruise Cook discovered New Caledonia farther west and gave the island the Roman name for Scotland. In 1606 the Spanish explorer Luis Vaez de Torres sailed through the strait that now bears his name, and may have sighted the continent of Australia. Also in 1606 the Dutch navigator Willem Jansz explored the Gulf of Carpentaria on the north coast of Australia. And, in 1642 Abel Janszoon Tasman explored the island (Tasmania) off the southeastern tip of Australia that now bears his name (although he named it Van Diemen's Land) and went on to circumnavigate the continent.

The CONTINENTAL SHELF around the periphery of the Coral Sea is of very limited extent except off the east coast of Australia—especially north of Brisbane—and along the southern coast of New Guinea. The sides of the volcanic Solomons and New Hebrides Islands are too precipitous to support a shelf of any great extent. The shallows do, however, support three of the world's largest coral reefs which contribute to the name of the sea. The largest of these is the Great Barrier Reef which runs for 1250 mi (2012 km) along the northeastern coast of Australia and lies from 10 to 150 mi (16 to 241 km) off the coast. The other two reefs are the Tagula Barrier Reef off the southeastern coast of New Guinea and along the Louisiade Archipelago and the New Caledonia Barrier Reef which rings that island and extends northward as far as the D'Entrecasteaux Reef. These reefs support an abundance of both plant and animal life. (See CORAL REEF.)

The deep basin of the Coral Sea is steep-walled, and the floor is contorted by huge trenches, basins, rises, and plateaus. The Coral Sea Basin on the north, the New Hebrides Basin on the northeast, and the Tasman Abyssal Plain thrusting up along the southwestern flank are separated by the Lord Howe Ridge which strikes north-south down the middle of the sea. The basins lie below 12 000 ft (3658 m), and the rise ranges from the surface to about 4000 ft (1219 m). The great Queensland Plateau with its many atolls and reefs lies to the southwest of the Coral Sea Basin. On the periphery of the sea there are two trenches: the San Cristobal Trench which runs around the southern flank of the Solomon Islands and the New Hebrides Trench which runs down the west coast of the New Hebrides Islands. These trenches are deeper than the basins, being on the order of 30 000 ft (9144 m) in depth. The deep floor of the Coral Sea is covered with red clay and globigerina ooze, whereas the shallow continental shelf is covered primarily with carbonate debris from the reefs and atolls.

The surface circulation in the Coral Sea is dominated by the EAST AUSTRALIAN CURRENT which flows southward along the coast of Australia. Its WATER is supplied by the Pacific between the Solomon and New Hebrides Islands.

CORALLIMORPHARIA is an order of SEA ANEMONE in the subclass Zoantharia. These animals resemble CORAL in their weak musculature, tentacles, and complex NEMATOCYSTS. They do not possess a skeleton, and the tentacles are arranged in radial rows on the disk, while the base has no muscular system. They occur in shallow WATER from the temperate zone to the tropics. Most species are solitary, although they may occur in large, asexually produced aggregations. Examples are *Corallimorphus* and *Corynactis*.

CORALLINE ALGAE See ALGAE.

CORERS See INSTRUMENTATION; OCEANOGRAPHY.

CORIOLIS EFFECT is the apparent deviation from a straight line of an object moving *freely* over the earth's surface, provided the earth itself is used as the reference system against which the profile of movement is measured. The Coriolis effect dictates that a freely moving object, no matter what its direction of travel, will appear to curve to the right in the Northern Hemisphere and to the left in the Southern Hemisphere. The phenomenon derives its name from the French mathematician Gaspard Gustave de Coriolis who first described it in 1835.

Basic to the understanding of the Coriolis effect is a recognition that the shape of the earth approaches that of a sphere, that the earth revolves about its own axis once every 24 hours, and that the distance —and therefore the speed—an object *fixed* on the earth's surface must travel in order to complete a revolution is a function of its latitude. For instance, in round numbers, an object fixed at the equator travels at a rate of 1000 mph (1600 km/h); at 30° latitude it travels at 866 mph (1400 km/h); and at 60° latitude it travels at only 500 mph (800 km/h), because the higher the latitude, the less far the object must travel in completing each revolution. It is also important to note that the *rate of change* of the fixed object's speed increases with increasing latitude. Therefore, the Coriolis effect—the apparent curvature—is greatest near the poles and falls to zero in the vicinity of the equator.

To demonstrate the interrelationship of the Coriolis effect with rotational speed as a function of latitude, we might visualize the flight of a rocket fired from the North Pole and aimed at Gabon on the west coast of equatorial Africa. Let us assume that the flight speed of the rocket is 1000 mph (1609 km/h) and that wind speed is neglected. The flight time of the rocket will be about 6 hours. During the time that the rocket is in flight, Gabon will have traveled 388 800 mi (625 579 km) as a result of the earth's orbital speed around the sun and the earth will have rotated 90° about its own axis. Therefore, observers will see the rocket curve continuously to the right and land, not at Gabon but in the vicinity of Ecuador on the west coast of South America. They will also note that the amount of curvature in the flight path is greatest in the high latitudes and becomes very gentle in the vicinity of the equator.

The Coriolis effect influences all our lives, but the effect is usually so small as to go unnoticed. In walking or driving, the friction of our shoes or tires is sufficient to prevent us from drifting from our intended track. An airline pilot compensates constantly for the Coriolis effect, but the correction for crosswinds is so much greater that the Coriolis correction is obscured. Long-range gunners long ago learned to use a Coriolis correction in their firing equations. And geologists have long speculated that the Coriolis effect causes rivers to erode at a greater rate at one bank than at the other. (This phenomenon is known as Baer's law, after the Russian scientist Karl von Baer.)

The circulation of the oceans and the atmosphere gives the Coriolis effect its greatest impact on our lives. Without this effect air would flow directly from high-pressure areas to low-pressure areas and quickly smooth any irregularities. In the vicinity of the equator where the Coriolis effect is near zero, this is exactly what happens —the result is the well-known doldrums. No hurricane or typhoon occurs within 5° of the equator. In higher latitudes, on the other hand, the wind between high- and low-pressure centers—due to the Coriolis effect—flows at right angles to the pressure gradient. The result is the strong and persistent "cells' of rotating high- and low-pressure air which provide the variability of our weather.

In the ocean, conditions are analogous to those in the atmosphere. In the North Atlantic, for instance, the WATER in the north is cold and dense, whereas that in the south is warm and less dense. As a result of this pressure gradient the southern water is induced to flow north. But, instead of simply flowing north where it would override the colder water and erase the clear distinction between the two, the Coriolis effect causes it to turn to the right. This process is aided by the near clockwise pattern of the wind in the North Atlantic—it blows west to east in the north and east to west in the south. The result is a clockwise pattern of ocean currents—often called the North Atlantic Gyre—made up of the GULF STREAM, NORTH ATLANTIC CURRENT, CANARY CURRENT, and North Equatorial Drift Current (see EQUATORIAL CURRENT SYSTEM). Since the Coriolis effect increases with increasing latitude, the Gulf Stream becomes a bold, swift current as it moves north. The southward-moving Canary Current, on the other hand, becomes increasingly sluggish as the Coriolis effect diminishes in lower latitudes.

One of the more obvious results of the Coriolis effect is to be found along the coast of California. The prevailing wind is from the northwest. The combined effect of the wind and the Coriolis effect causes the warmer surface waters along the coast to drift off to the southwest which, in turn, causes the colder water at greater depth to well up to replace it. In the summer the warm, moisture-laden wind blowing in from the northwest is chilled by the cold surface water; result: the famous California fog.

CORROSION (also called rust) is the deterioration or destruction of a metal or metal alloy by a chemical or electrochemical reaction with its environment. In the ocean environment, this reaction is usually electrochemical.

While there are complex and varying mechanisms connected with the electrochemical deterioration of metals, basically most marine corrosion may be represented by the equation

Metal + SEAWATER
→ hydroxide of the metal + hydrogen

This phenomenon is rooted in the fact that seawater, or a solution consisting of dissolved salts, possesses the property of electrical conductivity. It is therefore called an electrolytic solution. If a supply of oxygen is maintained at the corroding areas, metals will enter into solution and flake off or corrode in an electrolytic solution like seawater, which is fully saturated with air near the surface. The discharge of hydrogen ions also takes place, and both this (cathodic) reaction and that of the breakdown of the metal (anodic reaction) are accompanied by the flow of electricity. (See ION.)

Various metals have different tendencies to go into solution in the form of ions under the conditions briefly described, and, in essence, a measure of this tendency is termed the *corrosion potential.*

The rate and amount of corrosion of materials emplaced in the ocean depends upon the nature of the seawater. Since electric currents follow the path of least resistance, the conductivity of the seawater plays a major part in the mechanism. This conductivity increases with SALINITY and TEMPERATURE.

Oxygen in seawater can either inhibit or accelerate deterioration of materials depending upon the material and the situation. Primarily, the oceans are aerobic (contain oxygen) throughout. In the low-oxygen and anaerobic areas which are to be found in some quiescent ocean bottoms (e.g., BLACK SEA), sulfides exist because of large populations of sulfide-producing BACTERIA present there. When these sulfides exist in sufficient amounts, an acidic condition at and near the bottom causes corrosion to certain materials.

Corrosion also depends on the internal makeup of the material itself and its operational use. The rate of corrosion is influenced by the material's several internal factors (viz., chemical composition, distribution of secondary phases, residual stresses, voids, inclusions, and dissolved gases) as well as by environmental factors (viz., concentration, temperature, movement of the seawater, etc., plus the presence of electrochemical couples, surface films, and applied stresses). Some of these factors are understood. Many are not.

One method (first developed and applied to ships by Sir Humphry Davy in 1824) used to prevent corrosion of metals immersed in seawater is called *cathodic protection.* In cathodic protection, an outside anode is used to supply current that "neutralizes" the local corrosion currents. This technique reduces the rate of corrosion of metallic structures by making the corrosion potential of the metal more electronegative (cathodic). By this method, steel, aluminum, and several common structural materials, when immersed in seawater, can be protected by cathodic current supplied either from sacrificial anodes (magnesium, zinc, and aluminum) or an external direct current source.

Such schemes are especially adaptable for large ocean structures. However, one problem with cathodic protection is that caused by the effect of hydrogen which may be absorbed by the cathode metal. This hydrogen-embrittlement effect causes a loss of ductility and fatigue life. (Fatigue failure is a material failure due to cyclic imposition of a stress lower than the stress required to cause failure on one application.) Fatigue life is an important consideration in some materials such as the higher-strength steels used in submarine hulls and other structures designed to operate at greater depths in the ocean.

Other marine materials whose electrons are free to move (as in a storage battery) are vulnerable to corrosion. Stress corrosion, which may be defined as the deterioration that unexplainably takes place in many materials under the influence of a stress, has also long been a problem with some brass and steel alloys used in ordinary seawater applications.

The various forms of corrosion and the means employed to prevent this deterioration of marine engineering materials represent a multibillion dollar cost expenditure to nations of the world involved in OCEAN ENGINEERING endeavors.

COUSTEAU, JACQUES YVES (1910–), a French naval officer and marine explorer, won international acclaim for his film documentaries and books about the oceans, marine life, and activities of humans on and beneath the sea. He was also a prime contributor to the invention of the Aqualung, the first to take color pictures underwater, a pioneer in the use of television underwater, and one of the early investigators to demonstrate the feasibility of saturation diving. He was also very active in early programs to reduce ocean pollution.

Cousteau was born June 11, 1910, in St. André de Cubzac, France, and educated at Academy Stanislas, Paris, and the Brest Naval Academy. He served with the French Navy from 1930 to 1956 and with the French underground during World War II. Directly after the war, Cousteau helped clear mines along France's Mediterranean Coast. In 1950 Cousteau became the commander of the research ship *Calypso,* and in 1957 he became the director of the Museum of Oceanography at Monaco. It was through these two associations that Cousteau conducted most of the exploration and research that led to his film documentaries, books, and a television show entitled "The Undersea World of Jacques Cousteau." Three of his films, *The Silent World* (1956), *The Golden Fish* (1960), and *World Without Sun* (1966) won Oscars.

CRAB is the common name for a number of CRUSTACEANS in the order Decapoda, which includes SHRIMP, LOBSTER, the hermit crabs, and the true crabs. Crabs are members of the Anomura and Brachyura of the order Decapoda. Anomura comprises 12 families, which include about 1300 species of crablike or lobsterlike forms with a tail fan, while Brachyura is composed of 26 families and over 4500 living species of true crabs, most of which are marine with a worldwide distribution.

The anomuran crabs are very diverse crustaceans, but in nearly all anomuran crabs the abdomen is bent forward ventrally or is asymmetrical, soft, and twisted. Brachyurans are distinguished by the short,

The edible Alaskan king crab can have a leg span of over 36 in (91.44 cm). *(NOAA)*

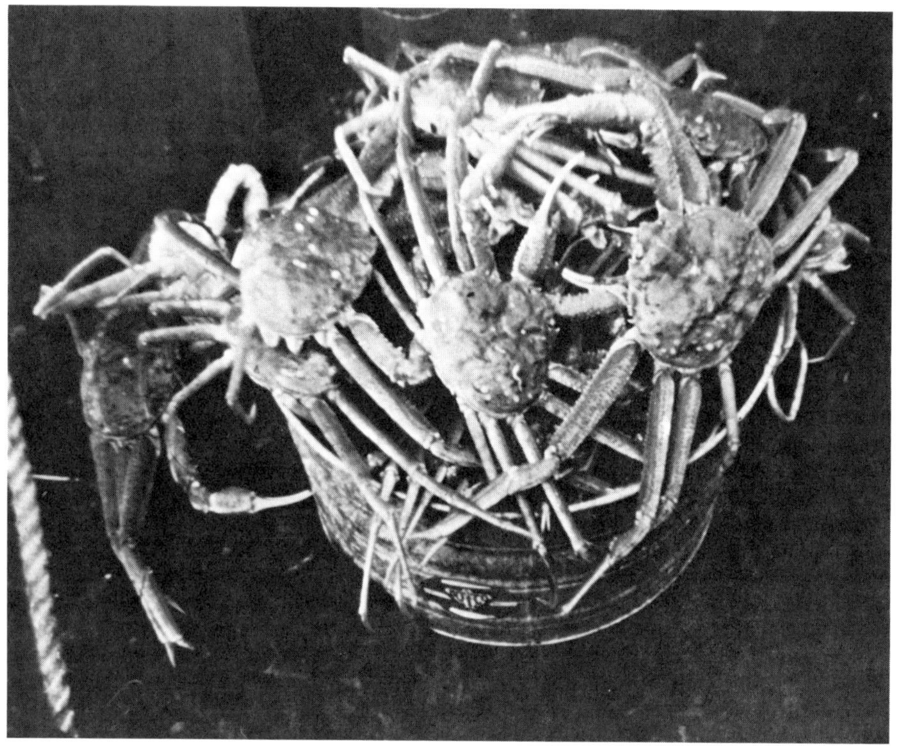

Spider crabs have long, thin legs, one species having a leg span of over 8 ft (2.44 m). Many cover themselves with seaweed so as to blend into the natural shore environment. *(NOAA)*

The scientific name for the edible blue crab means "savory, beautiful swimmer." *(Bureau of Sport Fisheries & Wildlife)*

A land crab strikes a defensive pose. These crabs can live and feed away from the ocean because they are equipped with modified gill chambers which permit respiration in air. *(Bureau of Sport Fisheries & Wildlife)*

broad, and generally flattened carapace (the case or shield-back covering), the small abdomen curled beneath the body and fitting into a depression under the thorax, the part between the neck and abdomen, and five pairs of walking legs with well-developed pincers or chelipeds on the two front legs.

Among the best-known of the many anomuran types of crabs are the king crabs (family Lithodidae), the hermit crabs (family Paguridae) such as the common species *Eupagurus bernhardus,* and the porcelain crabs (family Porcellanidae) represented by the common *Porcellana platycheles* of Europe.

The king crabs are unusually large crablike anomurans (inhabitants are mostly of cooler waters of the Northern Hemisphere) found off the coast of Alaska, Japan, and Siberia. Highly prized as a food, the large *Paralithodes kamtschatica* species often has a leg span of over 3 ft (91 cm). They somewhat resemble the spider crabs, which are true crabs.

The horseshoe crabs are often very erroneously referred to as king crabs. These horseshoe crabs are neither anomurans nor brachyurans but arachnids, members of a zoological class that includes spiders, scorpions, and ticks. The horseshoe crab, *Limulus polyphemus* has outlived all its closest relations and survived unchanged for hundreds of millions of years.

The hermit crabs, which are mostly marine occupy the empty shells of snails and other gastropod mollusks. (See MOLLUSK.) Some species live in straight-tubed shells, while others coil their large soft abdomen to the right into appropriately configured shells. When this animal senses danger, it withdraws completely into its shell. When emerging, the claws are put forth first.

In addition to the 650 living species of king and hermit crabs, there are the porcelain crabs. Like the hermits, these animals are found primarily in tropical and temperate waters, usually living in the shallow parts of these oceans. They match their coloration to the environmental background, thus providing themselves with excellent camouflage.

The pelagic California red crab, *Pleuroncodes planipes,* which translates to "bulgy-sided crab with flat feet," is a small [3.5–5 in (8.89–12.70 cm) long] anomuran of the family Galatheidae. This family consists of some 200 species of small lobsterlike crabs with abdomens and tail fans well-adapted for swimming, and their long first pair of legs are armed with pincers (chelipeds). Rows of hairlike setae are located on the front and hind margins of the walking legs. Each female may have two and sometimes three broods of eggs in one season. Females may carry up to 3650 eggs which hatch after about 2 weeks in deep water off the CONTINENTAL SLOPE. The larval and juvenile stages of these crabs are often carried to the beaches of Baja California in huge numbers by winds and currents.

The mole crabs, or hippas (the group Hippidea of the section Anomura), live in the surf zone of tropical and temperate shores of the world's oceans. Here they move up and down the shores, always staying in the active wash of the WAVES. As the water from one wave recedes, they burrow down in the sand, often disappearing until after the next wave has passed. Avoiding the impact of the breaking wave, they emerge to bury themselves bottom first in the wet sand and extend long, ornate antennas which filter plankton from the receding water. They are a link in the OCEAN FOOD CHAIN, a major connection between PLANKTON and larger animals.

The true crabs, or those 4500 species in the Brachyura, vary greatly in size from the minute [$\frac{1}{4}$ in (6 mm)] crabs of the family Pinnotheridae to the giant spider types of the family Majidae, as represented by the largest [5-ft (1.5-m) leg span] species, *Macrocheria kaempferi,* found in the deep waters off Japan. Some of the better-known brachyurans are the fiddler crabs with their distinctive enlarged, single claw, e.g., *Uca spp.,* common in areas of MANGROVES in the warmer sections of the world; the edible blue crab, e.g., *Callinectes sapidus* (from *Calli,* meaning "beautiful," plus *nectes,* "swimmer," and *sapidus,* "savory"), widely distributed along the Atlantic coast of the United States; the many species of dromiid-type crabs, e.g., *Dromia vulgaris,* which live in deep water; the ghost crabs (ocypode); and the stone crabs (*Menippe*). The latter is a highly esteemed seafood, especially the species *Menippe mercenaria* which is found in the waters off the southern coast of the United States and Cuba.

Crabs are chiefly omnivorous, although some are predatory, and a few are herbivorous. Their natural enemies are the OCTOPUS and various large FISH. Spawning habits differ within the two main sections since some crabs are strictly marine, while others have adapted to a terrestrial existence. However, most of the latter return to the ocean to release their eggs which hatch in the first larval stage as free-swimming zoea. See LARVA. In this regard, the true crabs have hard shells, or exoskeletons. Periodically, in order to grow, they shed this external armor or shell. This process is called molting. Before the molt starts, a new, soft exoskeleton forms inside and the crab backs out of the old shell as it loosens. The new shell is soft and elastic, allowing the crab to grow. It is particularly vulnerable to attack during the soft-shell stage and seeks refuge in a secluded spot until the new shell hardens. Chitin, an amorphous polysaccharide similar in nature to cellulose, is the structural material that forms the shells of crabs and other crustaceans. Crabs lose one or more legs during their lifetimes and are able to grow new ones through a regeneration process.

Crab is second to shrimp as the shellfish most preferred as food by many people of the world, especially those in the United States. Among the most important species of crabs used for food are the blue crab of the Atlantic and Gulf coasts of the United States (*Callinectes sapidus*), Dungeness crab of the Pacific Coast (*Cancer magister*), the giant king crabs of Alaska, the BERING SEA, Japan, and Kamchatka (*Paralithodes kamtschatica, P. platypus,* and *P. brevipes*), and the red crab and blue crab of Japan (*Chionectes opilio* and *Neptunus trituberculetes,* respectively). Primarily these are caught by means of triggered tunnel openings, baited traps or pots with buoy lines, and in trawls and dredges.

CROMWELL CURRENT is a subsurface current that flows eastward beneath the equator in the PACIFIC OCEAN. The current originates in the Western Pacific in the area north of the Admiralty Islands and disappears in the vicinity of the Galapagos Islands off the coast of Ecuador. The current is about 186 mi (300 km) wide and 656 ft (200 m) thick and flows at a speed of 1.9–2.9 kn. The current is named for Townsend Cromwell who discovered it in 1952.

CRUSTACEANS are the members of a large class of arthropods (phylum Arthropoda, subphylum Mandibulata). The crustaceans usually possess a segmented body, an exoskeleton (usually a horny chitinous outer shell), jointed feet and mandibles, and two pairs of antennae. Some 26 000 species exist within several subclasses and orders of which the malacostracans, or higher crustaceans, constitute the largest number and most familiar shellfish types.

Crustacea literally translates to animals with a crustlike shell. Some common examples of marine forms are LOBSTER, BARNACLES, SHRIMP, and CRAB. The crustaceans vary from other classes of related animals, such as insects, in that they have two pairs of antennae. The head consists of six fused segments, and the anterior segments of the body behind the head are collectively called the thorax, or cephalothorax. The posterior segments of the body are usually more distinct than the anterior, and they constitute the abdomen. Nearly all crustaceans have two sexes—male and female. The cirripedes are usually hermaphroditic, but the barnacles, or Cirripedia, have both male and female reproductive organs and are therefore always hermaphroditic.

Growth of crustaceans takes place when the outer shell is periodically molted or shed. This occurs at varying intervals, ranging from hours áfter hatching to weeks in some of the adults. The cycle is governed by a molt-inhibiting hormone present in an organ of the eyestalks as well as by a molt hormone, located in the body of crabs near the eyestalks.

Some of the adult crustaceans possess structures which are unlike those of the more typical forms. Barnacles, for example, attach themselves permanently to some solid object like a rock, secrete a limy shell of several plates, and gather food by sweeping the water with appendages resembling a net. *Sacculina* is a parasite on crabs which in the adult stage resembles the mycelium of a FUNGUS.

The hermit crabs have soft abdomens, not as hard and protective as the calcified CHITIN on the anterior part of the body. They seek out empty snail shells and back into these, fitting the abdomen into the space formerly occupied by the snail. As a hermit crab grows, it hunts around for a larger habitat. When it finds a shell that might be suitable, it explores the shell with its antennae. If it is satisfied, it pulls its abdomen out of the old shell and backs quickly into the new one.

Among the most numerous of ocean animals is a group of crustaceans known as COPEPODS. They eat ALGAE and PROTOZOA in great numbers and are eaten in turn by fishes and WHALES, forming an important link in the chain of food organisms from the smallest to the largest predators. They are small in size but are present in tremendous numbers in the world's oceans.

In the sea, crustaceans live at almost all depths and range in size from these tiny [$\frac{1}{10}$ in (0.25 cm) long] copepods to crabs having a leg span of 10 ft ($3\frac{1}{3}$ m).

Many crustaceans are filter feeders and screen PLANKTON and DETRITUS from the ocean waters, others are carnivorous, and still others convert DIATOMS or the microscopic marine plant life into food for larger animal life.

CURRENTS are those movements of ocean water characterized by regularity, either of a cyclical nature or, more commonly, as a continuous stream flowing along a definable path.

Three general classes of ocean currents can be distinguished: (1) those caused by differences in the density of seawater, and known as gradient currents; (2) wind-driven currents caused by the "pull" of the wind as it flows over the sea surface; and (3),

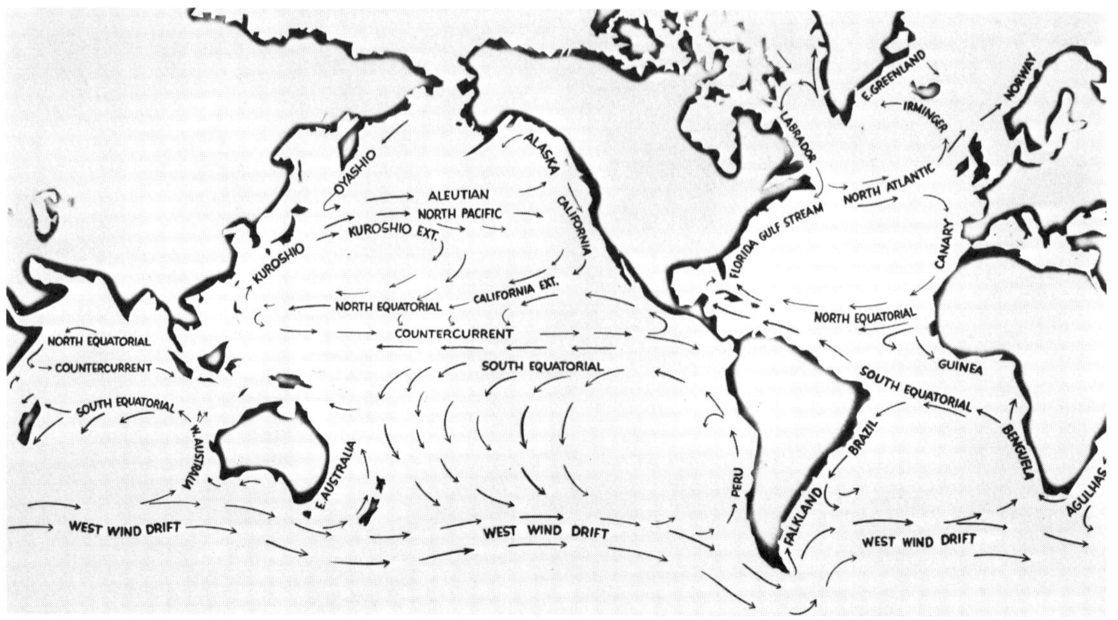

CURRENTS. The diagram represents all the major currents of the world's oceans. (*NOAA*)

those currents produced by long-wave motion. The latter class includes tidal currents (see TIDES) as well as those currents produced by internal WAVES, seiches, and tsunamis (see SEICHE; TSUNAMI).

Easily the most obvious, and certainly the most important ocean currents are the surface currents which circle the ocean basins to either side of the equator. Absorbing heat from the sun while they flow as broad, sluggish streams along the equatorial region, these currents turn poleward and release their stored heat in high latitudes. Thus, the coastal states of Alaska, Washington, and Oregon in the eastern PACIFIC OCEAN and England in the eastern ATLANTIC OCEAN have milder climates than their latitudes would ordinarily dictate.

These great "rivers in the sea" typically flow in a clockwise gyre north of the equator, and counterclockwise in the southern hemisphere. Those forces which generate and sustain the currents, as well as those which control their direction, are now fairly well understood by oceanographers. The first of these is the distribution of land masses such that they typically form a barrier on either side of the major ocean basins (Atlantic, Pacific, Indian). (See BASIN.) If no land existed and the earth were uniformly covered with water, the flow pattern of ocean currents would be greatly simplified. Under the influence of prevailing wind patterns the currents would circle the earth east to west in the tradewind belts (to either side of the equator) and in the vicinity of the poles, and west to east under the influence of the prevailing westerlies at intermediate latitudes. The latter current is at least partially illustrated by the ANTARCTIC CIRCUMPOLAR CURRENT that endlessly circles west to east about Antarctica because there is no land barrier between the tips of South America and Africa and the coast of Antarctica.

Given the turning effect of land barriers, the remaining factors that generate, sustain, and direct the major surface currents of the ocean are the wind, density differences, friction, and CORIOLIS EFFECT. The North Atlantic provides an excellent example of how these factors combine to produce a current gyre. Just north of the equator the trade winds, blowing steadily from east to west, produce a broad, sluggish current known as the North Equatorial Current (see EQUATORIAL CURRENT SYSTEM). As this current drifts across the Atlantic, its waters soak up heat from the tropical sun. By the time the water has reached the vicinity of the American coast its decreased density (because of increased temperature) has caused it to stand at a slightly higher level than the cooler water of greater density to the north. The tiny slope resulting from these different water levels causes the warm water of the North Equatorial Current to flow northward as soon as it is deflected by the American edge of the basin.

As the warm water of the North Equatorial Current flows northward through the CARIBBEAN SEA, the GULF OF MEXICO, and out through the Florida Strait to become the powerful GULF STREAM, it comes increasingly under the accelerating influence of the Coriolis effect. This "force," actually a deflection caused by the revolution of the earth, is zero at the equator but increases with increasing latitude until it reaches a maximum at the poles. Thus, the northward-flowing current is accelerated because the Coriolis effect is steadily increasing. Additionally, the Coriolis effect causes any body moving freely on or near the surface of the earth to be displaced to the right of its path in the northern hemisphere and to the left in the southern hemisphere. As a result, the Gulf Stream is continually deflected to the right as it reaches higher latitudes until it comes under the influence of the prevailing westerlies which drive the water back across the Atlantic as the NORTH ATLANTIC CURRENT.

The European edge of the North Atlantic basin and the Coriolis effect combine to turn the current southward for its long run down the European and North African coasts as the CANARY CURRENT. But now the Coriolis effect is decreasing, and the current gradually slows, broadens, and becomes a sluggish drift of water which still retains some of the heat it gained in the tropics. In the vicinity of the Cape Verde Islands the Canary Current turns west under the influence of the trade winds to become the North Equatorial Current and complete the North Atlantic gyre.

This pattern of a strong, swift current on the western edge of the basin, and a broad sluggish current on the eastern edge is repeated in each of the major basins of the Atlantic, Pacific, and Indian Oceans. In the North Pacific for instance, the powerful KUROSHIO CURRENT flowing northward past Japan is the equivalent of the Gulf Stream, and the CALIFORNIA CURRENT on the eastern edge of the Pacific basin is represented by the Canary Current in the Atlantic. Only in the northern INDIAN OCEAN basin is the pattern disrupted. There, partly because of the restrictive influence of the eurasian landmass to the north, and partly because of the seasonal reversing of the monsoons, the circular pattern does not develop.

In response to the piling-up of water in the western basin by the North and South Equatorial Currents, a return flow, called the Equatorial Countercurrent, develops between the two. Flowing west to east just north of the equator (ca. 7° N) this current only flows partway across the basin in the Atlantic, but crosses the entire basin in the Pacific.

The great surface currents of the world's oceans play an important role in the mixing of the surface layers and in the transport of heat from the tropical regions to higher latitudes. To gain some insight into the transport capacity of these currents it is only necessary to realize that the Gulf Stream carries about 100 times the amount of water transported by all the rivers of the world combined, and the Antarctic Circumpolar Current carries perhaps twice this amount.

CUTTLEFISH is the name for a marine decapod MOLLUSK of the genus *Sepia;* the decapod mollusks belong to the order Decapoda, an order of dibranchiate cephalopods, containing the SQUID and cuttlefish. Like the squid, the cuttlefish is distinguished by a hard internal shell, eight arms and two long retractable tentacles, and lateral fins capable of undulation.

The cuttlefishes of the family Sepioidea are night-hunting carnivores of SHRIMP and small fish. Species of the genus *Sepia* (e.g., the common cuttlefish *Sepia officinalis*) are very characteristic. They are usually broad-bodied and small, on the average of 6 in (15 cm) in length. The cuttlefishes are confined to the Old World where they live by day in the sands of coastal waters, ranging from near the surface to about 330 ft (100 m). The deeper bathypelagic inhabitants, or spirula-coiled cuttlefish (e.g., *Spirula spirula*), are found at depths of 650–5700 ft (198–1737 m).

These swimming animals, which, contrary to their name, are not fish, can change their buoyancy by varying the density of their shell—the cuttlebone (an internal skeleton made of horny matter covered with a chalky layer). This cuttlebone is composed of many chambers containing both liquid and gas, and when a cuttlefish needs to increase its buoyancy, it removes salt from the liquid in these chambers. This causes the liquid to flow into the more saline bloodstream by OSMOSIS (the process by which water from a weak salt solution passes into a stronger salt solution across a membrane). In this regard, they are unlike the squid (family Teuthoidea) which have within their bodies a feather-shaped shell called the *pen* or *gladius* that offers little buoyancy control. Hence, such squid have to swim actively to stay afloat as do the midwater fish of the bathypelagic zone that do not possess a SWIM BLADDER.

The cuttlebone is sold whole for feeding cage birds; it is also manufactured into a powder for industrial polishing purposes, as well as for use as dentifrice. The ink sac of the cuttlefish is a source of sepia, a natural brown pigment.

CYCLONE See HURRICANE.

DAMSELFISH (demoiselle) is a name for several species of carnivorous bony fish of the family Pomacentridae in the order Perciformes. Most species are small [less than 6 in (15 cm) long], somewhat flat-bodied fish that inhabit shallow tropical waters of the world's oceans, especially around coral reefs and rocky shores. See CORAL REEF. Many are bright-colored, and some such as the sergeant major (*Abudefduf saxatilis*) have stripes.

The damselfishes have many interesting characteristics. For example, the best known, the clown fish (*Amphiprion percula*) lives in close association with the SEA ANEMONE, whose stinging cell tentacles are used to kill other fish. The clown fish and some other damselfishes are able to establish rapport with the anemones, and they often raise their families among the tentacles of these animals. Another peculiarity of the damselfish family is that some species employ their pectoral fins in an oar-like or rowing fashion.

DARWIN, CHARLES (1809–1882) was an English naturalist whose major contributions to the study of marine biology and his theory of EVOLUTION, or the study of the effects of function and environment upon organisms, revolutionized the study of natural history.

Darwin was educated at Cambridge and the University of Edinburgh. Shortly after his graduation, Darwin, at the age of 21, readily accepted the non-paying job of "volunteer naturalist" on the *H.M.S. Beagle.* Commanded by Capt. Fitz-Roy, this ship, during the period 1831–1836, conducted an around-the-world scientific survey. Most of this time was spent in mapping the coasts of South America, especially the southern parts. On this voyage, Darwin made many original observations of a geological, botanical, and biological nature and, in doing so, established much of the framework of modern OCEANOGRAPHY.

Moreover, his studies on the fauna of the Galapagos Islands provided much of the basis for the famous Darwinian theory of evolution. Darwin's monumental work, *On the Origin of Species by Natural Selection or the Preservation of Favoured Races in the Struggle for Life,* was published in 1859. The book had profound social as well as scientific effects. Many interpreted Darwin's theory as a direct attack on all religious beliefs. Others challenged its scientific validity as well. However, the famous and respected scientist Thomas H. Huxley ably championed it, and Darwin's ideas gained widespread acceptance. Modern research also has supported and substantiated Darwin's general view that life began in very primitive forms and that, in the struggle for survival, organisms have ascended and differentiated, with new forms and species being constantly created.

While Darwin is best known to the world at large for his books *The Origin of Species* and *The Descent of Man,* he was equally eminent in geology, botany, and zoology. His theory on the formation of ocean coral reefs as set forth in his work *A Treatise on Coral Reefs, Volcanic Islands, Geological Observations* has held up in its major aspects to this day. (See CORAL REEF.) He was also one of the first scientists to study PLANKTON, the microscopic plant and animal life of the ocean that are the initial link in the ocean food chain. In his *Monograph of the Cirripedia* he described the parasitic marine CRUSTACEANS, including the BARNACLES which in their adult stage attach themselves to ship hulls. Darwin's other important publications include *A Journal of Research into Geology and Natural History* and *The Zoology of the Voyage of the Beagle.*

DAVIDSON CURRENT is a coastal countercurrent which flows north along the California coast inshore of the southward-flowing CALIFORNIA CURRENT. Developing during the winter months, the current can be identified as far north as 48°N LATITUDE.

DEACON, SIR GEORGE EDWARD RAVEN (1906–), British oceanographer, was one of the early scientists to do a comprehensive study of the temperature and salinity structure of the South ATLANTIC OCEAN and the SOUTHERN OCEAN. The results of these efforts were published in 1933 and 1937, respectively. During World War II, Deacon worked on UNDERWATER SOUND and WAVES. During this period his research team became the first to make a spectrum analysis of sea waves and to show the value of a concept of a wave spectrum. From 1949 to 1971, Deacon served as the director of the National Institute of Oceanography in England, an institution which has contributed much to our understanding of the oceans.

Deacon was born in Leicester, England, on March 21, 1906, and was educated at King's College, London. Early in his career, he sailed to the Southern Ocean as part of the *Discovery* investigations. For his many achievements, Deacon received the Polar Medal (1942), Alexander Agassiz Medal (1962), Royal Medal (1969), Founder's Medal (1971), and the Scottish Geographical Medal (1972).

DECIBAR, a metric unit of pressure, is defined as follows:

$$1 \text{ decibar} = 0.1 \text{ bar} = 10^5 \text{ dyn cm}^{-2}$$

The pressure exerted per square centimeter by a column of SEAWATER 1 meter tall is approximately one decibar. The depth in meters and the pressure in decibars, therefore, are expressed by nearly the same numerical value. (See HYDROSTATIC PRESSURE.)

DEEP SCATTERING LAYERS are widespread strata in deep ocean waters which cause sound to scatter and return echoes.

Marine organisms absorb, diffuse, reflect, and produce sound. Deep water [deeper than 650 ft (ca. 200 m)] generally has one or more well-defined layers of organisms, consisting of a variety of sea animals. These layers range between 660 and 2600 ft (ca. 200 and 800 m) in depth during the day, with most of them migrating to or near the surface during the night. Such animals as Myctophid (lantern) fish about 1.97–3.15 in (5–8 cm) long and the EUPHAUSIIDS (e.g., KRILL) are somewhat typical inhabitants of the layers. The former possess gas-filled swim bladders and because of this are excellent sound scatterers. The Euphausiids are less-efficient scatterers because they lack a SWIM BLADDER. However, they often occur in dense swarms and can be important scatterers at high frequencies of directed sound.

Deep scattering layers are easily detected by an ECHO SOUNDER operating in the 3–60 kHz frequency range.

See also DEEP-SEA FISH AND ANIMALS; FISH; SEA NOISE.

DEEP SOUND CHANNEL See UNDERWATER SOUND.

DEEP-SEA CHANNEL See SUBMARINE CANYON.

DEEP-SEA FISH AND ANIMALS include those forms of marine life that spend their life cycle in the ocean depths at 3300 ft (1000 m) or more.

The three generally recognized faunal zones of the world's oceans are: the littoral, on the continental shelves, down to approximately 660 ft (200 m); the bathyal, on the continental slope, about 660 ft (200 m) to about 6600 ft (2000 m); the abyssal, approximately 6600 ft (2000 m) to the deepest part of the ocean or approximately 36 000 ft (10 909 m).

As one goes deeper into the ocean, the SEAWATER cools rapidly. Below 6600 ft (2000 m), the temperatures can vary between approximately 3.6 and 1° C. At most depths, the SALINITY or salt content remains constant from the surface to the bottom and there is enough dissolved oxygen to support animal life throughout the column. However, the HYDROSTATIC PRESSURE, or the pressure exerted by the weight of the water, increases by one atmosphere (14.7 lb/in^{-2}) with every 33-ft (10-m) increase in depth. For example, at 32 800 ft (10 000 m), the hydrostatic pressure amounts to more than 6 tons in^{-2}. This factor seems to account chiefly for the changes in the forms of FISH and animal life in the deep parts of the sea.

Before the British CHALLENGER EXPEDITION (1872–1876), the common belief among biologists was that life could not exist in ocean depths greater than approximately 1800 ft (550 m). This particular expedition proved the existence of life down to the 16 500 ft (5000 m) level and subsequent investigations (e.g., GALATHEA in 1950–1952 and VITYAZ in 1953, among others) have found that a surprisingly large number of species exist in the deep sea.

Actually, what may be characterized as the deep sea, or those regions lying below the 3280-ft (1000-m) level, covers more than four-fifths of the world's ocean volume. In the top layers above this dark and cold environment, the animal life is very rich. These top layers are between 660 and 3300 ft (200 and 1000 m) deep, and many of the animals of this area make migrations up and down in the waters in search of food. Here, also are found several varieties of the commercially exploited PELAGIC FISH stocks that make excursions from time to time from the surface to these depths. However, most deep pelagic fishes are small [about 3–4 in ($7\frac{1}{2}$–10 cm)] and the area is especially distinguished by the presence of vast schools of luminiferous fishes of the genus *Cyclothone,* belonging to a group called the Stomiatoids. The Stomiatoids are carnivores having large jaws and caninelike teeth and all but a few are equipped with long filament appendages—called *barbels*—of an intricate variety of designs; the barbels hang from the chin region. These fishes resemble the BARRACUDA in their overall long slender design. However, unlike the latter, they possess barbels and light or luminescent organs on their body, tail, head, and fins (or even on the barbels). One species of Stomiatoid, the *Cyclothone signata,* is thought to be the most numerous fish in the ocean. The lantern fishes, the group Myctophidae, are also extremely numerous in this zone. These illuminated (primarily with green light) fish are also prevalent in the moderately deep (900–1700 ft) parts of the ocean, and they furnish the diet of TUNA in the PACIFIC OCEAN and of fur SEALS along the coast of Japan.

Other bathypelagic fish are numerous (viz., the hatchet fish, such as the species *Sternoptyx,* which resemble hatchets in appearance and have an iridescent skin; the bristlemouths, e.g., *Gonostoma elongatum;* the bigscales, e.g., *Melamphaes pumillus;* the pelican or gulper eel, *Eurypharynx pelecanoides,* a small nonmigrator, practically all mouth, which preys on organisms in the deep scattering layers), but the two groups previously mentioned are the most abundant. In addition, there are SQUID, EUPHAUSIIDS, COPEPODS, worms, JELLYFISH, CRUSTACEANS, CLAMS, and snails.

In the lower abyssal zone essentially the same kinds of animals are found as in the bathyal layer above. However, because there is less food available in this region, certain families, genera, and species are quite different, fewer in number and smaller in size.

In the HADAL or deep-trench areas, there is a higher density of ocean life although fewer species (approximately 370 as compared with a pelagical population of approximately 3000 species and the approximately 157 000 species of the benthos). In the hadal zone, the food supply is greater than it is on the abyssal plains since these great depths are closer to land. However, the community of ocean life is constantly disrupted due to sediment sliding off the trench walls, hence the fewer species.

The hadal fish consist of three predominant species: *Careproctus amblystomopsis, C. kermadecensis,* and *Bassogigas profundissimus.* These species show the same adaptations of color (grayish) and shape to life in eternal darkness as do many abyssal species, but in the hadal zone they are completely blind. Worms, crustaceans, clams, and sea cucumbers are present, and the worms and crustaceans are much larger in size than those found in the other depths.

A complete understanding of deep-sea life requires much additional research involving the origin and regulation of the diversity of species; the effect of high hydrostatic pressure and low temperature on rate processes and enzymatic systems and the transport routes of organic matter and other materials, such as man's wastes in these regions. See HABITABLE ZONES.

DEEP-WATER WAVE See WAVES.

DEMOSPONGIAE See SPONGES.

DENITRIFICATION is a process that takes place in the oceans whereby nitrate (NO_3^-) is reduced to a gaseous end product, nitrogen (N_2) or nitrous oxide (N_2O); the process is carried out by BACTERIA that use nitrate as a hydrogen acceptor when dissolved oxygen concentrations are at a very low level.

Denitrification takes place especially in such areas as enclosed bays, fjords, and shallow basins where the waters are inhibited from mixing with the adjacent open SEAWATER. Here the nitrate concentration falls to a low level, and sulfate (SO_4^{2-}) is used as a hydrogen acceptor in anaerobic respiration, with hydrogen sulfide (H_2S) as the end product. Denitrification occurs in the open ocean in the northeastern tropical PACIFIC OCEAN where UPWELLING takes place.

See also NITROGEN CYCLE.

DENSITY of SEAWATER is the mass of seawater per unit volume; its value at 77° F (25° C) temperature in water of 35 PARTS PER THOUSAND of dissolved salts, or 35 ppt SALINITY, is approximately 64 lb ft^{-3} 1.02412 g cm^{-3}).

Pure WATER (water without salts in solution) has a density of approximately 62.4 lb ft^{-3} (1.0029 g cm^{-3}) when measured at 77° F (25° C). The differences in density values between pure water and seawater are due to the differences in salinity, and a change in this salinity alters the density. Density of seawater is a function not only of the salinity or composition but also of TEMPERATURE and PRESSURE, and it may be computed from measurements of them. It may also be measured directly. For example, surface water can be measured directly by a hydrometer although *in situ* and surface measurements are difficult. However, since density increases with increases in pressure and salinity and decreases with a rise in temperature, the density for water beneath the surface can also be determined directly from those latter values by use of oceanographic tables. Such tables were first evolved by Martin Knudsen in 1902 and are used for computing density (in approximation) from CHLORINITY. Density values may also be derived with greater precision from the electric conductivity measurements of seawater of varying compositions.

The role of seawater density is especially important in OCEAN ENGINEERING and in naval work where underwater vehicles are involved. For example, strong density gradients may hamper the ascent or descent of such craft [i.e., a density change of 0.001 changes the bouyancy of a U.S. Navy fleet-type submarine by as much as 5400 lb (2440 kg)].

Density is also important in its relation to ocean CURRENTS. The greatest variations in seawater densities occur at the surface where the water is subjected to several conditions. Such conditions include the density being decreased by precipitation, runoff from land, melting of ice, or heating. As the surface water becomes less dense, it floats on top of the more dense water and there is little tendency for the water to mix. However, when the density of surface water is sufficiently increased (by evaporation, formation of sea ice, and by cooling), it sinks to the level of other water having the same density. Here, it tends to spread out to form a layer or to increase the thickness of the layer below it. The less dense water rises to give way to it, and the surface water moves in to replace that which has descended. In this manner, a convective circulation is established, and this circulation proceeds until the density becomes uniform from the surface to the depth at which a greater density occurs. When the surface water becomes sufficiently dense, it sinks all the way to the bottom. If this happens in an area where horizontal flow is unobstructed, this water spreads to other regions, creating a dense bottom layer. The greatest increase in density naturally occurs in polar regions, where the air is cold and great quantities of ice form. Here, the cold dense polar water sinks to the bottom and then spreads to lower latitudes. This process has gone on for long periods of time, and much of the ocean floor is covered with dense polar water, thus explaining the layer of cold water at great depths in the world's oceans.

In some respects, oceanographic processes are similar to those occurring in the atmosphere. The convective circulation in the ocean is somewhat the same as that in the atmosphere, and water masses having nearly uniform characteristics are analogous to air masses.

See also INSTRUMENTATION; SEAWATER.

DENSITY CURRENT See TURBIDITY CURRENT.

DESALINATION, is the process of removing salt from SEAWATER or brackish water so as to render potable (fresh) WATER.

Water is central to the history of human beings. Of the world's water supply, 97 percent is contained in the oceans, some 2 percent is held in the frozen icecaps, and the remaining 1 percent is found in lakes, the atmosphere, and in ground water. The fresh water of lakes and rivers, so necessary to all life, tends to flow toward the oceans where the earth's water is stored as salt water.

In view of the very meager amount of fresh water available to satisfy many personal, agricultural, and industrial needs, desalination of suitable portions of the 330 million (330×10^6) mi^3 [1.38 billion (1.38×10^9) km^3] of ocean water is the ultimate answer to the problem.

History The idea of desalting sea and brackish water is by no means new. In ancient times, men dug canals to channel seawater into shallow pools so that heat from the sun would evaporate the water and leave the salt, the substance which was then desired. In ancient Rome, peat and wood were burned under containers to distill or evaporate seawater and to recover salt. During the time of sailing ships, fresh water was carried in casks, and this supply was replenished by the use of simple evaporation techniques. However, with the advent of the oceangoing steamship which required large quantities of fresh and noncorrosive water for the boilers, the first seagoing desalination plants were developed. By the 1870s, such units were used on shipboard. During World War II, the requirement for better and larger seagoing desalination stills became essential. Techniques for the process were greatly improved upon and applied to large vessels, including aircraft carriers, that needed fresh water both for their crews and steam boilers.

A land-based solar distillation plant was first built and used in Chile's Atacama Desert in 1872 for supplying water to animals used in nitrate mining operations. This plant operated for 30 years on a process that consisted of passing the sun's radiation through a transparent cover onto a source of salty water. The water vapor condensed on the cover which was arranged to collect and store it.

This same process is used today, and small community-scale stills are now being utilized. Durable designs requiring little day-to-day attention and operating with minimal maintenance have been developed in the United States, France, Spain, and Australia. Solar distillation is also used on a small commercial scale to supply villages in isolated areas of Australia and small communities in the Mediterranean Basin and the CARIBBEAN SEA. Modern research on solar distillation is emphasizing new materials and designs for economical and durable construction to reduce costs.

Low-cost desalination of seawater would be extremely beneficial to arid lands bordering a SEA or a salt lake. Over the past few decades, many proposals to build huge distillation plants to produce water for agriculture have been widely advertised and intensively promoted. Although new and improved desalting methods using membranes and ION exchange have been developed, some authorities feel that the potential of desalting seawater for general agricultural use has been overpublicized. They argue that it appears that the process will continue to be uneconomical for field agriculture for many years, particularly in the light of increasing costs of energy, which is a significant component of the total cost of desalted water. On the other hand, for some special applications, e.g., controlled-environment installations, desalted seawater may be economical as a general solution for augmenting water supplies for irrigation. The possible alternative is irrigation of some selected crops with saline water containing 3000–5000 parts of salt per million parts of water, as opposed to seawater which averages around 35 000 parts per million (ppm) of salt. (See PARTS PER THOUSAND.) By contrast, fresh water is generally regarded as containing less than 1000 parts of dissolved salts per million parts of water (1 g/L).

In the early 1950s, serious consideration was given to the land-based desalination of seawater for domestic and industrial purposes. As a consequence, there are several commercial distillation plants throughout the world producing up to several million gallons per day for such usage. As an example, a very large single-unit seawater distillation plant went into operation in July 1967 at Key West, Florida. It is designed to consume 5–6 million gal (19–23 million L) of seawater per day and pass it in a boiling state through a maze of superheated pipes and chambers. At one prong of the output end, 2.63 million gal (9.956 million L) of water pour out; this water is so pure it must be passed over a bed of lime to give it body and taste. At the other output release, mineral-laden residue flows back to the sea.

Some of the other large plants (distillation or evaporation) of the world are located in the Caribbean—Aruba, Cuba, Virgin Islands, Curaçao, and Nassau; in the United States—Texas, California, and New Mexico; in the Soviet Union—Shevchenko on the Caspian Sea; in Israel, Italy, Venezuela, and at Kuwait and Qatar on the PERSIAN GULF.

Processes Basically, there are two ways for converting seawater into fresh water: (1) removing water from salt and (2) removing salt from water. In both approaches, the main problem is to be able to carry out the separation efficiently and economically.

Several methods are used for removing water from salt. These include in the main a variety of processes where the seawater is distilled (or evaporated and condensed) by heat energy. Such processes are called long-tube vertical distillation, multistage flash distillation, vapor-compression distillation, and solar conversion. Salt separation is affected by freezing techniques (e.g., butane secondary refrigerant, vacuum-freezing vapor compression). In another process, salt water under pressure is passed through a membrane causing pure water to flow out of the saline solution in the opposite direction to osmotic flow (reverse osmosis). In the electrodialysis method, an electrolytic cell which contains two different types of ion-selective membranes is utilized for salt removal. One type of membrane allows passage of positive ions, or cations, and the other allows passage of negative ions, or anions. The electric current imposed on the electrolytic cell provides the driving force for the ions. The cation-permeable membrane allows passage of the positive sodium ions, and the anion-permeable membrane allows passage of the negative chlorine ions, yielding fresh water between the membranes.

Piezodialysis is another method that utilizes special, stable selective filters or membranes and pressure as the driving force to produce fresh water from salt water.

While distillation has been generally accepted as a most practical method, the technique of freeze desalination, in principle, has several advantages over the other methods used. This is especially true since the estimated power consumption for carrying it out is less than that for the other processes. Moreover, the scaling and CORROSION problems are not nearly as severe as in the other techniques. However, in spite of these advantages, commercialization has been slow due to its relatively low level of development and to equipment problems that have

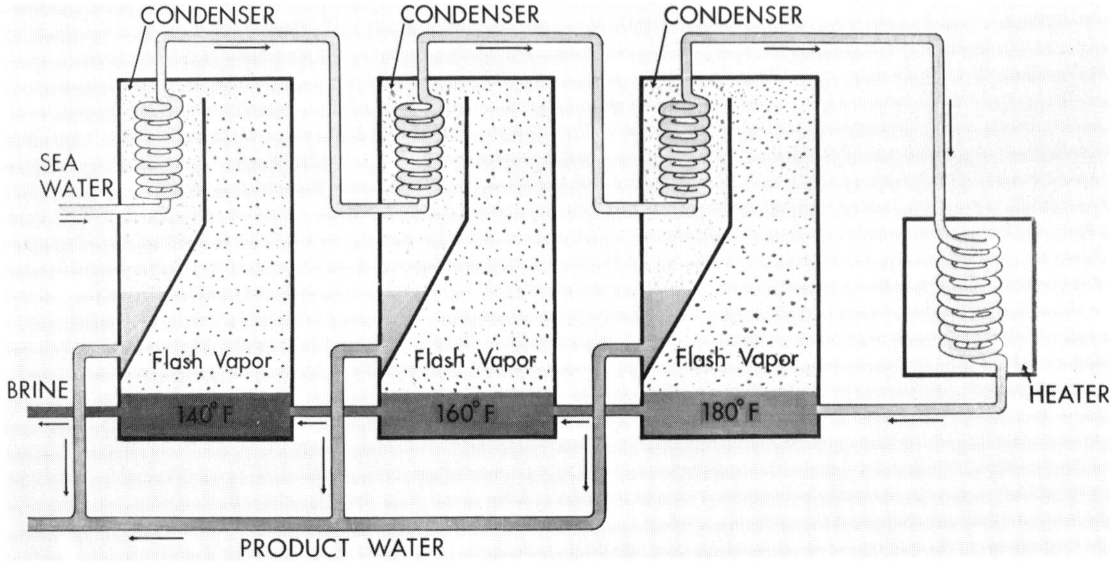

DESALINATION. Flash-distillation system—Operation of the multistage flash-distillation process is pictured schematically in this diagram. Seawater is progressively heated and then introduced into a large chamber where a pressure just below the boiling point of the hot brine is maintained. When the brine enters this chamber, the reduced pressure causes part of the liquid immediately to boil—or flash—into steam. The remaining brine is passed through a series of similar chambers at successively higher vacuums where the flash process is repeated at progressively lower temperatures. Progressive heating is accomplished by piping the incoming seawater through the flash chambers, starting at the low-temperature end. In each chamber, the flashed vapor condenses as it gives up its heat to the cooler seawater in the condenser. Final heating of the seawater before flashing takes place in a saltwater heater. By this arrangement, about 90 percent of the heat required for boiling is recirculated and only 10 percent is supplied by the saltwater heater.

been experienced with many of its components. Many such problems are being overcome, and freezing is approaching the threshold of commercial viability.

Reverse osmosis also shows much promise because of its simplicity of operation which does not require heating, boiling, or chemical additions. However, reverse osmosis for seawater does present problems more difficult than for brackish waters. In seawater, the concentration of salt is much higher, making higher operating pressures necessary and rendering more severe membrane composition and boundary-layer salt buildup. Seawater also contains many microorganisms and organic materials which can cause FOULING. In addition, the metallic components of the system are subject to corrosion.

DESALINIZATION See DESALINATION.

DETRITUS refers to the nonliving particles of matter suspended in the waters of the oceans, such as clay, mud, silt, dust, pollen grains, siliceous skeletons of dead micro-ALGAE, etc., as well as complete dead organisms and decaying particles from formerly living organisms.

Detritus in SEAWATER consists of a rich variety of PARTICULATE MATTER or nonliving debris (both inorganic and organic in origin) from many sources. Since this matter floats rather than sinking, it is generally characterized as fine material [about 2 mils (0.005 cm) average in size]. The size of detritus particles seems to decrease with increasing depth in the ocean.

Chemical analysis has shown that some types of detritus contain a diversity of chemical compounds which are useful as food for marine organisms. Moreover, the ingested particulates can act as a nutrient for beneficial bacterial growth in these organisms.

DIAPIR See CONTINENTAL SHELF.

DIATOM OOZE See DIATOMS; MARINE SEDIMENTS.

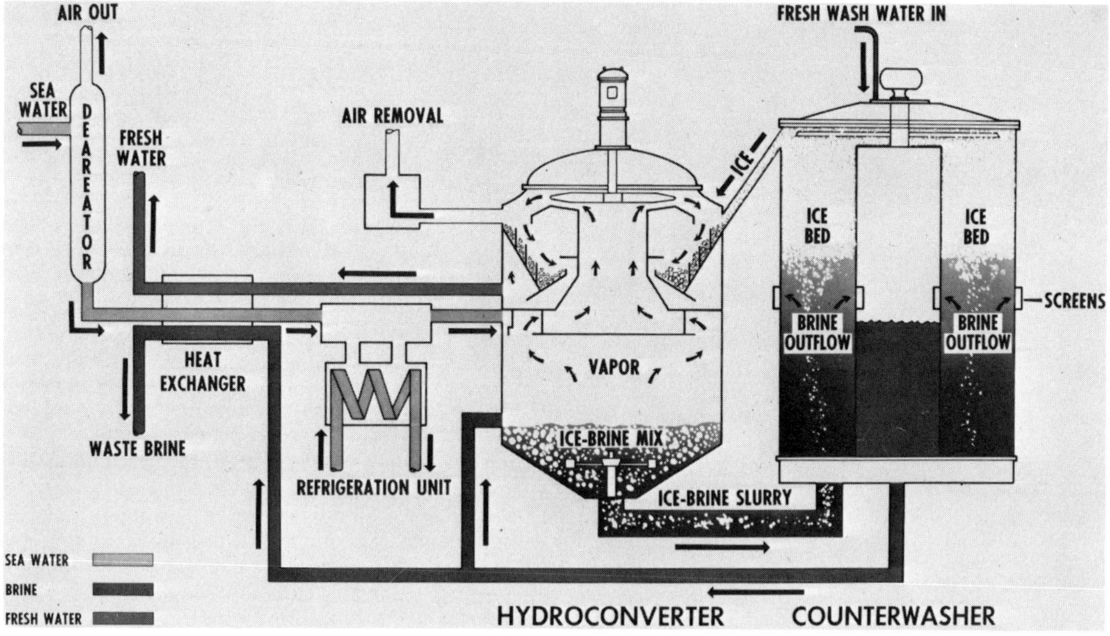

DESALINATION. Vacuum-freezing vapor-compression system—Seawater enters system at 60–70° F, passes through deaerator and then through heat exchangers where it is cooled by product water and waste brine. The feed is introduced into the freezer where it is converted into a 50 percent ice slurry at 3–4 mm Hg pressure. The slurry is pumped from freezer to counterwasher where brine is separated and the ice crystals are washed free of salt. From the counterwasher, ice is scraped into a melter where it is melted by the condensation of compressed water vapor from the compressor.

DIATOMS are microscopic, one-celled ALGAE belonging to the plant division Chrysophyta and the class Bacillariophyceae. Some taxonomists classify diatoms under the separate kingdom Protista and the phylum Chrysophyta. Diatoms constitute an important food supply in the ocean environment.

Marine diatoms are members of the yellow-green group of algae and are the most abundant of plants of the phytoplankton. There are many different species, and although all are a form of plant life, diatoms do not have leaves, stalks, or roots. Rather, their most distinguishing feature is a transparent boxlike shell of SILICA, filled with the protoplasm of the cell.

Encased in these shells of fantastic designs, which are bilaterally symmetrical in the order Pennales and circular or irregular in the Centrales, diatoms are rarely absent from any portion of the surface waters of the world's oceans. And, although diatoms are prevalent in the ARCTIC OCEAN and SOUTHERN OCEAN, they are also usually most numerous in some coastal waters where their nutrients are abundant. Reproduction can take place asexually or sexually.

When diatoms die, their shells sink to the ocean floor where such deposits (diatomaceous ooze) form a characteristic seabed feature. Over 10 million mi² of sea bottom, stretching across the north and south sides of the PACIFIC OCEAN, are diatomaceous. One bed off the California coast is over one-half mile thick.

The fossil remains of diatoms are found in large beds in many parts of the world. These friable porous silica deposits, which are the dried-out equivalent of diatomaceous ooze, are called *diatomaceous earth.* This is mined from land sources in such places as Denmark, France, and Finland. The material is chemically inert and has many physical properties suitable for industrial and scientific applications. It is used especially in the manufacture of polishing powders, toothpastes, pharmaceutical preparations, soap, waterglass, explosives, and fireproof packing material.

Diatoms constitute a vitally important food sup-

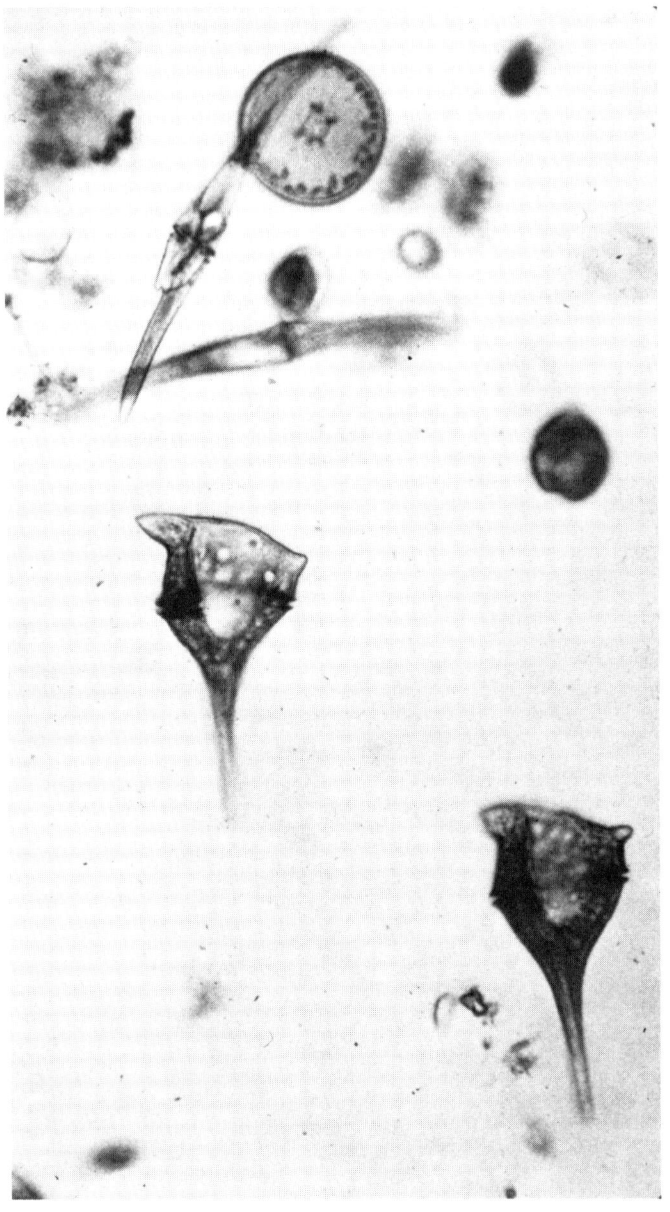

DIATOMS. One-celled diatoms appear in fantastic shapes beneath the microscope.

ply in the world's oceans where this most common form of plant life abounds. They furnish the basic food for zooplankton (such as COPEPODS), which, in turn, are eaten by other small marine animals in the OCEAN FOOD CHAIN. An ordinary circuit in this chain of events would be: diatom → copepod → HERRING → COD (or MARINE BIRDS) → BACTERIA → SEAWATER → diatom. Thus, diatoms supply the key connecting link and greatly surpass in bulk the productiveness of all the other aquatic plants.

See also PLANKTON.

DINOFLAGELLATES are small marine organisms which comprise a large portion of the ocean's PLANKTON, and have animal-like attributes (motion and food ingestion) as well as the ability, characteristic of plants, to carry on the process of PHOTOSYNTHESIS. It is because of these conflicting abilities that dinoflagellates have been classified traditionally in both the animal kingdom (order Dinoflagellida, class Phytomastigophorea, subphylum Sarcomastigophora, phylum PROTOZOA) and in the plant kingdom (class Dinophyceae, division Pyrrhophyta, subkingdom Thallobionta). This confusion can be lessened somewhat by considering a separate kingdom, Protista, as one containing unicellular organisms with a eukaryotic cell structure but not the true properties of plants or animals. In this kingdom (Protista) there are four phyla: Phylum Euglenophyta, composed of the euglenoids; Phylum Protozoa, unicellular animals—mostly heterotrophic; (a heterotroph is an organism that lives by ingesting and breaking down organic matter for its food); Phylum Pyrrophyta, the dinoflagellates and cryptomonads; and Phylum Chrysophyta, the autotrophic DIATOMS. (See also TAXONOMY).

Dinoflagellates are found in vast quantities in all the world's oceans, mostly in the top layers which receive sunlight. However, they can use low light intensities in synthesizing their food, and can live at lower nutrient levels than other microscopic forms such as DIATOMS. Some feed like animals while others live on decomposing organic matter. All contain chromatophores, pigment-containing bodies which give them a reddish color. Prolific "blooming" or excessive collections of some toxin-containing dinoflagellates (e.g., *Gymnodinium brevis*) cause what is known as the RED TIDE, a phenomenon which results in mass mortality of fish and other marine forms in the area. However, not all such fish kills connected with red tides are the result of the toxin produced by dinoflagellates. In enclosed bays and various near-shore ocean areas, fish kills following red water are due to the significant decrease in oxygen dissolved in the water that causes fish to suffocate. The marked decrease in oxygen after such blooms is due to the consumption of oxygen by microorganisms which consume the dying and dead dinoflagellates. Paralytic shellfish poisoning in man often occurs following the ingestion of certain mollusks and other organisms which have fed upon toxic dinoflagellates (e.g., *Gonyaulax catenella* or *G. tamarensis*). See MOLLUSK. The neurotoxin responsible for this poisoning is one of the most lethal biological toxins known, and it has been found in various MUSSELS and CLAMS.

Dinoflagellates move in the water in a characteristic spinning motion, and, if the water is agitated, many of these small organisms will luminesce or give off flashes of light. Some dinoflagellate cells are covered with an armorlike plate of cellulose; others lack such a covering, all are equipped with flagella, whiplike structures which they employ for locomotion. They reproduce by fission or the splitting of their parts. Shapes vary widely from tiny disks to bent threads. Flotation is accomplished chiefly by oil droplets in their structure as well as by their flagella. They form a very important part of the OCEAN FOOD CHAIN.

See also PLANKTON.

DISCONTINUITY LAYER See THERMOCLINE.

DISCOVERY I AND II were British exploratory investigations primarily concerned with biology (1925) and general oceanographic research (1930) in the SOUTHERN (or Antarctic) OCEAN.

Strictly speaking, the first Discovery expedition was carried out in 1792–1794 by George Vancouver in the *H.M.S. Discovery*. In his work on this voyage, which was a search for the Northwest Passage, Vancouver secured a place in history by his "Great Chart," the first complete survey of the North PACIFIC OCEAN and the northwest coast of America.

What is generally known as the Discovery I expedition was the 1925 voyage of the *R.R.S. Discovery*, a ship originally built for the famous Antarctic explorer Robert F. Scott. This ship was taken to the Antarctic for a scientific exploration of that area, but was concerned mainly with determining the whaling prospects of the region. In addition to the biological research done, some survey and charting of the island of South Georgia was made during the expedition.

The *R.R.S. Discovery II* was built in 1929 and specially designed to be capable not only of repeating the Discovery I survey of the biological and hydrological conditions of the Antarctic whaling grounds more efficiently but also of making a line of oceanic observations in the region of the South Sandwich Islands.

The oceanographic, geological, and biological research successfully undertaken by these two expeditions made the Antarctic waters among the best known and charted of any part of the ocean at that time. In Discovery II, some of the first work was done with the then new echo-sounding apparatus (see SONAR), and soundings were obtained at depths as great as 4421 fathoms (8085 m).

DIURNAL TIDE See TIDES.

DIVING is the act by a person of going beneath the surface of the water for purposes of recreation, scientific research, or the performance of useful work such as that involved with OCEAN ENGINEERING, fishery endeavors, etc.. In diving, specialized equipment includes SCUBA, diving helmets, diving suits, and other gear designed to protect and assist the diver in the ocean environment.

History Humans always have been heavily dependent upon the world's oceans for food and for transportation. They have traveled on the ocean surface in various types of craft, from hidecovered boats and man-powered galleys to wind-driven, steam-propelled, and nuclear-powered ships. They have cast nets, set lines, and otherwise caught food from the sea.

It can only be guessed when the first man or woman discovered that by holding their breath they could enter the world beneath the surface. However, since archaeological findings in and around Mesopotamia have included large amounts of mother-of-pearl shell ornamentation that could only have come from diving sources, it can be said that diving goes back 5000 years. Doubtless, in those very ancient times the type of diving done was what we now refer to as skin diving, that is, shallow-water [approximately 10–80 ft (3–24 m)] diving done without breathing apparatus. Such dives of a duration of 1 to 2 min, which is about the time the average person can hold their breath underwater, were carried out in a quest for food, SPONGES, CORAL, and mother of pearl.

Some of the first reliable written records of skin diving are from the Greeks, who, as early as the fifth century B.C., tell of divers active in military operations. Such missions included cutting anchor cables on enemy ships, boring holes in the hulls of these ships, and building harbor defenses at home and destroying them abroad.

The techniques of these primitive divers were followed by their successors for many centuries and are still being used today in a few areas of the world. For example, among the Polynesians, Japanese, and Koreans, whose underwater exploits are legendary, are skin divers who train from childhood, developing tremendous lung capacity and stamina. (See AMA.)

DIVING

An aquanaut wears an experimental heated suit, powered by batteries in the waist batterypack. He is carrying drift indicators, which he releases underwater, and anyone who finds and reports an indicator receives a reward and the indicator itself. *(U.S. Navy)*

A diver leaves the submersible decompression chamber "elevator" prior to entering the U.S. Navy's Sealab I undersea laboratory. *(U.S. Navy)*

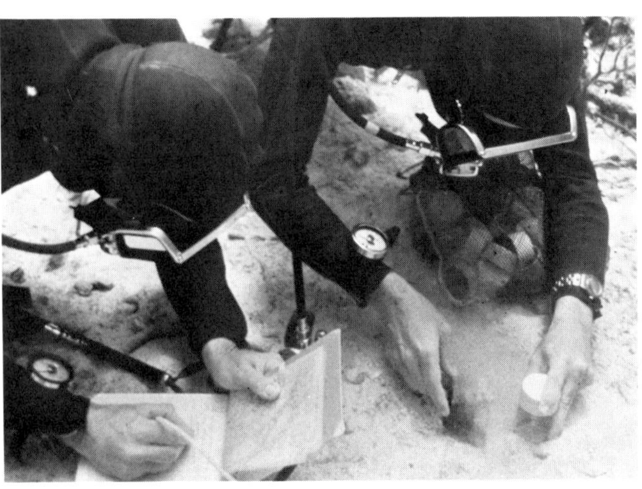

Divers work on the ocean floor near St. John, Virgin Islands. *(General Electric Co.)*

Two divers wait to ascend by a stage from 60 ft under water. *(U.S. Navy)*

Aquanauts break surface into the open hatchway entrance of Sealab II on the ocean floor near La Jolla, California. The entrance is open at all times, with an air pressure inside the lab equaling the ocean pressure and keeping the water from rushing in. *(U.S. Navy)*

While there is practically no period in human history during which people have not tried, in various ways, to penetrate the waters of the sea, these breath-holding excursions characterize most of the efforts before the nineteenth century. But, indeed, the origin of diving bells dates back to the time of Alexander the Great (332 B.C.), with premodern modifications and improvements made on them notably by the English astronomer Edmund Halley in 1690 and by another Englishman, John Lethbridge, in 1715. Moreover, the genesis of the submarine was largely due to the work of Cornelius Van Drebbel in England early in the 17th century and that of David Bushnell in America during the Revolutionary War. However, these kinds of devices sealed human beings in a metal structure where they were either vertically suspended by a cable from the surface, as in the case of the bell, or were provided only a limited means of transport through the sea, as with the early submarines.

Most historians credit the development of the first practical diving dress, or so-called hard-hat (or tethered) diving suit, to Augustus Siebe in 1839. Siebe's design consisted of a waterproof diving suit to which a removable helmet was securely attached and hermetically sealed on the diver's shoulders. A pump supplied the necessary breathing air from the surface via an umbilical-cord hose to the helmet. No really major changes have been made to this design since its invention.

Siebe's invention gave the diver the ability to remain underwater for extended periods of time with enough freedom of movement to perform significant work during this time. However, it remained for two Frenchmen, Jacques Yves Cousteau and Emile Gagnan, in 1943, to introduce a portable means of an air supply for the diver. (See COUSTEAU, JACQUES YVES.) Cousteau's aqualung device with its *s*elf-*c*ontained *u*nderwater *b*reathing *a*pparatus (or SCUBA) revolutionized diving by releasing the diver from a dependence on the surface hose, permitting increased freedom of movement without increased hazards.

Heat Loss Water allows a rate of heat exchange approximately 25 times that of air at the same TEMPERATURE. Thus, one limiting factor to the person who enters the world beneath the surface is exposure to cold. In fact, any water temperature that is lower than body temperature will cause a loss of heat as long as the diver is in the water. Also, the colder the water, the faster the loss of body heat and the shorter the period of time the swimmer will be able to remain in the water. Moreover, the high heat conductivity of water also causes the diver's metabolism to burn up calories to replace the heat lost.

The physical description of the human body in thermodynamic terms relates to the maximum temperature of the body, approximately 98.6° F (37° C), which is what physiologists call the "core" temperature. The skin temperature of the body varies with the environment and the part of the body concerned. In general, it is higher toward the central portion of the body and lower toward the extremities. The average human, by preference, maintains a mean skin temperature of 91° F (33° C) but can accept a minimum mean comfort temperature of 87.9° F (31° C). When the skin temperature drops to approximately 55.4° F (13° C), the cold becomes painful and nerve damage can take place. The rates of heat loss per unit area of hands and lower arms are surprisingly high when compared with the trunk and head surface.

Diving Dress In water below 78° F (25.5° C) some form of clothing must be worn. The low temperatures that prevail in many areas of the ocean create serious diving problems, which illustrate that really efficient insulating materials must be used to fabricate the needed wet suits required by free swimmers in the future exploration and exploitation of the CONTINENTAL SHELF and in the sea beyond.

The thermal environment in which the underwater swimmer or diver must work is, in general, that of the continental shelves of the world's oceans. Here the maximum PRESSURE depth is reached around 600 ft (180 m), and the SEAWATER temperatures in certain areas may reach 28° F (−2° C), although 41° F (5° C) is a more usual temperature encountered.

Various underwater life-support systems require the capability both for free swimming as well as "hookah" umbilical diving gear in which the air supply and/or electric heat is supplied through a hose. The capability to support a work period of 4 hours underwater is also desirable. The activity level of work may vary from that of the relatively inactive scientific observer to the vigorous work of the underwater construction diver.

It is, therefore, apparent that some mechanism is needed to maintain the body temperature when the diver is working in cold water. The principal method used to limit the rate of heat loss is by some form of protective insulated clothing. This clothing also affords protection from possible abrasions and minor bites from various forms of sea life.

Wet suits used in scuba are made of foam neoprene [normally $\frac{3}{16}$ or $\frac{1}{4}$ in (4.7 or 6.3 mm) thick] and constructed so that a layer of water is trapped next to the diver's skin. This layer is warmed by the body, and the body heat loss to the water is lessened. The basic wet suit consists of neoprene pants and jacket, with boots, gloves, hood, and vest being optional. When using a wet suit, the diver usually requires an additional weight such as a leaded belt to compensate for buoyancy.

A special type of wet suit, designed for use in very cold waters, is similar to the standard one except that it is configured to accept a hot-water supply from an external source. This open-circuit hot-water wet suit, as it is called, has a system of tubes fastened to the outside of the suit and allows hot water to be distributed evenly through holes within the suit. The hot-water supply may originate from a surface support ship or platform and may be pumped directly to the diver or passed to the diver from a diving bell, submersible, or underwater habitat.

While dry suits are used to a limited degree in scuba, they are the regular dress in so-called hard hat, the historic deep-sea method of contained diving. The first dry suits for the skin diver were made of a sheet of thin gum rubber which cover the diver from head to foot. Only the hands and face are exposed, and insulation from the cold is obtained by wearing warm clothing underneath to trap the molecules of warm air emanating from the surface of the body. The hard-hat diver's outer dress is made of vulcanized sheet rubber between layers of cotton twill. Underneath this, ordinary working clothes or several sets of underwear can be worn, depending on the water temperature and the individual diver's preference.

The evolvement of the modern-day wet suits used in scuba came about because of several difficulties encountered in the dry suits. Early suits (dry) were difficult to keep waterproof, and when they remained waterproof, they were difficult to manage. Invariably they required two additional people to assist the diver into the suit and to fold and close the back with a rather large U clamp. Once inside, it required a great deal of movement and bending to expel the air from the suit. If the air were not expelled, the swimmer would be buoyant in the water and would be subject to injury in rough seas or in the surf line. Also, the face pieces with these suits were made originally of thin rubber which demanded some skill in matching the material of the suit with that of the face mask in order to maintain a tight waterproof seal. However, dry suits or variations of them have been improved upon and now find their greatest use in regions of extremely cold water such as the Arctic and Antarctic.

There are two types of modern dry suits: the closed-circuit hot-water suit and the variable-volume dry suit. In the first type, the suit consists of the dry suit and a special type of underwear through which heated water is circulated. The water is pumped from a heater, through the specially designed underwear, and back to the heat source. The hot water can originate either from a heater carried by the diver or from a surface support heater. In the variable-volume dry suit, a lightweight one-piece dress, constructed of closed-cell foamed neoprene, is worn with thermal underwear. The hood and boots are an integral part of the suit. It can be inflated via an inlet valve which is connected to the diver's air supply. Air inside the suit can be exhausted by another valve, and by controlling these valves, the properly weighted diver can maintain complete buoyancy control at any depth.

Pressure In both methods of diving, namely, hard hat or tethered and scuba or free diving, it is necessary to deliver a breathable gas mixture to the diver's lungs and body cavities at a pressure equivalent to that exerted by the overlying water column (see HYDROSTATIC PRESSURE).

In the marine environment, anything so placed must be capable of withstanding pressures which increase 14.7 psi (1 atm. of pressure) or 10 N cm^{-2} for every 33 ft (10 m) of seawater, or nearly 2.7 psi for every 6 ft of depth (1.9 N cm^{-2} for every 180 cm). Interestingly, people can withstand enormous hydrostatic pressure on their outer bodies since human tissue has about the same DENSITY as seawater and is virtually incompressible. However, gas in the human body obeys Boyle's law which states that at a constant temperature, the volume and pressure of a gas are inversely proportional, or the volume decreases as the pressure increases. Thus, for long dives below 33 ft (10 m) (depth equivalent to 1 atm or 14.7 psi), adjustments are made in the diver's breathing gas mixture. Such an adjustment is predicated upon maintaining the number of oxygen molecules close to the number that is breathed at sea level. Too much oxygen can be toxic, and too little can cause asphyxiation. Nitrogen forced into the blood of an air-breathing diver begins to produce a state of intoxication at depths of 100 ft (30 m) and incapacitates the diver by nitrogen narcosis at about 300 ft (90 m). In this latter regard, it should be noted that many combinations of breathing gases are used in diving. Air, a mixture of gases (and vapors) containing nitrogen (78.084), oxygen (20.946), argon (0.934), carbon dioxide (0.033), and other rare gases (0.003), is most common. This air is compressed for diving applications. However, many combinations or mixtures of breathing gases are used for special diving situations.

Oxygen is the sole life-supporting gas used by the human body. The other gases breathed by a diver in a gas mixture serve only as carriers and diluents for the oxygen, which is dangerous to the diver when excessive amounts are breathed under pressure.

Nitrogen is commonly used as a diluent with oxygen in a diving gas mixture, although it has several disadvantages as compared with some other gases, e.g., helium. For instance, when used under increasing partial pressures, nitrogen produces the disorder known as "nitrogen narcosis," characterized by a loss in diver judgment and disorientation.

Helium is used extensively in deep diving mixtures as an oxygen diluent. It has several advantages over nitrogen because of its low density and lack of narcotic effect at depth. However, it has a high thermal conductivity, which results in rapid loss of body heat, and helium-oxygen breathing mixtures cause temporary speech distortions (Donald Ducklike voice).

Argon, neon, and hydrogen are gases which have been used experimentally as diluents with oxygen to form diver breathing gases; however, these gases are not normally used in diving operations.

If the diver returns too rapidly to the surface from a deep dive, the gases dissolved in the blood and the tissues are released as bubbles, thereby producing decompression sickness known as "the bends." To avoid this, the diver must undergo decompression, a controlled process that allows time for inert gases to be removed from the body by respiration and other natural processes. The causes of the bends were not fully understood until J.B.S. Haldane's work. Haldane first established a set of rules in the form of decompression tables. Subsequent developments and changes have been based on this work as well as that by Bert and Lambertsen. Decompression is not required after dives of 33 ft (10 m) or less, regardless of the time spent on the bottom. As dive depth increases beyond 33 ft, the time that can be spent on the bottom without decompression decreases, diminishing to 5 min at approximately 190 ft (60 m). However, once the diver is saturated for a particular depth, no additional inert gases are absorbed, and it is possible to remain at that depth and pressure indefinitely without increasing required decompression time.

Saturation Diving This fact led to the concept of saturation diving that has been used by the U.S. Navy in its sea-bottom living experiments conducted in the 1960s by Captain George Bond of the U.S. Navy. Saturation diving has also been used by the American investigator Edwin A. Link, the French explorer Jacques Yves Cousteau, and by the Soviets in their underwater living experiments.

In this concept, the diver is provided with a shelter on the seafloor which is pressurized to the outside water pressure [water pressure increases approximately 1 lb in^{-2} for every $2\frac{1}{4}$ ft of depth (0.689 N · cm^{-2} for every $67\frac{1}{2}$ cm of depth)]. The shelter is provided with a suitable breathing gas mixture, and after about 24 h of exposure under pressure, all tissues of the diver's body have a gas saturation equivalent to the surrounding environment, and the diver is considered to be "saturated." Divers then stay pressurized indefinitely without further increasing their decompression requirements [e.g., they need 27 h or decompression time to ascend from a 200-foot (60-meter) depth after saturation]. Thus, requirements for decompression are based on depth rather than the duration of the dive; e.g., a diver saturated to 300 ft (ca. 90 m) requires the same amount of decompression time (approx $2\frac{1}{2}$ days) whether his bottom time is 1 day or 1 month. By way of contrast, the time of decompression, or the gradual elimination of the inert breathing gas (nitrogen or helium) which dissolves in the tissues of the human body when it is subjected to gas under pressure, required under conventional diving techniques increases with the length of time spent on the bottom. For instance, a diver who spends 20 min at a 200-ft (60-meter) depth breathing a helium-oxygen mixture requires approximately 1 h to return to the surface safely. The same diver with 60 min working time at this depth needs $2\frac{1}{4}$ h to ascend.

However, in the saturated diving situation, the diver lives in a chamber or habitat that provides an exit directly to the surrounding ocean environment for work and research. After hours of useful work at depths, a return to safety and comfort of the underwater habitat is made. Since there is no appreciable difference between the pressure of the habitat and the outside water, decompression of a diver entering the undersea chamber is not required. Rather, decompression of the saturated diver for the total time spent at depth is accomplished in a single phase upon return to the surface after days or weeks of useful work on the ocean bottom. Such decompression may be accomplished in a chamber located on board a ship.

The offshore oil industry has utilized saturation diving to greatest practical advantage thus far.

Scuba Diving Types Scuba is of three basic types: open circuit, closed circuit, and semiclosed circuit.

In the open-circuit type, air is taken from a supply tank used by the diver, and the exhaust is vented directly to the surrounding water.

The basic closed system uses a tank of 100 percent oxygen for breathing, and the gas being used by the diver is recirculated in the apparatus. It passes through a chemical filter which removes carbon dioxide waste, and additional oxygen is added from the tank to replace that which is consumed in breathing. The closed-circuit systems employ a mixed gas for breathing and electronically sense and control oxygen concentration. This type of apparatus retains the bubble-free characteristics of 100 percent oxygen recirculators while significantly improving depth capability.

The third basic type, semiclosed apparatus, combines features of the other two systems. A mixture of gases is used for breathing, and the apparatus recycles the gas through a carbon dioxide removal canister, continually adding a small amount of oxygen-rich mixed gas to the system from a supply tank. The supply gas flow is preset to satisfy the body's oxygen demand, and part of the recirculating mixed gas stream, equal to the supply gas flow, is continually exhausted to the water. Because the quantity of makeup gas is constant regardless of depth, the semiclosed scuba provides significantly greater endurance than open-circuit systems in deep diving.

Importance of Diving In summary, until the introduction of the self-contained underwater breathing apparatus (scuba) in the early 1950s, even the shallow waters bordering the landmasses of the world were inaccessible to most people. However, production and use of this diving equipment has increased at a fantastic rate. Although precise figures are unavailable, it is estimated that there are over a half-million civilian sport divers in America employing scuba gear. With such equipment these underwater enthusiasts have the freedom of fish in the top layers of the ocean. And anyone who has once ventured beneath the surface must admit to a feeling of atavistic "at homeness" there.

Whether this sensation is real or engineered by subliminal influences is unimportant. It is important that it does exist—since it will no doubt assist in providing a needed future supply of divers for underwater engineering operations and scientific research. The ability of human beings to make decisions and to manipulate things with their hands cannot be duplicated in the same way by any machine. However, to safely accomplish the increasing number of underwater mechanical tasks, both divers and machines working together are needed. The types of diver/machine combinations required include free divers with hand tools; work packages controlled and operated by free divers; manipulators controlled by operators in untethered, self-propelled submersibles or in tethered, unmanned submersibles; and surface-controlled manipulator systems equipped with television.

The kinds of work the diver can do are as diverse as the ocean itself. Such classes of pursuits include a wide range of efforts, from underwater parks, underwater archaeological investigations, salvage operations, underseas warfare, and underwater storage, to oil and hard mineral extraction from the ocean floor, MARICULTURE, advanced COMMERCIAL OCEAN FISHING techniques, intelligent waste disposal, and many more.

Developments to enable human beings to perform useful work at greater depths will continue unless it is found that definite and insurmountable physiological or behavioral phenomena limit the diver's ability to go deeper. As all working depths increase, humans will be forced to become less directly involved in actually carrying out the underwater tasks, and remotely controlled manipulators will be required to augment human capabilities.

DOLPHIN is the common name for about 33 species of cetacean mammals included in the family Delphinidae and distinguished by a distinctive beak-shaped mouth. Dolphins of the family Platanistidae are found in rivers (e.g., the Ganges dolphin, *Platanista gangetica,* and the Amazon dolphin, *Inia geoffrensis*) as well as in estuaries and coastal waters (e.g., the La Plata dolphin, *Pontoporia blainvillei*). In some areas the terms PORPOISE and dolphin are used interchangeably. For example, the bottlenosed dolphin, *Tursiops truncatus,* a widely distributed species in temperate and tropical parts (chiefly coastal) of the world's oceans' is commonly called a porpoise. Also two species of FISH (*Coryphaena hippurus* and *C. equiselis*) are referred to as dolphins.

See also DOLPHIN FISH; PORPOISE; MARINE MAMMALS.

DOLPHIN FISH (*dorado, dolphin*) is the name for two species (*Coryphaena hippurus* and *C. equiselis,* or pompano dolphin) of large [approximately 6 ft (1.8 m) long in some adults of the first species and about 1–2½ ft (0.3 to 0.7 m) in length for the pompano dolphin] bony fishes that live in the very top layers of the ocean; the name *dolphin* is also applied to some cetaceans that are not fishes but mammals.

Dolphins are pelagic or open-ocean fishes that live singly or in schools. Although they predominate in top layers and warmer waters of the world's oceans, dolphins extend over a wide geographic range. An extremely active and fast-swimming fish, the dolphin exist on a diet of young SAILFISH, flying fish, CRAB, SQUID, and SHRIMP. Fishes that prey upon the dolphin are the TUNA, marlins, and SHARK.

The dorado's beautifully multicolored iridescent body is due to light-reflecting skin crystals and to saclike skin cells (or chromatophores). These pigment-containing cells take on a different shade of color as the sac is expanded and contracted. The body of the dolphin is streamlined and blue in color above and bright yellow below, with an even-bordered undifferentiated dorsal fin that almost covers the full length of the back. The color is lost in death, but the dying fishes undergo several dramatic color changes. The two species of dolphins, *Coryphaena hippurus* and *C. equiselis,* when of equal size, are so similar in appearance that they can only be told apart by counting the number of rays on the dorsal fin. The *C. hippurus* has 55–65, while the *C. equiselis* has 48–55.

DOWNWELLING See UPWELLING.

DUGONGS See SEA COWS.

EAST AUSTRALIAN CURRENT is a warm [59–77° F (15–25° C)], swift [12–20 in/s (30–50 cm/s)], relatively narrow [62–124 mi (100–200 km)] current that forms between the Great Barrier Reef and Chesterfield Reef in the CORAL SEA and carries approximately 1.06 billion (1.06×10^9) ft^3/s [30 million (30×10^6) m^3/s] of tropical waters south along the east coast of Australia. From January to March the WATER is supplied by the South Equatorial Current (see EQUATORIAL CURRENT SYSTEM) as it flows into the Coral Sea from the north and northeast. For the remainder of the year the supply is from subtropical water flowing in from the east.

Once the East Australian Current passes the southern tip of Australia, it begins to broaden and break up into eddies. Much of the water turns and flows eastward across the TASMAN SEA and to the north of New Zealand. It apparently makes no contribution to the eastward-flowing Circumpolar Current which appears to be self-contained. (See CURRENTS.)

EAST CHINA SEA (Tung Hai) is a marginal sea of the western PACIFIC OCEAN which lies off the east coast of China. It is bordered by the YELLOW SEA on the north, the JAPAN SEA on the northeast, the Ryukyu Island chain on the east, and the SOUTH CHINA SEA on the south. The boundary between the East and South China Sea is a line running from Fuki Kayu, the north point of Formosa (Taiwan), to Kiushan Tao (Turnabout Island), then to the south point of Haitan Tao, and then westward along the parallel of 25° 24′N to the coast of China at Fukien. The East China Sea covers an area of 290 272 mi^2 (752 000 km^2), occupies a volume of 63 096 mi^3 (263 000 km^3), and has a mean depth of 1145 ft (349 m). The maximum depth is 8914 ft (2717 m).

With the exception of a trough that runs from Kyushu to Taiwan along the western edge of the Ryukyu Island chain, the East China Sea lies on the CONTINENTAL SHELF. The trough slopes to the south, where it attains a maximum depth of 8914 ft (2717 m). The sediments in the trough are a combination of muds derived from the surrounding land and of foraminiferal oozes. The continental shelf sediments are made up of sands, silts, and muds introduced by such rivers as the Yangtze and Min Kiang. Rock outcroppings are exposed in the Formosa Strait because of the strong current.

The tide range in the East China Sea is somewhat complicated by the island chain on the east, but ranges from 16 ft (5 m) around Taiwan to 36 ft (11 m) in Hangchow Bay southwest of Shanghai. A small branch of the warm KUROSHIO CURRENT flows into the sea around Taiwan and moves northward at speeds of 18–40 nautical mi/day (33–74 km/day). Otherwise the water motion in the sea is under the influence of the southwest (summer) and northwest (winter) monsoons. During the summer the area is subjected to the typhoons that sweep up out of the PHILIPPINE SEA.

EAST GREENLAND CURRENT represents the main outflow of the ARCTIC OCEAN. It flows southward through the Davis Strait between Greenland and Iceland. Part of the current joins the IRMINGER CURRENT and the general eastward flow of the North ATLANTIC OCEAN. The remainder of the current rounds the tip of Greenland to flow north along the west coast as the West Greenland Current. (See CURRENTS.) In the surface layers the East Greenland Current flows at the rate of about 0.5 kn.

EAST SIBERIAN SEA, a sea of the ARCTIC OCEAN, is located off the extreme northeastern coast of the U.S.S.R. On the north the sea is bounded by a line from the northernmost point of Wrangel Island,

EAST SIBERIAN SEA. On June 14, 1881, the *USS Jeannette,* under the command of George Washington De Long, sank under the growing pressure of the Arctic ice pack. The crew made their way across the ice to the coast of Siberia. *(U.S. Navy)*

through DeLong and Bennett Islands, to the northern extremity of Kotelni Island. On the east the sea is bounded by a line from Wrangel Island, through Cape Blossom, to Cape Yakan on the mainland. On the west it is bounded by the LAPTEV SEA, or a line connecting the islands of Kotelny, Malyi, and Great Liakhov to Cape Sviatoy on the mainland. The East Siberian Sea covers an area of 350 487 mi² (901 000 km²), has a volume of 12 715 mi³ (53 000 km³), and has a mean depth of 190 ft (58 m).

The first chart and scientific exploration of the East Siberian Sea was made by Ferdinand von Wrangel, a Russian naval officer, polar explorer, and (later) governor of Alaska, who led an expedition to the region in the years 1820–1824. In 1878–1879, Nils Adolf Erik Nordenskjöld, the Swedish polar explorer, studied the sea as he passed through on his historic west-to-east transit of the Northeast Passage. After the three-masted bark *Jeannette* was crushed in the pack ice and sank on June 14, 1881, the crew, under the leadership of George Washington De Long, the American naval officer and polar explorer, trekked to Bennett Island and then Kotelni Island on the edge of the East Siberian Sea before beginning the long trip (which De Long did not survive) to the Siberian Coast. Thirty-seven years later, in 1918, Roald Amundsen, the Norwegian polar explorer, crossed the East Siberian Sea aboard the *Maud* in a successful transit of the Northeast Passage but an unsuccessful attempt to reach the North Pole.

The climate of the East Siberian Sea is typically Arctic and one of the harshest in the world. The mean temperature of January and February is −14.8 to −27.4° F (−26 to −33° C), with an absolute minimum temperature of −54.4 to −58° F (−48 to −50° C). The mean temperature for July is 37.4–44.6° F (3–7° C) along the Siberian Coast and 32–35.6° F (0–2° C) along the northern stretches of the sea. In August the WATER temperature off the mouth of large rivers may reach 42.8–44.6° F (6–7° C) but sinks to 33.8–35.6° F (1–2° C) near the ice edge. The East Siberian is ice-covered for most of the year, with floe ice persisting off the coast throughout some summers.

As is the case with the Laptev Sea to the west, the inflow of warmer, sediment-laden river water, combined with the rafting and grinding effect of SEA ICE, has a profound effect on coastlines and near-shore features of the East Siberian Sea. Two of the largest rivers flowing into the sea are the Kolyma and the Indigirka which annually deposit 8.3 and 16.7 million (16.7×10^6) tons of sediment, respectively, in shallow coastal areas. The warm river water prevents coastal and shallow water sediments from freezing so that shores may erode up to 50 ft (15 m) in a single year. The eroded sediment, combined with that brought in by rivers, results in the formation of bars and shallows near shore.

Mainly because of severe ice conditions, the East Siberian Sea is poor in plant and animal life. However, whitefish, grayling, Arctic char, pole flounder, and the like are found in the coastal areas. The mammals are represented by the WALRUS, SEALS, and POLAR BEARS, and the birds by cormorants, sea gulls, murres, and guillemots.

EAU DE MER NORMAL See NORMAL WATER.

ECHINODERM is a common name for any of the more than 9000 species of exclusively marine animals in the phylum Echinodermata (a Greek word meaning "spiny skin"). Echinoderms are distinguished from all other animals in that they characteristically possess a complex external skeleton composed of crystalline calcite plates, and a unique water-vascular system. This latter system of fluid-filled vessels is utilized to serve the needs of nutrition, respiration, locomotion, and sensory perception.

The echinoderms are all marine animals which metamorphose from bilateral free-swimming larvae to stationary or slow-moving adults. All adult living echinoderms become radially symmetrical, although there are marked differences in appearance among various representatives of these animals. In general, limy plates are imbedded in the skin, and often these bear spines which project through the skin and protect them from some predators. In the classification of echinoderms, a chart of their TAXONOMY shows some 4 subphyla, 10 classes, 8 subclasses, 4 superorders, 37 orders, and 3 suborders. This is not surprising in view of the fact that these animals have existed for some 600 million (600×10^6) years, and during such a time span several divergent structural patterns have evolved. Variations are almost limitless. No other INVERTEBRATE animal outranks echinoderms in the complexity of skeletal structure, which may include more than 2 million components in an individual. Some living examples of the echinoderms are

1. The starfishes (subclass Asteroidea in the subphylum Asterozoa). These have a central disk from which five or more arms are radiated. There is no distinct boundary between the disk and the arms. Limy plates are loosely embedded in the skin.
2. The SEA URCHIN (class Echinoidea in the subphylum Echinozoa). These have a compact and globular-shaped body enclosed in a hard shell, or test, formed from plates which bear movable spines.
3. The sea cucumbers (class Holothuroidea in the subphylum Echinozoa). These have a cylindrical and muscular body in which five "arms" are incorporated. The skin is soft and fleshy. Most live buried in the sand or mud at the bottom of the ocean near the shore. Around the mouth is a circle of tentacles. Two internal gill-like respirator organs connect with the anus, and SEAWATER is pumped into these, aiding in gaseous exchange.
4. The SEA LILIES and FEATHER STARS (class Crinoidea in the subphylum Crinozoa). These echinoderms are the only surviving members of the subphylum Crinozoa. The sea lilies have five upwardly directed arms which are highly branched and resemble long coarse feathers projecting from a small central disk. Ciliated grooves on the arms are used for transporting food particles into the mouth. They have closely joined plates in the skin. Most adult forms are attached to rocks or some other substrate by stalks or stems. Sea lilies live in deep ocean waters. Most of the feather stars are stemless and are adapted for a free-swimming existence.

Echinoderms are widely distributed in the world's oceans at all bottom depths. Some have been dredged from the Mariana Trench in depths as great as 35 120 ft (10 705 m). Others are known to be common on the ocean floor at average depths of 13 000 ft (3962 m), and still others are encountered in near-shore areas. In addition to their importance in the OCEAN FOOD CHAIN, they have a paleontological importance because of their many diverse types of well-preserved and abundantly available fossils. These furnish the means by which evolutionary trends generally can be defined.

ECHO SOUNDER, a SONAR-type device (also called sonic depth finder), is used to measure ocean depth and to locate underwater objects. A sound pulse is transmitted vertically downward into the water by means of a transducer, and the time required for the pulse to return after reflection is measured electronically. See INSTRUMENTATION; UNDERWATER SOUND.

ECOLOGY, or environmental biology, is the study of the various interactions between living and nonliving parts of the environment; it is concerned primarily with the assessment of the linkages between the living organisms, their structure, function, and life cycles, and how these affect people; it is also concerned with living organisms and the balance of nature.

Ecology is a relatively new science that had its beginnings about 100 years ago. The biologist Haeckel coined the term about 1870 by combining

ECHINODERM

Five arms radiate from the body of this echinoderm, giving it the immediately recognizable (even to the nonprofessional) shape of the starfish. *(U.S. Navy)*

This large orange "starfish" recovered near Nicaragua bears remarkably long, thin, and numerous arms. *(NOAA)*

the Greek roots *oikos,* "house," and *logos,* "study"; literally, the word means the study of organisms "at home." Frequently it is used to refer to the environment itself rather than to its study. However, as is the case with OCEANOGRAPHY, ecology is a synthesizing science which draws heavily upon many other disciplines: botany, geology, chemistry, meteorology, biology, and several others. Since the interrelationships of nature are many in number and diversity, ecological research is equally diverse, although it is basically concerned with a better understanding of the biology of groups of organisms and with their functional processes on the lands, in fresh waters, and in the world's oceans.

Such research has as one of its primary purposes to provide authoritative information that can be used in organized environmental planning and management. This planning is not an easy task since it has a direct impact not only on the relations between humans and their physical and biotic environment, but also on their social and economic environment. For example, conservationists argue with evidence that certain areas of the world and various species of organisms there and other places must be kept untouched by people so that the organisms would be protected and preserved at all costs. Industrialists, as well as many others, at times also present rather convincing ecological information to show that new facilities (i.e., nuclear power plants, oil drilling sites, housing, etc.) will not have an adverse impact upon these environments. Coupled with this dichotomy of thought, there are other factors that must be considered and somehow dealt with intelligently. These are the urgent need for new sources of energy to supply the growing needs of human beings; the world's population explosion with its attendant needs for food, housing, etc.; the cultural differences among the world's population; and the indisputable fact that in the past much of the ecosystem has not been wisely managed so as to maintain its quality or make it more habitable.

In this total ecosystem, or what is defined as "a functional system which includes the organisms of a natural community together with their environment," are the world's oceans. The manner in which this vital resource is ecologically managed in the future directly affects the survival not only of the living organisms who inhabit it but also of humans on this planet.

Many global effects or changes in the climate, habitability, fauna and flora, ocean, atmosphere, landmasses, or other large elements of the Earth, resulting from human activities, have been postulated. Such theories and speculations have included assertions that the buildup of carbon dioxide from fossil-fuel combustion might warm up the planet and cause the polar ice to melt, thus raising the SEA LEVEL several hundred feet and submerging coastal cities.

Serious questions have been raised about what the effects on ocean and terrestrial ecosystems will be from systematically discharging into the environment such toxic materials as heavy metals, oil, and radioactive substances; or from such nutrients as phosphorous which can over-enrich coastal waters. (See also OCEAN POLLUTION.)

Induced, uncontrolled pollution resulting from human activity presents the greatest threat to the ocean system. Toxic liquid wastes and sewage dumped into rivers all find their way to the oceans, a huge mixing pot of national and international waters, where they are joined with lumps of material from OIL SPILLS, ship's wastes, pollutants from noxious industrial gases in the atmosphere that enter the ocean by rain and snow, and a variety of other induced foreign substances such as hydrocarbons.

EEL is a common name for a number of unrelated fishes included in the orders Anguilliformes (the true eels) and Cypriniformes; all have an elongate, serpentine body without scales or pelvic fins.

The true eels consist of several hundred species, most of which are marine; these are somewhat typified by the morays (e.g., *Muraena helena*) and members of the genus *Conger.* They are usually found in the shallow parts of tropical and subtropical waters of the world's oceans, although a few inhabit the depths (e.g., the Gulper eels), and some can be found in cold waters.

The American eel (*Anguilla rostrata*) and the European eel (*A. vulgaris*) are representative of true eels that are catadromous—an unusual characteristic, meaning that such eels live in fresh water or estuaries but move to the open ocean to spawn. This behavioral pattern is exactly opposite to that of the ANADROMOUS FISH which migrate to fresh water from the oceans. The precise life history of catadromous eels is not well understood. Aristotle was positive that all eels originated in "the bowels of the earth." It is now generally believed that these mature eels spawn at considerable depths in the SARGASSO SEA. The female eel is capable of producing 3–10 million ($3-10 \times 10^6$) eggs during spawning and dies a short time afterward.

The larvae are called *leptocephalus* because of their similarity to thin transparent leaves. Most eel larvae are small in size, although the *Leptocephalus giganteus* species often attains a length up to 6 ft (1.8 m). This peculiar ribbonlike leptocephalus larval form is also

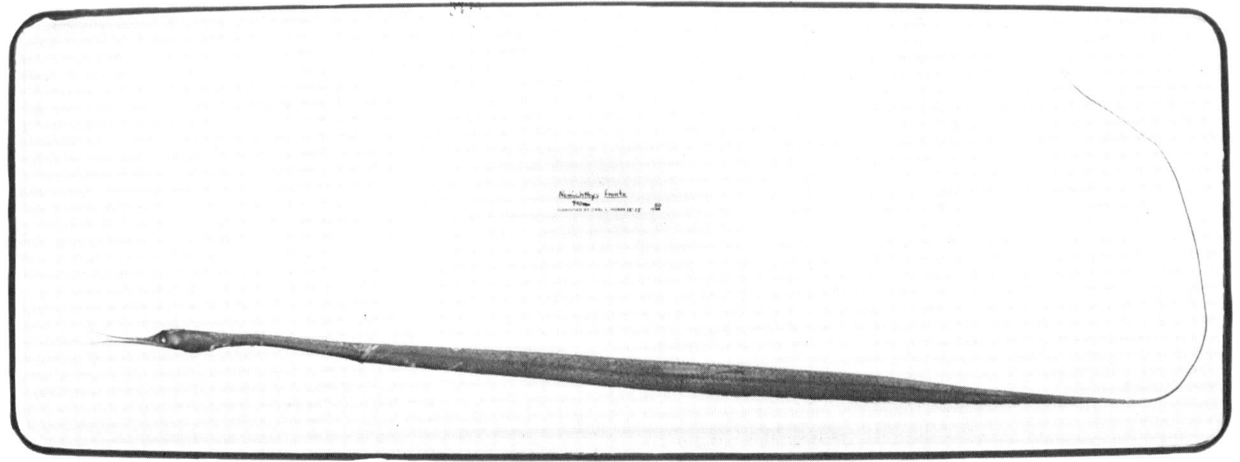

EEL. This deepwater threadeel is distinguished by a peculiar beaklike snout and a long, threadlike body. *(Scripps Institute of Oceanography)*

characteristic of the young stages of tarpon, bonefish, ladyfish, and in some of the FLATFISH.

From the time of spawning, it is estimated that it takes about a year for the young eels—called elvers—to develop and to return without exception to the very continental waters of their parents. They move into bays and inlets after this remarkable navigation, and in late spring, great numbers of them begin their journey upstream. The males tend to stay in brackish water, while the females travel much greater distances. They live and grow in fresh water for long periods—5–20 years—before returning to the sea to spawn.

When mature, the American eel, an inhabitant of east coast waters of the United States, averages about 3 ft (1 m) in length, with a maximum of about 4 ft (1.21 m). The European eel is larger, often attaining a length of 5 ft (1.5 m) and a weight of 20 lb (9 kg). In the Central American form, the American eel grows larger, sometimes measuring 10 ft (3 m) in length.

Eel is a fish seldom eaten in the United States but known and appreciated in Scandinavia, eastern Europe, The Netherlands, Germany, Italy, Japan, and England (where it is often served as a smoked fish and called *Blind Robin*).

EELGRASS, or *Zostera marina,* is a flowering plant (of the family Potamogetonacae) that grows fully submerged in shallow SEAWATER. This unusual plant is native to the coasts of the Northern Hemisphere from the subarctic to the subtropic. Related species are found in the Southern Hemisphere, and another marine-flowering plant, *Posidonia caulini,* grows extensively in the MEDITERRANEAN SEA. Also, there are abundant beds of *Thalassia testudinum,* or TURTLE GRASS, found in the tropical parts of the ATLANTIC OCEAN. All such grasses act to support and provide a natural habitat for huge quantities of marine life.

In addition to its many intricate ecological functions, dried eelgrass leaves have been used as a fuel, a packing and upholstering material, insulation, fodder, and fertilizer. However, the only recorded case in which the ocean has been used for grain production is that of the *Zostera marina* harvested by Seri Indians on the west coast of Mexico. These Indians prepared *Zostera marina* grain by harvesting the grain-bearing part, sun-drying it, threshing the plants with wooden clubs, and loosening the seeds by rolling the seed heads between their palms. The product was winnowed (tossed in the air to remove the hulls), then the grain was toasted, rewinnowed, and ground into a flour. Cooked in water into a thick gruel, the flour has a bland flavor. Like wheat flour, *Zostera* grain flour can be variously flavored; its protein and starch contents compare favorably with those of wheat, rice, and other grains: 13.2 percent protein, 50.9 percent starch, and 1.0 percent crude fat.

Zostera marina, in addition to its potential as a grain crop, is an important food for some sea turtles and water fowl. Its most important functions, however, are ecological, as a shallow-water mud-flat stabilizer and as a sustainer of the productivity of estuarine and other coastal areas.

See also SEAGRASS.

eH See REDOX POTENTIAL.

EL NIÑO refers to a set of oceanographic conditions that, at intervals of about 5 to 8 years, results in a biological catastrophe in the coastal waters of Ecuador and Peru. Hardest hit are the ocean birds, tens of thousands of which literally starve to death when the fish on which they feed migrate to other areas. El Niño (Spanish for the Christ Child) gets its name from the fact that the phenomenon usually occurs about Christmas time.

The relatively cold PERU (HUMBOLDT) CURRENT flows northward along the coast of Chile and Peru until it turns westward to join the South Equatorial Current (see EQUATORIAL CURRENT SYSTEM). In the vicinity of this current, the CORIOLIS EFFECT and the prevailing wind combine to push the surface water to the west. The displaced water is replaced by the steady UPWELLING of cold, nutrient-laden water along the coast. The steady enrichment of the surface layer produces an abundant PLANKTON crop that, in turn, produces one of the most prolific fishing grounds in the world. Anchovies and TUNA are particularly abundant, as are the sea birds that feed on the anchovies. The sea birds, in turn, produce commercial deposits of guano that is used for fertilizer.

On occasion, during the Southern Hemisphere summer, the trade winds that drive the South Equatorial Current westward along the equator weaken and sometimes fail. During such periods, upwelling of the cold, rich water ceases, and a shallow layer of warm, high-salinity water from the eastward-flowing Equatorial Countercurrent moves south to replace it. Under these rather sterile conditions the plankton crop is destroyed and the anchovies migrate temporarily to other feeding grounds, leaving the sea birds to starve.

Hydrogen sulfide produced by the rotting sea life that results from El Niño tends to blacken the hulls of ships. This effect is known as the Callao Painter (El Pintor) after the harbor at Callao, Peru where it is most noticeable. A more local and less destructive version of El Niño, occurring almost every year in the period April through June, is known as Aguaje.

ELASMOBRANCHS is the name for the sharks and RAYS. (See SHARK.) These cartilaginous fishes belong to a subclass (Elasmobranchii) of the class Chondrichthyes and are distinguished by separate gill openings, the absence of a SWIM BLADDER, and the presence of clasper organs in the male for internal fertilization and sensory organs (AMPULLAE OF LORENZINI) in the head region.

It is believed that the sharks and rays, as well as the Holocephali (the ratfishes), evolved 150 million (1.50×10^8) years ago from the Placodermi a group of extinct fishes armored with bony plates and having imperfect skeletons.

Elasmobranchs are characterized especially by their cartilaginous skeleton and by body tissues whose osmolality is brought to that of SEAWATER by the presence of a large concentration of urea in addition to concentrations of salts in the body fluids. This osmoregulatory system enables the elasmobranchs to absorb water through the gill and oral membranes in small amounts sufficient for the formation of urine. This system also requires that the embryo of the female must be either internal or in a tough-shelled egg within her body. Thus, internal fertilization by copulation is essential. Such copulation is carried out via a modified pelvic fin, or clasper, used by the male for the transmission of sperm.

By way of comparison, the TELEOSTS, or true fishes with real bone, underwent a different evolutionary history than the elasmobranchs, and the structure of their kidneys and their protein metabolism are quite different. Also, unlike the elasmobranchs, such fishes rarely possess seminal vesicles, the temporary storage organs for spermatozoa, and true sperm ducts are not found in some forms, such as in some SALMON. In others the sperm is passed to the exterior through modified kidney tubes, and in others the sperm does not pass through the kidney at all. Moreover, these marine fishes maintain a blood osmolality considerably below that of seawater and close to that of the higher invertebrates. (See INVERTEBRATE.) While many waste products are passed out of the body via the skin and the anus, most are eliminated by the kidneys. The seawater ambient is more concentrated than the body fluids of these fishes, and so water is removed or "drawn out" of the body and the seawater salts are drawn in according to the principles of osmoregulation (the physiological mechanism for the maintenance of an optimal and constant level of osmotic activity of the fluid in and around the cells). (See OSMOSIS.) The marine fish consumes large amounts of seawater by mouth. Freshwater fishes, on the other hand, have adopted special devices to prevent the intake of water and the loss of salts which tend to move out of their bodies. Freshwater fishes take in water through the gills and not through the mouth. While the saltwater fishes seldom urinate, the freshwater fishes utilize their kidneys to carry away waste products in large quantities of urine.

ELEMENT is one of the basic components of matter; it is a substance made up of atoms with a unique atomic number. Some common elements are oxygen, nitrogen, hydrogen, iron, copper, gold, silver, and aluminum.

The elements are classified in families or groups in the periodic table. This table, which lists the elements, sequentially according to increasing atomic number or atomic weight and arranges them in horizontal rows (periods) and vertical columns (groups), is designed to illustrate similarities in the properties of the elements as a periodic function of the sequence.

Elements are also frequently classified as metals and nonmetals. A metallic element is one whose atoms form positive ions in solution, and a nonmetallic element is one whose atoms form negative ions in solution. (See ION.) Approximately 75 percent of all the elements are metals, and the remainder are nonmetals. Most are solids at room temperature, although mercury, gallium, cesium, and bromine are liquids, and the rest are gases.

Most of the elements found in nature are combined with other elements in the form of compounds. A few such as oxygen, nitrogen, helium, neon, argon, krypton, radon, sulfur, gold, silver, and copper occur in the free or uncombined state. The most abundant element on Earth is oxygen, and the next most abundant is silicon. In the universe, hydrogen is the most abundant element and helium is second.

It is very probable that all the elements are present in the world's oceans, although some have not been detected. At least 84 of the basic elements have been identified in SEAWATER, and some of their concentrations are highly variable. In this regard the major seawater components are almost constant in their relative proportions, while the trace elements may vary with geographical location and water depth.

Since most of the chemical elements have been found in seawater, the chemical and geological oceanographers particularly are provided with a countless number of problems concerning the source, function, speciation or molecular structure, and significance of the elements and their interactions.

ELVER See EEL.

ENERGY (from the oceans) is the intrinsic ability of the oceans to perform work. Various natural energy sources, distributed in several forms throughout the world's oceans, represent forces that can be extracted and utilized. Basically, these forces have the capacity for doing work by being applied in one way or another to transfer energy or motion to a body, with the resultant action used to operate appropriately designed engineering systems.

Petroleum, natural gas, and coal are some of the energy-producing substances that can be obtained from off-shore ocean locations. Such mining endeavors have been carried out for several years by the larger nations of the world. However, several other natural forms of energy exist in the oceans as alternative sources for exploitation: one is solar energy, another is geothermal energy, and a third is the energy contained in SEAWATER because of its chemical composition.

Solar energy is an inexhaustible resource. More than 200 000 times the energy consumed in all fossil and nuclear fuels is delivered daily to this planet by the rays of the sun. Geothermal energy, or heat conducted from the interior of the earth to the surface, also represents a huge resource. Geothermal or "geopressured" water resources are also known to occur along the Texas and Louisiana Gulf coasts. Unlike other geothermal areas, the energy potentials of the waters in these geopressured areas are not limited to thermal energy. The extremely high fluid pressures of these hot waters are a potential source of mechanical (hydraulic) energy. In addition, dissolved natural gas, primarily methane, contributes significantly to their energy potential.

Both solar and geothermal energy are absorbed by the oceans, and the heat transmitted is responsible, either directly or indirectly, for several ocean phenomena. Such phenomena, in turn, can be employed as kinetic energy sources. In the main these sources are ocean thermal differences, ocean WAVES, and ocean CURRENTS.

The very chemical composition of ocean water itself—basically a compound of hydrogen and oxygen—represents another plentiful energy source, namely, hydrogen, the cleanest-burning fuel on earth. Hydrogen can be dissociated from seawater by electrolysis.

Ocean Thermal Gradients Various nonpolluting mechanisms can be designed and built to generate power from differences in ocean water TEMPERATURE. Thermal gradient conversion, a variant of solar power, was suggested as early as 1881 when the French physicist Jacques d'Arsonval predicted that humans could use heat from the ocean.

The concept of electrical generating systems is based upon the fact that in ocean equatorial and tropical waters, the heat of the sun on the sea surface (a daily average of 1400 Btu/ft^2) produces temperature differences of over 40° F (22° C) between the surface waters and depths of 1000 ft (300 m) or more. The warm surface water can be used to boil and vaporize a volatile liquid, like propane or ammonia, and the resulting gas can then be used (much like steam) to drive a turbine-bladed electrical generator. The cold waters below the surface would be pumped up and used to condense the gas back into a fluid, and this fluid would be recycled through the turbines. It is postulated that these turbines would be capable of generating as much as 160 000 kW which would be transmitted ashore by cable.

Ocean Currents Ocean currents, especially western boundary currents, are a source of extractable energy. Such currents derive their energy by a complicated process involving the adsorption of solar heat in the oceans and atmosphere. By transformations and redistributions of this heat from the earth's equatorial zones toward the polar areas via moving currents of seawater and air by the earth's rotation, the strongest sea currents are ultimately focused to the western edges of the ocean basins (or the eastern coasts of the continents). Among such powerful current systems of the world's oceans are the GULF STREAM off the United States, the KUROSHIO CURRENT off Japan, and the Agulhas-Somali system off the African Coast.

Ideas for tapping this source of ocean energy include emplacement of appropriately designed propeller-type turbines (water "windmills") near the surface in the current. Ocean currents usually are strongest close to the surface.

Ocean Waves As in the case of oceanic currents, ocean waves (advancing crests and troughs of seawater propagated by the force of the wind) contain a significant amount of kinetic energy which may be converted to usable form (e.g., electricity, compressed air, etc.).

It has been estimated that wave power off the coast of Great Britain, if harnessed, could provide that country's total annual consumption of electricity. Several schemes for converting such a form of ocean energy to electricity have been proposed. One of the most advanced of these is a device that consists of a large system of oscillating vanes that move up and down with the wave action. Such an induced movement would produce pulses of high-pressure water to drive turbines. This energy could, in turn, be used to dissociate water into its components, hydrogen and oxygen, with the hydrogen transported to land as fuel for power stations and automobiles.

Although such a concept has not yet been put to work, some steps have been taken to convert wave motion to electrical power in navigational buoys. A turbine generator system may be utilized for such a purpose at some ocean site after detailed wave distribution information for the site is acquired.

Tides The ocean's tidal cycle (about $12\frac{1}{2}$ hr) provides an enormous amount of seawater placed in motion, especially in coastal areas.

Some utilization of this energy source has been achieved, and in Brittany (near St. Malo), France, for example, the French, in 1967, constructed a large tidal electric power-generating complex across the tidal flow. In this area, there are two low and two high tides every day, and the tidal range averages more than 37 ft (11 m) in height. A novel generator shaped like a submarine has been designed to efficiently deal with the peculiarities of the tidal flow during both flood and ebb. The Russians also have tidal power stations in bay locations of the ARCTIC OCEAN. In the Passamaquoddy estuaries between the United States (Maine) and Canada, the bay has a tidal range of 50 ft ($15\frac{1}{4}$ m).

Hydrogen Hydrogen, a constituent of water (H_2O), is in plentiful supply in the earth's lakes, oceans, and streams. This efficient fuel can be made by electrolysis of the seawater, a process used to break it down to hydrogen and oxygen. Suggestions have been made to carry out this electrolysis at sea by converting ocean current and/or thermal gradient energy to electricity which is required for the electrolysis process. Other schemes for power for the process have also been made (e.g., nuclear reactors).

Hydrogen generation (from seawater) for use as a fuel is attractive in concept. The gas can be bottled, stored, piped, pumped, burned, and liquefied. Hydrogen as a fuel can be substituted for petroleum and coal in various industrial processes; i.e., for gasoline in internal-combustion engines. The element H_2 is also easily converted to other fuel forms such as ammonia, hydrazine, and methanol. In addition, hydrogen may be conveniently stored as a metal hydride; these materials release hydrogen when heated.

Other Sources Other possible approaches to solar energy, broadly defined, are wind power, conversion of biomass into energy, thermal conversion to electricity and photovoltaic (direct) conversion.

The wind, for example, has been used to power ships since human beings first used sails; and, on land the wind has been used for pumping water, for irrigation, and for grinding grain from about the twelfth century A.D. Wind-conversion systems were also generating usable amounts of electricity in Denmark as early as 1910.

A problem with wind is that its force is not always available when needed. Thus, one important requirement for utilizing this energy source is the capability to store its energy in the form of electricity or to combine wind generators into a larger electrical network. Improvements in storage batteries, such as the high-temperature lithium and sulfur systems, would assist greatly in using wind-generated electricity at sea and on land on a larger scale.

Direct use and conversion of solar energy also has been used in DESALINATION of seawater and domestic heating for many years. The idea of using high-temperature solar steam plants for the conversion of electric energy holds promise. Steam boilers used in generating electricity require temperatures of approximately 1000° F (540° C). By concentrating the sun's rays by some lensing techniques such as those imitating the eye of the HORSESHOE CRAB (an ideal concentrator of light), these required temperatures could be attained for steam boiler operations.

Another resource involving the ocean and land environment is obtainable via PHOTOSYNTHESIS, or bioconversion. Tremendous amounts of plants and trees cover over two-thirds of the world's landmasses. Such vegetation uses sunlight to fix carbon in the form of fats, proteins, and carbohydrates, as does the phytoplankton which abounds in one-tenth of the world's oceans. (See PLANKTON.) If portions of this ocean biomass (organic content), including, for example, seaweeds, especially KELP—which is the fastest-growing known plant, sometimes growing as much as 2 ft (0.6 m) per day—could be harvested and methodically subjected to controlled decomposition (anaerobic) or digestion, the production of methane, a fuel, would be significant.

Summary It has been estimated that in the United States alone, the requirement for power, as expressed in measures of heat energy or British thermal units (Btu's), may reach something like 200 quadrillion Btu, or "quads," in the early twenty-first century. (A quad, or 1 quadrillion Btu = 10^{15} Btu.) Moreover, at the current pace of usage, the nations of the world will consume, in the period 1977–2000, an amount of energy greater than that used by human beings in all past recorded history. Energy from the world's oceans can help supply some of these energy needs.

EQUATORIAL COUNTERCURRENT See EQUATORIAL CURRENT SYSTEM.

EQUATORIAL CURRENT SYSTEM consists of a North and South Equatorial current, measuring 621–932 mi (1000–1500 km) across, which flows east to west along either side of the equator in the ATLANTIC OCEAN, PACIFIC OCEAN, and INDIAN OCEAN at relatively low speeds of around $\frac{1}{2}$–2 kn. These currents are relatively shallow [approx 1640 ft (500 m)] and are strongly influenced by the northeast and southeast trade winds which dominate the wind system in the equatorial region. In the relatively narrow [310 mi (500 km)] region between the two westerly flowing equatorial currents is an easterly flowing equatorial countercurrent which, among other factors, is related to the region of low winds between the two trade wind systems. The equatorial countercurrents are about 984 ft (300 m) thick, with speeds of 1–3 kn, and occupy a belt about 248 mi (400 km) wide which is centered on the equator.

In the Atlantic the North Equatorial Current originates to the north of the Cape Verde Islands and flows almost due west at an average speed of about 0.7 kn. In the western part of the Atlantic this current joins the branch of the South Equatorial Current that has crossed the equator and turns northward to terminate in the current through the Yucatan Channel and the ANTILLES CURRENT. The continuation of these currents represents the beginning of the GULF STREAM which dominates a large part of the circulation of the North Atlantic.

The South Equatorial Current in the Atlantic begins off the west coast of Africa just south of the Gulf of Guinea and flows westward at an average speed of about 0.6 kn, although it may reach a value of 2.5 kn off the coast of South America. As the current approaches Cape São Roque, the eastern extremity of South America, it divides, the southern limb curving south along the coast of Brazil to feed the BRAZIL CURRENT, with the northern limb joining the North Equatorial Current as it turns northward.

In the Pacific the North Equatorial Current begins where the waters of the Equatorial Countercurrent turn north off Central America and flows westward across the Pacific until it turns northward in the vicinity of the Philippines to feed the KUROSHIO CURRENT which dominates much of the circulation of the western and northern Pacific. The North Equatorial Current, measured at about 160°W LONGITUDE, has a volume transport of about 1.59

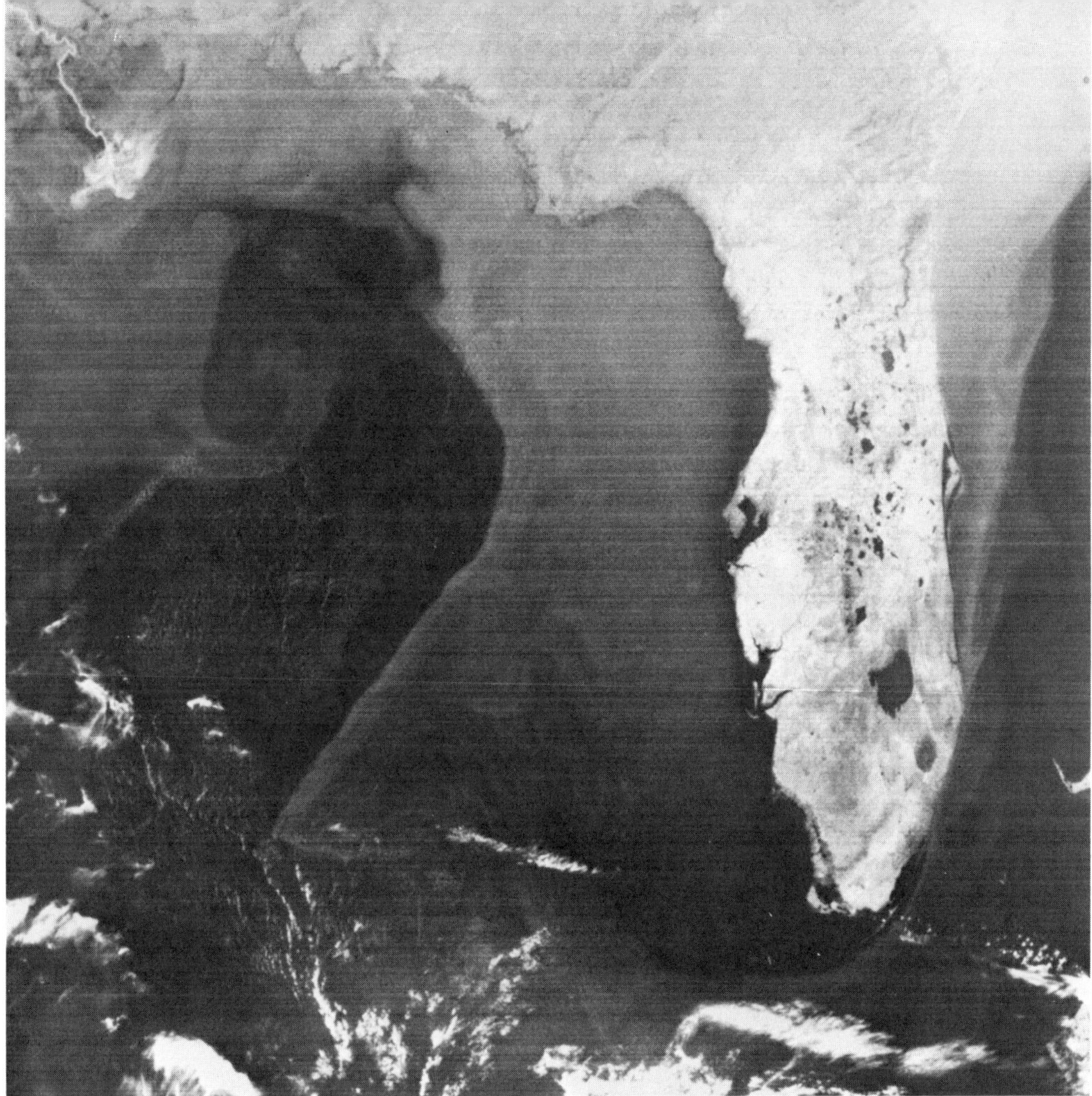

EQUATORIAL CURRENT SYSTEM. The satellite photograph shows a looping current which is an extension of the Atlantic South Equatorial Current into the Gulf of Mexico. The birdfoot delta of the Mississippi River is shown in the upper left corner. *(NOAA)*

billion (1.59×10^9) ft^3/s [45 million (4.5×10^7) m^3/s], which, in turn, is comparable to that carried by the North Equatorial Current in the Atlantic.

The South Equatorial Current in the Pacific begins as a northward flow parallel to the coast of South America, which gradually turns to the west in the vicinity of the equator. After crossing much of the Pacific, the current turns southward through the CORAL SEA to feed the swift EAST AUSTRALIAN CURRENT. The southward flow is particularly pronounced during the period from January to March.

In the Indian Ocean the North and South Equatorial currents are somewhat different due to the greater variability of prevailing wind patterns. The North Equatorial Current, for instance, is well developed in an east-west flow during February and March under a prevailing northwest monsoon. However, in August to September, when the southwest monsoon blows, the North Equatorial Current disappears and is replaced by the west-east–flowing Monsoon Current. The South Equatorial Current, on the other hand, flows east-west throughout the year. Approaching the east coast of Africa, it turns south to feed the strong AGULHAS CURRENT which flows to the tip of Africa before most of its waters gradually turn south and then back to the east.

ESTUARY is a distinctive body of water in which fresh water flowing from the land mingles with the salt water of the ocean. Most oceanographers accept a more rigid definition as follows: a semi-enclosed coastal body of water having free connection with the open sea and within which the seawater is measurably diluted with fresh WATER derived from land drainage. Based on topography, this definition leads to the identification of three major types of estuaries.

Coastal Plain Estuaries During the last Ice Age, the lowering of the SEA LEVEL induced rivers to deepen their valleys, particularly in that region where they previously met the sea. With the melting of the glaciers, the sea level returned nearly to its former level, thus drowning the newly-cut river valleys. Where sedimentation did not keep pace with the inundation, an estuary was formed, and the typical features of a river valley were retained. The estuary is triangular in shape—broadening toward the sea—the depth rarely exceeds 180 ft (60 m), and the floor is covered by recently deposited sands and muds. An example is the Chesapeake Bay estuary system.

Fjords In some locations such as the coast of Norway, the Pleistocene ice sheet during the last continental glaciation significantly deepened the existing river valleys near the coast. This down-cutting—often through hard rocks—resulted in a unique valley shape. A typical fjord is roughly rectangular in shape, has steep sides and a U-shaped bottom, except where the bottom has been flattened by sedimentation. A sill typically exists at the seaward end of the drowned valley. Sill depths typically range from 131 to 492 ft (40 to 150 m), although some are as shallow as 16 ft (5 m). The fjord itself may reach 2625 ft (800 m) in depth and be up to 62 mi (100 km) in length. The Sogne Fjord in Norway is an example.

Bar-built Estuaries This type of estuary is typical of tropical areas and along coasts having a high rate of sediment accumulation. Bar-built estuaries resemble the Coastal Plain Estuary in that they occupy drowned river valleys. They differ in that the sediment load carried by the river is so high that wave action in the breaker zone (see BEACH) has built an offshore bar that acts to restrict water flow. During flood stages the bar may be swept away, but it is quickly reestablished when the river resumes normal flow. The estuary at the mouth of the Roanoke River on the Virginia coast is an example.

Anomalous estuaries are, of course, formed by such processes as faulting, landslides, and volcanic eruptions. For instance, San Francisco Bay has been formed by the drowning of the valleys of the Sacramento and San Joaquin rivers as a result of faulting along the San Andreas Fault system.

Classification Based on the fact that the inflowing fresh water is lighter than the relatively static salt water, estuaries may be classified on the basis of stratification and flow characteristics.

Salt-Wedge Type Here, in its simplest form, the salt water lies in the estuary as it would if the river did not exist; i.e., its perimeter is everywhere at sea level. The fresh water, being lighter, flows into the sea over the top of the saltwater wedge. In actual practice the friction between the two layers results in the point of the salt wedge being blunted and the upper surface of the wedge being roughened. If the freshwater velocity is great enough, this surface roughness may grow into waves which actually "break." When this happens, the salt water is entrained in the upper fresh layer and results in brackish water flowing to the sea. The salt wedge must now compensate for this loss of water, and a landward flow along the bottom results. If the tidal range is great enough, both layers may flow toward the sea during ebb tide. The Mississippi River estuary is an example of this type.

Fjord Type Such estuaries are similar to the salt-wedge type except that the saltwater layer is much deeper (thicker) and is to some degree trapped by the sill. The freshwater layer tends to be of a thickness equal to the sill depth. The inflow of salt water is the result of tidal action, and the entrainment of salt water by the outward-flowing fresh water—especially in the vicinity of the sill where velocities are higher. Where sill depths are very shallow, the replacement of the lower salt layers is so slow that anoxic conditions similar to those in the BLACK SEA may develop near the bottom.

Partially Mixed Estuary Here the distinction is based on the existence of a significant tidal range which causes the contents of the estuary to oscillate with the tidal period. The amount of water that is added or subtracted by this oscillation is called the tidal prism. The tidal oscillation increases the turbulence in both the salt and fresh layers such that appreciable water is exchanged between layers. This increases the salinity of the surface layer, decreases the salinity of the lower layer, and increases the flow rate of water in both directions. This can result in flow rates in both directions within the estuary which may be up to 20 times the rate at which the river flows. Due to the resulting turbulence the salinity increases toward the sea in the upper layer and decreases toward the head of the estuary in the lower layer. The James River estuary in Virginia is an example of the partially mixed type.

Summary In general, aquatic life is sensitive to the salt content of the water in which it lives. The estuary represents a unique environment in which salinity varies all the way from that of river water to that of the sea and fluctuates frequently in any given location within the estuary. A number of species have adapted to and thrived within this unstable environment, although they have not distinguished themselves by a diversity of species as have those in more stable environments. The estuary also serves as a nursery for many animals which spend their lives in the open sea. Unfortunately for the animals, human beings have seen fit to build many of their major cities on estuaries (New Orleans, San Francisco, New York, etc.), and the dumping of municipal and industrial waste, channel dredging, and land filling has rendered survival in the estuarine environment more difficult.

EUPHAUSIIDS are a few species of planktonic ocean animals in the order Euphausiacea of the class Crustacea. These CRUSTACEANS, or euphausiids, are large for planktonic animals, and because of their size, wide distribution, and abundance in the world's oceans, they represent a key element of the marine biomass.

Like COPEPODS, euphausiids are SHRIMP-like in appearance. They are found in all the world's oceans, especially in Antarctic waters. The Antarctic species, *Euphausia superba,* is called KRILL, and this vitamin-rich euphausiid furnishes the principal food for the BALEEN whales. They are also the diet of many other animals such as SEALS, HERRING, some MARINE BIRDS, and even *Homo sapiens.*

Euphausiids, which, on the average, range in size from approximately $\frac{3}{4}$–2 in (2–5 cm), feed chiefly on ALGAE, DETRITUS, and copepods. They are filter feeders, and both their feeding mechanisms and mouth structure are complex.

During the day most of these animals live at considerable depths [e.g., 16 000 ft (5029 m) as a possible maximum], and many make long-distance vertical migrations diurnally. (See DEEP SCATTERING LAYERS.) They possess luminous photophores which give off a blue-green light. (See BIOLUMINESCENCE.)

Swarming is common in many species and is usually associated with reproduction. The fertilized eggs of the female develop into larvae (see LARVA) and an adult form is reached in 1 year in the tropic oceans and 2 years in the colder waters such as those of the SOUTHERN OCEAN.

EUPHOTIC ZONE, also called *photosynthetic zone,* is the upper level of the oceanic region from the surface down to the limits of effective light penetration for PHOTOSYNTHESIS.

See also ALGAE; HABITABLE ZONES; OXYGEN CYCLE.

EVOLUTION is the theory, or "fact," most widely accepted by biologists which states that all the different species of living organisms are genetically related to each other. This is based upon the thought that biological changes in these organisms have produced descendants that now differ from their ancestral species.

The subject of evolution has been one of the liveliest topics of biology since Charles Darwin, in the 1850s, first published his book *On the Origin of Species by Means of Natural Selection, or the Preservation of Favored Races in the Struggle for Life.* (See DARWIN, CHARLES.)

In this work, Darwin argued that evolution could be primarily accounted for from the following observations and postulates:

- That there are variations in organisms.
- That a survival-of-the-fittest situation exists in nature, and only those organisms that can adapt to their environment will survive.
- That these adaptive characteristics are inherited and the surviving forms then pass them on to their successors.
- That new species are developed as environments change so that with a change in environment a different organism should be formed and the characteristics which define the species should likewise change.

Evolution began, in the opinion of many scientists, when a small population became isolated from the rest of the species. With this isolation, gene combination changes took place, and the isolated groups were no longer able to mate with the parent stock. Thus, a new species was born. The birds, bony fishes, the flowering plants, and all other kinds of life, including *Homo sapiens,* all sprang from a single ancestral species in this way.

This one mode of speciation is a gross oversimplification both of the major processes that make the diversification of species possible and the research tasks involved in the study of living systems.

Evolution has resulted by many small changes that take place continually. Such changes occurred over millions of years to bring about the major divergences of phyla. All have come about by the cumulative effect of many small evolutionary steps in the succession of life as we know it.

For example, probably all life originated in the oceans, and at sometime in the distant past the fishes were less specialized than those of today. From a study of fossils, it has been determined that 225 million (225×10^6) years in the past (the boundary between the Paleozoic and Mesozoic eras) some event occurred that brought about a worldwide transition from largely marine-based life forms to land-based forms. Some of the fishes evolved into amphibians, and still later reptiles evolved from the amphibians. Both birds and mammals sprang from the reptiles. All owe their structure and survival to successful adaptation to the particular environmental niches in which they are found as a result of changing global CLIMATE and flora. Also, the impetus to conquer other realms due to competition accounted for such an early migration and hence subsequent development.

The total number and kinds of such systems that have emerged in the tangle of evolution over the past millions of years and that now constitute the biosphere are fantastically great. There are probably over 5 million species of plants and animals now living on Earth. Each is not only unique in its own way but, as previously alluded to, subtle differences have developed even within species. For example, one species of MARINE BIRDS will fly over 2000 mi (3220 km) of ocean to colonize a new home. Yet an almost identical species stubbornly refuses to travel a few miles from one island to another. Then, too, the HORSESHOE CRAB, the fairy SHRIMP, and the SHARK have remained basically unchanged for more than 200 million years, while some species have been known to have evolved from a parent group in a few thousand years.

Such questions, or anomalies of behavior, do not deny the validity of the theory of evolution, which is now almost universally accepted by biologists as a working theory. Evolution furnishes an explanation for a great body of factual data from the fields of comparative anatomy (the study of similarities in structure among different species of living organisms), comparative embryology (the study of similarities in the embryonic developments in different species), comparative physiology (the study of physiological processes—locomotion, sensation, coordination, and excretion—in organisms), TAXONOMY, and biogeography (the study of geographical distribution of organisms correlated with variations in characteristics in order to determine the relationships of different environments and the effect of geographical separation and isolation on the change of hereditary characteristics).

Coupled with this, perhaps the most direct evidence of evolution is provided by the fossil record of a billion years. Such a record shows profound changes in the structure of plants and animals and very recognizable trends, adaptations, and extinctions correlated with changes in environment. These evolutionary changes and adaptations most often show that the structures radiated by divergence from a common ancestral form.

EWING, WILLIAM MAURICE (1906–1974), an American geophysicist, is best known for the development and application of geophysical instruments and techniques to the study of the ocean floor. His name is particularly associated with studies of turbidity currents, abyssal plains, mid-oceanic ridges, and crustal thicknesses. (See ABYSSAL PLAIN; CONTINENTAL DRIFT; TURBIDITY CURRENT.) He helped to found the Lamont-Doherty Geological Observatory at Columbia University in 1949 and served as its director until 1971, at which time he moved to the University of Texas.

Ewing was born in Lockney, Texas, on May 12, 1906. He received his bachelor's (1926), master's (1927), and doctoral (1931) degrees from Rice Institute in Houston. He was an instructor in physics at the University of Pittsburgh from 1929 to 1930. He took a similar post at Lehigh University in 1930, becoming an assistant professor in 1936 and associate professor of geology in 1940. In 1935, a committee composed of distinguished geologists asked Ewing to undertake the task of applying geophysical techniques to the study of the oceans. Thus was born a lifelong pursuit. Among his early studies were a classic refraction study of the CONTINENTAL SHELF off the east coast of the United States and gravity measurements using the gravity pendulum newly developed by F.A. Vening Meinesz.

Ewing spent the war years of the early 1940s at the WOODS HOLE OCEANOGRAPHIC INSTITUTION where he was the leading physicist in the development and application of underwater photography and UNDERWATER SOUND for the Navy. He introduced the long-range sound transmission studies which resulted in the sofar system and, ultimately, the Navy's long-range surveillance and detection system. (See SOFAR CHANNEL.)

After World War II Ewing joined the faculty of Columbia University where he helped found, in 1949, the Lamont-Doherty Geological Observatory. During the two decades that he served as director, the observatory grew into one of the most influential research groups in the world. The major contributions of this period were the development and improvement of oceanographic instrumentation, a system of geophysical measurements accepted by most other research institutions, the development of a worldwide system of seismographs for earthquake detection, and an active program of radio and isotope geology for measuring the ages of marine sediments.

During his career Ewing wrote or was co-author of 280 papers and three books. He received 10 honorary degrees from universities in four countries and 26 medals and awards from institutions and scientific societies of eight nations.

FALKLAND CURRENT originates near the tip of South America and flows north between the Falkland Islands and the coast of Argentina. In the vicinity of the Rio de la Plata it merges with the southward-flowing BRAZIL CURRENT and turns eastward to cross the ATLANTIC OCEAN as the South Atlantic Current. (See CURRENTS.)

FAN VALLEY See SUBMARINE CANYON.

FATA MORGANA is a complex mirage, usually appearing over WATER or ice, in which images may be so distorted and magnified that a small boat appears as a lighthouse, a distant cliff as a Disneyland castle, or with no apparent source image, an island with towering mountains appears where none exists.

The fata morgana takes its name from a mirage that for centuries has periodically appeared in the Strait of Messina between Sicily and the toe of the Italian boot. To the Italian poets who wrote of it, the mirage reminded them of the submarine crystal palace that legend attributed to Fata Morgana (Italian for Morgan le Fay, the sorceress and half-sister to King Arthur), who was able to conjure huge and fantastic apparitions.

The simple mirage is that commonly seen on roads and in the desert as a shimmering pool of water. Here a layer of air just above the ground surface is heated and rarefied as compared with the layer above it. At certain viewing angles light waves do not strike the ground but are reflected to the eye by the heated layer of air. In such cases one sees the sky and not the road or desert sand.

The fata morgana, on the other hand, requires different and more complex conditions. Essential is a level, evenly illuminated surface, such as water or a broad expanse of ice, and stratified air in which the surface layer is colder than that above it. Under such conditions, light waves reflected from some distant object may be temporarily trapped (the cold surface layer acts as a waveguide) so that a distant observer perceives the image to be quite close by. In addition, vertical and horizontal stratification within the cold surface layer of air causes it to act as a poor-grade lens—as a result, the image is often grossly distorted in all its dimensions.

Early Arctic explorers referred to the fata morgana type of mirage as "looming." Perhaps the most famous case of looming was Crocker Land, a land of towering mountains and deep valleys first sighted by Admiral Robert E. Peary in 1905 during his unsuccessful attempt to reach the North Pole. In 1913 an expedition set out to relocate Crocker Land, and found it as Peary had described it some 400 miles (644 km) west of the northern tip of Greenland. But, as the expedition approached the landmass later in the day, it simply disappeared to be replaced by an unbroken sea of ice.

Tobias Gruber, in 1781 and again in 1786, demonstrated the role of atmospheric density layers in producing mirages by bending light waves. Independently, Gaspard Monge arrived at the same explanation for the mirages he witnessed in Egypt while accompanying Napoleon on his expedition against the Mamelukes in 1798.

FATHOMETER See UNDERWATER SOUND.

FEATHER STARS See ECHINODERM.

FIJI SEA, though occasionally so-called in the literature, is not an officially recognized sea. Some oceanographers, however, believe that the wedge-shaped area of the western South PACIFIC OCEAN defined by a line down the Melanesian Borderland (New Hebrides Islands) to New Zealand and north to Tonga Island is sufficiently distinctive to warrant its identification as a separate sea.

FINFISH See FINS; FISH; MARICULTURE.

FINS refer to appendages on FISH and other aquatic animals that are used for propulsion, guidance, and balance. About 97 percent of the fishes in the world's oceans are the ray-fin fishes (Actinopterygii). A characteristic of these fishes are their fins. Basically, most have two sets of paired fins—the pectorals, immediately behind the gills on the side of the head, and the pelvics, found farther back. The dorsal fin is located along the midline of the top of the body, and this fin may be subdivided into a spiny and soft part. The anal fin is along the underside behind the vent and before the caudal, or tail, fin. In different types of fishes, the fin placement and structure varies. In the typical PELAGIC FISH the dorsal and anal fins are employed as stabilizers, while the tail, pectorals, and pelvics are used for propulsion and stopping.

FISH (of the oceans) is the common name for the cold-blooded aquatic vertebrates (or backboned animals) belonging to the following classes: (1) Cyclostomata (the jawless ones), such as the lampreys and hag fishes, living representatives of a primitive line of fish; (2) Chondrichthyes (the cartilaginous fishes), e.g., the RAYS and sharks; and (3) Osteichthyes (the bony fishes), consisting of those with fleshy fins or arm- and leg-like appendages and those with rayed fins. Rayed-fin fish are divided into three groups: (1) the Teleostei, a very large and diverse assemblage constituting about 90 percent of all the living fish; (2) the Holostei which consist of bowfin and freshwater garpikes; and (3) the Chondrostei, the freshwater paddlefish and the freshwater and marine sturgeon which is large, primitive, and sharklike in many ways. (See LAMPREY; SHARK.)

In one system of animal classification, all 17 phyla are to be found in the oceans and five of these are exclusively marine. So extensive is marine life that it has been estimated that the HABITABLE ZONES in the ocean are about 300 times greater than those of the land and freshwater areas combined.

This is not too surprising due to the vast area covered by the world's oceans and their great depths. Moreover, the amazingly varied and abundant life in the sea is found under more favorable physical and chemical conditions than exist in any other known environment. TEMPERATURE variations are small, and since water has a high SPECIFIC HEAT value, whatever fluctuations take place happen slowly. Also, PHOTOSYNTHESIS, one of the basic food-forming processes of green plant cells, requires water as one of its basic materials. SEAWATER can carry and circulate the nutrient food supply for its inhabitants. In addition, the ionic strength of this solution facilitates osmotic regulation, and various phenomena act so as to "buffer" the solution. This means that changes from acid to alkaline, and the reverse, are resisted, and a nearly constant pH is maintained. (See PH.) Further, just as with the commonly prescribed swimming exercise for humans suffering from osteoarthritic problems, the buoyant support provided by the water makes the requirement for skeletal support much less stringent than the environment imposed on land creatures. Incidentally, some forms of marine life have no skeleton, but even large ocean WHALES, which do have a skeleton, can live only with the support and moisture of the surrounding water.

About 90 percent of the fish types thriving in the world's oceans can be described typically as those aquatic, backboned, cold-blooded animals possessing gills and fins. Such a description thus excludes other marine animals such as sea mammals, mollusks (e.g., CLAMS, etc.), and CRUSTACEANS (e.g., CRAB, SHRIMP, etc.). (See also MOLLUSK.)

Ichthyologists estimate that there are approximately 25 000 different species of ocean fishes. This educated estimate of 25 000 includes some 20 000 + species of bony skeleton fish, over 600 with skeletons of cartilage (e.g., the sharks and rays), with unknowns making up the difference.

Also, many fish species have not been named as yet, while others have sometimes been named more than once because of inadequate descriptions and variations due to geography. One source of such confusion has been caused by the fact that, in some species of fish, the male has been described as belonging to a different species than the female because of striking differences in color and body configuration. (This is referred to as sexual dimorphism.)

Ocean fishes vary greatly in length, from adult gobies who are ½ in (1.25 cm), to swordfish of more than 14 ft (4 m) and the sharks, such as the 25- to 30-ft (7.6–9 m) basking shark and the 50–60-ft (15–18 m) whale shark. However, the majority of ocean fish species are small, and the estimated overall average length of all the existing species would be about 6 or 7 in (15 or 17.5 cm) in length.

With the exception of such pelagic (surface or near-surface) fish as the abundant HERRING, SARDINES, and MENHADEN that feed directly upon PLANKTON (tiny floating organisms), the majority of ocean fishes are predators that feed upon other fishes smaller than themselves. However, a few of the larger fishes also feed upon plankton and various plant organisms. Interestingly, the two largest sharks (the basking shark and the whale shark) live wholly on plankton. Some experts also believe that the large OCEAN SUNFISH feeds principally on JELLYFISH.

The CONTINENTAL SHELF areas, including shallow seas, make up 7.6 percent of the oceans. Generally, these areas have fertile waters since in the shelf locations there is an effective recirculation of the slowly sinking nutrients back into the euphotic layer which is the place where the most organic matter is produced. Approximately two-thirds of all the ocean fish (or 17 000 species) live in these areas of the ocean. The majority of these fish usually swim within 20 mi (32.2 km) of the shore, and in summer some ANADROMOUS FISH (e.g., the SALMON) migrate inland to freshwater estuaries and rivers to spawn. Others that inhabit the close-in shore waters move offshore for that function. Epipelagic fish, like the TUNA, migrate widely to many areas of the open ocean and especially favor those areas wherever temperature conditions are most favorable for their metabolism.

There are probably no parts of the ocean, with the possible exception of some of the extreme 20 000 ft (6096 m) depths and beyond, that do not contain fish. Some midwater species live in the upper regions of the bathypelagic environment. While the tunas and marlins often frequent this depth zone [but at depths no greater than about 3000 ft (914 m)], the residents (e.g., the lantern fish) of this depth zone bear little resemblance to these fishes. For, unlike the tunas, and marlins, the inhabitants of this region have a gas-filled organ called a SWIM BLADDER. This makes it possible for them to sustain a firm skeleton and well-knit muscles at neutral buoyancy. Thus, they can live at such depths and also make migrations to the food-rich surface water. Such fish often inhabit what scientists refer to as the DEEP SCATTERING LAYERS, or those mid-depth portions of the ocean composed primarily of marine animals that migrate vertically to the surface at sunset and then descend at sunrise. These bands of migrating marine life are known to scatter the sound waves transmitted by SONAR. However, marine biologists do not yet have a sufficient detailed understanding as to how this scatter takes place nor do they have adequate information on either what kind of organisms live in the layers or how these organisms behave within the layers.

Below the 3000-ft (914-m) depths are the so-called "sunless" fishes (such as the ceratoid anglers and gulper eels) which, in contrast, have sparsely developed tissue systems. In these dark regions are to be found various luminous fishes and other animals, but below 4500 ft (1371 m) even this chemical light is scarce.

At the 6000–18 000-ft (1828–5485 m) depths, a region of uniform darkness, SALINITY, and cold both at the equator and at the poles, fish specimens are also to be found. However, these fishes are essentially not too different from those in the ocean layer immediately above.

Below this 18 000-ft (5485-m) level, fish species with backbones are very rarely found.

The application of names to the wide variety of fish in the ocean has been complicated over the years by the confusion which still exists because of the scientific nomenclature applied to them, as well as the common names, folk names, and invented names that have been given to ocean life.

The scientific (Latin) names are governed by voluntary adherence of zoologists to rules laid down by the International Commission on Zoological Nomenclature. The code of this Commission states that any zoologist who discovers a fish species lacking a scientific name may appropriately describe it and within certain rules then give it a latinized name consisting of two parts—the generic and specific (i.e., *Thunnus obesus* or bigeye tuna). Linnaeus, the Swedish biologist, in his book published in 1758, furnished the working foundation for this worldwide system of scientific names. (See LINNAEUS, CAROLUS.)

This latinized, two-word system, while subject to changes, is invaluable to the scientist, but invariably has little meaning or is too cumbersome for the average person to use. Thus, common names usually take precedence over Latin names, for it is much easier for the nonprofessional to speak of a marlin when actually referring to the *Makaira nigricans* (the Pacific blue marlin) or of a swordfish for *Xiphias gladius,* of wahoo for *Acanthocybium solandri,* DOLPHIN for *Coryphaena hippurus,* and so on. However, such common names as these can also be complicated by geography and accepted usage. For instance, the species *Oncorhynchus tshawytscha* is known as the king salmon in California and as the chinook salmon in Alaska.

Some varieties of fish, usually those of the same species or fish of the same size and feeding habits, travel in groups or schools (schools are sometimes called shoals). Not all fishes school. The reasons for this behavioral difference among species are not well understood. Several theories regarding the advantages of schooling have been advanced. These theories include such ideas as the safety-in-numbers concept, or the larger a school, the less the probability of any individual member of the group being eaten by a predator. Another theory is that by swimming in schools, many members of the group can ride along in the wake of the school leaders and exert less swimming effort and thus consume less oxygen than do single fish swimming alone. Moreover, schooled fish, which feed upon smaller fish and small invertebrates, do have a very definite feeding advantage since such prey cannot easily avoid the crowd. (See INVERTEBRATE.) Perhaps another basic reason for schooling is that via this traveling pattern successful reproduction is made more certain. For in the presence of spawning, most female fishes freely release their eggs into a mass of milt discharged by the males. In schooling fishes, the milt is very dense, and this denseness ensures that its contact with the eggs will result in fertilization, as contrasted with the situation where individual males and females spawn together. Here, the milt may well drift away from the eggs with the current.

Fish pigments, the coloring matter in fish cells, are chiefly the carotenoids (also the principal pigments in plants and animals). Carotenoids are of two chemical types: the carotenes, which are hydrocarbons; and the xanthophylls which, in addition to hydrogen and carbon, also contain oxygen. In fish, there are primarily three compounds of the xanthophyll type: astaxanthine, taraxanthine, and lutein. Most species contain lutein and either astaxanthine or taraxanthine, but not both.

Astaxanthine is the pigment responsible for the color of salmon flesh. It is also present, generally combined with protein, in crustaceans; in this form it has a blue or green color. Upon heating with water, the protein bond is broken, releasing the red astaxanthine. In nearly all fishes, these pigments are found in the skin, where they occur in the esterified forms.

Many facets of fish characteristics and behavior are still not well understood. However, it is of more than casual interest and importance that the basic metabolic functions of fishes and mammals are similar. Fishes, while they are more vulnerable to diseases and biochemical changes than human beings, may well serve as an early warning system to humans. By studying linkage effects of various chemical environmental pollutants (viz., pesticides, residues of DDT, aldrin, dieldrin, and toxaphene, plus polychlorinated biphenyls—PCBs, and others that are often released to the major chemical sink, the world's oceans) on fish, some useful clues can result for the protection of other life forms. For example, it is only recently known that DDT attacks the nervous system of fish. Stored up in the liver, it can send a lobster under stress into convulsions. In the laboratory, low levels of toxic compound polychlorinated biphenyls (PCBs) affect the thyroid gland in fish. In higher amounts, PCBs cause the fins of fish to rot away. Lower levels of the insecticides aldrin and dieldrin alter the amino acid metabolism in fish and produce signs of mental retardation. Toxaphene, another insecticide, attacks bone composition.

See also BIOLOGICAL OCEANOGRAPHY; OCEAN POLLUTION.

FISHING INDUSTRY See COMMERCIAL OCEAN FISHING.

FJORD See ESTUARY.

FLATFISH is the name for a number of demersal (or bottom-dwelling) FISH having a laterally compressed asymmetrical body with a twisted mouth and both eyes on the same side of the head, the upper side. Such fish comprise the order Pleuronectiformes which consists of several diversified families and many genera and species such as the halibut, sole, plaice, flounder, and turbot.

Flatfish is a demersal fish, without a SWIM BLADDER, that spends most of its long life (e.g., 12–35 years in some types such as the halibut) lying partially buried on the bottom. Lying on one side, most species are capable of changing their body color and pattern to blend with their surroundings, such as sand, mud, or rocks. This nervous control phenomenon is governed through the eyes which in the adult fish are located in the same side of the head and are able to move independently. Most species are right- or left-sided, in that some lie and swim on the right side and others on the left.

The halibut has been a prized food fish since ancient times in the Scandinavian countries and British Isles. The English called these fishes "holy-day butte" and ate the butte (the middle English word for flatfish) on holy days. The term holy butte changed eventually into the word halibut.

The North Atlantic halibut (*Hippoglossus hippoglossus*) is the largest of the flatfish, sometimes attaining a length of 12 ft (3.65 m) and a weight of 700 pounds (318 kg). The other species is the North Pacific halibut (*H. stenolepis*). Both live in the inshore waters of these areas as young fish and migrate to deeper [approx 1000 ft (305 m)] waters as they mature. Here, after they reach an age of 8 years or older, they spawn in the ATLANTIC OCEAN from May to June and in the PACIFIC OCEAN during the winter. A large female can lay over 2 million (2×10^6) eggs, and these buoyant eggs float to the surface where they hatch in a few days. The very young fishes have a symmetrical body and swim upright. As they grow, however, one eye, usually the left, moves over the top of the head toward the right one. As this takes place, the fish turns more and more on its left side and develops fringing fins on its brown and black upper body. The time required for the left eye to migrate 120 degrees to join the right is only a few days. When this cycle is complete, the fish swims with its blind side parallel to the bottom, and the dark top side of its body permits it to hide in the sand and rocks. The CRAB, MOLLUSK, COD, SKATE, and especially the HERRING form the main part of their diet.

Sole are small [1–2 ft (30.5–61 cm) in length] fringe-finned flatfish that, like the halibut, rest and swim with their right side upward. Most species such as the naked sole (*Gymnarchirus williamsoni*) inhabit the shallow waters of the warmer parts of the world's oceans, although a few species (e.g., *Bathysolea profundicola*) live in deep water, and some such as the common sole (*Solea solea*) enter the northern temperate seas of the Atlantic off the coast of Europe. Some species migrate into freshwater streams. Unlike the halibut, the sole exists almost entirely on bottom-living animals, although the young soles feed upon COPEPODS and fish larvae. See LARVA.

In the flounders, which are widely distributed and found, for example, along almost every coastline of the United States, the blackback, or winter flounder (*Pseudopleuronectes americanus*), is a right-eyed fish, while the fluke, or summer flounder (*Paralichthys dentatus*), is left-eyed. The former species is small [$\frac{1}{2}$–5 lb ($\frac{1}{4}$–$2\frac{1}{4}$ kg) in weight], and the fluke weighs up to 15 lb (6.80 kg). In swimming, these fish glide gracefully through the water, and when resting on the bottom, they cover themselves over with sand or silt with only their eyes protruding. When small fish or other prey are sighted, flounder launch themselves quickly toward them with a squirt of water from the gill on the underside.

In the United States most flounders are caught in large quantities by the use of trawls, while halibut fishermen use long lines since nets are prohibited so that the immature (less than 8 years old) fishes are protected.

Turbot and plaice are foremost among other economically important types of flatfish, with the plaice being one of the best known and most important. They live on the seafloor off the coasts of Western Europe from the Mediterranean countries to Scandinavia, and Iceland. Spawning occurs at different times of the year, depending upon the part of the sea inhabited. Each female plaice lays a large number of eggs (50 000–400 000), and in some areas millions of these fishes come together each year to spawn.

Almost all the flatfishes are good eating, and plaice, halibut, sole, and flounder are of great economic value. The turbot are known as the left-eye flounders and are extensively fished in the MEDITERRANEAN SEA and BLACK SEA.

FLORES SEA, located in the western PACIFIC OCEAN, is separated on the south from the INDIAN OCEAN by the islands of Flores, Komodo, Banta, and the easternmost point of Sumbawa; is bounded on the west by a line from Sumbawa to the Paternoster Island Group and then to the southernmost point of Sulawesi (Celebes); on the north by the south coast of Sulawesi; and on the east by the western limit of the BANDA SEA between Flores and Sulawesi. The Flores Sea covers an area of 46 706 mi^2 (121 000 km^2), occupies a volume of 53 260 mi^3 (222 000 km^3), and has a mean depth of 6000 ft (1829 m).

The Flores Sea basin is largely occupied by the Flores Trough, an elongate depression with a maximum depth of 16 863 ft (5140 m). Two submarine ridges run southward into the trough from Sulawesi. The bottom sediments consist of terrigenous and volcanic muds and globigerina ooze. The shallow Gulf of Bone (Annaba) joins the Sea along its northern rim.

Pacific water flows southward into the Flores Sea through the Molucca and Banda depressions, and continues westward to the BALI SEA and JAVA SEA. Surface WATER temperatures in the Southern Hemisphere range from 78.8° F (26° C) in winter to 84.2° F (29° C) in summer.

FLORIDA CURRENT may be defined as all the moving WATER from the Straits of Florida to a point off Cape Hatteras where the current ceases to follow the CONTINENTAL SLOPE. It is one of the swiftest of ocean currents, flowing at a rate of 2–5 kn.

The Florida Current can be traced directly back to the Yucatan Channel where the CARIBBEAN CURRENT enters the southern GULF OF MEXICO. Most of the current takes the most direct route to the Straits of Florida. A smaller branch loops around the Gulf and rejoins the main current in the east. After passing through the straits, the current is reinforced near the Bahamas by the ANTILLES CURRENT. The main Florida Current is retained as far north as Cape Hatteras where it becomes the GULF STREAM.

FLOUNDER See FLATFISH.

FOAM See SEAFOAM.

FOG is a visible aggregate of minute water droplets suspended in the atmosphere near the earth's surface. It results from the condensation of atmospheric moisture in the form of liquid droplets or ice

San Francisco's Golden Gate Bridge disappears into the fog. The famous California fog is produced by warm, moist air flowing over cold water that wells to the surface just off the coast. *(NOAA)*

Fog smothers the Golden Gate Bridge. Angel Island lies in the foreground, within the San Francisco Bay, and San Francisco on the left, while the Pacific Ocean extends into the distance. *(National Archives)*

crystals. Fog differs from clouds only in that the base of the fog is at the earth's surface while that of the cloud is at some altitude. It is distinguishable from haze by its gray color and its dampness, and from mist by the fact that it reduces visibility to a greater degree. The most notorious fogs originate over the ocean and, in some localities, drift in over the adjoining land. The most significant characteristic of fog is the hazard it poses to ships, aircraft, and highway travel because of the reduction in visibility. By definition fog decreases visibility to a maximum of 0.62 mi (1 km).

Fog forms when the temperature of the air becomes the same, or nearly the same, as the dew point (i.e., the temperature to which a parcel of air must be cooled at constant pressure and constant water-vapor content in order for saturation to occur). Therefore, fog may form by lowering the air temperature to the dew point or by increasing the moisture content of the air to the saturation point. Given the first condition, three types of fog may form:

1. *Advection Fog.* This type of fog forms when warm, moist air flows over a cold land or sea surface. The water chills the air to its dew point. Examples of this may be found in the LABRADOR SEA where the warm moisture-laden air flowing in from the GULF STREAM is chilled by the cold Labrador Current. (See CURRENT.) The heavy fog combined with floating icebergs makes this an especially hazardous area for navigation. The same process is responsible for the frequent fogs of late summer and winter along the California coast. Here the cold water is provided by UPWELLING.
2. *Radiation Fog.* This is a type of fog that is characteristic of land rather than sea. It occurs when the land surface cools down after the sun has set. The land, in turn, cools the air to its dew point. Therefore, radiation fog forms typically at night and persists until the sun has again warmed the land surface and the land, in turn, has warmed the layer of air above it. This type of fog does not form at sea because the water retains most of its heat throughout the night.
3. *Upslope Fog.* Such fogs are found when moisture-laden air flows over a land surface that gradually increases in elevation. Since the temperature commonly drops with altitude, the rising air eventually reaches its dew point in a process known as adiabatic cooling.

In the case in which moisture is added to the air until saturation occurs, two types of fog commonly form:

1. *Steam Fog.* This type is also called "sea smoke," "arctic fog," and "sea mist" and occurs when very cold air flows over relatively warm water. This air chills the moisture-laden air just above the surface and produces a relatively thin layer of fog. Since the fog forms immediately as the evaporated moisture leaves the sea surface, it often appears that the sea is smoking or steaming. Steam fog rarely exceeds 100 ft (30 m) in height.
2. *Frontal Fog.* Where a warm front is advancing over a layer of cold, unsaturated air, the formation of a frontal fog is possible. Rain from the warm front falls through the leading edge of the cold layer, providing the moisture it lacks. The greater the temperature differential between the rain and the cold air wedge, the more rapidly the fog will form.

FOOD PYRAMID See OCEAN FOOD CHAIN.

FORAMINIFERANS is the name applied to marine protozoans. (See PROTOZOA.) They are distinguished by the possession of a secreted shell, or test, which encloses a body with continually changing shape.

Foraminiferans are very numerous and widely distributed in many parts of the world's oceans. However, in spite of their abundance, only some 30 species are known to be pelagic, i.e., to inhabit the open oceans. Of these, 22 species are found only in the warm ocean waters of the world and a few are in the Arctic and Antarctic waters. Most foraminiferans are benthonic, living on the ocean floor or upon benthic ALGAE or other organisms. These types range in habitation from the intertidal zone to the oceanic depths.

They feed mainly on DIATOMS, although the fragments of the larger algae, BACTERIA, other protozoans, and other organic matter are also used. The food may be transported into the shell, or it may be digested outside by a network of temporary extrusions (pseudopodia) from the body and only the dissolved nutrients ingested. These pseudopodia stretch for amazing distances and are not only employed for the capture and digestion of food but also serve for locomotion and for anchorage of the organism. In operation the pseudopodia, which are long, slimy threads of protoplasm, weave around each other to form a sort of spider's web. Such a web, in the case of foraminiferans of the genus *Astrorhiza* can cover an area with a radius of 2.4–2.8 in (6–7.1 cm). This represents a net some 10 times larger than the shell diameter.

Foraminiferans range in diameter from 20 μm 4.7 in (11.9 cm), although the average is less than 0.04 in (1 mm).

These organisms, possessing shells of varying compositions, also have the ability to concentrate titanium, and contain about 4 million (4×10^6) times as much titanium as the SEAWATER they inhabit. Their early stages are spent in the upper, sunlit layers (the photic zone) of the sea. The castoff shells of forminiferans make up the sediment in certain areas of the ocean floor, especially in the shallower oceans. The most important of the forms is that of the planktonic genus *Globigerina.* Its bottom shells make up globigerina ooze, a pelagic sediment consisting of 30 percent calcium carbonate. All foraminiferans are excellent stratigraphic markers for petroleum exploration or as tools for paleoecologic analysis of ancient sediments. They have been invaluable in oceanographic studies of the earth's history.

FOULING is a term commonly used to mean the adhesion of various marine organisms to the hull and other underwater parts of ships as well as to equipment placed in the ocean. This type of marine attachment and growth, which the Swedish botanist Carolus Linnaeus called "calamitas navium," is especially serious in the case of ships since it inhibits their efficient operation. (See LINNAEUS, CAROLUS.) It has been estimated that fouling organisms cost world shipping over $500 million annually.

Written records concerning this problem and how to cope with it date from the fifth century B.C. In those times, for example, the Phoenicians took their ships out of the water every year or two, hit the BARNACLES with a hammer, and placed the ships back in the water. In ancient times, it was not too unusual for a ship to have its bottom encrusted with more than 300 tons of marine growth. Some merchant ships of today have reported collecting over 200 tons of barnacles on their bottoms in a 2-month period. Other structures in SEAWATER, such as screens around seawater intakes, fencing for MARICULTURE farming, components in offshore platforms, buoys, platform grading, ladder steps, etc., can be overloaded by the extra weight of fouling.

Most typical of all the fouling organisms which can attach themselves to ships and underwater equipment are the barnacles. As a group, these CRUSTACEANS have a tolerance to a wide range of seawater TEMPERATURE and depth, although some varieties of barnacles grow more abundantly at greater depths than others.

However, the fouling organisms of ship hulls consist not only of barnacles but also of various species of hydroids, ALGAE, calcareous worms, and sea squirts. In their larval stages, the animal organisms are free-swimming. (See LARVA.) They attach themselves to the hull while a ship is in port, and unless detached by friction when the ship is underway, they remain in place and grow.

The fouling process is initiated when a clean surface placed in seawater becomes coated with a slime film of dead organic matter. This type of coating is formed within minutes or hours from the time of immersion. Shortly thereafter motile spores or reproductive cells of fungi and algae settle and grow, producing living films which may damage the surface directly (as in the case of acid-secreting BACTERIA) or indirectly by providing a host surface for other organisms, such as the larvae of barnacles and MUSSELS. (See FUNGUS.) After metamorphosis these larvae grow to macroscopic sizes.

Three groups of marine organisms are responsible for most of the destruction wrought each year to wharves, piers, ferry slips, and other harbor structures built wholly or in part of wood. These are the Teredinidae, Pholadidae, and Limnoriidae. Probably the most destructive are the Limnoriidae, although the shipworms, or TEREDO, are responsible for considerable marine hardware damage. Marine borers have permanently damaged lead sheathing at depths of 7000 ft (ca. 2120 m) in the INDIAN OCEAN. These borers may grow to a length of more than 5 ft (1.2 m) while attaining a diameter of less than an inch (2.54 cm). Oceanographic research has shown that, to varying degrees, they are sensitive to SALINITY, TEMPERATURE, food supply, action of CURRENTS, POLLUTION, dissolved oxygen concentration, pH, and the amount of dissolved hydrogen sulfide (H_2S) in the water.

An effective method of preventing fouling which has been successful with modern ships has been through the use of toxic paints. Such paints, or plastic coatings containing ions of copper, tin, lead, or mercury, poison the organisms during their larval stages.

As a result of improvements in coatings, larger modern ships are now able to remain out of dry dock for more than 2 years with no reduction in speed or increase in fuel consumption due to fouling, compared with approximately only 6 months of satisfactory performance prior to World War II. Copper additives and toxic arsenic compounds coupled with creosote are also good preservatives for wood. A 30- to 60-mil (0.030- to 0.060-in) plastic sheet protective barrier is also used.

However, even with such commendable progress, the control of fouling, the impact of toxic compounds on the environment, and the CORROSION of materials (which is often started by fouling microorganisms) are still of major concern. Thus, it may be said that practical antifouling methods must rest on increased understanding of the toxic effects of paints or coatings now in use as well as that of the

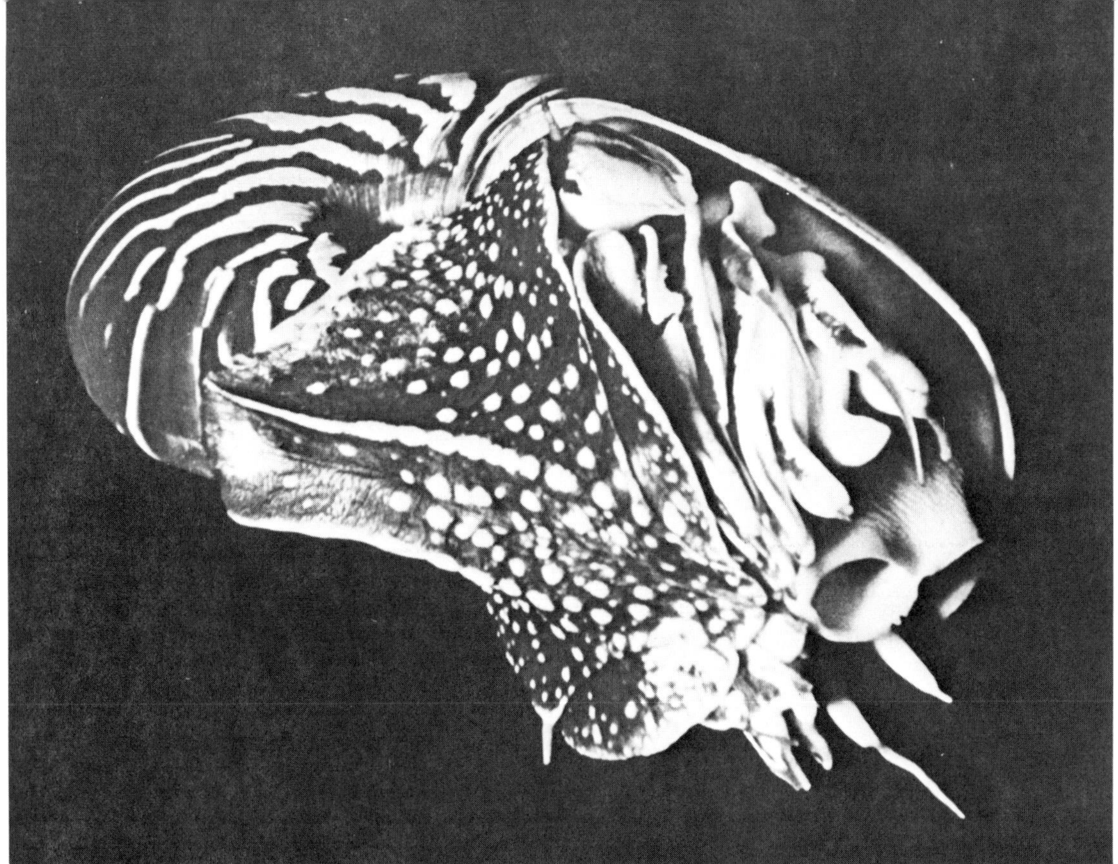

geographic distribution, life histories, physiology, and behavior of the different species of organisms, particularly those that infest ships in tropical ports.

FUNGUS. Fungi have, along with bacteria, a place in the ocean food chain, but they can destroy ocean plants and animals in catastrophic proportions. The gastropod in this photograph is massively infected with a fungal parasite.

FRINGING REEF See CORAL REEF.

FRONTAL FOG See FOG.

FULMAR See MARINE BIRDS.

FUNGUS (pl., fungi) is a nucleated, usually filamentous, spore-bearing organism having no chlorophyll but a complex type of cell construction called eukaryotic, which is identical to that of true plant and animal cells. The most widely held view on the phylogeny of fungi is based upon the concept that they originated from algal or protozoan ancestors and through successive EVOLUTION gave rise to their present classification as plants.

The fungi constitute a phylum (Eumycophyta) of the plant kingdom, and several thousand species are known. Most of these occur on land. However, fungi and yeasts are indigenous to the waters and sediment of the world's oceans. Together with BACTERIA, they are the principal agents of decay or the decomposition of organic matter. In that saprophytic role they serve an important function in the nutrition of marine ALGAE and animal life. However, it is not known with any degree of preciseness what the real importance of bacteria and fungi is in the OCEAN FOOD CHAIN, especially since much of the dissolved organic matter (e.g., glucose, glycolate, and amino acids) is consumed by these microorganisms.

Marine fungi commonly attack many ocean animals, especially certain SPONGES and mollusks. (See MOLLUSK.) In the latter case, a fungus disease of oysters caused by *Labyrinthomyxa marina* is of significant importance in the OYSTER-growing areas of the United States. This fungus can destroy virtually all the seed oysters along the Louisiana coast.

Fungi such as *Labyrinthula* and *Cochliobolus thalas-*

siae also cause disease in certain ocean-flowering plants. In 1931, the EELGRASS of the North Atlantic coast and Pacific Coast as well as that of Europe was infected by *Labyrinthula* in catastrophic proportions. In the Danish seas, the eelgrass has never come back to its former abundance since that epidemic. Both types of fungi are especially destructive to the Thallassia or TURTLE GRASS beds in semitropical areas. The major site of the infection occurs on blades exposed at low tide, and it is believed that the fungi are supported in their activity by other microorganisms.

In the marine sedimentary environments fungal filaments provide a source of food for the unsegmented worms (Nematodes) which feed by inserting a feeding stylet into the fungal cells and remove protoplasm.

Marine fungi also have the ability to break down wood fibers so that when organisms that specialize in wood boring, such as the TEREDO, attack the hull of a wooden ship or a piling, their job is already partially completed. See FOULING.

GALATHEA is the name of a Danish research vessel which made a voyage around the world lasting from 1950 to 1952; during this voyage, oceanographic observations were made in five OCEAN TRENCHES at depths between 19 800 and 33 000 ft (6035 and 10 058 m). The observations made and samples collected by bottom dredging were primarily concerned with marine life. Anton Bruun, the leader of the *Galathea* expedition, first proposed the word *hadal* for the deep-trench zone and its fauna. In dredging the Philippine Trench, scientists on board the *Galathea* proved that life can exist in the greatest ocean depths; they also showed that representatives of all the major INVERTEBRATE groups thrive at depths of more than 33 000 ft (10 058 m).

GALEOID is the name for a major SHARK group; the galeoids consist of the isurids, which include the mackerel sharks, and the carcharhinids, which include the tiger sharks.

The other major shark group is the SQUALOID (e.g., the spiny dogfishes). Sharks belonging to these different groups (galeoid and squaloid) vary in such features as the size of their jaws, shape of their teeth, and structure of FINS. For instance, the galeoids possess an anal fin, and the squaloids have no such fin.

GASTROPODS is the general name for some 30 000 species of diverse land, freshwater, and marine mollusks belonging to the class Gastropoda. (See MOLLUSK.) This is the largest of the seven classes in the phylum. Three subclasses of gastropods—the Prosobranchia, the Opisthobranchia, and the Pulmonata—are recognized. Of these, the first and second groups consist primarily of marine organisms such as the ocean snails, while the Pulmonata comprise land and freshwater groups.

The gastropods, or univalve mollusks, are characterized by a single spirally coiled shell or, as an exception, with a shell that is not spirally coiled, as is the case of the slugs. As a class, these animals represent a very diverse group morphologically and include slugs, limpets, and conches. (See CONCH; LIMPET.) Typically, there is a distinct head, one or two pairs of tentacles, and a flattened fleshy foot. Marine snails of the family Conidae, the cone shells, are potentially dangerous to human beings, although they have long been collected because of the attractive patterns their shells display. Of more than 400 species, however, most contain a fully developed venom apparatus. The venom apparatus of cone shells lies near the shell opening. The radular teeth are thrust into the victim, and the venom is believed to be forced under pressure into the wound opening. Cone shells may be seen crawling along the sand or may be found under rocks or near CORAL where they feed on small fishes and worms. Some of the more dangerous tropical species are:

- Court cone, *Conus aulicus,* ranges from Polynesia to the INDIAN OCEAN.
- Geographer cone, *C. geographus,* inhabits the Indian Ocean and the PACIFIC OCEAN from Polynesia to east Africa.
- Marbled cone, *C. marmoreus,* ranges from Polynesia westward to the Indian Ocean.
- Striated cone, *C. striatus,* inhabits the area from Australia to east Africa.
- Textile cone, *C. textile,* ranges from Polynesia to the RED SEA.
- Tulip cone, *C. tulipa,* ranges from Polynesia to the Red Sea.

In addition to these, many forms of marine gastropods exist in all the world's oceans from the coastal zones to the great depths. Certain groups live on rocks, on mud and sand, and among coral or on the large ALGAE. Some feed on this algae, while others are carnivorous predators feeding on other mollusks and echinoderms. (See ECHINODERM.)

GASTROPODS. As a class, the gastropods consist of over 30 000 diverse species of snails, slugs, limpets, and conches among others. There is wide variation in the colors and forms of the shelled gastropods, some of which are very beautiful. (*NOAA*)

GASTROZOOID is a nutritive polyp of colonial COELENTERATES characterized by the possession of tentacles and a mouth. See PORTUGUESE MAN-OF-WAR.

GEOLOGICAL OCEANOGRAPHY is the study of the topography or shape of the floors of the world's oceans, the distribution and character of the bottom sediments, the structure and composition of the underlying rock formations, and the geologic processes that have taken place during the millions of years of the ocean floor's history.

To make such studies possible, the geological oceanographer employs such tools and techniques as echo-sounding equipment and other methods for mapping the ocean floor. Underwater cameras, various bottom samplers, coring devices, submersibles, etc., are all valuable tools used for marine geological work.

See also INSTRUMENTATION; OCEANOGRAPHY.

GEOLOGICAL OCEANOGRAPHY. Volcanic eruptions and formations are a subject of interest to marine geologists. Here the camera peers into Mt. Mihara, an active volcano on the island of Oshima, Japan. *(U.S. Navy)*

GLOBIGERINA OOZE See FORAMINIFERANS; MARINE SEDIMENTS.

GLOMAR CHALLENGER is a United States oceanographic drilling vessel with a dynamic positioning capability; that is, the ship's position over a drill hole is maintained by thrusters controlled by a bottom-placed SONAR beacon and hydrophones on the ship's hull. The information gained in this way is fed into a computer which automatically controls any ship movement over the sonar beacon. This 400 by 65 ft (121 by 19 m) ship is equipped with a drilling derrick that towers about 200 ft (61 m) above the water line. The ship, built in 1968, has a capability of drilling cores from the ocean bottom in up to 20 000 ft (6060 m) of water.

The *Glomar Challenger* has made many voyages since 1968, taking and analyzing geological samples or cores from various parts of the world's oceans. Some MARINE SEDIMENTS were obtained from oceanic basins in the deepest parts on both the American and African sides of the North ATLANTIC OCEAN. (See BASIN.) These sediments were dated at about 155 million (155×10^6) years.

Deep-sea drilling operations on the *Glomar Challenger* began in August 1968 in the GULF OF MEXICO. The drilling of cylindrical cores of the ocean-floor sediments was carried out first in parts of the Atlantic Ocean, the PACIFIC OCEAN, as well as in the Gulf of Mexico to test the hypothesis of CONTINENTAL DRIFT by seafloor spreading. Later, cores were taken from all the oceans for a general investigation of the sediments of deep ocean basins, and for investigating the crustal rocks beneath the sediment.

The cores obtained and the investigative work done by the *Glomar Challenger* in the Deep-Sea Drill-

GLOMAR CHALLENGER. The *USS Glomar Challenger* has, since, 1968, set a number of records in drilling and coring operations on ocean floors.

ing Project have proved invaluable to scholars in all the earth sciences. Moreover, the *Glomar Challenger* drilled and cored in 11 720 ft (3550 m) of water on the Sigsbee Knolls (Gulf of Mexico) and later set a record (which it subsequently broke) in the North Atlantic, drilling 2759 ft (836 m) below the ocean floor in 16 316 ft (4944 m) of water.

Subsequent accomplishments of the *Glomar Challenger* have been many and varied. This ship was the operating heart of the Deep Sea Drilling Project, widely recognized as one of the most significant and successful programs in the history of geological research, and helped to accumulate much new scientific data, contributing to PLATE TECTONICS, geochemistry, paleoclimatology, paleontology, and paleo-oceanography. Among its other accomplishments and discoveries are:

- The verification of seafloor spreading
- Discoveries of salt domes having oil exploitation potential at the bottom of the Gulf of Mexico and off the coast of West Africa
- Discovery that glaciers started forming in the Antarctic as early as 15 million years ago
- Discovery that about 12 million years ago the MEDITERRANEAN SEA was a huge "Dead Sea" full of salt and gypsum
- Discovery that, at times, TURBIDITY CURRENTS have transported coarse sand from the Amazon River to the Mid-Atlantic Ridge
- Various technical innovations in deep-sea drilling now used by the offshore drilling and mining industry, e.g., dynamic positioning, sonar re-entry techniques, improved core bits, and improved coring techniques

The spectacular work of the Deep Sea Drilling Project/*Glomar Challenger* has led to a proposal for a 10-year follow-up program in ocean margin drilling (OMD). Its objectives would be the furtherance of increasing earth scientific knowledge, resource exploration, and technological development.

As evidence of the rapid development in drilling technology, a breakthrough was achieved by the *Glomar Challenger* when, on June 14, 1970, a deep-water hole was successfully reentered. The crew changed the worn-out bit and succeeded in finding and reentering the hole in 10 000 ft (3030 m) of water. Several months later, a similar operation was successfully completed in 13 000 ft (3040 m) of water; the bit wore out after drilling 2300 ft (697 m) into the seafloor, was changed, and reentered the same hole to drill 200 additional ft (60 m). Reentry was accomplished with the aid of a high-resolution scanning sonar system protruding beyond the drill bit and guiding it into a funnel-shaped receiving cone mounted on the ocean floor. This achievement heralded a new era in offshore drilling technology.

GOLDBERG, EDWARD D. (1921–), an American chemist, is best known for his work on the geochemistry of natural waters and sediments, the application of radioactive dating technologies to sediment age determination, marine pollution, and the use of glaciers in studying recent geological events at the earth's surface. He is professor of chemistry at the Scripps Institution of Oceanography.

Goldberg was born in Sacramento, California, on August, 2, 1921. He received his bachelor of science degree from the University of California, Berkeley, in 1942, and his doctorate in chemistry from the University of Chicago in 1949. During World War II he served with the Navy for 4 years. Goldberg joined the Scripps Institution of Oceanography as an assistant professor in 1949 and became full professor in 1961. In 1960 he spent a year at the Physikalisches Institut, University of Bern, Switzerland, as a Guggenheim Fellow, studying the rate of accumulation of glaciers. And, in 1970, he served as NATO Fellow, studying North Sea pollution at the Institut Royal des Sciences Naturelles de Belgique.

As author and editor, Goldberg has contributed many articles and books to the general field of oceanography. He has served as editor of a technical series in oceanography entitled *The Sea: Ideas and Observations.* He was coeditor of the two volumes *Earth Sciences and Meteorites* and *Man's Impact on Terrestrial and Marine Ecosystems,* and recently completed a book for the United Nations entitled *The Health of the Oceans.* In 1975, as part of his continuing interest in marine pollution, he initiated a surveillance program on U.S. coastal marine pollution for the Environmental Protection Agency.

GRAVITY WAVE See WAVES.

GREAT AUSTRALIAN BIGHT lies off the south coast of Australia and occupies the arcuate indentation in the continent from a baseline that stretches from West Cape Howe at the southwestern tip of Australia to Southwest Cape on Tasmania. The Bight covers the fairly extensive CONTINENTAL SHELF that rims the southern coast, and the deep seafloor that contains the Great Bight Abyssal Plain with depths of over 18 000 ft (5486 m). Except for a few seamounts, the deep basin is relatively smooth. (See SEAMOUNT.) The Great Australian Bight covers an area of 186 824 mi^2 (484 000 km^2), occupies a volume of 110 118 mi^3 (459 000 km^3) and has a mean depth of 3117 ft (950 m).

GREENLAND SEA, a sea of the ARCTIC OCEAN, is located between Greenland, Iceland, and Spitsbergen. Its northern boundary is a line joining the northernmost point of Spitsbergen to the northernmost point of Greenland. The eastern boundary is the west coast of Spitsbergen, and the western boundary is the east and northeast coasts of Greenland between Cape Nansen and the northernmost point. On the southeast the sea is bounded by a line joining the southernmost point of Spitsbergen to the eastern extreme of Gerpin in Iceland. The southwest boundary joins Straumness in Iceland to Cape Nansen in Greenland. The Greenland Sea covers an area of 465 130 mi^2 (1 205 000 km^2), comprises a volume of 417 443 mi^3 (1 740 000 km^3), and has a mean depth of 4737 ft (1444 m).

The first explorer to sight the Greenland Sea may have been Pytheas of Massalia during a voyage of discovery which took place around 320 B.C. Unfortunately, the personal accounts of his voyage have been lost, and we are left with only the statements of such classical writers as Polybius and the learned geographer Strabo. According to their accounts, Pytheas sailed through the Strait of Gibraltar and turned northward along the coasts of Spain and France where he spent some time studying the tides. Crossing over to the British coast—the limit of the known world at that time—he appears to have circumnavigated Britain via Land's End, Kent, and the Orkneys. From Britain, the accounts indicate, Pytheas sailed north for 6 days to discover a land that lay one day's sail from the frozen sea and where during the summer the day is only 2–3 hr long. This may have been Iceland on the southeastern edge of the Greenland Sea. However, some scholars believe the coast of Norway to be more consistent with the evidence. If Pytheas' voyage is to be believed at all, then he must be counted as one of the greatest explorers of all time since his voyage was not surpassed for over a thousand years.

During the second half of the ninth century, Iceland was colonized by a group of Irish monks in search of solitude and escape from the Norsemen who had begun to plunder the British Isles. Unfortunately for the monks, the Vikings arrived shortly thereafter on the first leg of their westward exploration and colonization. From parts of Iceland the tops of the Greenland mountains can be seen on a clear day. Responding to the pull of this new land, two Icelanders guided a party across the Danish Straits in 980 to spend the winter on Greenland's east coast. In 982 Eric Thorvaldsson—better known as Eric the Red—was exiled from Iceland for 3 years for murder. Eric spent the exile years exploring the coasts of Greenland in preparation for the colonization that followed. Later—probably 986—he led a large group of Icelanders to Greenland in what has been termed the largest polar expedition before that of Richard E. Byrd. He established two settlements in the vicinity of Julianehaab and Godthaab. Iceland annals record that around 1194 the Vikings, fanning out over the Greenland Sea in search of SEALS, WALRUS, and WHALES, discovered Spitsbergen—thus competing the navigation of the Greenland Sea.

The sediments on the floor of the Greenland Sea are mixed, ranging from rock stripped bare by ice rafting and currents, through terrigenous material deposited by melting glaciers, to globigerina and diatom oozes. In some areas, particularly around Iceland, volcanic ash contributes significantly to the bottom sediments. These sediments are draped over a complex bottom topography characterized by ridges and basins. Two large basins, the Iceland Basin with a maximum depth of 9163 ft (2793 m), and the Greenland Basin with a maximum depth of 15 899 ft (4846 m), are separated by the West Jan Mayen Ridge which rises to a depth ranging from 3280 to 4921 ft (1000 to 1500 m). An extension of the mid-oceanic ridge runs northward in a disjointed fashion from Iceland, through Jan Mayen Island, and to the west of Spitsbergen. To the west of Spitsbergen two smaller basins, the North Spitsbergen Basin (Lena Trough) with a maximum depth of 13 123 ft (4000 m) and the South Spitsbergen Basin with a maximum depth of 12 303 ft (3750 m), are divided by the Barentsberg Ridge in the south and the Spitsbergen Ridge in the north.

A belt of sea ice is present off the east coast of Greenland for the entire year. Navigation along the southern part of the Greenland Sea is possible during September with specially constructed ships. The tidal range is about 6.2 ft (1.9 m) on the east coast of Greenland and 6.5–16.4 ft (2.0–5.0 m) at Spitsbergen. Two tides are recorded daily.

The waters of the Greenland Sea are composed of four distinct water masses: the Polar WATER which flows southward over the Greenland shelf to join the Greenland Current; the Intermediate Water which is contributed by the Spitsbergen Current and sinks below the Polar Water further south to form a layer between depths of 656 and 1640 ft (200 and 500 m); the Norwegian Deep Water which comes from around Jan Mayen and flows beneath the Intermediate Water layer; and the North Atlantic Water which comes from the IRMINGER CURRENT and is heavier than the Polar Water but lighter than the other two water masses.

GREENLINGS See LINGCOD.

GROIN See BEACH.

GROUPER (often called jewfish, sea bass, sea perch, soapies, or rock cod) is the name for 400 species of fishes of the family Serranidae, most of which are similar in appearance—large mouths, strong backward-pointing teeth, and heavy-bodied. The family Serranidae has several members that are hermaphrodites (that is, an individual possesses both male and female reproductive organs). These fish inhabit tropical oceans mostly, although a few live in temperate waters and some in freshwater areas.

The groupers, or serranids, vary in size from those that are about 1 in (2.5 cm) as adults to the 12-ft (3.6 m) long Queensland grouper (*Epinephelus lanceolatus*) of the Indo-Pacific. Most are rather similar, having two dorsal fins, the front spiny and the rear one with soft rays. Groupers are noted for their color patterns and their ability to change colors to match with environmental changes.

The red grouper, *Epinephelus morio*, is the most abundant and commercially important grouper in the U.S. waters. This near bottom–dwelling fish is easily identified by the straight-edged dorsal fin membrane which, in other groups, is notched between the spines. It is restricted to the tropical and subtropical western ATLANTIC OCEAN from Cape Hatteras to Brazil and is found in the GULF OF MEXICO with less-abundant groupers: the snowy grouper (*Epinephelus niveatus*), Warsaw grouper (*E. nigritus*), rock hind (*E. adscensionis*), Kitty Mitchell (*E. drummondhayi*), red hind (*E. guttatus*), yellowfin grouper (*Mycteroperca venenosa*), scamp (*M. phenax*), gag (*M. microlepis*), yellowmouth grouper (*M. interstitialis*), and black grouper (*M. bonaci*).

Like the snappers, most groupers prefer hard bottoms of broken relief with CORAL heads and outcrops of rocks. (See SNAPPER.) Off Florida's west coast, the red groupers frequently occupy the crevices, ledges, and caverns formed by the limestone reefs. (See CORAL REEF.) Some such as the wreckfish, *Polyprion americanus*, common in the MEDITERRANEAN SEA and the warm-water areas of the Atlantic, live in and around sunken ships.

The name *jewfish* is often applied to several fish but really belongs to one species of grouper, *Epinephelus itajara*, which attains a length of 6 ft (2 m) and a weight of 600 lb (274 kg). This name, it is believed, was given to the fish because it is one with distinctly sharp fins and scales. Therefore, it was characterized as a clean fish according to Levitical law.

GRUNION is the name for the best-known FISH of the family called Atherinidae which is related to the mullets. One species, the *Leuresthes tenuis*, is a moonphase spawning fish, while the *Hubbsiella sardina* spawns by day.

The grunion is a small silvery fish [approximately 5–7 in (12.7–17.7 cm) long] that comes ashore to spawn in sandy beaches of the West Coast of the United States and Mexico, especially during spring periods of the year for 3 to 4 nights following a full moon and high TIDES. A second species (*Hubbsiella sardina*) that occupies the coastal waters farther north spawns by day. Both species come up on the beaches in pairs, a male and female; and swim and slither toward the high-water mark where they bury themselves in the sand. Here, their eggs (1000 to 3000 per individual) are laid and fertilized in less than a minute. The male wraps himself around the female and fertilizes the eggs as they are laid. Then the spawning fish attempt to return to the ocean. The sand burial of the eggs prevents them from being shaken and hatched prematurely. In nature the correct rhythm is maintained, and hatching of the eggs takes place at the next high tide, usually some 14 days later. At that time, the waves wash away the sand and carry the young hatchlings out to sea.

The grunion "run" or mass spawning event and the seasonal sport of gathering these fish, which are flavorful eating, has many followers in California.

GUINEA CURRENT originates in the vicinity of the Cape Verde Islands and flows southeastward along the west coast of Africa, and then turns to the east to flow into the Gulf of Guinea. The current later curves south and then west to join the South Equatorial Current. (See EQUATORIAL CURRENT SYSTEM.) The Guinea Current is a continuation of the eastern branch of the CANARY CURRENT and is augmented by the Equatorial Countercurrent. The Guinea Current flows at the rate of about 0.5–0.9 kn. The temperature of the surface WATER ranges from 77 to 82° F (25 to 28° C).

GULF See OCEAN.

GULF OF CALIFORNIA is a long, narrow arm of the PACIFIC OCEAN which lies between the Lower California Peninsula (Baja California) and the State of Sonora on the mainland of Mexico. The gulf was given the name Vermilion Sea (because of the red color imparted by plankton blooms similar to those in the RED SEA) by Francisco de Uttoa who explored the region in 1539. It has also been called the Sea of Cortez after Hernando Cortez who sailed its waters in 1536. The first European to sight the gulf was Núñez de Guzmán during a voyage in 1532. The Gulf of California is about 700 mi (1126 km) long and ranges from 30 to 150 mi (48 to 241 km) wide. It covers an area of 68 322 mi^2 (177 000 km^2), occupies a volume of 34 985 mi^3 (145 000 km^3), and has a mean depth of 2684 ft (818 m).

Scientists believe that the formation of the Gulf of California is related to CONTINENTAL DRIFT, specifically to the interaction between the floor of the Pacific Ocean and the North American continent. The evidence indicates that it is intimately connected with the complex of faults and fractures along the West Coast, of which the San Andreas Fault is the better known. Approximately 25 million years ago, Baja California presumably nestled against the mainland of Mexico. Then, about 3 million years ago, the gulf began to open as the peninsula moved northwestward—a process that is still going on. The separation of the peninsula from the mainland created tension cracks on the floor of the gulf which filled with lava. These northeast-southwest trending fracture zones may be responsible for the several present-day basins along the axis of the basin which have a similar trend.

The present configuration of the Gulf of California has a modest CONTINENTAL SHELF along the east side and precipitous slopes along the western edge. The eastern continental shelf ranges up to 31 mi (50 km) in width, and breaks at a depth of about 361 ft (110 m). The head of the gulf is dominated by the delta of the Colorado River. Here, mud and sand extend offshore for as much as 20 mi (32 km), with depths as little as 60 ft (18 m). The deeper floor of the gulf consists of a series of basins which become progressively deeper to the south. The gulf contains a number of islands, with Ángel De la Guarda and Tiburón being the largest. A number of permanent rivers enter the gulf from the east, but none come from the arid peninsula of Baja California.

The sediment of the Gulf of California is largely mud and silt from the Colorado River in the north. However, since the building of Hoover Dam in 1935, this source of sediment has become insignificant. The rivers along the eastern coast also contribute considerable quantities of terrigenous sediment both to the shelf and to the deeper basins by means of several submarine canyons. (See SUBMARINE CANYON.) In the deep basins the sediment consists of alternating layers of PLANKTON remains and clay.

Water circulation in the Gulf of California is largely controlled by the prevailing wind pattern. During the winter the wind blows from the northwest, moving the surface water out to be replaced by deep nutrient-rich Pacific WATER. During the summer the wind shifts to the south and moves surface water from the Pacific into the gulf, forcing the deeper water to flow out. The temperature of the

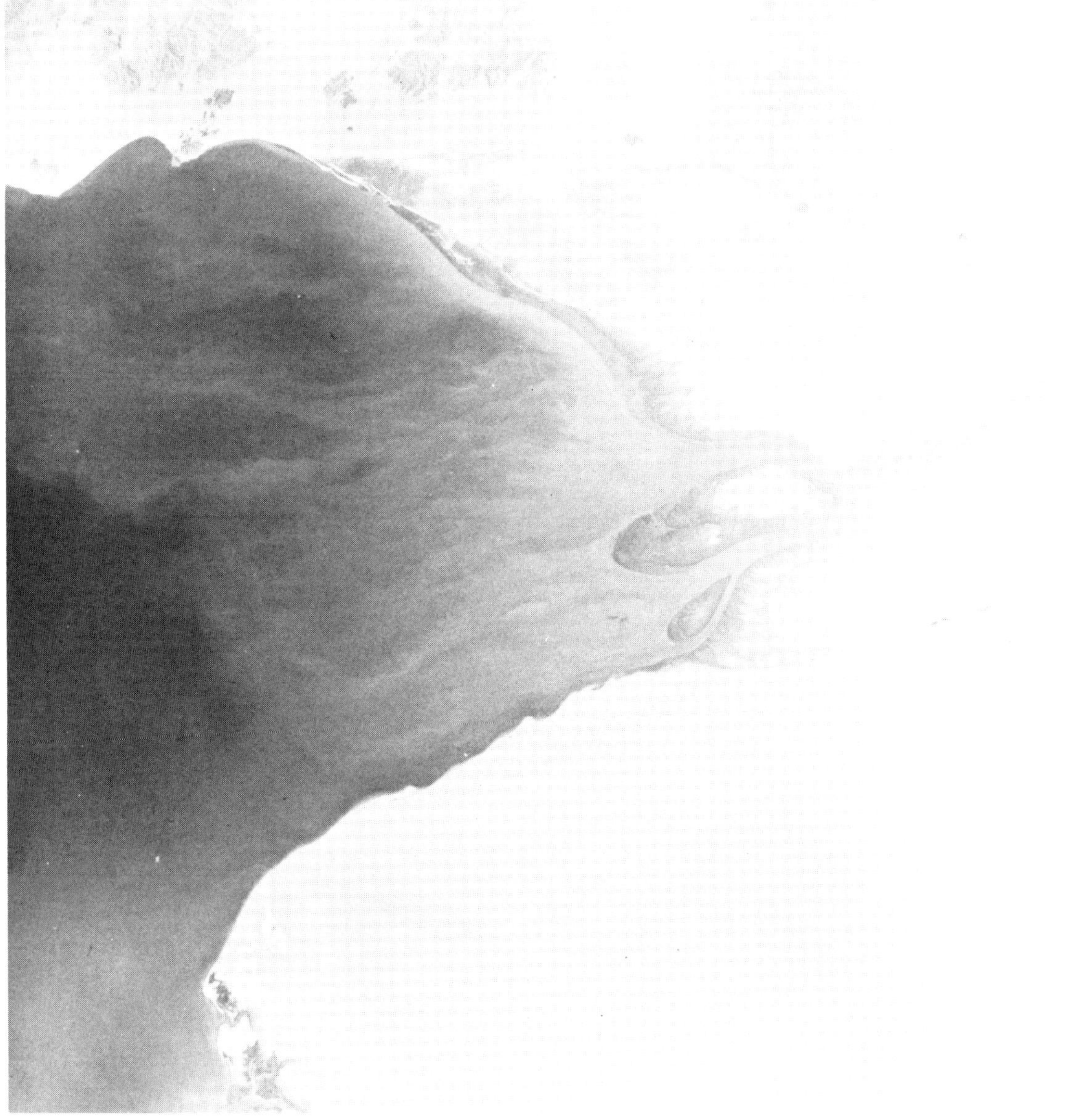

GULF OF CALIFORNIA. Scientists consider the Gulf of California to be related to the complex of faults and fractures along the western coast of North America, theoretically created by the process of continental drift. Here, the Colorado River pours into the Gulf of California. (*NASA*)

surface water ranges from 86°F (30°C) in summer to 60.8°F (16°C) in winter. The combination of temperature and the nutrient-laden water results in an abundance of sea life. Plankton, the basic link in the food chain, are prolific, the FISH population (including sport fish) is extraordinary, and oyster beds and sponge fisheries are to be found along the east coast.

GULF OF MEXICO, an arm of the western ATLANTIC OCEAN, is a near-circular body of water bordered by the United States, Mexico, and Cuba. The recognized boundary line on the south and southeast begins at Cape Catoche on the Yucatan Peninsula and runs across the Yucatan Passage to Cape San Antonio on Cuba's west coast. The line then runs through Cuba to the island of Dry Tortugas, Rebecca Shoal, and up the Florida Keys to the mainland at the eastern end of Florida Bay. The gulf connects with the CARIBBEAN SEA through the Yucatan Passage, and with the Atlantic Ocean through the Florida Strait. The Gulf of Mexico covers an area of 595 598 mi² (1 543 000 km²), occupies a volume of 559 447 mi³ (2 332 000 km³), and has a mean depth of 4960 ft (1512 m). The greatest depth is 12 700 ft (3871 km), which occurs in the Sigsbee Deep near the coast of Mexico.

The CONTINENTAL SHELF of the Gulf of Mexico is continuous around the entire rim of the gulf except for the Yucatan Passage and the Strait of Florida. The shelf is narrowest along its western rim (East Mexico Shelf) and broadest off the Yucatan Peninsula (Campeche Shelf), the coast of Texas and Louisiana (Texas-Louisiana Shelf), and off the west coast of Florida (West Florida Shelf). The sediment cover ranges from carbonate debris off Florida, to muds and silts off Mississippi and Louisiana (resulting from the Mississippi River), to sand and silts around much of the remaining shelf. The sediment off the coast of Texas rests in a depositional trough containing up to 50 000 ft (15 240 m) of sediment. The shelf on the northern and southern rim contains vast oil deposits.

The CONTINENTAL SLOPE is also continuous around the rim of the gulf and has one of the steepest known slopes as it plunges toward the deep basin. Beginning with a break in the shelf as shallow as 225 ft (69 m), the slope falls away at the rate of around 10 ft/mi (2 m/km), increasing to as much as 300 ft/mi (57 m/km). Near the contact between the continental slope and the CONTINENTAL RISE the bottom slope on the north, east, and west is steep enough (39°) to be called an escarpment (West Florida Escarpment, Sigsbee Escarpment, and Campeche Escarpment). The slope contains many irregularities which may have been caused by slumping of the sediment. From the base of the escarpments the continental rise merges smoothly with the ABYSSAL PLAIN—one of the smoothest in the world—that is often called the Sigsbee Deep.

Dominating the eastern portion of the gulf basin is the Mississippi Cone which extends downslope from the West Florida Escarpment to the Sigsbee Deep. This gigantic wedge of gray silt was undoubtedly deposited by the Mississippi River during a lower stand of sea level—probably during the last Ice Age. Today the Mississippi River, beginning as a small outflow from Lake Itasca in northcentral Minnesota, incorporates some 250 tributaries and drains an area of 1 250 000 mi² (3 236 259 km²). Each year it dumps 350 million (350×10^6) yd³ (27 million m³) of topsoil into its 15 000-mi² (38 835-km²) delta and into the Gulf of Mexico. During the retreat of the continental glacier that scoured as far south as Cargo, Illinois, the sediment available for transport by the Mississippi River system must have been far greater than today.

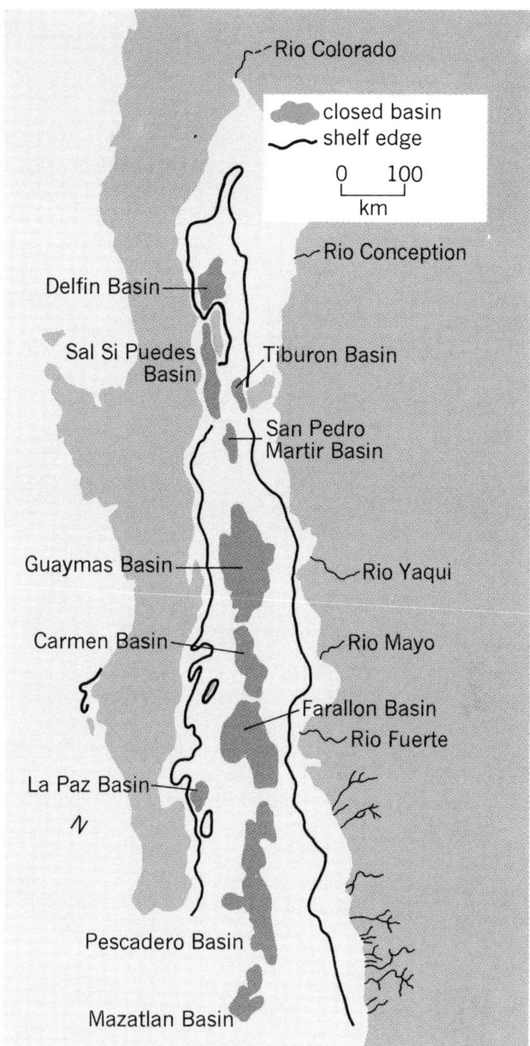

MAP 6. Gulf of California.

Turbidity currents have transported material from the Mississippi Cone out onto the flat floor of the abyssal plain which is interrupted only by a series of hills known as the Sigsbee Knolls, some of which rise more than 1200 ft (366 m) above the bottom. (See TURBIDITY CURRENT.) These knolls are all thought to be due to salt domes. Such domes occur where the weight of overlying sediment has forced a deep layer of salt to flow plastically upward to de-

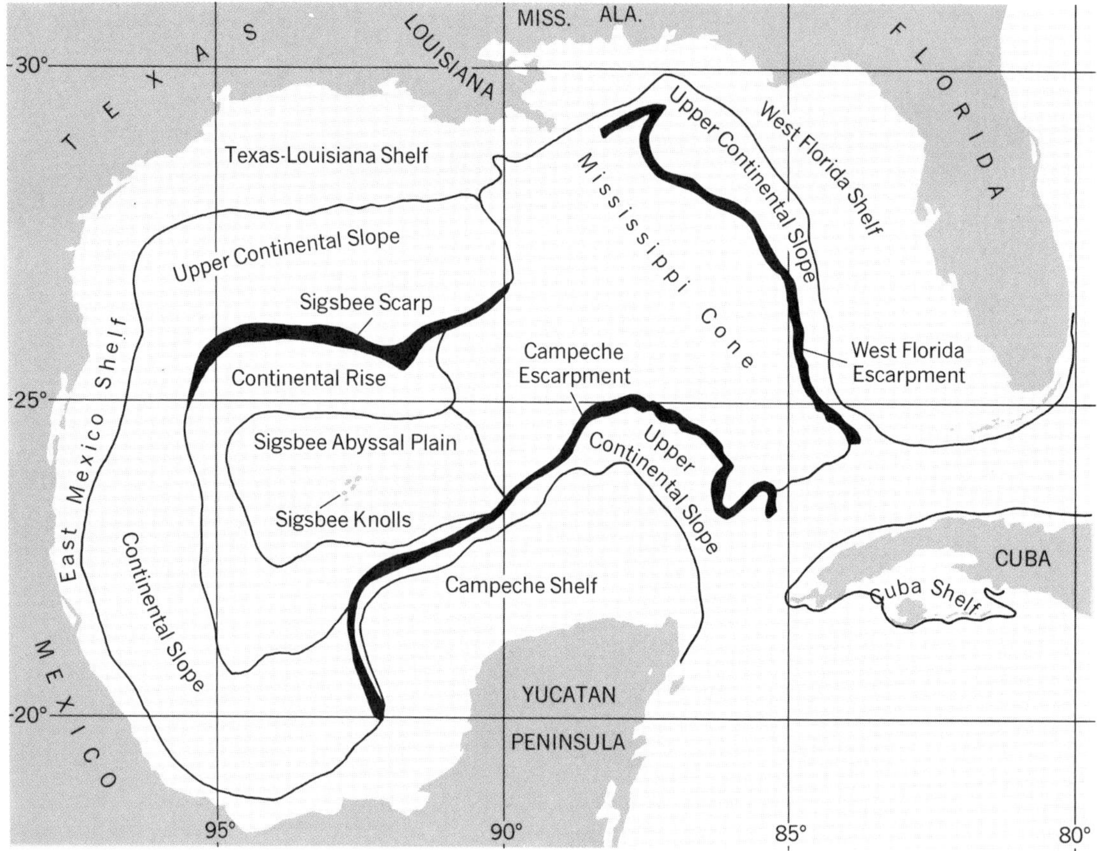

MAP 7. The Gulf of Mexico.

form the sediment. Such salt domes are common along the Texas-Louisiana coast and along the coast of Mexico. They are often the source of oil, gas, and sulfur.

The main WATER circulation in the Gulf of Mexico is produced by the Yucatan Passage. After entering the gulf, the current curves eastward to flow through the Florida Strait as the FLORIDA CURRENT, which later becomes the GULF STREAM. The Yucatan Current is made up of Atlantic water flowing into the Caribbean Sea through the several passages in the Greater and Lesser Antilles chains. The gulf experiences only one high and one low tide each day. The tide range is from 1 to 2 ft (0.3–0.6 m).

GULF STREAM, one of the mightiest of ocean currents, flows northward from Cape Hatteras to the Grand Banks off Newfoundland. It is generally considered to be a part of the Gulf Stream system which includes the FLORIDA CURRENT, which flows from the Yucatan Channel through the Florida Strait and northward to Cape Hatteras, and the NORTH ATLANTIC CURRENT, which continues across the North Atlantic to Europe. The ANTILLES CURRENT flowing north along the Greater Antilles adds its WATER to the system in the vicinity of the Bahamas, and considerable water flows from the SARGASSO SEA to the east. The result is a current that often attains speeds of 4–5 kn and transports as much as 3.99 billion (3.99×10^9) ft^3 [113 million (113×10^6) m^3] of water per second—greater than all the rivers of the world combined.

The core of the Gulf Stream, the jet, is about 50 mi (80 km) wide and appears to follow the contour of the CONTINENTAL RISE. As a result, there are many meanders in its course which may swing 40 mi (64 km) from the main current direction and not rejoin the main route for 200 mi (321 km). These meanders are not stable but are in a continuous state of evolution. Often the meanders become so curved

that they break away as independent eddies that persist for some time in the Sargasso Sea to the east. Since most of its water comes out of the tropical regions, the temperatures of the Gulf Stream are relatively warm, at 78.8° F (26° C) on the surface and around 50° F (10° C) at a depth of 800 m. The water inshore of the current is perhaps 10° colder, whereas that to the east is very similar to the current's temperature. The inshore boundary of the current is very sharp and can often be seen as it is approached from the west.

The Gulf Stream is the most extensively studied current in the world's oceans, but many questions are yet to be answered.

GULF OF MEXICO. The great bowl-like Gulf of Mexico is clearly depicted in this satellite photo of North and Central America. The northern and southern rims of the gulf's continental shelf contain rich oil deposits. *(NASA)*

GUYOT, ARNOLD (1807–1884), a Swiss geologist and geographer, was born in Switzerland. His boyhood days were spent at La Charix-de-Fonds, a village located at the foot of a narrow gorge. The mountainous terrain of the area inspired in Guyot a love of nature and inquiry.

At the age of 18, Guyot went to Germany to complete his education. However, in 1827 he returned

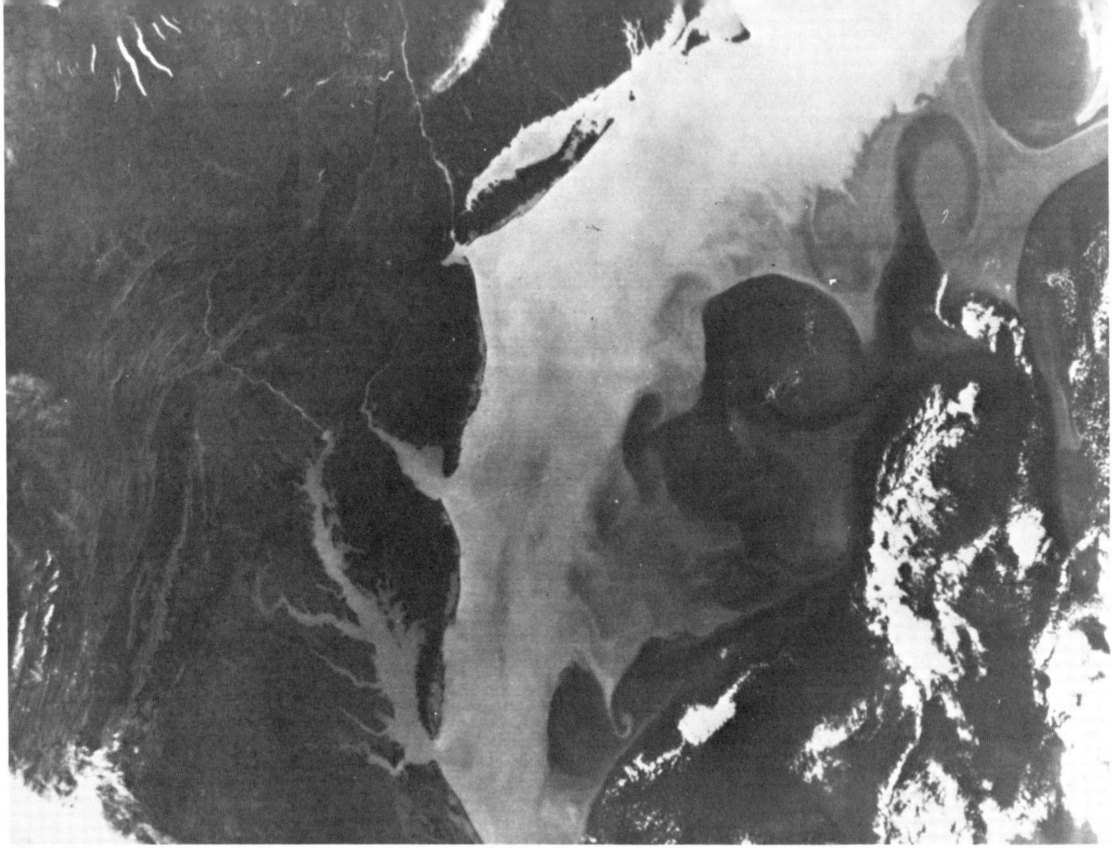

GULF STREAM. A NOAA satellite captured this very-high-resolution infrared image of the ocean adjacent to the Atlantic coast of the United States in May of 1977. To the left, the land area is dark, and the cold coastal water is much lighter. The dark-toned swirls in the ocean reveal the warm Gulf Stream. Note the irregular interface of the two waters of different temperature as the Gulf Stream meanders and sheds eddies. *(NOAA)*

to his homeland to study theology. In 1829, he went to Berlin to attend a series of lectures by historians and theologians at the University of Berlin. While there, he met several other professors on the scientific faculty and also made the acquaintance of Humboldt. (See HUMBOLDT, ALEXANDER VON.) These men, and especially the geographer Carl Ritter, made a strong impression on Guyot, and he abandoned theology for the natural sciences. In 1834, he graduated from the University of Berlin with the degree of doctor of philosophy. His graduating thesis was entitled, "The Natural Classification of Lakes."

After graduation, he went to France and became a tutor to the children of Count de Pourtales-Gorgier. In this job he visited the Pyrennes and traveled in Italy, Belgium, and Holland. In 1838, he was encouraged by his friend Agassiz to study the glaciers of the Alps. (See AGASSIZ, LOUIS.) He did so and made several important original discoveries regarding their structure and action, although his findings were not published until 40 years later.

In 1839, Guyot returned to his home in Switzerland where he joined the Society of Natural Sciences and accepted the chair of history and physical geography at the postgraduate school known as the "Academy." He spent the next 10 years in this position and also carried out extensive research of a meteorologic, barometric, orographic (relating to mountains), and glacialistic nature.

When a revolution in 1848 closed the Academy, Guyot was urged by Agassiz to come to the United States. He did so, and in the winter of 1848 he delivered a series of lectures in French at the Lowell Institute in Boston on "Comparative Physical Geography." These were translated into English and published under the title of *Earth and Man.*

For the next 6 years, Guyot worked for the Massachusetts Board of Education in lecturing to teachers on geography. He also prepared a series of geographies and maps for schools which were used extensively throughout the United States. In addition, he prepared instructions for meteorological observations for the Smithsonian Institution in 1850 and a volume of meterological and physical tables which was published in 1852.

Appointed professor of physical geography and geology at Princeton in 1854 (where he remained until his death in 1884), Guyot, in addition to his distinguished work at Princeton, also served for 30 years as a lecturer and consultant to the Smithsonian Institution. With the latter's strong encouragement, Guyot made barometric investigations of the mountain ranges of the eastern United States from the White Mountains of New Hampshire to the mountains of North Carolina. His barometric measurements of altitudes of high mountain peaks were remarkable for their exactness. Incidentally, the oceanographic term *guyot,* which refers to a flat-topped submarine mountain that rises to 0.5 mi (0.8 km) or more above the surrounding ocean floor, was originated in honor of Professor Guyot.

In 1863, when the National Academy of Sciences was established in the United States, Guyot was one of its original members or incorporators.

GUYOT is a flat-topped SEAMOUNT whose crest lies at least 600 ft (183 m) below the surface of the sea. The existence of such features was first discovered in 1946 by Harry H. Hess while serving as a naval officer in the Pacific during World War II. Hess named truncated seamounts in honor of Arnold Guyot, a nineteenth century Swiss geologist who served for 30 years (1854–1884) as a member of the faculty at Princeton University. (See GUYOT, ARNOLD.)

Guyots have been identified in every ocean except the ARCTIC OCEAN. Evidence indicates that these features are the relics of volcanic islands whose tops were cut to sea level by the eroding action of waves. The structure subsequently subsided due to a combination of structural adjustments both on the seafloor and in the guyot, and to the raising of sea level following the last Ice Age (11 000–8000 years ago). Dredging from the tops of guyots has recovered the basalt rubble and reef remains that would be expected from the eroded top of what was once an island. The tops are not truly flat but slope gently upward to a low pinnacle at the center.

By definition all guyots lie at least 600 ft (183 m) beneath the sea surface, and some have been noted as deep as 8200 ft (2500 m). Most lie between 3280 and 6560 ft (1000 and 2000 m). Only about 160 guyots have yet been identified in the world's oceans.

GYMBRETOXIN is a potent neurotoxin found at certain times in a species of the DINOFLAGELLATES, *Gymnodium brevis.* See RED TIDE.

HABITABLE ZONES are the various areas that marine organisms occupy in the ocean environment.

In the world's oceans, life is concentrated in about 4 percent of the total body of water. The basic reason for this is that sunlight is needed to permit PHOTOSYNTHESIS for the marine PLANKTON on which all FISH and WHALES directly and indirectly base their subsistence. In the tropics, the zone of photosynthesis extends down to a maximum depth of 260–330 ft (80–100 m). In the northern latitudes, this zone reaches no more than 50—66 ft (15–20 m) below the surface. (See LATITUDE.)

The two primary divisions of the marine environment are the benthic and the pelagic. The benthic is the environment of or near the ocean floor, while the pelagic division includes that of the water mass.

Benthic The benthic can be divided into the littoral (coastal) system, and the deep-sea system. The separation of these systems is based on distinct faunal (animal) and floral (plant) change connected with light penetration.

The animals of the benthic division are those which live on, in, or close to the ocean floor. They can be divided into two major groups: (1) the epifauna (those animals that live upon hard substrata such as rocks or pilings, or on other organisms), and (2) the infauna (those animals that inhabit soft ocean bottoms).

The infauna occupy something like 90 percent of the world's ocean bottoms and the epifauna reside in the remaining 10 percent. However, there are about four times as many epifaunal species as there are infaunal species. This is due to the more favorable factors for adaptation of life in the ocean regions nearer to the water surface. On the other hand, some forms of marine life, such as CLAMS, STARFISH, snails, and worms exist in both groups. However, many species and genera of these same animals are quite different and such differences distinguish their benthic grouping.

The littoral system is composed of that portion of the ocean environment from the shoreline to a depth of 600 ft (ca. 181 m). In this system, there are three zones: the supralittoral, littoral (intertidal), and sublittoral. The littoral (intertidal) zone includes the ocean floor area exposed between high and low tides and extends to the lowest depth at which the more abundant ocean plants can grow, approximately 40–60 ft (ca. 12–18 m). In the upper portion of the littoral zone and the lower portion of the supralittoral zone, the action of WAVES, wind, and TIDES make life difficult. Here, organisms must be firmly attached to the bottom, burrow into the mud, or have developed anatomical structures which enable them to survive periods of exposure to the air. Still others adapt by using tidal pools or rock caves as temporary homes. Owing to the great variations in type of substratum, character of shoreline, degree of exposure, salinity and TEMPERATURE of water, the littoral zone as a whole contains more habitable environments and associated living forms than any other zone in the benthic region.

The sublittoral one extends to a depth of the outer ledge of the continental shelf. Here, plant life is not as abundant as in the littoral zone, although great numbers of different species of organisms feed on each other and on the depositional materials from the overlying waters.

The ocean floor animal life of the littoral systems, in general, consists of three groups: (1) sessile (immobile) forms, (2) creeping forms, and (3) burrowing forms. Animals belonging to the sessile or immobile group include SPONGES, MUSSELS, limpets, chitons, adult BARNACLES, corals, tube worms, bryozoans, hydroids and anemones. (See CORAL; LIMPET; SEA ANEMONE.) Some animals of the creeping group are the sea urchins, CRUSTACEANS, flatworms, and brittle stars. (See SEA URCHIN.) Burrowing animals include many bivalves, snails, and worms.

In the deep-sea system, there are the bathyal, abyssal, and hadal zones. The bathyal zone extends from the sublittoral zone (600 ft or 180 m) to a depth between 3200 and 4000 ft (970–1212 m). In the bathyal zone, light is found in diminishing quantity and plant life is rare. However, the animal population may be abundant. Beneath the bathyal zone, or below 4000 ft (1200 m), lies the abyssal zone and beneath it is the hadal zone. These areas receive no sunlight and the water temperature is near freezing at all times. The animal population of these zones is less numerous than in the upper zones. Because plants that require sunlight cannot live in this environment, animals depending on plant food must live chiefly on the sinking organic material derived from near-surface plants. Animals of the lowest (hadal) zone are chiefly mollusks, GASTROPODS, hydroids, alcyonarians, special types of SPONGES, crinoids, and sea urchins.

Pelagic In the pelagic division of the ocean, which includes the entire water mass, are the neritic and oceanic systems or provinces. Distinguished from life on the ocean floor, the plants and epifauna of the world's oceans are found in the neritic province. This region includes all the water mass from the lowest tide line to the outer edge of the continental shelf. Thus, it corresponds to the littoral system of the ocean floor. In this environment, the greatest abundance of ocean life is to be found. Such life consists of NEKTON—the free-swimming animal forms such as squids, fishes, and WHALES—and PLANKTON—floating animal and plant forms (zooplankton and phytoplankton). DIATOMS are a major component of the phytoplankton. It can be said that practically all marine life depends on plankton or related life forms of organic materials. Of these, the phytoplankton, chiefly microscopic ALGAE which reproduce asexually and are extremely sensitive to temperature, must live in the upper 600 ft (181 m) or less because of their dependence on photosynthesis which, aided by enzyme action, produces sugar and free oxygen from carbon dioxide, water, and light. Phytoplankton are consumed by zooplankton, which, in turn, are consumed by the larger marine animals.

The depths from the outer continental shelf to the deepest parts are called the oceanic province; here the waters contain the ocean fauna and flora, i.e., many marine animals, vast numbers of fish, great masses of plankton and floating algae (e.g., sargassum) in the upper (or epipelagic) zone and vast numbers of plankton and fish in the middle (or mesopelagic) zone. Below this, in the deepest waters are BACTERIA, and some animals, as well as the remains of plants and DETRITUS.

HADAL pertains to the environment of the deep ocean trench zone and its fauna; these TRENCHES range up to 36,500 feet (11,000 m) in depth. The word *hadal* was first proposed by Anton Bruun, leader of the GALATHEA Expedition (1950–1952).

See also HABITABLE ZONES.

HADDOCK (*Melanogrammus aeglefinus*) is the name for an ocean fish belonging to the family Galidae in the order Gadiformes; it is related to the COD and very similar to it although somewhat smaller in size. The haddock is one of the more important commercial fishes and is found on both the North ATLANTIC OCEAN and in the ARCTIC OCEAN. It is sometimes called the finnan or finnan haddie.

HALIBUT See FLATFISH.

HALMAHERA SEA occupies an area of about 36 670 mi^2 (95 000 km^2) and is located in the western PACIFIC OCEAN between the islands of Halmahera and New Guinea. It is bordered on the northeast by the Pacific, and on the south by the CERAM SEA. Kau Bay, between the outstretched arms of Halmahera Island, lies on its northwestern rim. The seafloor is a complex of basins—the Halmahera Basin being the more pronounced—separated by ridges and covered by terrigenous and volcanic muds and globigerina ooze. (See MARINE SEDIMENTS.) Pacific WATER flows southward through the sea and into the Ceram Sea to the south. Surface currents depend on the prevailing winds and generally flow westerly during the summer and swing more to the south during the winter. Surface temperatures range from 84 to 79° F (29 to 26° C).

HALOBATES are members of the genus *Halobates*, the only known insects found in the world's oceans; there are 39 species of these unique insects found in the oceans, seas, and lagoons.

Halobates (from the Greek words *hal,* meaning "salt," and *bates,* meaning "walker") are insects that have been known to marine biologists and oceanographers since the early 1800s, although these marine organisms were not extensively studied until the 1960s and 1970s. In general, they are found in tropical and subtropical waters; their ranges extend from LATITUDE 40°N to 40°S in the open ATLANTIC OCEAN and the PACIFIC OCEAN, but is more restricted in the INDIAN OCEAN. Five species are found in these areas: *Halobates micans, H. sericeus, H. germanus, H. splendens,* and *H. sobrinus.* All occur in the Pacific Ocean; the first two are the only species found in the Indian Ocean; only *H. micans* inhabit the Atlantic. Since all live exclusively on the air-sea interface and cannot fly, their distribution is related to such ocean surface properties as salinity, TEMPERATURE, and CURRENTS. As adults, they are dark gray or black in color and about 1 in (25 mm) in body length; they have a leg spread of some $1\frac{1}{2}$ in (38 mm). The eggs are laid on any available piece of flotsam or jetsam in the ocean.

These active and agile insects possess a water-repellent coating of body hairs which assists them in retaining a supply of oxygen when they are submerged by waves or spray. They have the ability to "skate" at considerable speeds (approx 1–2 kn) on the surface and occur together in large numbers in some areas.

They feed primarily on PLANKTON. MARINE BIRDS appear to be the main documented predator of the halobates whose biology and role in the OCEAN FOOD WEB are not yet well understood.

HALOGENS are the elements fluorine (F), chlorine (Cl), bromine (Br), iodine (I), and astatine (At). The abundance of the halogens occurring in SEAWATER is estimated to be: chlorine = approximately 19 000 parts per million (ppm); bromine = approximately 65 ppm; fluorine = approximately 1.4 ppm; and iodine = approximately 0.05 ppm. (See PARTS PER THOUSAND.) Markedly higher concentrations of iodine and fluorine occur in rain water than in seawater, and the relative abundances of chlorine, bromine, fluorine, and iodine in the earth's crust (the lithosphere) are 550 ppm, 1.6 ppm, 770 ppm, and 0.3 ppm, respectively. Astatine, a radioactive element, occurs naturally only by the continuous decay of elements such as uranium 238 and thorium 232. The total amount of astatine present at any one time in the outer mile of the earth's crust has been approximated at 2.4×10^{-4} oz (6.86 mg).

HEEZEN, BRUCE CHARLES (1924–1977), an American marine geologist, made many notable contributions to a better understanding of the floor of the oceans, its topography, its sediment characteristics, and the processes by which it is formed and altered. Even as a student, Heezen took the position that a SUBMARINE CANYON results from the erosional effects of a TURBIDITY CURRENT and went on to argue persuasively that the many Atlantic Cable breaks occurring over a span of a few hours in 1929 were caused by an earthquake-triggered turbidity current. He was among those who first proposed that the mid-Atlantic Ridge (see CONTINENTAL DRIFT) is part of a worldwide system of ridges that run through every ocean and was one of the first to observe that a rift valley runs along the center of the ridge.

Heezen wrote extensively on sedimentary processes in the deep sea, on tectonics, seismicity, and on ocean-bottom currents as interpreted by bottom photographs. His book, *The Face of the Deep,* was nominated for a National Book Award in 1972. His maps of the topography and physiography of the ocean floor have been hailed as accomplishments with a profound effect on the development of oceanography and marine geology.

Heezen was born in Vinton, Iowa, on April 11, 1924. He graduated from the University of Iowa in 1948 and received his master's degree (1952) and doctorate (1957) from Columbia University. He joined the staff of the Lamont-Doherty Geological Observatory at Columbia University, where he became associated professor of geology in 1964.

HERRING is the common name for about 70 genera of FISH of the family Clupeidae (a major family which includes such species as herring, sardine, shad, and MENHADEN) in the order Clupeiformes. The herrings are one of the most common and abundant of the open ocean (pelagic) fishes and the most important of all food fishes in the world's oceans.

Herrings occur in all the world's oceans and flourish in a range of water temperatures from 43° F to 59° F (6–15° C). (See TEMPERATURE.) They are found in extensive compact schools, especially at spawning time which takes place close to shore in the winter months for some species and in deeper waters during the winter for other types. A female lays about 10 000 to 30 000 eggs at each spawning, and the eggs, which are covered by a very sticky substance, sink to the bottom as soon as they are laid. Here they are relatively immune from attack by predators as compared with floating eggs. When hatched, the larvae develop rapidly into baby fish. (See LARVA.) Maturity is reached in 4–5 years, and spawning takes place after 5 years. Some herrings live 20 years.

While they are the most primitive of the higher bony fish, herrings present a good example of the typical fish-form, with its foot-long torpedo-shaped (fusiform) body, single dorsal fin, pectoral FINS, single anal fin, forked tail, and pelvic fins. However, the fins have no supporting spines and are soft-rayed. Scales present on the body are blue-green to light brown on the back and silvery on the lower flanks and belly. They are deciduous and are readily

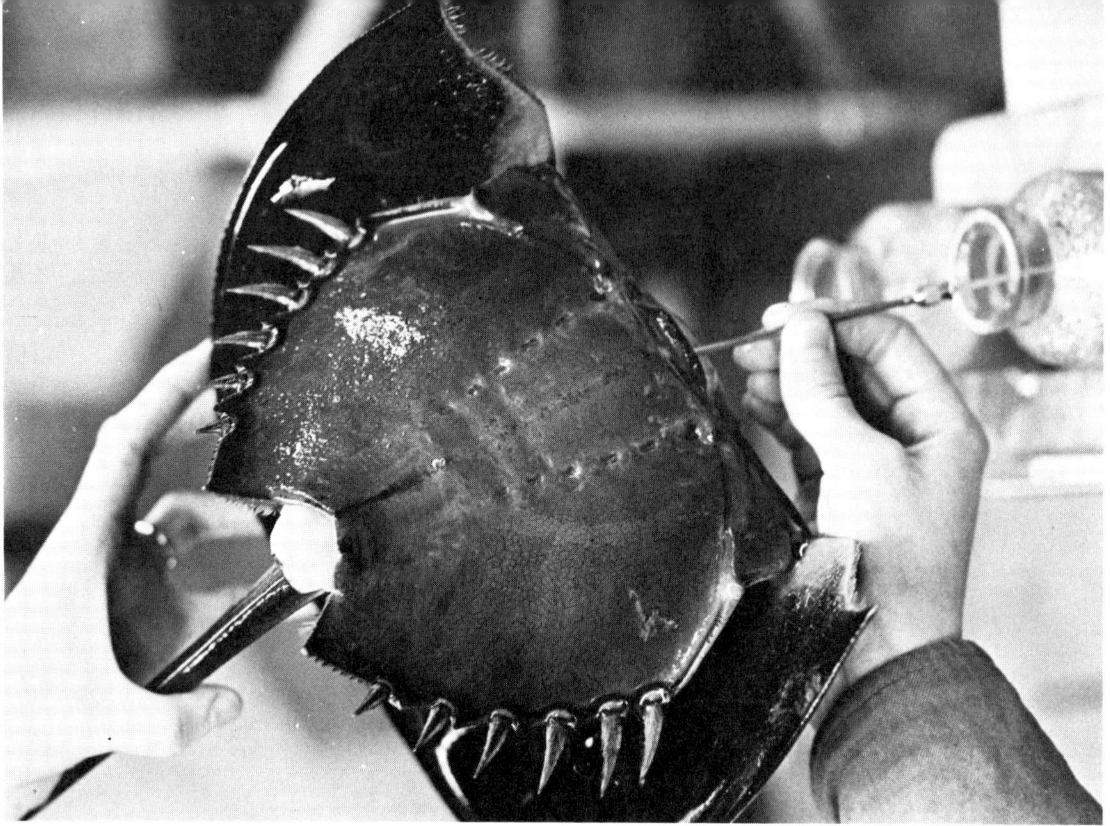

HORSESHOE CRAB. This marine animal (not a true crab) lives on the ocean bottom and feeds on a wide variety of both plant and animal life. The North American horseshoe crab is common in the coastal waters of the Atlantic Ocean off the eastern part of the United States. (*Woods Hole Oceanographic Institution*)

rubbed off. There are no scales on the head, and the SWIM BLADDER may be absent. The teeth are feeble and small or lacking. Most herrings feed upon PLANKTON. In the North ATLANTIC OCEAN herring especially favors a diet of the copepod *Calanus*. (See COPEPODS.)

The herring *Clupea harengus* is the most common herring of the North Atlantic, although it has a circumpolar distribution. About eight other species of this same genus exist (e.g., the sprat or brisling, *C. sprattus,* found especially in the Mediterranean, and the North Pacific herring, *C. pallasii*). The herring *Sardina pilchardus* (commonly called sardines or pilchards) is distributed in large schools mostly along the European coasts of the Atlantic, while *S. sagax* occurs along the coasts of Japan, Chile, California, and South Africa.

Shad, which are relatively deep-bodied ANADROMOUS fish of the herring family and members of the genus *Alosa* (e.g., the white shad or *Alosa sapidissima*) occur along the Atlantic seaboard of the United States. Spawning is carried out in rivers from spring to early summer. Some, such as the allis shad (*A. alosa*), are found along the coasts of Europe. They attain a length of about 30 in (76 cm) and a weight of 8 lb (3.6 kg).

For many centuries, the herrings have been a mainstay of commercial fishermen, who catch them in large drift nets which hang down in the water like a wall. While there is no great demand for fresh herrings they are very popular in other forms—smoked, pickled, salted, and kippered. Many of the young are processed in oil, canned, and sold as sardines.

HEXACTINELLIDA See SPONGES.

HIGH SEAS See OCEAN.

HIGHER HIGH TIDE See TIDES.

HIGHER LOW TIDE See TIDES.

HORSESHOE CRAB is the common name for arthropods making up the subclass Xyphosurida, especially the subgroup Limulida. See CRAB.

HUMBOLDT, ALEXANDER VON, (1769–1859) German scientist, was born in Berlin, Germany. He was educated at the Universities of Frankfort, Berlin, and Göttingen, as well as the Academy of Mining at Freiberg. After serving for 3 years (1792–1795) as the Director of Mines in Bayreuth and Anspach, Humboldt, perhaps inspired by his friend Georg Forster, who had served on James Cook's expeditions to the South PACIFIC OCEAN, planned a scientific expedition to South America. In 1799, accompanied by the French botanist Aimé Bonpland, he made a 5-year voyage (on a ship called the *Pizarro*) to Cuba, Mexico, and many places along the coast of South America. The findings of this scientific journey were published over a 10-year period in a 17-volume work titled, *The Travels of Humboldt and Bonpland in the Interior of America.* These books were divided into five general sections: Zoology and Comparative Anatomy; Geography and the Distribution of Plants; Political Essays and Descriptions of Peoples and Institutions in the Kingdom of New Spain; Astronomy and Magnetism; and Equinoctical Vegetation. In this coverage was a record of the first organized observations of the cold ocean current that flows along the west coast of South America (the current that has been called the HUMBOLDT CURRENT although modern oceanographers prefer the name PERU CURRENT).

Humboldt, between 1845 and 1858, made many scientific research journeys over Europe and Asia, and in this period he delivered some 61 lectures. The main subjects covered in these were the History of Science, the Nature of Volcanoes, the Crust of the Earth and Earthquakes, Physical Geography, Astronomy, Mountains, the Nature of the Oceans, the Distribution of Matter, the Atmosphere as an Elastic Fluid, the Geography of Animals, and the Races of Man. The lectures were published in Humboldt's *Cosmos,* an encyclopedic account and explanation of the physical universe.

These works have established Humboldt as one of the greatest scientific contributors of his time to physical geography, meteorology, and OCEANOGRAPHY.

HUMBOLDT CURRENT See PERU CURRENT.

HURRICANE is a tropical storm with a cyclonic (circular) wind pattern (counterclockwise in the Northern Hemisphere and clockwise in the Southern Hemisphere) which exceeds 74 mph (119 km/h). Often called the "Meteorological Monster of the Sea," hurricanes, because of their size, are the most destructive of all storms. Such disturbances have a diameter that ranges from less than 100 mi (161 km) to as much as 1000 mi (1609 km). Winds with speeds that range from 74 to 225 mph (119 to 362 km/h) revolve about an area of relative calm known as the "eye."

Hurricanes form in the tropical regions of the ATLANTIC OCEAN, PACIFIC OCEAN, and INDIAN OCEAN. They occur most frequently from June through October, infrequently in May and November, and rarely during the other months of the year. The southern part of the North Pacific east of the Philippines is the most prolific spawning ground in the world for hurricanes, having an average of 20 per year as compared with 7 for the North Atlantic, 7.5 for the North Indian, and 6 for the South Indian oceans. No hurricanes are known to occur east of about 140°W LONGITUDE in the South Pacific and South Atlantic.

The term *hurricane* is customarily used for storms of the above description which occur in the Atlantic Ocean. The same storm is called a *typhoon* in the Pacific, a *tropical cyclone* in the Indian Ocean, and a *willy-willy* in Australia. Hurricane comes from the Spanish word *huracán,* and the Spanish are thought to have adopted it from the Caribbean Indian tribes. For instance, the Mayans called their storm god *Hunraken,* whereas *Huraken* is the thunder and lightning god of the Quiché of Guatemala. Other tribes of the region used such words as *aracan, urican,* and *huiranvucan* for hurricanes.

The oldest technique for designating hurricanes to distinguish one from another was to use the latitude and longitude coordinates of the storm track. This proved to be rather cumbersome. For several hundred years the hurricanes of the Caribbean area were named for the saint's day on which they struck a populated area (e.g., the San Ciriaco hurricane of 1899 that wreaked great destruction on Puerto Rico). During World War II the military services began the practice of using the phonetic alphabet to designate typhoons in the Pacific (i.e., Able, Baker, Charlie, etc.). The first written use of a girl's name for a hurricane was in George R. Stewart's novel *Storm* published by Random House in 1941. Toward the end of World War II the U.S. Navy, Army, and Air Force began increasingly to use girl's names in preference to the already overworked phonetic alphabet. The practice is now widely accepted, and, in recent years, male names have also been added. Each year the U.S. Weather Bureau, the Navy, and Air Force weather services select an alphabetical list of names; the first storm of the season is given a name beginning with *A,* the second a name with *B,* etc. It is customary to use different names each year for about 10 years and then begin repeating. How-

HURRICANE

The winds of Hurricane Gladys were whirling at a speed of 80 knots when this photograph was taken by Apollo 7. The view is toward the southwest, with Cuba hidden in the background. The wind pattern is counterclockwise since the storm is in the northern hemisphere. *(NASA)*

Twin waterspouts touch down at sea. Seawater only rises a few feet up the spout; the remainder is caused by condensation from the rotating air. Waterspouts rarely cause damage, even when they pass directly over a ship. *(NOAA)*

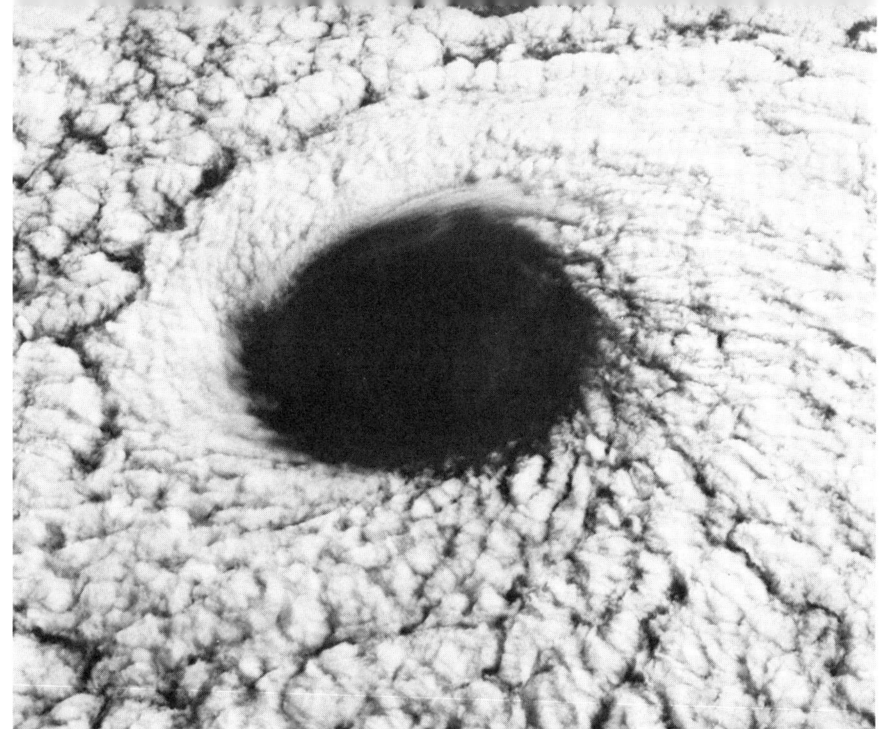

This aerial oblique shot shows an unusual cloud formation near Catalina Island off the California coast. The formation did not become a hurricane, but a well-developed "eye" common to all hurricanes is present. *(U.S. Navy)*

The leading edge roll cloud formation of a pre-frontal squall line moves in over Jacksonville, Florida. *(U.S. Navy)*

ever, the name of an exceptional hurricane is retired for a longer period of time.

Local usage has led to some confusion over such terms as *cyclone, anticyclone, tornado, waterspout,* and *whirlwind.* For instance, in the Midwest a tornado is often called a cyclone, and tornado and hurricane are often used interchangeably in other areas.

The word *cyclone* is from the Greek *kyklos,* meaning "coil of the snake," and was coined in about 1839 by Henry Piddington while he was investigating tropical storms in India. A cyclone is any system of rotating winds (except a tornado, waterspout, or whirlwind) in which the winds spiral inward and typically cover a circular or elliptical area having a diameter of 50 to more than 1000 mi (80–1600 km). Within this area the barometric pressure diminishes progressively toward the center, or eye, of the disturbance. Cyclones are classed as tropical or extratropical, depending on the characteristics of the surrounding air mass. A hurricane, then, is a tropical cyclone.

An anticyclone, as the name implies, is the reverse of a cyclone; i.e., an area of high pressure exists at the center, and the wind spirals out in all directions. The winds from an anticyclone are not of storm intensity and usually signify good weather due to relatively dry, descending air. The large, nearly stationary anticyclones, such as the Azores-Bermuda High of the North Atlantic, have a profound effect on global weather.

The tornado is characterized by a funnel-shaped cloud that may or may not reach the ground. Of those tornadoes in the United States which do reach the ground, the average path is 400 yd (366 m) wide and 16 mi (26 km) long. The winds, which turn in a counterclockwise direction, are estimated to blow at speeds of 450–500 mph (724–805 km/h), and the tornado speed over the ground ranges from 0 to 40 mph (64 km/h). Understandably, the tornado is considered the most violent storm in nature.

A waterspout is similar to a tornado, but forms over water and is less violent. Waterspouts have passed over ships without causing damage and have crossed the coastline with only minor damage. Over the ocean the salt water only rises a few feet in the funnel; the remainder is fresh water produced by condensation.

A whirlwind (dust devil) may be likened to a tornado in miniature, except that it most often occurs in clear weather and is only made visible by the dust it entrains. The whirlwind is a few inches to a few feet in diameter and rarely causes even minor damage.

Although Benjamin Franklin recognized in 1743 that storms move progressively from one spot to another, often in opposition to prevailing surface winds, and William C. Redfield first recognized the rotary nature of such disturbances in 1831, scientists are still not sure of the full details of hurricane formation. It is known that a number of meteorological and oceanographic parameters are required to generate and sustain a hurricane, but even when all these parameters are present, a storm does not always result. Fortunately, the process by which a hurricane is formed is extremely sensitive, and the disturbance begins committing suicide the day it is born.

The dominant feature in hemispheric weather is the existence of relatively stable high-pressure cells, or anticyclones, centered at about 30° north and south of the equator. In the North Atlantic this cell is called the *Azores-Bermuda High,* is centered southwest of the Azores, but extends from Spain to the southeastern United States. The High is bounded by wind patterns that are similar to the ocean circulation about the SARGASSO SEA; the prevailing westerlies on the north and the northeast and the east trade winds on the south are the more fully developed. The High is best developed in summer and reaches its peak in August. The position of the High is responsive to the sun's inclination and moves some 6–10°N during the summer. It also oscillates north-south in response to forces around it, but any persistent deviation from its normal position seems to contribute to the formation of hurricanes. The High extends to elevations of about 50 000 ft (15 240 m), and its axis is tilted about 5° to the south.

Between the high-pressure cells of the Northern and Southern Hemispheres lies the Intertropical Convergence Zone (ITC), or doldrums. It is along this line that the prevailing winds of the two hemispheres converge; the northeasterly and easterly winds of the Northern Hemisphere converge with the southeasterly and southwesterly winds of the Southern Hemisphere. The position of the ITC, also responsive to the sun's inclination, lies about 12°North of the equator in August and within about 1–5° of the equator in February. The ITC is an area of ascending air, cloudiness, and precipitation, with clouds often towering to 40 000 ft (12 192 m). It appears that many disturbances that later mature into hurricanes originate in this belt of moist, rising air.

The temperature of the ocean water is critical to the formation and intensification of hurricanes. The surface temperature of the tropical North Atlantic Ocean ranges from 77° F (25° C) off the coast of Africa to 85° F (29.4° C) in the GULF OF MEXICO in August. These temperatures drop to about 74° F (23.3° C) during the winter. As the dry, heavy air from the Azores-Bermuda High descends along its periphery, it is heated by compression and its relative humidity drops. Along the tropical boundary of the High this air moves for hundreds of miles across the tepid water surface. The air gradually assumes a temperature that is within about 1° F of that of the water and, because of high evaporation from the ocean, becomes heavily laden with moisture to a height of several thousand feet. This process provides the moisture and heat that is so essential to a hurricane: as soon as a hurricane moves into an area of colder water, its energy begins to diminish. There is some evidence that hurricanes will not form over water with a temperature of less than 79° F (26° C).

Above the moisture-laden layer of surface air, and separating it from the dry, descending air of the High, is a temperature inversion in which the air is warmer than that above or below it. This inversion layer sets in at an altitude of 5000–8000 ft (1524–2438 m) and acts to stabilize the layer below.

These appear to be the conditions which provide the basic ingredients for hurricane formation. What is now needed is a trigger which will disrupt the inversion layer and allow the moist surface air to rise. This might be provided by the westward progression of a low-pressure trough known as an *easterly wave* or by the injection of an unusually strong cell of polar air (Polar Trough). Once the inversion layer is disrupted, the moisture-laden surface layer begins to rise through the break and new air spirals in counterclockwise to replace it. If conditions are such that the air, after rising vertically through the vent, is dispersed with sufficient rapidity that it does not impede the air rising behind it, a hurricane has been born. Whether or not it matures and persists depends upon an adequate supply of energy.

A hurricane is actually a heat engine. As the warm, moist air rises, it expands, cools, and reaches its saturation point so that some of its water condenses. This condensation represents latent heat of condensation which, in turn, is added to the surrounding air causing it to rise faster. The now dryer air escaping at the top of the vent mixes with the surrounding air and begins to sink. As it sinks, it is again heated by compression and spirals back into the disturbance. This process results in a temperature gradient in which temperature increases to a hot core at the center of the disturbance. This gradient produces the kinetic energy necessary for high wind speeds.

Once a hurricane is born, it tends to move clockwise around the periphery of the high-pressure cell; first west, then north, and finally eastward (Northern Hemisphere). This storm track, so familiar to those along the Gulf and East Coast of the United States, is largely due to the CORIOLIS EFFECT. As an average storm matures, its hurricane-force winds will have a diameter of about 100 mi (160 km), and gale-force winds of 40–74 mph (64–119 km/h) will stretch across a diameter of 350–400 mi (563–644 km). The relatively calm eye, with wind speeds of 10–20 mph (16–32 km/h), will have an average diameter of 14 mi (23 km), although diameters as small as 4 mi (6.4 km) and as large as 25 mi (40 km) have been observed. Although the efficiency of the hurricane in converting heat to energy is low (approx 3 percent), it typically produces the equivalent of 16 trillion (16×10^{12}) kWh of energy each day of its life, as compared with the total of 2 billion (2×10^{9}) kWh of electricity produced each day by the entire United States. Said in another way, a hurricane with a diameter of 500 mi (805 km) will release latent heat energy at the rate of 10 trillion horsepower and will do so for as long as 10 days. Or, in more modern terms, an average hurricane will lift about 20 billion tons of water every 24 h—the equivalent of 50 000 atomic bombs of the Nagasaki type.

The average life of a hurricane is 9 days, although the average for August hurricanes is 12 days. Much longer periods, of course, have been recorded. For instance, the San Ciriaco hurricane of 1899 originated off Cape Verde on August 3, crossed Puerto Rico with terrible destruction on August 8, passed the Bahamas and was off Cape Hatteras on August 16 and 18, and was still being carried on weather maps in the vicinity of the Azores on September 6—a life of 5 weeks. In a like manner, some hurricanes are rejuvenated along the way and have extraordinarily long tracks. The Galveston hurricane of 1900 originated off the coast of Africa, crossed the Atlantic, entered the Gulf of Mexico, and went ashore at Galveston, Texas, killing some 6000 people. The storm was rejuvenated in the vicinity of the Great Lakes, recrossed the Atlantic, and died in Western Siberia.

The damage caused by hurricane winds and those of the tornadoes formed along their leading edge is, of course, enormous. However, the most severe damage and the greatest loss of life results from the storm surge (see WAVES) which precedes the leading edge of the hurricane. For instance, in 1737 a 41-ft (12.5-m) storm surge killed 300 000 people near Calcutta, India, and 50 000 in the same locality in 1864. The 6000 lives lost in the Galveston hurricane of 1900 were due to drowning, as were the 500 in Louisiana who died as a result of Hurricane Audrey in 1957.

The U.S. National Weather Service provides a hurricane warning service for the Gulf and East

Coast states and the islands of the Caribbean. Information used in detecting and tracking the disturbance is provided by ships, commercial and weather service aircraft, instrumented balloons, and radar. More recently the satellite has provided a valuable new dimension to hurricane detection and tracking. In April 1960, Tiros I, an experimental weather satellite equipped with television, ushered in this new era by detecting a fully developed typhoon 800 mi (1287 km) east of Brisbane, Australia.

The Weather Bureau issues its warnings in one of three categories; (1) *Hurricane watch* alerts the population that wind and water damage may result from a hurricane within 36 h. (2) *Gale warning* is issued to coastal states and maritime interests that winds of gale force [40–74 mph (64–119 km/h)] may be expected within 24 h. Frequently a gale warning will be issued for areas that are expected to lie along the periphery of a hurricane track. (3) *Hurricane warning* is issued to alert the population to expect winds of 55 mph (88 km/h) or higher within 24 h. Emergency action is justified.

HYDROGEN SULFIDE See SULFUR CYCLE.

HYDROGRAPHY is the measurement and description of the physical features of the oceans, seas, lakes, rivers, and their adjoining coastal areas, with particular reference to their use for navigational purposes.

HYDROIDS are small (usually a few millimeters in size) animals in the order Hydroida, the largest order of the class Hydrozoa in the phylum Cnidaria (Coelenterata).

The order Hydroida is comprised of the freshwater hydras, the attached (sessile) and generally colonial-type hydroids, and several of the smaller JELLYFISH.

In the typical life cycle of these animals (e.g., *Obelia geniculata*) the fertilized egg usually develops into a free-swimming ciliated LARVA, which, when attached to a substrate (viz., a shell), develops into a polyp with a mouth surrounded by tentacles at its free end. This polyp then produces stolons (runners), stems, and other polyps all of which are connected to make up a hydroid or hydroid colony. The stems of a colony are covered by a stiff, secreted substance, the perisarc or periderm, which supports and protects it. (Such outer integuments of the hydroids are also called hydranths.) Buds develop on the colony and are liberated as jellyfish or medusae which produce the sperm and eggs.

Most hydroids are found near the shore attached to various supports such as rocks, wharves, boats, MUSSELS, BARNACLES, CRAB and snail shells, and seaweed. See COELENTERATES.

HYDROLOGIC CYCLE See WATER CYCLE.

HYDROPHONE See UNDERWATER SOUND.

HYDROSTATIC PRESSURE is the force caused by the weight of a water column acting on a unit area; such a force, or pressure, is exerted uniformly in all directions. The atmospheric pressure at the surface of the ocean is equal to 14.7 pounds per square inch (psi), which is equivalent to the pressure exerted per square centimeter by a column of mercury 29.9 in (760 mm) high at a temperature of 32° F (0° C), where the acceleration of gravity is 980.665 cm/s².

With each descent of 33 ft (10 m) of depth in the ocean, the pressure increases by 1 atm. Pressures can be given in what is referred to as absolute pressure. That is, the atmospheric pressure or surface pressure is added to the depth or gage pressure. However, in most physical oceanographic work other than in SCUBA diving (see DIVING), the atmospheric pressure at the sea surface is usually neglected. Conversely, pressure is calculated as an essential function of water depth, and the pressure values are given in decibars (1 atm equals 14.7 psi, or slightly over 1 million dyn/cm², which is equal to 1 bar). A tenth of a bar, or decibar, is approximately equal to 1.47 psi of pressure, which is nearly the pressure exerted by the weight of 3.280840 ft (1 m) of water. Therefore, the ocean pressures are often given in decibars. For instance:

- Pressure at surface: 0
- Pressure at approx 3.28 ft (1 m): 1 decibar or approx 1.47 psi
- Pressure at approx 32.8 ft (10 m): 10 decibars or approx 14.7 psi
- Pressure at approx 65.6 ft (10 m): 20 decibars or approx 29.4 psi
- Pressure at approx 98.4 ft (30 m): 30 decibars or approx 44.1 psi
- Pressure at 3280 ft (1000 m): 1000 decibars or approx 1470 psi
- Pressure at 10 000 ft (3048 + m): 3048 decibars or approx 4480 psi

When depth is measured in fathoms [1 fathom = 6 ft (3.3 ft = 1 m)], the (gage) pressure at various depths is expressed in more familiar terms for the mariner.

As a ready rule of thumb for calculating these ocean pressures, one may use the figure of 0.448 psi increase for each foot of descent. Thus, for a depth of 32.8 ft, the figure 0.448 multiplied by 32.8 equals 14.7 psi of pressure at this depth.

It should be noted that all the figures given above are for SEAWATER having a DENSITY (weight per unit volume) of 64 lb/ft^3 (1.025 g/cm^3). However, the density of seawater is more precisely a function of its TEMPERATURE, SALINITY, and pressure; this density ranges from 63.6 to 67.4 lb/ft^3 (1.02 to 1.08 gm/cm^3). Water with a mean salinity, mean potential temperature, and located at a depth of 6224 ft (1897 m) (half the mean depth of the ocean) would have a density of 64.6 lb/ft^3 (1.036 gm/cm^3).

For a solid or liquid, the number assigned on the basis of the ratio of the density of the solid or liquid to the density of fresh water, which has a density of 62.4 lb/ft^3 (g/cm^3), is called its specific gravity. Thus, in most calculations of hydrostatic pressures in the ocean, a seawater specific gravity of 1.025 is used as a recognized standard. Since seawater is approximately 1.025 times as heavy as fresh water, it requires 34 ft (10.4 m) of descent in fresh water rather than 33 ft (10.1 m) to give a pressure of 1 atm or 14.7 psi. (See INSTRUMENTATION.)

Interestingly, human beings theoretically can withstand enormous hydrostatic pressures on their outer bodies. However, the effects of pressure together with the "inert" breathing gases (nitrogen and helium) that must be used at high underwater pressures are not yet completely understood by physiologists. Even less understood is the fatigue effect on the lungs that exists due to the increased breathing resistance that takes place when a diver breathes gas under pressure. (See DIVING.)

Many engineering materials also suffer fatigue under pressure. For example, when a submerged vessel is subjected to increased pressure, combined axial and radial loadings are created. These cause compressive and flexural stresses to develop within the material of construction. Such stresses, when severe, lead to two possible modes of failure. First, if the vessel is thin-walled relative to its diameter, is long, and is not rigidly reinforced by rings or bulkheads, it can fail by buckling. In this case, the vessel first loses its shape, and as even greater stresses are introduced, exceeding the strength of the material at the point of maximum deflection, a localized yield failure takes place. Second, if the vessel is thick-walled, short, and otherwise rigidly supported, the stresses will build up to a point where they exceed the strength of the material without instability deformation, and yield failure results.

ICEBERG See SEA ICE.

INDIAN OCEAN, the third largest of the world's oceans, is bracketed by Africa and Australia on the west and east and by Asia and Antarctica on the north and south. It differs from the ATLANTIC OCEAN and the PACIFIC OCEAN in that most of its WATER lies below the equator. Including its marginal seas the Indian Ocean covers an area of 31 498 372 mi² (81 602 000 km²), has a mean depth of 14 055 ft (4284 m), and occupies a volume of 83 872 000 mi³ (349 600 000 km³). The maximum depth of 24 442 ft (7450 m) is found in the Java Trench south of Java.

The official boundaries of the Indian Ocean are the following: on the north, the southern limits of the ARABIAN SEA, the LACCADIVE SEA, the BAY OF BENGAL, the East Indian Archipelago, and the GREAT AUSTRALIAN BIGHT; on the west, from Cape Agulhas to the Antarctic continent along the 20°E latitude line; on the east, from South East Cape, the southern point of Tasmania, down the meridian of 146°15′E, to the Antarctic continent; and on the south, the Antarctic continent.

The waters of the Indian Ocean wash some 35 independent nations and one continent uninhabited by indigenous peoples. Some 12 marginal seas and identifiable water bodies interface with the ocean (clockwise from the southern tip of Africa): the Mozambique Channel between Mozambique and the offshore island of Madagascar; the RED SEA and the Gulf of Aden between Africa and Saudi Arabia; the PERSIAN GULF and the Gulf of Oman between Saudi Arabia and Iran; the Arabian Sea lying off the curve formed by Africa, Saudi Arabia, Pakistan, and India; the Laccadive Sea off the southeastern tip of India; the Bay of Bengal off the east coast of India; the ANDAMAN SEA between the Andaman and Nicobar islands and the Malay Peninsula; the SAVU SEA off the Lesser Sunda Islands; the Great Australian Bight off the southern coast of Australia; and the Davis Sea off Antarctica.

The major rivers flowing into the Indian Ocean are the Limpopo and Zambezi of Africa, and the Irrawaddy, Brahmaputra, Ganges, Indus, and Shat-el-Arab of Asia. Numerous volcanic islands and several large islands that are obviously continental fragments are found in the ocean. Madagascar off the east coast of Africa, parts of the Seychelles to the north of Madagascar, Socotra off the Gulf of Aden, and Sri Lanka off the southeastern tip of India are granitic in composition and thought to be continental fragments. Other islands include Mauritius, Réunion, and Rodrigues to the east of Madagascar, and Prince Edward, Crozet, Amsterdam, St. Paul, and the Kerguelen islands in the southern part of the ocean.

Early Exploration The Indian Ocean was the first of the major oceans to serve as a trade route, but until the fifteenth century it was the exclusive province of the Egyptians and Greeks. About 600 B.C., according to Herodotus, the Greek "father of history," King Necho of Egypt sent an expedition of Phoenician sailors out into the Indian Ocean through the Red Sea and along the east coast of Africa. According to the account the ship returned 3 years later through the Straits of Gibraltar and the MEDITERRANEAN SEA—a remarkable voyage for that period. The expedition of Alexander the Great to the Indus River in Pakistan added considerably to the knowledge of such peripheral waters as the Persian Gulf, the Gulf of Oman, and the Arabian Sea. By 150 A.D. Ptolemy was able to construct a map which showed the Indian Ocean as an enclosed sea stretching from Africa to China and bordered in the

INDIAN OCEAN. India juts forward between the Arabian Sea and the Bay of Bengal in this satellite photograph, with Ceylon trailing off to the right. (*NASA*)

south by an unknown land. (A legend from Greek antiquity held that a vast continent existed in the Southern Hemisphere.)

The trade routes across the northern part of the Indian Ocean were adversely affected by the fall of the Roman Empire in 476 A.D., but the knowledge was retained by the Egyptians and recovered by the Europeans during the Crusades (1096–1300). Marco Polo returned from his 24 years in China through the Malacca Strait in 1292. In 1497 Vasco da Gama sailed down the west coast of Africa, rounded the Cape of Good Hope, and sailed into the Indian Ocean to open a new era of trade and exploration.

Over the 300-year period between 1500 and 1800 several notable voyages extended the knowledge of the Indian Ocean. Del Cano crossed the central part of the Ocean in the *Victoria* in 1521. Sir Francis Drake, in 1580, sailed westward through the Indian Ocean to complete his 3-year voyage around the world aboard the *Golden Hind*. Abel Tasman, in 1642, circumnavigated Australia and discovered Tasmania. And James Cook, in 1772 and in 1776, became the first to cross the southern Indian Ocean in his exploration of the Antarctic Continent. (See COOK, JAMES.)

In spite of this unprecedented early history of

trade and exploration, the Indian Ocean remained until recently the least explored of the three major oceans. The establishment of the Pan-Indian Ocean Science Association in the early 1950s and the International Indian Ocean Expedition of 1960–1965 has added significantly to our knowledge of the ocean, its climate, its waters, and its basin.

Climate In general the winds in the Indian Ocean follow the global pattern of air circulation, with the major exception of the monsoons which dominate the weather north of the equator. The controlling factor in the overall wind pattern is the existence of semipermanent subtropical high-pressure cells in both the Northern and Southern Hemisphere at about 30° latitude. Air moves outward from these cells toward a semipermanent low-pressure belt known as the doldrums or the intertropical convergence zone in the vicinity of the equator; air moves in the opposite direction toward low-pressure zones known as the subpolar lows at about 60° latitude. As the air moves toward the equator, it is deflected to the right in the Northern Hemisphere and to the left in the Southern Hemisphere by the CORIOLIS EFFECT. As a result, the wind approaching the equator from the north and south is induced to flow from east to west. The high- and low-pressure cells move northward and southward with the seasonal march of the sun.

During the summer in the Northern Hemisphere the doldrum belt moves northward and extends from the southern part of the Arabian Peninsula across Pakistan and northern India. In this position the low-pressure trough is deepened by intense heating from the land. The air flowing down from the high-pressure cell to the north is northerly over the Red Sea and Persian Gulf, while that flowing up from the southern high-pressure cell is deflected after crossing the equator and becomes the strong, steady southwest monsoon that blows over the Arabian Sea and the Bay of Bengal. During the long trip across the Indian Ocean the air picks up moisture which results in the prolonged torrential rains for which the summer monsoon is famous. Typically, some 14 in (360 mm) of rain falls on Amini in June alone, and even more falls on the windward slopes of Sri Lanka. Air temperature during the season is around 86–95° F (30–35° C) in the daytime and 75.2–80.6° F (24–27° C) at night. The southeast monsoon lasts from May to September.

During the winter in the Northern Hemisphere the doldrum belt shifts to a position 10–15° south of the equator. The air flowing toward the doldrum belt from the north comes from a great high-pressure cell over Siberia. As the air flows toward the Indian Ocean, it is deflected and slowed by the lofty Himalayan Mountains. The result is a dry, subsiding air mass that forms the relatively weak northeast monsoon that brings very little rain to the area between November and January.

The doldrum belt—called the intertropical convergence zone by scientists—is an ill-defined zone of calm to light winds, ascending air, and low pressure which is best defined during the period from December to March. The inflow of air from the high-pressure cells brings moist, unstable air which results in showers and occasional thunderstorms. Due to the enormous inflow of moist air from the high-pressure cells, the doldrum belt is one of the rainiest zones on earth. For instance, from 90 to 196 in (2285 to 4290 mm) of rain falls on the Seychelles and the Chagos Archipelago each year. The lowest rainfall in the doldrum belt is 35–48 in (890–1225 mm) in the vicinity of the island of Rodrigues.

The horse latitudes to the south are centered at about 35°S in January and 30°S in July. This is a belt of relatively warm, dry, subsiding air and high pressure. Mild, fair weather persists year-round. During the early days of sail, ships were often becalmed in the horse latitudes for long periods of time. If horses were being transported, they were thrown overboard to conserve water—hence the name.

South of the horse latitudes are the westerlies—the so-called roaring forties and furious fifties—which continually circle the Antarctic continent. With no land to obstruct them, these winds blow with a yearly average of 20–25 kn, and the winter maximum often exceeds 70 kn. The interaction of the cold Antarctic air with the warm air from the southern border of the horse latitudes often produces severe storms. Precipitation ranges from 30 to 55 in (750 to 1400 mm) annually, and the temperature in the westerly belt ranges from a daily high of 104° F (40° C) in Amsterdam Island to 68° F (20° C) at Heard, some 994 mi (1600 km) to the south.

Origin The Indian Ocean basin was formed by the ongoing process of CONTINENTAL DRIFT which began with the breakup of the massive continent of Gondwana at an estimated 180 million years ago. As late as 75 million years ago both India and Australia were joined with Antarctica in a large continent that lay off the southeastern coast of Africa. Then, responding to forces still not fully understood, India broke away and "sailed" northeast and then north for 3100 mi (5000 km) at an average rate of 3 in (7.5 cm) per year. About 45 million years ago India collided with the continent of Asia. This great collision crumpled the thick sediments of the Asian continental shelf and forced them up into what are now the lofty Himalayan Mountains.

The northward movement of the subcontinent of India was made possible by two lateral cracks in the seafloor, which allowed the Indian plate to slide past the plates on either side. These transform faults are now represented by the Ninety East Ridge which runs south from the Bay of Bengal and the Owen Fracture Zone which runs south from the western Arabian Sea.

Forty-five million years ago, at the time India was colliding with Asia, Australia broke free of Antarctica and moved northeastward to its present position, while Antarctica moved southward. About 20 million years ago, Arabia broke away from Africa and moved eastward to form the Red Sea and the Gulf of Aden.

As a result of these several processes the Indian Ocean is a comparatively young ocean, and its basin is one of the most complex in the world.

Basin The topography of the floor of the Indian Ocean is dominated by four north-south trending ridges. By far the more massive of these ridges is the mid-Indian Ocean Ridge which joins the Antarctic, African, and Indian plates and is part of the 40 000-mi (64 000-km) long mid-Oceanic Ridge that circles the globe. The mid-Indian Ocean Ridge is spreading out from its center (i.e., adding new seafloor) at a rate of 4 in (10 cm) per year. To the north the Ridge turns westward to enter the Red Sea where its spreading widens that sea year by year. To the south the Ridge forks to form the Southwest Indian Ocean Ridge and the Southeast Indian Ocean Ridge. Most of the earthquake activity in the Indian Ocean occurs along the mid-Indian Ocean Ridge and along the Java Trench to the east.

Just to the east of the northern half of the mid-Indian Ocean Ridge is the Chagos Laccadive Plateau which terminates in the Arabian Sea off the west coast of India. In places this plateau rises above the water surface to form the Maldive and Laccadive islands. Further to the east is the remarkably straight, 3100-mi (5000 km) long Ninety East Ridge which represents the now stable transform fault produced by the northward migration of the Indian subcontinent. The Ninety East Ridge towers from 8900 to 12 000 ft (2712 to 3658 m) above the surrounding seafloor and lies from 3000 to 7000 ft (914 to 2134 m) below the water surface. At the southern extremity of the Ridge the Diamantina Fracture Zone, which produces the Ob Trench, runs eastward to Australia. To the north of the Diamantina Fracture Zone is Broken Ridge which is thought to be a submerged continental fragment. And, further to the north, along the coast of Sumatra and Java, lies the Java Trench which plunges to a depth of 24 437 ft (7450 m).

To the west of the mid-Indian Ocean Ridge is a complex area made up of the Owen Fracture Zone in the north and several islands and submerged plateaus which are thought to be continental fragments. Notable among these is the large island of Madagascar and the submerged Madagascar Plateau to the south. The arcuate Seychelles-Mauritius Plateau lies between the Owen Fracture Zone, Madagascar, and the mid-Indian Ocean Ridge. Further to the south lie the Mozambique Plateau and the Agulhas Plateau.

The ridges and plateaus divide the Indian Ocean basin into a series of basins which range from 186 to 5591 mi (300 to 9000 km) in width, and into smooth abyssal plains—among the smoothest in the world—which occur at depths of 9842–19 685 ft (3000–6000 m). In the north, the Indus and Ganges rivers on either side of India form conspicuous alluvial fans on the seafloor; these are known as the Indus Cone and the Ganges Cone, respectively.

The continental shelves of the Indian Ocean are somewhat narrower than those of the Atlantic Ocean. (See CONTINENTAL SHELF.) They range in width from a few hundred meters off some islands to 124 mi (200 km) off Bombay, India. The break in the shelf occurs at an average depth of 460 ft (140 m). Numerous submarine canyons slice through the continental shelves, with the most notable being off the Indus and Ganges rivers. (See SUBMARINE CANYON.)

The floor of the Indian Ocean is covered primarily by pelagic sediments (see MARINE SEDIMENTS). Approximately 25 percent of the seafloor is covered by red clay and about 54 percent by calcareous oozes derived from the shells of animals of the globigerina type. Siliceous oozes from the shells of DIATOMS cover about 20 percent of the seafloor below 50°S latitude. The remaining sediment is derived from and occurs near land.

Water and Its Circulation Surface water circulation in the Indian Ocean, as in other oceans, is largely determined by the prevailing wind pattern. Unique to the Indian Ocean, however, is the influence of the monsoons which cause the equatorial circulation to reverse with the seasons. The Northeast Monsoon Drift Current flows generally east to west from November to April. Off the coast of Somalia the current turns south to cross the equator and then turns eastward to form the Equatorial Countercurrent (See EQUATORIAL CURRENT SYS-

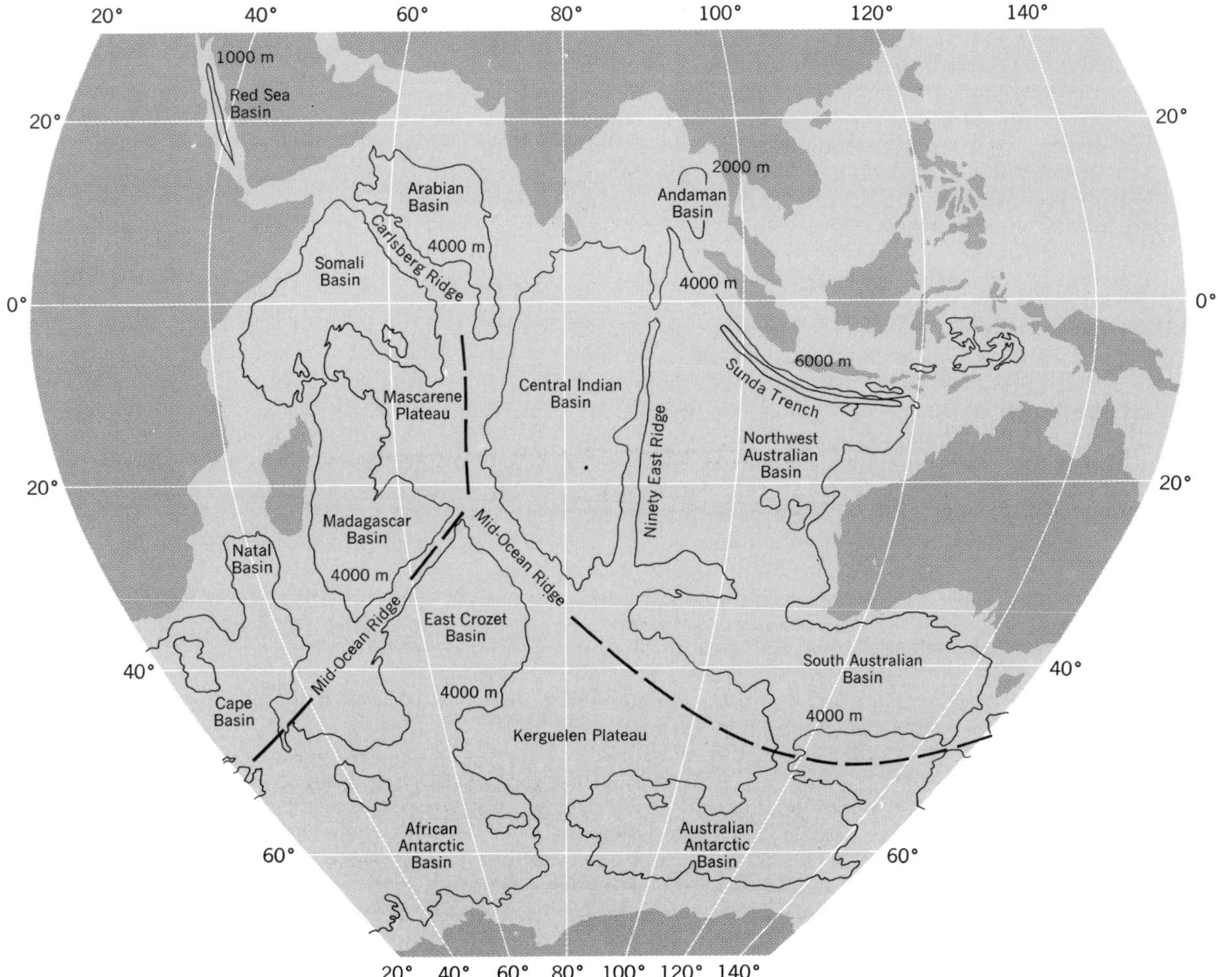

MAP 8. The Indian Ocean.

TEM). With the onset of the southwest monsoon in April the circulation reverses directions to form the Southwest Monsoon Drift. The Somali Current begins to flow northward, and by May the principal currents north of the equator start to flow in an easterly direction. (See CURRENTS.) By July the monsoon current is fully developed and the Equatorial Countercurrent has shifted north of the equator to join the eastward flow.

South of 10°S latitude the circulation is generally constant throughout the year and forms a counterclockwise gyre about the southern Indian Ocean basin. The Mozambique Current joins the AGULHAS CURRENT and flows southward along the east coast of Africa where it joins the eastward-flowing WEST WIND DRIFT CURRENT off the coast of Antarctica. In the southern summer the West Wind Drift, which is 124–149 mi (200–240 km) wide and flows at a rate of about 18 in/s (45 cm/s), turns northward before reaching Australia and is joined by a current flowing in from the Pacific south of Australia. In the winter the West Wind Drift is joined by a southward-flowing current along the west coast of Australia and continues on into the Pacific. The West Australian Current is the eastern leg of the gyre and flows steadily northward during the summer. During the winter it becomes southerly and very much weakened. The South Equatorial Current (see EQUATORIAL CURRENT SYSTEM) represents the northern leg of the gyre and is the westward continuation of the West Australian Current under the influence of the southeast trade winds in the vicinity of the Tropic of Capricorn. After reaching Madagascar, the current splits; the northern arm splits again, with one branch turning westward and then south to form the Mozambique Current, and the other turning eastward. The southern branch joins the Mozambique Current south of Madagascar to form the Agulhas Current.

Extensive upwelling occurs along the Arabian and Somali coasts when the wind shifts the relatively sterile surface water further out to sea. This cold, deep water carries 10–20 times the nutrients carried by the water it replaces. This rich water produces spectacular PLANKTON blooms which make ideal feeding grounds for fishes. Occasionally the plankton population seems to explode with disastrous results. The oxygen depletion, the toxicity, and the fouling of FISH gills causes great ecological damage. In 1957 a plankton bloom in the Arabian Sea was estimated to have killed the equivalent of the entire world catch of fish for a year.

The salinity of the Indian Ocean water varies from 32 to 37 parts per thousand (ppt). The waters of the Red Sea and the Persian Gulf are considerably more saline than the average because of excessive evaporation. The Bay of Bengal and the area between Sumatra and Australia, on the other hand, tend to be less saline than the average because of high precipitation and the inflow of lower-salinity water from the Pacific. The same is true of the water off Antarctica because of melting ice.

Ice formation occurs in the Southern Hemisphere winter around Antarctica. During the summer this ice, along with glacial ice from the continent, is carried out to sea by winds and currents. Some ice reaches as far north as 40°S latitude where it becomes a navigational hazard.

Food Resources The annual fishing harvest of the Indian Ocean is about 2.5 million tons (2.3 million metric tons). Ninety percent of the catch is by small nonpowered, family-owned boats belonging to the people of the littoral states. Very little processing of this catch is attempted: half is marketed fresh from the boat, about one-third is cured by simple means, about 5 percent is frozen, another 5 percent is processed as fishmeal, 1 percent is canned, and the remainder is used in other ways.

The remaining 10 percent of the annual catch is by the modern fishing fleets of the U.S.S.R., Japan, Taiwan, and South Korea. These fleets seek mostly the high-value open-ocean fish such as TUNA and the BILLFISH. In 1973 the South Korean fleet took 53 400 tons (48 400 metric tons), the U.S.S.R. 44 200 tons (40 100 metric tons), Japan 33 300 tons (30 200 metric tons), and Taiwan 23 400 tons (21 200 metric tons). The much sought after tunas such as yellowfin, bigeye, bluefin, and albacore, and the billfishes are presently being overfished. Nearly all such fishes are being taken by the longline technique, consisting of a surface line several kilometers long to which numerous lines with baited hooks are attached.

The smaller tuna—skipjack—is presently being harvested in the vicinity of the Maldives and Sri Lanka at the rate of 50 000–70 000 tons (45 400–63 500 metric tons) per year. About 1.5 million tons (1.4 million metric tons) of bottom fish such as croakers, snappers, groupers, skates, and grunts are harvested annually by local fishermen. (See GROUPER; SNAPPER.) Out of a potential harvest of 6–7 million tons (5.4–6.4 million metric tons) annually, only about 800 000 tons (726 000 metric tons) of sardines, anchovies, and MACKEREL are taken by local fishermen. Sharks are also underutilized. (See SHARK.) A potential harvest of 400 000 tons (363 000 metric tons) for use as both food and leather has been estimated. India, Pakistan, Indonesia, and Thailand account for most of the estimated 265 000 tons (240 000 metric tons) annual harvest of SHRIMP.

Improvement in the local fishing industry of the Indian Ocean is badly needed to reduce the estimated 3 million tons (2.7 million metric tons) annual protein deficiency of the peoples of that area.

Mineral and Petroleum Resources About 40 percent of the world's offshore oil production comes from the Indian Ocean area, and the Persian Gulf fields continue to lead the world in production. Considerable reserves of natural gas are also present, but in 1975 only 1 percent of the world production came from this area. Extensive exploration for additional offshore fields is being conducted in the Gulf of Kutch, Gulf of Kambat, Bay of Bengal, Andaman Sea, the northwestern continental shelf of Australia, Mauritius, and the South African continental shelf.

MANGANESE NODULES containing manganese (14.7 percent), iron (13.5 percent), nickel (0.42 percent), cobalt (0.25 percent), and copper (0.22 percent) have been found at depths of 11 483–16 400 ft (3500–5000 m). However, no economical method of recovery has yet been demonstrated.

In 1947–1948 the Swedish research vessel *Albatross* discovered hot brines near the bottom in the Red Sea. Later investigations indicated that the bottom 65–130 ft (20–40 m) of seawater is being heated to around 133° F (56° C) by the hot rocks that continually well up in the Red Sea extension of the mid-Indian Ocean Ridge. The heating of the water causes a concentration and deposition of such minerals as manganese, iron, zinc, and lead. Cores indicate that a rich deposit of mineral-bearing muds nearly 100 ft (30 m) thick exists in some areas. But again, no economic method of recovery has been developed and the ownership is being contested by Saudi Arabia and Sudan.

INFRAGRAVITY WAVE See WAVES.

INLAND SEA is defined as a body of water that is surrounded by land except for a narrow STRAIT by which it is connected to an OCEAN or another SEA. Excellent examples of inland seas are the shallow Sea of Azov which flows through Kerch Strait into the BLACK SEA, which flows through the Dardanelles into the MEDITERRANEAN SEA, which, in turn, flows into the ATLANTIC OCEAN through the Strait of Gibraltar. Another example is the BALTIC SEA which flows into the NORTH SEA through the Skagerrak and Kattegat.

Many "inland seas" are actually lakes since they do not connect with a sea or ocean by way of a strait; instead, their waters are derived from runoff from the surrounding land, and their outflow, if any, is by way of a stream. The best example of such lakes is the Caspian Sea—the largest inland body of water in the world—which lies below SEA LEVEL in the Caucasus Mountains of southwestern Russia and northern Iran. Roughly the size of the state of California, the Caspian covers an area of 163 500 mi² (421 800 km²), is 750 mi (1200 km) long, and ranges from 130 to 300 mi (208 to 480 km) in width. Lying some 92 ft (30 m) below sea level, the Caspian is fed by such rivers as the Volga, Kura, Terek, and Ural of Russia. Because, in part, some of the water from these rivers is being diverted to irrigation projects in the Caspian Basin, the inflow of water no longer equals that lost to evaporation, and the level of the Caspian Sea is dropping. There is no outlet for the waters of the Caspian, except through evaporation, and the water is salty, though less so than typical ocean water (see SALINITY).

Approximately 175 mi (280 km) east of the Caspian Sea lies another large saltwater lake, the Aral Sea. Covering an area of 24 630 mi² (63 545 km²), the Aral is fed by the Amu Darya and Syr Darya. Like the Caspian, it has no outlet.

In the United States an example of an "inland sea" is the saltwater lake of southeastern California known as the Salton Sea. Lying between the Chocolate and Santa Rosa Mountains, and some 240 ft (80 m) below sea level, the Salton has a length of 30 mi (48 km) and measures 8 to 10 mi (13 to 16 km) in width. The Salton Sea was formed between 1905 and 1907 when the Colorado River broke through irrigation head gates and flowed northward to fill the Salton Basin. Since the repetition of such floods was rendered unlikely by the construction of the Hoover Dam in 1936 and since it receives little water from its arid drainage basin, it is doubtful that the Salton Sea can survive.

INSTRUMENTATION, OCEANOGRAPHIC includes instruments and devices used in oceanographic work for collecting samples of the ocean environment and for measuring the various biological, chemical, and physical characteristics of the ocean as well as other marine environmental features directly above the ocean surface and the seafloor below.

Equipments that are used systematically for executing this work fall under three categories: (1) collection, (2) measurement, and (3) data handling (or the computer methods of making measurement information available for utilization). Various other ancillary tools (e.g., shipboard handling gear, calibration equipment, buoy release devices, navigational aids, certain hardware for instrument repair, etc.), although also used in such work, cannot be properly characterized as instrumentation.

The types of collecting devices are widely varied. They include such equipments as geological corers, grabs, and dredges for sampling portions of the ocean bottom, nets for PLANKTON and nekton and other biological acquisitions, and bottles for SEAWATER collection. Historically, representative devices such as these have been classified as a primary part of oceanographic instrumentation. However, instrumentation is mainly concerned with measurements, and to the extent that collecting or sampling devices are used to perform measurements, they are instruments. Actually, in advanced oceanographic work, any instrument is seldom used by itself; rather, it is a part of a network of others. For example, computer systems are employed to process and reduce much measured data, and at times they are used simultaneously with the event being observed. This automated recording and processing method permits conclusions to be drawn and appropriate experimental corrective actions to be implemented immediately. Computers also make data taken on board oceanographic ships available faster to the scientific community. Data processing systems are especially useful in physical and GEOLOGICAL OCEANOGRAPHY.

Oceanographic measurements can be taken (1) *in situ* (in the environment), (2) on board a surface ship or at a "field" station (ocean platform), and (3) at a land laboratory.

More than 120 marine parameters have been identified by ocean data collectors and users as being important and necessary for the support of their particular oceanographic or ocean-oriented programs. More than 21 000 oceanographic instruments of about 34 generic types have recently been inventoried among various agencies and institutions in the United States alone. This proliferation of instruments and advances in the technology did not take place without its historical antecedents, and even the first seafarers of the ancient Mediterranean world accumulated measurements and observations on TIDES, CURRENTS, depths, and other oceanographic phenomena. (See MEDITERRANEAN SEA.) Over the years and up until the very recent past, most good instruments have come from relatively few seafarers, seagoing oceanographers, and others interested in observing the varied behavior of the world's oceans.

Some Historical Examples To give but a few examples illustrative of this point, we may begin with the little-known but significant work of the Italian Count Luigi Ferdinando Marsigli. Marsigli, in the early 1700s, was probably the first to build and use a dredge not too unlike those of today that are employed to collect marine animals from the ocean bottom. He also built a current meter equipped with a propeller to measure subsurface currents in the BLACK SEA and the Bosporus.

James Cook, who was sponsored by the Royal Society of Great Britain, made a remarkable scientific ocean expedition (1768–1771), and, during his three voyages to the South PACIFIC OCEAN, using rather primitive but effective devices, Cook compiled an impressive amount of information about winds, currents, and the TEMPERATURE of those waters. He also made soundings to depths of over 2000 ft (609 m). On Cook's second voyage, Johann Reinhold Forster and his son Georg attempted some of the earliest deep-sea soundings. They were successful in this and brought up samples of blue mud from the Pacific bottom from depths of approximately 4600 ft (1402 m). (See COOK, JAMES.)

Alexander von Humboldt of Germany was inspired by the work of his contemporaries Cook and Forster, and using instruments of his time, he made several noteworthy contributions to OCEANOGRAPHY by documenting his observations and measurements of currents (the current off the western coast of South America is often referred to as the Humboldt Current), marine life, tropical storms and the mappings of volcanoes in the Western Hemisphere. (See HUMBOLDT, ALEXANDER VON.)

Charles Darwin, the famous English naturalist, invented and used a tow-net device for collecting plankton while he was on the South American expedition of the *H.M.S. Beagle* (1831–1836). Darwin's work in marine biology and his theory of how a CORAL REEF forms represent landmark studies in oceanography. (See DARWIN, CHARLES.)

During the 1839–1842 U.S. Exploring Expedition, headed by Charles Wilkes, copper wire was used in making sounding of the SOUTHERN OCEAN depths. This was a superior technique compared with previous methods that used rope lines primarily for the purpose.

The great American statesman Benjamin Franklin, who also had a fervent interest in understanding nature in its many forms, evolved (with the help of Timothy Folger, an American whaling captain) a chart of the surface currents of the north ATLANTIC OCEAN. In this late eighteenth century work, Franklin used an oaken bucket with a simple thermometer for sampling the water temperature —a method that remained essentially unchanged for almost 200 years.

The Englishman Edward Forbes made notable contributions to the fields of marine biology in particular and oceanography in general. In his work, Forbes also improved upon the existing dredge designs used for collecting samples of deep-sea bottom life. He employed these designs in 1835–1840 off the Isle of Man and in the Aegean Sea. Prior to Forbes, the first really effective device intended for this kind of work in the deep sea is credited to Sir John Ross. Ross, some 20 years earlier, while exploring Baffin Bay, designed and had built what was termed a "deep sea clamm." This was used successfully to depths of around 1300 ft (396 m).

The American Matthew F. Maury is acknowledged by many as the "father of modern oceanography" because of his epic work in the correlation of data on the surface phenomenon of the oceans, his bathymetric map of the North Atlantic, and his famous book *The Physical Geography of the Sea.* (See MAURY, MATTHEW FONTAINE.) Interested in deep-sea soundings and the nature of the sediments on the ocean floor at these depths, Maury in the 1850s first used a novel device, whose invention and design he strongly influenced, to bring up samples of the ocean bottom at the 12 000-ft (3658 m) depth. This consisted of a hollow iron pipe to which a large cannonball was attached at the top. The cannonball–rod configuration was connected by a line arrangement so that when the rod reached and penetrated the bottom, aided by the weight of the ball, the sediment-filled rod was then brought back by the line to the surface.

The most significant early oceanographic instrument work was done on board the famous oceanographic voyage of the *H.M.S. Challenger* (1872). (See CHALLENGER EXPEDITION.) Moreover, due to the influence of such men as Wyville Thomson, T.R. Huxley, and Sir John Murray, the pioneering deep-sea investigations that took place on this expedition formed the basis for scientific oceanography. Although the instrumentation then used (water samplers, dredges, trawls, plankton nets, thermometers, current meters, etc., and various supporting gear) was primitive by modern standards, it was the very best of its time, and oceanographic measurements that were obtained by these tools were surprisingly accurate.

Alexander Agassiz, in 1877–1880, conducted several oceanographic research voyages on the *Blake* in the CARIBBEAN SEA, the GULF OF MEXICO, and off the Atlantic coast of Florida. While on these expeditions, he and Charles D. Sigsbee invented a new and more effective type of dredge. Agassiz also improved upon the design of the trawl then commonly in use.

The instrument package for sampling seawater, and one that has been extensively used over the years, was invented at the turn of the century by the Norwegian explorer and zoologist Fridtjof Nansen. This water sampler, called the Nansen bottle, is lowered by cable into the sea along with a sounding device to determine the depth of the waters being collected. The bottle is lowered in the open position, and water is collected by sending a weighted "messenger" down the cable which triggers a closing mechanism on the sampler upon impact. The device is so constructed as to also measure the temperature of the collected water by means of reversing thermometers. Although more sophisticated and expensive instruments are now used for *in situ* determinations of seawater SALINITY, TEMPERATURE, and PRESSURE, the Nansen bottle, or Nansen cast, is a method still often employed.

Various oceanographic expeditions (e.g., the 1889 German Plankton Expedition and the 1886–1889 Russian *Vityaz* voyage), men such as Prince Albert of Monaco, several worldwide laboratories and marine stations (e.g., Marine Biological Laboratory at Woods Hole, Massachusetts, the Marine Biological Association Laboratory in Plymouth, England, and SCRIPPS INSTITUTION OF OCEANOGRAPHY in California), governmental agencies, and universities have all contributed to the improvement and development of oceanographic instrumentation over the years.

Instrumentation Today The relatively unsophisticated tools and devices evolved in the past have assisted their users in performing certain tasks and have been responsible for providing some good preliminary information on ocean-surface phenomena and marine biology, as well as on the behavior of tides, currents, temperature, and salinity. However, oceanographic instrumentation, a comparatively young technology, now has taken on a much more complex meaning.

Instrumentation plays an important part today both in oceanography and OCEAN ENGINEERING. Data derived from the oceans must be collected, measured, and then these measurements assimilated so as to obtain a better understanding of the ocean in all its facets and to apply this knowledge to its fullest extent. This is a difficult task due to the vast scope covered by both the fields of the science of the oceans (oceanography) and the engineering of the oceans (ocean engineering), and because of the enormous size and nonhomogeneity of the ocean environment. Consequently, instruments must be so constructed as to withstand the hostile surging and corrosive sea environment, and they must be designed and made with a high degree of reliability, maintainability, and standardization. This requires considerable design and manufacturing expertise. Obviously, international cooperation is also essential before such a goal as standardization can be fully met.

In the wide spectrum of work done at sea by the oceanographer using instruments, the following areas are of prime importance: ocean currents (subsurface and surface); tides and the motion of the WAVES; ocean temperatures, densities, and depths; ocean salinity and chemistry; sediment structure and ocean-floor topography; marine life and meterological information. Since these represent parts of the total coverage of oceanography, in the subsequent discussion, some of the main types of instruments and analytical techniques used are delineated under the generic headings of physical, chemical, geological, biological, and meteorological.

Instrumentation for Physical Oceanography

Among the most important parameters to be measured under physical oceanography—which is a branch of oceanography that treats physical properties, the movements of the ocean, and the variability of these properties in time and space—are those of current speed and direction; tides; wave height, speed, and direction; density; temperature and depth.

Currents In the case of ocean currents, both surface and subsurface, probably more types of instruments and devices are used than in any other

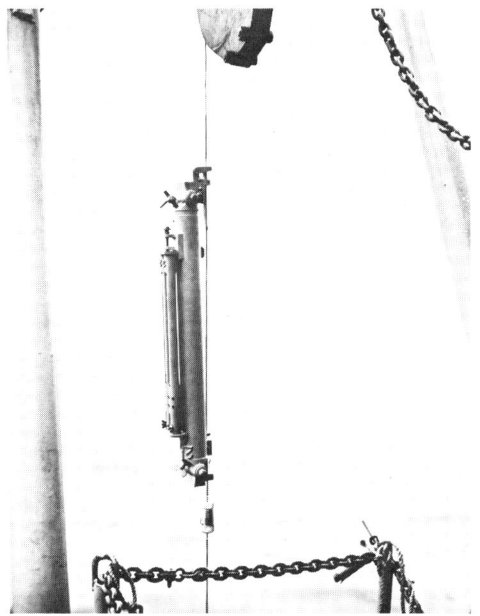

A Nansen bottle attached to a lowering wire descends into the sea. The Nansen bottle is a device used by oceanographers to obtain subsurface samples of seawater. The bottle is equipped with a rotary valve at each end so that when it is rotated at the desired depth, the valves close, trapping a water sample and setting the reversing thermometers. *(Oceanographer of the Navy)*

Buoys like these, which carry a radar reflector and beacon, are used as anchored reference points during surveys. They can also be allowed to drift freely with surface currents or dragged by a submerged drogue parachute. Current velocity is obtained by tracking the float with the ship's radar. *(NOAA)*

A Navy oceanographer prepares to lower a current meter from aboard ship in the Kara Sea of the Arctic Ocean. Information gathered by such instruments is used to compile oceanographic charts of currents, temperature, salinity, etc., for use by the Navy and the worldwide scientific community. *(Oceanographer of the Navy)*

The lower half of this instrument comprises electronic sensors for measurement of temperature and electrical conductivity, from which salinity is determined, and pressure, from which distance below the surface is determined. The upper half is a series of bottles used to obtain a sample of seawater for chemical tests and for standardization of the electronic sensors. *(Arnold Gordon)*

Geological oceanography requires remote-sensing equipment such as the gravity meter. The earth's magnetic field is constantly changing, and data are needed for ship navigation, submarine detection systems, and the study of the earth's geologic history. *(NOAA)*

The orange-peel bottom sampler brings to the surface deposits which will subsequently be studied for such qualities as size distribution, bearing strength, shear strength, and pullout resistance as well as chemical properties. *(NOAA)*

Study of the formation of the ocean bottom yields valuable information to the oceanographer. Here a corer, which will retrieve a vertical section of the ocean bottom, is lowered over the side of a Naval research ship. *(Oceanographer of the Navy)*

A box corer with samples from the bottom rises from the waters of the New York Bight. *(NOAA)*

Engineers unload current meters during POLYMODE 1976. Current measurements are important because ocean circulation affects the distribution of the physical and chemical properties of the ocean water masses and the distribution of plankton, and because ship navigators need to know about strong currents. *(Woods Hole Oceanographic Institution)*

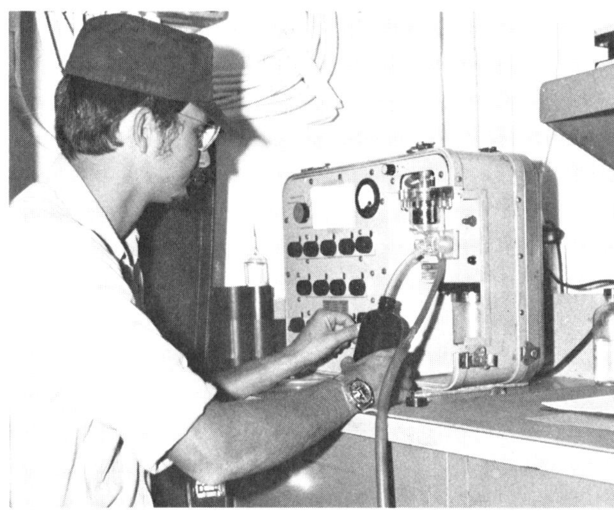

A marine science technician analyzes the salt content of seawater with the aid of the salinometer in the oceanographic laboratory on board the Coast Guard cutter *USCGC Dallas*, 480 mi (768 km) off the coast of Dakar, Africa. *(U.S. Coast Guard)*

Scientists examine a core sample from the New York Bight. *(NOAA)*

Geologists examine deep-sea core sample from the seafloor. Sediment layers help tell the story of the early history of Earth and its climates. *(Woods Hole Oceanographic Institution)*

An Isaacs-Kidd trawl net is used to sample plankton and fish at mid-water depth (hundreds of meters deep). The ship slowly tows the net through the water. *(Arnold Gordon)*

Plankton collect at the base of this net as it is slowly raised from the ocean depths to which it has been dropped. *(Arnold Gordon)*

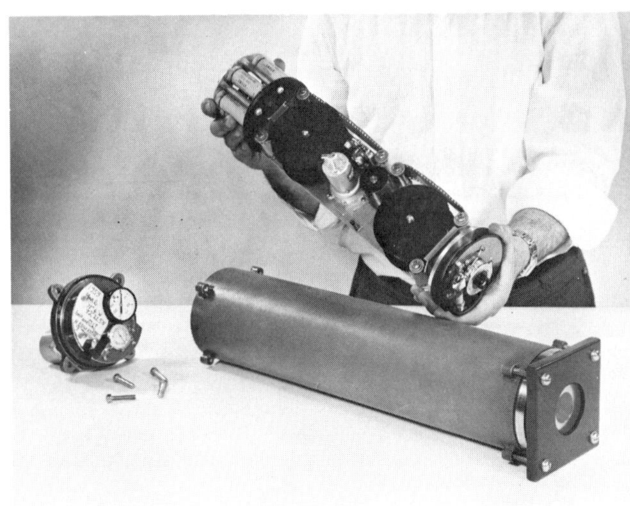

The underwater camera is a valuable tool in the study of ocean life and is used extensively in all sorts of underwater probes. *(Dr. H.E. Edgerton, MIT)*

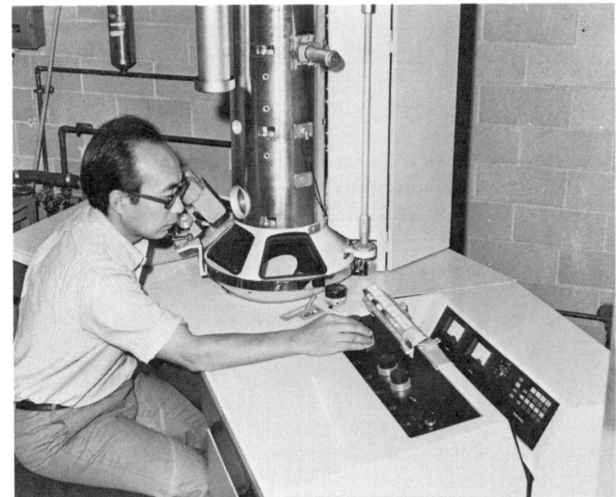

The electron microscope is among the highly sophisticated instruments used in chemical and biological oceanography. It greatly magnifies images electronically, and is used to study plants, animals, and sediments recovered from the seas. *(Woods Hole Oceanographic Institution)*

single oceanographic measurement. The methods and techniques of their employment vary greatly also, and these depend upon the use of floating or tethered measuring devices, the depths of the desired data, and the requirement for detailed measurements at a given locality or the need for several measurements over various oceanic areas. Currents can be measured directly or indirectly. Among the direct measuring devices are the drogues, drift bottles, dyes (e.g., Rhodamine-B tracers), mid-depth neutrally buoyant pingers (e.g., Swallow float), the buoy-operated Roberts radio current meter, the Ekman propeller-type meters, the Carruthers meter, and the Savonius rotor. The operating principle involved in the use of some of these is that the speed and direction of a free-floating body is determined by observing—either by tracking by radar or by appropriate sound gear—the distance and direction that the object, dye, or drift bottle goes from the ship or launching platform in a given time interval. In others, such as the fixed-body current meters (e.g., the Ekman and Savonius rotor), the velocity of the current is measured by correlating it with the speed of rotation of a vane-configured wheel or propeller which is actuated by the impinging current. The revolutions are recorded for a given length of time along with the direction by compass reading. A Roberts-type current meter employs a unique rotor actuated by the current to be measured. The difference in this regard from the Savonius type is that the rotor reverses with reversing current flow and the device also automatically compensates for oscillatory flow which can be caused by the strumming of mooring lines due to surface wave action and other forces. An optically scanned compass and a magnetic tape recording unit are used in the system. In still other types of meters, current can be determined indirectly by the rate of cooling on hot wires; by a device using radioisotopes wherein seawater flow rates are measured by radioisotope drift time over a fixed course; and also by a method of measuring (e.g., the geomagnetic electrokinetograph—GEK) the electropotential difference between two separated electrodes spaced given distances apart in the ocean. The resulting difference is due to the movement of the conducting seawater (an ELECTROLYTE) through the Earth's magnetic field. The induced voltage is related to current speed. Current meters may be rigged for mooring in either deep or shallow water, and the data sensed by the meter may be telemetered back to the ship or buoy, or the instrument may be designed to optically record current speed and direction data on photographic film. This film information is automatically converted to computer-compatible data (magnetic tapes) and the data then processed for use in reports and atlases.

The fluid dynamics or movement of deep, slow-moving ocean currents also can be determined indirectly from temperature and salinity profiles which, in turn, determine the vertical density structure of ocean waters. Ocean circulation additionally can be inferred from the distribution of various properties of seawater. For instance, the high-salinity–temperature characteristics of the subsurface outflow from the MEDITERRANEAN SEA permits this water to be identified and traced for several thousands of miles across the North Atlantic Ocean. General ocean circulation is a long-time average, global-scale flow occurring in an oceanic environment rich in flow phenomena on many scales. New techniques have identified a phenomenon that appears to be intimately linked physically to the general circulation and even to dominate the circulation energetically in many regions. This phenomenon is comprised of slow medium-sized fluctuations in the circulation itself. These fluctuations are either called low-frequency mesoscale variability or the eddying of the flow. This variability can take several forms, such as meandering of the Gulf Stream, large ring vortices that are snapped off from the currents during intense meandering events, and the mid-ocean eddies. To be ultimately useful for the study of CLIMATE, ocean circulation models are being developed so as to represent correctly the temperature distribution and transports of heat.

Tides The periodic variation in the level of ocean waters, resulting from the gravitational attraction of the moon and sun upon the Earth, are measured essentially by three types of devices. These are the nonregistering gauges, which require the observer to record the tidal changes; the registering or automatically recording gauges, which are equipped to plot the rise and fall of the tide; and the gravimetric instruments, which involve the measurement of gravity variations with tidal response.

The gravimeter method requires relatively few observations on the seafloor since the tidal displacement of the earth and the oceans can be observed over a wide area. This information includes all tidal components simultaneously, the displacement of the earth due to the weight of ocean waters, and bottom friction and resonance effects due to the shape of the coastline. The combination of such factors together with radar altimetry furnishes a rapid, reliable tidal prediction capability across great areas of the ocean while using a minimum of seafloor stations.

Waves The surface of the ocean is in constant motion, and the most obvious cause of this is wind-generated waves. These irregular moving bumps, or waves, in the open ocean are formed by a transfer of energy from the air to the water. Instruments to measure waves vary in design and in deployment. For example, measurements may be made from ships, buoys, aircraft, satellites, and stationary platforms or towers. From stationary positions on the bottom or along the shore, waves are measured by electronic wavestaffs (based on a variety of principles of operation), pressure transducers, sonic, infrared, and radar sensors, upward-looking sonar, and mechanical wave indicators. From moving platforms, waves can be measured with sonic surface viewers, inverted echo sounders, accelerometer devices, radar, infrared beams, photography, and lasers.

In regard to the last group, wave height–generated graphs or plots showing wave energy versus wave frequency can be made via infrared sensors mounted on the bows of ships. Continuous profile records of the ocean surface may also be made by a downward-directed laser beam from an aircraft. The more conventional sensing means employed for wave measurements are (1) the step resistance gauge, an electrical device supported vertically in the water and so designed to measure surface changes by means of changes in electric current and resistance as the water heights vary and the water contacts are shorted out along the submerged length of the gauge; (2) surface floating mechanisms connected mechanically to a recorder; (3) stereophotographs; (4) precise aircraft altimeters in aircraft flying over the ocean at a fixed elevation; (5) subsurface pressure gauges that can convert ocean pressure to electrical response by means of pressure transducers, with this value thus translated or computed by hydrodynamic theory to surface wave heights; (6) echo sounders or fathometers so emplaced below the surface that they "bounce" an echo signal off the ocean surface; (7) devices that are capable of recording the buoyant force on a vertical cylinder; and (8) surface buoys, having vertical accelerometers that actuate recorders.

The recording of wave conditions and the prediction or forecasting of these conditions are of obvious importance to ocean navigation, commercial shipping, DIVING, and salvage operations, resource exploitation, ocean animal and plant activity, OCEAN POLLUTION, and beach erosion.

Temperature, Density, and Sound Velocity Ocean temperature measurements are essential in all phases of oceanographic work. The common method employed for ocean temperature measurements at any desired depth in the sea has been the reversing thermometer attached to Nansen and/or Nisken bottles. When the bottle closes, the thermometer measures the temperature (to within a hundredth of a degree), thus providing a reading for a particular time and point. Within about 500 ft (150 m) of the surface, where the principal temperature changes occur, a continuous record of temperature can be obtained by an instrument called a bathythermograph, invented by Athelstan Spilhaus in 1938.

Ocean density (or the weight of a given volume of seawater at a specified temperature) can be computed from temperature, pressure, and salinity values, although it is possible to measure density *in situ* by densitometers or by hydrometers. When computing densities, the temperatures are usually determined either by means of mercury thermometers, bathythermographs (BT), expendable BT's (XBT), infrared sensors such as the airborne radiation thermometer (ART), or electronic sensors such as the salinity-temperature-depth recorder (STD).

The measurement of sound velocity in the oceans is highly useful. Underwater sound travels almost 5 times faster than sound in air, or about 4850 ft/s (1480 m/s). Also, sound velocity varies in the oceans, depending upon temperature, pressure, and salinity. In general, an increase in any of these results in an increase in sound velocity. Sound velocities decrease with depth in both a temporary and a permanent THERMOCLINE. There is very little change in temperature below the permanent thermocline. Thus, pressure is the most important factor, causing sound velocities to increase with depth. Because of its rapid drop in temperature, the thermocline bends any sound rays passing through it downward. Therefore, the ocean has sound channels above and below the thermocline—sound originating below the thermocline may not be audible above it, and vice versa.

Understanding the effects of biological organisms on the transmission and reception of sound is another important phase of underwater ACOUSTICS. It is thought that animals arranged in strata are responsible for much of the volume reverberation detected in the deep sea. The geographical and vertical distribution of these animals, the spectrum of sound scattered by them, their identities, and the changes in the frequencies of sound scattered by them during their vertical migrations are among the variables studied. Animals produce sounds themselves, and, to separate their sounds from the signal of interest, it is important to know which animals generate which sounds and where these sound producers are likely to be. Through these studies some insight may be gained as to why animals make SEA NOISE. The knowledge of how environmental variables in known areas affect sound transmission enhances the use of acoustic techniques to investigate many areas in order that the fisherman or marine biologist may be able to rapidly detect concentrations of particular species of organisms. Also, the geologist can more effectively use this method to identify rocks and sediment regions, and the physical oceanographer can apply the technique to the identification and tracking of water masses.

The sound velocity can be computed or may be measured directly by an STD with a sound velocimeter attached (SVSTD) or by sound velocity/depth profilers designed to make measurements while emplaced on the ocean bottom. This latter instrument contains a velocimeter and pressure sensor along with a digital display.

Basically, the bathythermograph, or BT, is an instrument capable of sensing temperature by means of a thermistor and sensing depth by a vibrating wire transducer. The temperature is converted electronically to a varying frequency, and the frequencies for both temperature and depth are transmitted to the ship or platform by a conductor cable. These frequencies are then fed to a recorder, digitized, and converted to temperatures, or they may be converted to voltages and fed to an *x-y* recorder to give a continuous profile of the ocean temperature with depth. The expendable BT (XBT) telemeters send back data to be processed in a like manner.

The salinity-temperature-depth recorder (STD) is an electrometric instrument used *in situ* to provide a continuous reading of the three variables on a recorder. The device consists of a conductivity cell, a temperature sensor, and a pressure-operated depth element. It is connected by a conductor cable to a unit, on board ship or platform, containing amplifiers, a salinity computing circuit, and a recorder.

The conductivity temperature indicator (CTI) consists of a two-electrode conductivity cell and a thermometer. The temperature values and conductivities are indicated on a pair of counters mounted on the housing for the amplifier and servomechanism.

Temperature measurements not only map thermoclines but assist in determining major current patterns in the ocean. By combining temperature (T) and salinity (S) measurements, TS diagrams are used in labeling large bodies of water and a relationship is traced by comparing TS characteristics.

Some controversy exists among ocean scientists concerning the equation of the state of seawater. Laboratory measurements on the SPECIFIC VOLUME of seawater, using a high-pressure magnetic float densimeter, appear useful in deriving such an equation of state from directly measured specific volume data on seawater over a wide salinity, temperature, and pressure range. Such data can be of special value in the interpretation of the movement and stratification of deep-water ocean masses, the calculation (made by thermal expansible values as a function of pressure) of the adiabatic temperature gradient, and the calculation of other thermodynamic properties of the ocean. Specific volume information is also of value in assessments of partial molal volumes of seawater based on temperature, and salinity since these volumes can be employed to determine the effect of temperature and pressure on the equilibria of importance in the world's ocean environment.

Depth The probing or sounding of the ocean depths has long been carried out by seafarers. Basically, the early method was to use a line of rope, twine, piano wire, or some other material to which a weight was attached and then lowered into the ocean. After the line was fully played out, the depth from the surface to the bottom was measured by the amount of line immersed. Naturally, this method had built-in errors caused by the uncertainty of the angle taken by the line from the ship to the bottom.

Ocean depths are still difficult to measure with precise accuracy. Sound is employed to perform this measurement function. Sound is also the only practical means of underwater communications, because radio waves are not transmitted very well in the sea and light beams are also attenuated very quickly in most waters. Various forms or kinds of sonar are used in undersea exploration to determine the depth of water as well as the location and contour of seamounts, trenches, shoals, and reefs. (See SEAMOUNT; CORAL REEF.) Permanently planted sound sources and receivers also are used in acoustics telemetering systems for reporting data from other oceanographic instruments like salinometers and velocimeters, as well as for acoustic monitoring. However, echo sounders which are used to bounce a sound signal off the bottom and then measure the time such an echo takes to return to the surface, thus measuring the depth, suffer from inadequacies due to reflection, reverberation, refraction, and absorption phenomena. Reflection takes place when sound waves are obstructed by solid objects or discontinuities in the ocean near the ocean floor; reverberation is that portion of the scattered sound that returns to the tranducer, the scattering being caused by irregularities in the ocean surface, bottom, and the water itself; refraction is the bending of sound rays due to variations in changes in velocity caused by variations in seawater density, while absorption reduces the echo intensity when part of the sound energy is changed to heat energy by friction caused by water or vibration of suspended PARTICULATE MATTER or bubbles in the water. The echo intensity can also be attenuated by soft forms of biological growth, and the echo interpretation is further changed by ambient noises of biological, mechanical, and wave-action origin. Thus, while echo sounders are widely used in various depth measurements, another reliable way to measure deep depths is to take advantage of the direct relationship between pressure and depth [seawater pressure increases roughly $\frac{1}{2}$ psi per foot (0.3 N/cm^2 per 30 cm of depth)]. In this way, ocean pressure is measured and converted to depth. The previously mentioned STD and BT instruments employ transducers (a transducer converts energy from one form to another; most pressure transducers change pressure to motion, then use the movement to produce an electrical output which can be calibrated and monitored) to utilize this principle directly, as do the present depth sensors.

The measurement of ocean depth is especially important since all meaningful oceanographic measurements must be positioned along coordinates of depth, LATITUDE, and LONGITUDE.

Instrumentation for Chemical Oceanography

Seawater, a unique mixture of almost everything, including salts, gases, trace metals, nutrients, and radioisotopes, presents a challenge to the analytical chemist. A large part of this challenge revolves around the difficulties encountered in carrying out chemical analyses at sea in real time. Here, working on a rolling and pitching platform with a large number of samples collected from several locations, the associated problems are formidable. Moreover, seawater samples obtained at depth under certain temperature and pressure conditions may or may not be the same in chemical character when analyzed on board ship or later at a land laboratory. In short, the goal in chemical oceanography has long been for reliable *in situ* instrumentation systems.

Up until quite recently, for example, most measurements of salinity were carried out by means of collected seawater in Nansen bottles, with the water so collected and then analyzed at a shore-based laboratory or at sea at a laboratory on board a research vessel. This analysis, in the case of salinity, was done by titrating the water with a standardized silver solution, with silver halides being precipitated chiefly as silver chloride. Thus, the chlorinity, or amount of chloride in the sample, was determined. This value provided the means of determining salinity. Salinity can also be found by the relation of conductance and salinity after the conductance of the sample has been measured by salinometers and conductivity cells. However, with *in situ* electrometric instruments that measure salinity-temperature-depth (STD) simultaneously and on a continuous reading by means of probes and temperature and pressure sensors, the long delays between field work and laboratory analysis are eliminated.

Another common chemical measurement of seawater is the determination of oxygen content. The amount of oxygen in various parts of the ocean is a basic quantity that must be measured before many other kinds of oceanic research can be carried out. The oxygen concentration determines the life-support capability of seawater, and its values are also used in deep-water mixing measurements. Traditionally, dissolved oxygen has been measured in seawater samples by Winkler titration. This classical technique is based on oxidation-reduction reactions between added manganous ions and the dissolved oxygen in the samples. (See ION.) The oxidized manganese is reacted with iodine in an acidic medium, and the amount of the iodine consumed is determined by titration with a standardized thiosulfate solution.

Gas-chromatography–instrumented techniques are also now being used at sea for measuring dissolved oxygen and other gases in seawater. In this method helium is usually employed to strip the dissolved gases from solution, and these gases are measured in a gas chromatograph equipped with thermal conductivity detectors.

Two *in situ* methods of ocean oxygen determinations include devices (1) that are capable of diffusing the oxygen through a semipermeable membrane, then reducing the oxygen molecule on a cathode (a polarographic measurement), and (2) that rely on the quantitative oxidation by dissolved oxygen of thallium metal containing a known amount of radioactive thallium 204. In this method a seawater sample is passed through a lined thallium cylinder. The thallium is oxidized and enters into solution after which it is sent through the radiation counters. These record the level of beta radiation from the thallium 204, and since the oxidation rate, and therefore the rate of thallium given off into solution, is proportional to the amount of dissolved oxygen in the seawater, the device can be instrumented to show oxygen content.

Research programs have led to the development of specialized methods in marine chemistry. For example, to scavenge trace metals and radioisotopes from seawater, an extraction technique has been developed by the U.S. Navy, using manganese-impregnated acrylic fibers. Large volumes of surface waters can be sampled by towing the fibers behind a ship; subsurface sampling is accomplished by flushing water through a specially designed sampler lowered to the required depth. Although originally intended to extract radium from the water, the manganese-impregnated fibers also will remove other trace elements (for example, lead and mercury). This fiber technique is potentially useful also in the treatment of waste waters to remove undesirable trace elements and radioisotopes.

Some of these determined seawater properties, such as the salt content, may be used with temperature to define various water masses. The distributions of the other properties may be used to trace the movement and mixing of water masses. Absolute rates of mixing may be determined by studying certain radioisotope concentrations. The chemistry of the water is important to the organisms living in it, for they extract vital nutrients, calcium carbonate, and various elements to support their existence. Because some organisms, especially corals, also incorporate radium isotopes, growth rate studies of these organisms are made easier. (See CORAL.) Chemical techniques are also vital to studies of pollution in the water or in sediments. Moreover, it can be said that in addition to the standard equipment generally encountered in an analytical chemistry laboratory (i.e., UV-visible spectrophotometer, gas chromatograph, fluorometer, pH meter, etc.), the use of sophisticated instrumentation, such as scanning electron microscopy, nuclear magnetic resonance, and Mössbauer and laser Raman effects are also necessary for work in chemical oceanography. This usage is expected to expand in both seagoing and land-based marine laboratories.

Instrumentation for Geological Oceanography

The largest percentage of geological oceanographic instrumentation is concerned with measurements related to nearshore and open ocean depths (bathymetry), sediment composition, magnetism, gravity, seismism, volcanism, erosion, and the other geological and geographical features of the ocean world.

This is a very broad field which draws upon the work of other oceanographic endeavors. However, it also should be recognized that perhaps in the past the geological oceanographer has used the greatest variety of remote sensing equipments. The geological oceanographer has also used acoustic, electromagnetic, geomagnetic, gravity, and seismic types of measurements to describe the ocean world environment, including the process of seafloor spreading. As a result, the knowledge of the shape of the ocean floor and the character of the subbottom structure that exists has been accumulated by the employment of such equipments as compressed-air guns, ocean-bottom seismometers (OBS), echo sounders, sonobuoys, magnetometers, and instruments for gravity measurements.

Both aircraft and ships equipped with magnetic airborne detection devices are employed for the survey and determination of accurate magnetic data. (The Earth's magnetic field changes constantly). Such information is of great value for improving navigation at sea, for submarine detection systems, and for the study of the Earth's geological history. Gravity measurements made by gravity meters having an extremely sensitive spring balance system are used aboard ships to measure gravity continuously while underway at sea. These meters can also be lowered to the ocean bottom where information is sent by cable to recorders on the ship. Gravity anomalies—the differences between observed and theoretical gravity values—assist the geological oceanographer in pin-pointing mineral and oil deposits in the ocean.

In acoustic measurement work, or the interaction of an acoustical field with the ocean bottom, the bottom reflections are dependent on the composition of the sediments, roughness, topography, and subbottom structures. In general, reflections are good over hard sandy bottoms and very poor over muddy bottoms.

Surveys designed to learn more about the ocean floor in the past and to a lesser extent, at present have been primarily based on a study of sediments. Samples of these deposits that have been laid down throughout time are collected by means of graps, mud samplers, piston and gravity corers, and dredges. When they are brought to the surface, the samples can be analyzed by a range of measurements—moisture content, size distribution, bearing strength, shear strength, pullout resistance, etc. Also the chemical nature can be determined and other features of the sediment can be characterized by x-ray techniques, scanning electron microscopy, and neutron activation analysis.

In simple terms, neutron activation analysis consists of bombarding a sample such as seawater or an ocean sediment with neutrons. In this way, radioisotopes are produced from the resident elements. The irradiation source can be either a nuclear reactor, an accelerator-type of fast neutron generator, or a neutron-emitter radioisotope such as Californium 252. The energy spectrum of the resulting x-rays from the decaying radioisotopes is recorded, and the energies of the spectral peaks and the areas under the peaks identify and relate to the amount of the elements present, respectively.

Modern systems completely automate the process of this analysis. The sample enclosed in a tube is pneumatically transferred to the neutron source for irradiation. It then proceeds to a station where unwanted short-lived activities are allowed to decay away. The activated sample is then shuttled to a detector site for counting and finally to a waste station for disposal. Computer data acquisition and reduction completes the analytical cycle.

Although rarely used, an interesting *in situ* measuring method of determining the character of bottom sediments in various parts of the ocean is done by the sediment density probe. The probe consists of a 26-ft (8-m) long by 4-in (10-cm) diameter tube containing a gamma ray emitting source, a lead shield, and a radiation detector. As in the case of the typical coring device, the tube is forced into the bottom sediment, and, while in place, the gamma ray source, shield, and detector are programmed to travel up and down at measured intervals inside the probe. In this way, the rays are absorbed in the sediment according to its density. Thus, a low radiation count at the detector indicates a high-density sediment since more radiation is absorbed and less reflected back to the detector. The reverse is true for high counts at the detector, which show a low density. This information is recorded on a film.

While the greater percentage of geologic oceanography is concerned with the study and analysis of nearshore and open-ocean depths (bathymetry) and sediment composition, many other fields constitute the lot of the ocean geologist, e.g., magnetics, gravity, seismism, erosion, and volcanism. All these analyses of the geological oceanographer are important since they give information concerning the rate and processes of sedimentation, the changes in ocean life and fast climatic conditions, the presence of bottom currents, engineering information such as the load-bearing capacity of the sediments, the ecological effects of POLLUTION on the ocean floor environment, and the various mineral and other resources available for use by human beings.

Instrumentation for Biological Oceanography

In marine biology, the life processes and distributions of organisms in the ocean—from BACTERIA and the smallest phytoplankton (microscopic plant life) (see PLANKTON) to the largest cetaceans (see MARINE MAMMALS) and fishes—are studied, as are the interrelationships between these organisms and the effects of the physicochemical environment upon their life processes, distributions, and interrelationships.

Biological instrumentation used in these studies is primarily mechanical. Ocean life existing in its many forms, from microscopic organisms to WHALES, and at all depths is collected by a variety of ways—different configurations of samplers, nets, trawls, etc. Most of these devices are highly specialized so that they are capable of obtaining specimens horizontally and vertically, capturing the more active swimmers, securing undamaged specimens, recording the amount of water sampled and the depth, and sampling over a long distance and at several depths simultaneously.

Underwater photography is a useful tool when employed in conjunction with biological samplers as well as other instruments for probing the sea. Many forms of marine life can be photographed in their natural environment, allowing the scientist to gain insights into ecological relationships.

In addition to the use of deep-sea cameras, remotely controlled vehicles, and pressurized organism-recovery systems, there are some other techniques that aid in the analysis of certain marine life.

In spite of the strong arguments that have been put forth by some in behalf of submersibles and underwater habitats for augmenting the arsenal of tools used by the biological oceanographer, these methods are expensive and the costs involved for their usage may, according to several scientists, be better expended on needed new and improved sampling devices and other techniques. On the other hand, high-resolution multiple-frequency acoustic systems used on submersibles hold promise as a major means for describing distributions of marine animals (especially planktonic and slower moving schooling nektonic forms) and thus provide methods for assaying these living resources of the ocean. Such a technique may also be employed by using a large towed body from a surface ship.

In biological sampling work, marine chemists have traditionally provided valuable independently obtained information for the ocean biologist. Most such data have been in the form of salinity, oxygen, and nutrient determinations. Now, with the advent of more capable instrumentation for *in situ* measurements, the correlations of collected biological information with chemical characteristics of the place where this information is taken can be made more straightforwardly and faster. In the laboratory special equipment is used for culturing oceanic bacteria, COPEPODS, and ALGAE. Oxygen production and consumption by marine organisms are measured by polarography. Plankton populations are assessed by use of Coulter counters and by reflectance spectrophotometry of filter deposits. Several important applications in marine biology have been found for the use of the electron microscope, a device for forming greatly magnified images of objects by means of electrons.

One particularly important problem area facing the marine biologist relates to the development of effective sampling equipment and methods for forecasting marine FOULING conditions in various parts of the world. Buoys and bottles set adrift in the open ocean can be used to study fouling organisms, and plates of several materials emplaced in various parts of the sea give indications as to the nature and extent of the attack of these materials. Other devices for use in marine biology are a towed turbidimeter which measures backscattered light, and a bar with a controllable temperature gradient for studying the effect of temperature on biological fouling.

The very nature of much of the work in biological oceanography suggests that many new and improved devices are needed. Several present instruments have been developed to meet a specific oceanographic need. Many other specific needs still remain.

Instruments for Meteorological Oceanography

Instrumentation in meteorological oceanography is primarily designed to obtain quantitative information about various aspects of the ocean and the ocean-air interface as these pertain to the atmospheric weather conditions and its predictions. In the execution of such work, typically, measurements of ocean surface and air temperatures, ocean currents, heat budget, barometric pressure, air-sea interactions, and waves and wind speed and direction all assist in studies having this objective.

Information on many of these parameters is provided to the meteorological oceanographer by others, especially by physical and chemical oceanographers.

In addition to the use of anemometers (for wind speed and direction), barometers (for atmospheric pressure), temperature sensors (for air and water temperatures), rain gauges (for precipitation), humidity sensors (for degree of wetness in the air), much data can be obtained by the use of satellite and aircraft remote-sensing equipment. This technique of measuring ocean properties from a distance is especially useful in extending the oceanographer's view to cover much wider areas than is economically possible by ships. Satellite sensors can measure ocean surface layers directly, or the measurements can be inferred by the employment of both microwave and multiple-frequency infrared radiation and visual sensors. For example, sea surface temperatures show up as variation patterns in infrared radiation upward from the ocean. Such designs can be utilized to find and track and map thermally different surface waters, currents, heat flow, evaporation, wind stress and major oceanic fronts (places where waters of different origins meet), and eddies (circular water movements). By optics, analysis of color, and the use of gated laser systems, useful information can be provided about the seafloor in shallow water areas. These satellite and aircraft systems can be used together with research ships, ocean-moored instrument arrays, buoys and aircraft to intensively study an ocean region.

Sensor-instrumented buoys, such as the MONSTER BUOY [a 40-ft (12-m) diameter, $7\frac{1}{2}$-ft ($2\frac{1}{4}$ m) thick buoy capable of operating unattended for a year at a time] also are used to measure air temperature at various heights, water temperature at several depths, wind velocity, barometric pressure, humidity, and precipitation. This collected information can be either stored on tape in the buoy or relayed upon demand to a shore station.

Studies of marine atmospheric physics utilize airborne devices for collecting marine aerosols and for measuring atmospheric moisture. A dewpoint hydrometer that eliminates water droplets and permits measurement of the dewpoint within clouds is also used. Another special instrument measures cloud condensation nuclei and determines the fraction arising from sea salt.

The importance of obtaining more accurate basic data in order to understand and predict oceanic and atmospheric conditions of the world's oceans and the resulting weather patterns over the world's continents is obvious.

INTERAMERICAN TROPICAL TUNA COMMISSION is an international fishery research organization that operates under the authority and direction of a convention originally negotiated between the United States of America and the Republic of Costa Rica. The convention entered into force in 1950. It is open to adherence by other nations that participate in the fisheries covered by it.

Five member governments contribute funds for operating the commission according to a formula outlined in the convention which is based on the amount of their TUNA catch and use. The principal duties of the commission under the convention are:

1. To study the biology, ECOLOGY, and the population dynamics of the tunas and tuna-bait fishes of the eastern PACIFIC OCEAN in order to establish how fishing affects the stock.
2. To recommend conservation and management measures to maintain the tuna stocks at a level that affords maximum sustainable catches, when commission research demonstrates that such measures are necessary.

INTERMEDIATE WAVE See WAVES.

INTERNAL WAVE See WAVES.

INTERNATIONAL DECADE OF OCEAN EXPLORATION (IDOE) was a long-term multipurpose oceanographic research program carried out on an international cooperative basis during the 1970s; major IDOE projects included environmental forecasting, seabed assessment, living resources, and environmental quality.

The idea for an IDOE was first proposed in March 1968 by the President of the United States. In December 1968, the United Nations General Assembly endorsed "the concept of an international decade of ocean exploration to be undertaken within the framework of a long-term programme of research and exploration. . . ."

In late 1969, the National Science Foundation (NSF) of the United States was charged with the responsibility of planning, managing, and funding the United States part of the IDOE.

The goals of the IDOE were to:

- Provide the scientific basis needed to improve environmental forecasting.
- Determine the potential resources of the ocean floor.
- Determine the quality of the ocean world environment by scientific observations of the natural state of that environment, evaluate the impact of human activities on that environment, and establish a scientific understanding of the basis for corrective actions needed to preserve the ocean environment.
- Improve worldwide data exchange through updating and standardizing national and international marine data collection, processing, and distribution
- Provide basic knowledge of biological processes needed for the intelligent utilization of living ocean resources.

In the pursuit of these objectives, several cooperative projects were carried out. For example, the Geochemical Oceans Sections Study (GEOSECS) involved geochemists at 14 United States universities and geochemists from Belgium, Canada, France, Germany, India, Italy, and Japan. In their environmental quality study, research teams were charged with making detailed measurements of ocean characteristics along Arctic longitudinal sections and in the ATLANTIC OCEAN and PACIFIC OCEAN, at all depths. In April 1978, GEOSECS completed its program on precise chemical measurements in the major oceans. These provided, for the first time, physical and chemical data measured on identical WATER samples. This was done by several U.S. laboratories along with scientists, laboratories, and ships of Canada, France, West Germany, India, Italy, and Japan. The samples obtained are filed at the WOODS HOLE OCEANOGRAPHIC INSTITUTION, and the data published in a 12-volume atlas. The results of the project have provided new and valuable insights regarding the general importance of PARTICULATES MATTER in marine chemistry and the rate of deep-water formation in the North Atlantic.

The environmental forecasting work consisted of a study of the relation between CLIMATE and the oceans. Wind and current recorders, ships, aircraft, and free-floating instruments of many types were used in the Mid-ocean Dynamics Experiment. (See CURRENTS; MODE). Other studies included the North Pacific Experiment (NORPAX), whose objective was to understand the fluctuations in the upper layers of the North Pacific Ocean and the relation of these fluctuations to the overlying and adjacent atmosphere. In addition to these studies, investigations of the Southern Ocean and their relation to atmosphere circulation patterns were carried out, plus a study (CLIMAP) was made on climate from a long-range investigation, mapping, and prediction basis. The focus here was on a description and explanation of climate changes over the last million years, with the deep-sea sediments used as the primary source of data.

The Seabed Assessment Program had as its principal objective the the resolution of questions concerning geologic processes along continental margins, mid-oceanic ridges, and deep-ocean basins. These projects included investigations that can be broadly grouped as continental-margin studies, plate tectonics and metallogensis studies, and manganese nodule studies. In the continental-margin work, attempts were made to better understand the interaction of continental plates, especially the aspects of this process that may relate to the formation of hydrocarbon and mineral deposits. Emphasis was especially on the African Atlantic margin, the Southwest Atlantic margin, and the Caribbean margin. An important seafloor spreading investigation of the plate tectonics and metallogenesis work was that concerned with the Nazca Plate in the Peru-Chile Trench area. Here, seismic refraction and seismic reflection observations taken together with analysis of bathymetry, geomorphology, volcanism, and seismicity provided a detailed chronological orientation for seafloor spreading patterns. Studies of the sediments in this area indicate that SEAWATER hydrothermal systems extract and transport metals initially disseminated in newly erupted underwater volcanic rocks, and that this process not only influences the composition of seawater, but it has led to the formation of many important continental ore bodies. Also in the plate tectonics and metallogenesis studies, the Project Famous (French American Mid-Ocean Undersea Study) scientists in their work using submersibles to explore the rift valley of the mid-Atlantic Ridge made the first direct observations of faults and lava flows of a newly formed ocean floor. Such studies have made it possible to construct accurate geologic maps showing both small and large scale geologic seafloor features. The photographs and samples taken during Project Famous have provided new information on seafloor spreading, the formation of new crust, and the hydrothermal emplacement of minerals.

The Living Resources Program concentrated on the study of complex marine ecological systems, with the goal being the ability to provide the scientific basis for the improved management and wiser utilization of the ocean's living resources. Included in this program were investigations such as the Coastal Upwelling Ecosystems Analysis (CUEA) and the Seagrass Ecosystem Study (SES).

INTERNATIONAL MARINE SCIENCES ORGANIZATIONS are those groups or bodies whose efforts are directed on an international scale for promoting studies of oceanographic problems and coordinating oceanographic research.

In recent years in the marine sciences, there has been a proliferation of committees, commissions, institutions, projects, etc., that either directly or indirectly perform the above mission. To name all these activities and to describe their detailed work would require an entire book. Several such directories have been compiled in the past. For example, the *World Directory of Hydrobiological and Fisheries Institutions,* (published by the American Institute of Biological Sciences, Washington, D.C., Library of Congress Card Number 63-18290) lists over 2000 worldwide organizations having to do with marine biology alone. In addition, the *International Directory of Marine Scientists* [published periodically by the Food and Agriculture Organization (FAO) of the United Nations] not only contains the names of 5745 marine scientists (arranged alphabetically and by area of oceanographic specialization) from 91 countries of the world but also lists the worldwide institutions involved both in OCEANOGRAPHY and OCEAN ENGINEERING. Another information source is the U.S. National Oceanographic Data Center (an activity sponsored by various U.S. goverment agencies having an interest in the marine environment). This center has published a listing of some 115 international organizations whose activities relate to the field of marine science.

Thus, the selection of those 16 organizations mentioned below is merely intended to represent a sample cross section of some of the primary international activities and centers involved in marine sciences or various facets of oceanographic endeavors.

• ***Advisory Committee on Marine Resources Research (ACMRR)***

Address: Secretariat
Advisory Committee on Marine Resources Research
Via delle Terme di Caracalla
Rome, Italy

Founded in 1962, the ACMRR is a nongovernmental body of the Food and Agriculture Organization (FAO) of the United Nations. The membership of the ACMRR is made up of 15 individuals appointed in their personal capacity for 1-year period and eligible for reappointment for the purpose of advising the Intergovernmental Oceanographic Commission (IOC). This advisory committee counsels the FAO director-general on the formulation and execution of the organization's program of work concerned with the research on marine fisheries resources. It also acts as an advisory body to the IOC on the fisheries aspects of oceanography. Reports of the work of this committee are published periodically.

• ***Commission for Maritime Meteorology (CMM)***

Address: President
Commission for Maritime Meteorology
Meteorological Branch
315 Bloor Street West
Toronto 5, Canada

The idea for such a commission as this emerged when a group of maritime nations met in 1853 to develop a program for collecting weather observations. Originally established as the Commission for Weather Signals and Marine Chart Projections and Scales in 1910, its name was changed to this International Commission for Maritime Meteorology and Weather Signals. From 1919 the commission has functioned under its present name, and in 1951 it was incorporated into the World Meteorological Organization (WMO) as one of that organization's eight technical commissions.

The work of the commission, whose membership is composed of representatives of 56 countries, is concerned with the improvement of weather service, the collection of meteorological data for use in preparing weather forecasts and warnings, the organization of meteorological networks for observations at sea, and the promotion of studies of the meteorological aspects of ocean WAVES and sea ice.

• *Engineering Committee on Oceanic Resources (ECOR)*

Address: The Secretary
Engineering Committee on Oceanic Resources
2101 Constitution Avenue, N.W.
Washington, D.C. 20418 U.S.A.

The Engineering Committee on Oceanic Resources (ECOR) was formally established in March 1971 as an international, nongovernmental, professional engineering body. The purpose of ECOR is to provide an international focus for professional engineering interests in marine affairs with particular emphasis on (1) establishing and maintaining international professional engineering communications in marine affairs; (2) providing advice, from an engineering viewpoint, on policy, program, and organizational marine matters to international and intergovernmental organizations as well as to individual nations; and (3) assisting the engineering profession in the use of the ocean and in the enhancement of the quality of the marine environment. The interest of ECOR includes all aspects of engineering practice (such as design, management, operation, planning, and research) and all engineering disciplines (such as biological, civil, chemical, electrical, mechanical, mining, naval architectural, and transportation) as they relate to the marine environment. The Engineering Committee on Oceanic Resources publishes a newsletter on a quarterly basis as well as several reports on various marine engineering subjects of international interest.

ECOR, in 1976, was composed of some 13 national members and 5 international members (e.g., International Association of Dredging Companies, International Association for Hydraulic Research, International Association on Water Pollution Research, International Ship Structures Congress, and Fédération Internationale de la Précontrainte).

• *International Association of Biological Oceanography (IABO)*

Address: Secretary
International Association of Biological Oceanography
National Institute of Oceanography
Wormly, Godalming
Surrey, United Kingdom

Founded in 1954, this association is a nongovernmental organization of the International Union of Biological Sciences (IUBS), Division of General Biology. The National Institute of Oceanography at Godalming also provides a standard seawater service for oceanographic activities throughout the world. (See NORMAL WATER.)

• *International Association of Physical Sciences of the Oceans (IAPSO)*

Address: Secretary
International Association of Physical Sciences of the Oceans
Woods Hole Oceanographic Institute
Woods Hole, Massachusetts 02543 U.S.A.

Established in 1919 as an oceanographic section of the International Union of Geodesy and Geophysics (IUGG), in 1967 the name was changed from International Association of Physical Oceanography to its present one.

The association promotes scientific study of oceanographic problems through publications (e.g., *Procès-Verbaux* and other scientific publications) and initiates and coordinates international research and international scientific meetings. Member countries of IUGG accredit delegations to IAPSO meetings through their national academies of sciences.

• *International Council for the Exploration of the Sea (ICES)*

Address: Secretary-General
International Council for the Exploration of the Sea
Charlottenlund Slot
Charlottenlund, Denmark

This council, established in 1902, operates a laboratory and encourages all research connected with the exploration of the oceans and coordinates the activities of the participating governments. Various reports, bulletins, and data lists are published periodically. Membership is composed of two delegates from each of the 17 member governments.

• *International Council of Scientific Unions (ICSU)*

Address: Secretary-General
International Council of Scientific Unions
Via Cornelio Celso 7
00161 Rome, Italy

This council was established in 1919 as the International Research Council (IRC), and the present name was adopted in 1931.

Primarily, the ICSU coordinates and facilitates activities of International Scientific Unions, acts as a coordinating center for national adhering organizations, encourages international scientific activity, and maintains relations with the specialized and related agencies of the United Nations. A quarterly bulletin, *Year Book,* and various other reports on its work are published.

• *International Geographical Union (IGU)*

Address: Secretary-General
International Geographical Union
Blumlisalpstrasse 10
8006 Zurich, Switzerland

The International Geographical Union was founded in 1922. It promotes the study of geographical problems, initiates and coordinates research requiring international cooperation, provides for scientific discussion and publications (e.g., *The IGU Newsletter*—semiannual reports of commissions, and bibliographies), arranges international congresses and appoints commissions for study of special matters during the interval between congresses. Membership is composed of adhering organizations, national committees, and associate members in 62 countries.

• International Hydrographic Bureau (IHB)

Address: Director
International Hydrographic Bureau
Avenue President J.F. Kennedy
Monte Carlo
Principality of Monaco

The International Hydrographic Bureau was founded in 1921. It establishes a permanent association between hydrographic services of various maritime countries, coordinates their work with a view toward rendering navigation easier and safer in all seas, endeavors to obtain uniformity in hydrographic documents, advance the science of hydrography, and facilitate the free exchange of hydrographic charts and information between nations. Various hydrographic bulletins and literature are published by this activity. Members are government representatives of 40 states.

• Intergovernmental Oceanographic Commission (IOC)

Address: Chairman
Intergovernmental Oceanographic Commission
UNESCO Headquarters
Place de Fontenoy
Paris 7e, France

The Intergovernmental Oceanographic Commission was formed in 1960 within UNESCO to promote the scientific investigations of the oceans, with the aim being to learn more concerning their nature and their resources through the concerted effort of all member states. The commission publishes summary reports of sessions. Government representatives of 65 states make up the membership of the commission.

• International Union of Geodesy and Geophysics (IUGG)

Address: Secretary-General
International Union of Geodesy and Geophysics
Geophysics Laboratory
University of Toronto
Toronto 5, Canada

Founded in 1919 as a constituent union of the International Research Union (now the International Council of Scientific Unions), this activity promotes the study of all problems concerning the configuration of the earth and the physics of the globe, oceans, and atmosphere. It initiates, facilitates, and coordinates research and investigation of those problems of geodesy and geophysics which require international cooperation. Monthly chronicles and occasional monographs are published. Membership is composed of representatives of seven international unions and 58 adhering countries represented by governments, scientific academies, or departments.

• Permanent International Association of Navigation Congresses (PIANC)

Address: Secretary-General
Permanent International Association of Navigation Congresses
60 rue Juste Lipse
Brussels 4, Belgium

Founded in 1900, this group fosters progress in the maintenance and operation of inland and maritime waterways, and compiles and publishes information about subjects in its field (e.g., quarterly bulletins, technical papers, proceedings, technical dictionary, bibliographical notes, and reports). Research into particular problems is initiated, and international congresses through which members may share their experience and research are organized. Membership is composed of 50 governments, corporate and individual members in 69 countries, and national branches in 25 countries.

• The Permanent Service for Mean Sea Level (PSMSL)

Address: The Permanent Service for Mean Sea Level
The Observatory
Birkenhead, Cheshire
England

The Permanent Service for Mean Sea Level collects and disseminates data (monthly and annual values of mean sea level are published triennially). The group conducts research to ensure a higher standard of accuracy and reliability in observations of sea level. It also encourages permanent installation and maintenance of new sea level gauges, provides a computational service, and modernizes methods of data storage and publication.

• Scientific Committee on Oceanic Research (SCOR)

Address: President
Scientific Committee on Oceanic Research
Scripps Institution of Oceanography
P.O. Box 109
La Jolla, California 92037 U.S.A.

Founded in 1957 by the International Council of Scientific Unions (ISCU), this body acts as a scientific advisory to UNESCO and the Intergovernmental Oceanographic Commission (IOC). It is a joint sponsor with the Indian Government and UNESCO of the Indian Ocean Biological Centre at Cochin, and also coordinates activities with the International Association of Physical Sciences of the Oceans, the International Council for Exploration of the Sea, and the Advisory Committee on Marine Resources Research. *Proceedings* is the title of the publication of the Scientific Committee on Ocean Research. Members are representatives of eight international unions (members of ICSU) and 28 national committees.

- ***United Nations Educational, Scientific and Cultural Organization (UNESCO)***

Address: Director-General
United Nations Educational, Scientific and Cultural Organization
Place de Fontenoy
Paris 7e, France

The constitution of UNESCO was adopted in 1945 based upon the idea that by promoting collaboration definite contributions to peace and security would be carried out among the nations through education, science, and culture. This organization develops international scientific cooperation in various disciplines including OCEANOGRAPHY by organizing meetings among scientists, assisting activities of international scientific organizations, and promoting the exchange of scientific information. Examples are the International Hydrological Decade (IHD), 1965–1975, and the program in oceanography. UNESCO publishes a wide variety of bulletins and publications. A catalog of publications is printed each year. Membership is composed of states of the United Nations; other countries can be admitted by vote.

- ***World Data Center (WDC)***

Address: World Data Center A, Oceanography
Washington, D.C. U.S.A.

World Data Centre B, Oceanography
Molodezhnaya 3
Moscow B-296, U.S.S.R.

The international exchange of data through World Data Centers (WDC's) was first organized during the International Geophysical Year (IGY) in 1958. At the end of the IGY, the responsibility for the exchange of data through the WDC's was assigned to Comité International de Géophysique (CIG) and later to the International Council of Scientific Unions' Panel on WDC's. The primary responsibility of World Data Centers is the collection and distribution of data, with the responsibility for processing of oceanographic data at the national level. Catalogs of data and newly received publications are published every 6 months by these centers.

In addition to these above-mentioned groups, which are intended to give the reader an insight into some of the types of international organizations involved in oceanographic endeavors, the work of the many national academies of sciences throughout the world, technical societies, and ocean science and ocean engineering institutions, laboratories, and centers should not be overlooked.

For example, many academies of sciences and national research councils have international programs that relate strongly to oceanography. To illustrate this point, the U.S. National Academy of Sciences (2101 Constitution Avenue, N.W., Washington, D.C. 20418, U.S.A.) has published many valuable studies on ocean science and engineering, and most of these have received wide international distribution. In addition, a number of international programs over the years have been carried out under its aegis, such as the International Biological Program, the International Hydrological Decade, and the International Decade of Ocean Exploration. Also, the Academy of Sciences of the Union of Soviet Socialist Republics (Moscow, V-71, Leninskii Prospekt 14, U.S.S.R.) has as part of its functions "to guide the development of fundamental research in marine science and to lay down the lines for international liaison on scientific matters." *Okeanologiya* (Oceanology), the journal published by this organization, is considered to be the leading scientific bimonthly publication in the Soviet Union. The publication deals with the physical, chemical, geological, and biological aspects of oceanography. It has been translated into English in its entirety and published by the American Geophysical Union of the United States.

Most other national academies of sciences of the world establish and maintain relations between the scientists of their countries and international scientific groups as well as between their scientific activities and those of scientists in other countries.

Regarding other scientific and engineering acitivies concerned with the oceans, over 90 countries have formal institutional and/or laboratory programs. Of these, the United States has over 275 centers, Japan about 164, the United Kingdom about 120, and the U.S.S.R. about 60.

INVERTEBRATE is a term applied to any animal not possessing a backbone (e.g., CRUSTACEANS, mollusks, worms and insects).

ION is an atom or molecule that by the loss or gain of one or more electrons has acquired a net electric charge. If the ion is formed from an atom of hydrogen or an atom of a metal, it is usually positively charged; if it is formed from a nonmetal, it is usually negatively charged.

SEAWATER salts, such as sodium chloride (ordinary table salt), are usually composed of an orderly arrangement of ions which are not free to easily move in the solid. However, when sodium chloride, which is designated by its chemical formula as NaCl (Na = sodium and Cl = chlorine), is placed in WATER, the chemical bond between these two elements is weakened. (see ELEMENT.) As a result, the sodium and chlorine are separated (or ionized) in the solution, and the resulting particles are called ions. The sodium has a positive (cation) electric charge (Na^+), while the chlorine ion has a negative (anion) charge (Cl^-). Of significance in chemical OCEANOGRAPHY studies is the fact that the electric charge of an ion is not neutral and it can combine with ions of opposite charge. Such ionic interactions between water molecules and the ions of the dissolved salts in seawater, as well as between ions, take place in the ocean. A simple example of one of the results of ion separation in seawater is CORROSION. Since the ions are relatively free to move about in this solution, they are subject to the influence of external forces. For instance, the presence of metallic elements immersed in the solution will provide a barrier to the motion of the ions and, on contact, a chemical reaction will usually result. In the event that two metals such as iron and copper are so immersed in this sodium chloride solution, the molecules of the salt separate, or dissociate:

$$NaCl \xrightarrow{\text{solution}} Na^+ + Cl^-$$

In the elementary process, the chlorine ions will eventually come in contact with both the iron and the copper to form ferric chloride and cupric chloride, respectively:

$$Fe + 3Cl^- \rightarrow FeCl_3 + 3 \text{ electrons}$$
$$Cu + 2Cl^- \rightarrow CuCl_2 + 2 \text{ electrons}$$

Normally, the ferric chloride would deposit itself on the iron and eventually action would cease. The water, however, has also dissociated:

$$H_2O \rightarrow H^+ + OH^-$$

so that the hydroxyl ion OH^-, on coming in contact with the ferric chloride which is extremely soluble in water, causes ferric hydroxide to be formed, which subsequently forms ferric oxide (reddish brown). The net result is that some of the ferric chloride remains as a yellow solution, the ferric oxide forms a reddish brown deposit on the iron, and the hydrogen which was left from the dissociation of water is liberated at the copper. The process continues until the iron is covered with ferric oxide (one of the main components in rust), at which point complex and unexplained reactions occur. Eventually the iron structure is physically weakened and, if given a long enough time, will be converted to rust.

In this example, the medium through which the various ions moved was water in which common salt (sodium chloride) was dissolved. The use of the word *salt* is a descriptive term applied to a distinct family of chemical substances which are defined as the products resulting from the displacement of hydrogen from an acid by a metal. Generally, salts which are soluble in water possess the property of electric conductivity, in which case the resulting solutions are called electrolytes. The strength of the electrolyte is determined by the degree of dissociation of the solute (the compound or element dissolved) or, loosely speaking, by the ion-carrying capacity of the solution.

IRISH MOSS, or carrageenin (*Chondrus crispus*), is the name for a species of red ALGAE. See PHARMACEUTICALS FROM THE OCEAN.

IRISH SEA is an arm of the eastern ATLANTIC OCEAN that lies between Ireland on the west and England and Scotland on the east. The northern boundary is the line joining the Mull of Galloway in Scotland to Ballyquintin Point in Northern Ireland. The southern boundary is a line from Wooltack Point in Wales, through Skomar Island and the Smalls, to Carnsore Point in Ireland. The Irish Sea communicates with the Atlantic through North Channel between Scotland and Northern Ireland in the north, and through St. George's Channel between Wales and Ireland in the South. The two main islands in the Irish Sea are the Isle of Man and Angelesey. The sea is about 230 mi (370 km) long and 140 mi (225 km) wide at its widest point. Oceanographers have given the name Celtic Sea to the body of WATER that lies to the south of the Irish Sea and which has as its southern border a line running from Land's End to Ushant.

Both the Irish and Celtic Seas lie wholly on the CONTINENTAL SHELF, and their floors are devoid of major irregularities. A trough, having a number of closed basins along its axis, can be traced southward through the Irish Sea and into the Celtic Sea. The eastern half of the sea is shallow, but the western half reaches a depth of about 850 ft (259 m). Sediments range from muds to sand and shell fragments.

Circulation in the Irish Sea appears to be mainly due to a wind-driven current that moves northward through the Celtic Sea. A number of eddy currents are present. The tidal range in the area is variable, being greater on the east than on the west. Tides of about 4 ft (1.2 m) occur along the Irish coast, whereas spring tides (see TIDES) may reach 40 ft (12 m) along parts of the English coast. Tides of 20 ft (6 m) are common. Surface water temperatures range from 41–50° F (5–10° C) in winter to 55.4–62.6° F (13–17° C) in summer.

IRMINGER CURRENT is a branch of the NORTH ATLANTIC CURRENT that curves north in the vicinity of Iceland. A minor part of the current splits off to flow north along the west coast of Iceland, but the main body continues to curve back to the west until it joins the southward-flowing EAST GREENLAND CURRENT. Both branches eventually rejoin the North Atlantic Current. The Irminger Current flows at the rate of about 0.1–0.2 kn at the surface.

IRMINGER SEA is not officially recognized as a separate body of water, but oceanographers find it a useful distinction. As loosely defined, the sea lies between the east coast of Greenland and the west coast of Iceland. The LABRADOR SEA lies on its southwest corner, and the GREENLAND SEA on the northeast. The southern border is not well defined, being marked by the boundary of dissimilar water masses rather than geographical features.

The basin of the Irminger Sea is largely occupied by the eastern part of the large Labrador Basin which ranges to depths of 15 000 ft (4572 m). The northeastern limit of the basin which underlies the Irminger Sea is about 7800 ft (2377 m) deep. On the east the basin is bordered by that portion of the Mid-Atlantic Ridge known as the Reykjanes Ridge with a depth of around 2160 ft (658 m). The southern part of the basin is largely covered with globigerina ooze (see MARINE SEDIMENTS), whereas the northern part shows significant admixture of volcanic and glacial debris.

The circulation in the Irminger Sea is dominated by the IRMINGER CURRENT which splits off from the NORTH ATLANTIC CURRENT south of Iceland and runs west and southwest into the Labrador Sea. The twice daily (diurnal) tides in the sea range from 2.9–8.2 ft (0.91–2.5 m).

IRON-MANGANESE NODULES See MANGANESE NODULES.

ISAACS, JOHN DOVE, III (1913–), an American engineer and oceanographer, has made many important contributions to a wide range of marine research areas such as varved sediments, halophytes, the marine food web, the deep scattering layer, instruments, seafloor photography, climatology, wave power, wave dissipation, harbors, sand transport, high-pressure effects, etc. Recent contributions have dealt with new ways to determine the growth and mortality rates in some pelagic marine species; the uncovering of a sedimentary record that permits the reconstruction of a calendar of recent prehistorical oceanographic, biological, and climatic events; the development of a number of new instruments, particularly collectors; deep-moored, unmanned instrument stations; a wave-powered generator; and a floating breakwater system.

Isaacs was born in Spokane, Washington, on March 28, 1913. He received a bachelor of science degree from the University of California at Berkeley in 1944 and served there as a research engineer from 1944 to 1958. He was appointed associate oceanographer at SCRIPPS INSTITUTION OF OCEANOGRAPHY in 1948, associate professor of oceanography in 1955, and professor in 1961. At Scripps he also served as assistant to the director (1948–1958), as director of the Marine Life Research Group (1958–1974), as acting director of marine resources (1961–1962), and as director beginning in 1971. He also served as the acting chairman of the Scripps Department of Oceanography (1966–1967).

ISELIN, COLUMBUS O'DONNELL (1904–1973), an American oceanographer, was born in New Rochelle, New York, on September 25, 1904. He graduated from Harvard University (A.B. degree in 1926 and A.M. in 1928) where he majored in mathematics. While at Harvard, he met Henry Bigelow, professor of zoology and oceanography. Under Bigelow's influence and coupled with his great love of the sea, Iselin decided while at Harvard to make a lifetime career of studying the oceans.

In 1929 he accepted the position as assistant curator of oceanography at the university's Museum of Comparative Zoology, a post he held for 19 years. When the WOODS HOLE OCEANOGRAPHIC INSTITUTION was founded in 1930, Iselin was appointed by Bigelow, the Institution's director, to captain the Institution's first research vessel, *Atlantis.* This specially built ship was the first full-time oceanographic research vessel in the United States to be operated by a private institution.

During the next decade, Iselin and the *Atlantis* traversed the Atlantic Ocean many times, gathering valuable data on ocean depths, on water temperatures, and bottom samples. Although Iselin's own field work and published papers almost exclusively deal with the western half of the North Atlantic and the Gulf Stream system, he was one of the first to point out that together the atmosphere and the oceans constitute a vast and complex heat engine powered by the sun.

During the periods 1940–1950 and 1956–1958, Iselin served as the director of the Woods Hole Oceanographic Institution. Between these terms he was a senior physical oceanographer at Woods Hole and concurrently filled a number of other posts at Harvard and the Massachusetts Institute of Technology.

The recipient of many honors for his outstanding

work in marine science and his major role in the development of one of America's principal oceanographic institutions, Iselin is recognized as a man who greatly influenced modern oceanography.

ISOPODS is a common name for members of an order of CRUSTACEANS of the class Malacostraca which includes the SHRIMP, LOBSTER, CRAB, and sow bugs. The isopods are generally characterized by a head (cephalon) bearing a pair of legs, in addition to mandibles and mouthparts posterior to the mandibles. The appendages and trunk limbs are sharply differentiated into thoracic and abdominal series.

Although many of the 3000 known species of isopods are terrestrial (e.g., the sow bugs) and are usually found in moist environments under decaying wood and leaves and under rocks, the majority are marine. In the world's oceans they range from the intertidal areas to the greatest depths of the oceans. (See HABITABLE ZONES.)

These isopods are usually less than $1\frac{1}{2}$ in (3.8 cm) long, and in the most common genera their food consists of wood (genus *Limnoria*), seaweeds (genera *Limnoria* and *Idothea*), and animal flesh (genera *Cirolana* and *Cymothoa*). (See ALGAE.)

One $\frac{1}{2}$-in (1.27-cm) isopod called *Sphareroma* attacks the roots of MANGROVES. As a result, in some parts of Florida the red mangrove swamps are shrinking, the peat that had been formed under these trees has eroded away, and several islands have already disappeared.

JAPAN SEA (*Sea of Japan*), a marginal sea of the western PACIFIC OCEAN, is bounded on the east by the three main islands of Japan (Kyushu, Honshu, and Hokkaido), on the west and southwest by Korea, and on the north and northwest by the Soviet Union. To the northeast the Sea has access to the Sea of Okhotsk through the Soya Strait, and in the south it has access to the YELLOW SEA and EAST CHINA SEA through the Korea Strait. (See OKHOTSK, SEA OF.) Access to the Pacific Ocean is through the Tsugaru Strait between Honshu and Hokkaido. The Sea of Japan has an area of 377 508 mi^2 (978 000 km^2), a mean depth of 5748 ft (1752 m), and a volume of 410 966 mi^3 1 713 000 km^3). The greatest depth to be found is 13 284 ft (4049 m) in the Japan Basin.

The earliest people to cross what is now the Sea of Japan probably walked across. Evidence indicates that during the low stand of sea level which accompanied the last Ice Age (11 000 to 8000 years ago), the Korea Strait and the Soya Strait were exposed and served as land bridges between Japan and the continent of Asia. The fossilized remains of elephants similar to those in India are found on the southern Japanese islands, and the remains of woolly mammoths are found to the north. In a similar manner the early inhabitants of China, Korea, and Siberia probably migrated to Japan.

The written history of Japan begins in 660 B.C. with Jimmu Tenno who invaded the main island of Honshu and became the first emperor of Japan. In 550 A.D., a Korean Buddhist sailed to Japan to bring a different culture and a written language. However, it was not until Portuguese sailors landed in 1543 that Japan became known to Europeans, although Marco Polo brought back a report of its existence when he returned from China in the late 1200s. Even so, a reasonably good map of the Sea of Japan was not published until 1660, and the name by which it now goes did not appear until 1815. The name is credited to the Russian navigator, A.J. von Krusenstern.

The continental shelves of the Sea of Japan are abbreviated and rather poorly developed. (See CONTINENTAL SHELF.) They terminate at a depth of 656 ft (200 m) along the west coast of Japan and at a depth of 459 ft (140 m) along the east coast of Korea and the Soviet Union. The shelves are cut by numerous submarine canyons whose mouths terminate at a depth of about 6562 ft (2000 m) along the western rim of the sea, and at about 2625 ft (800 m) along the eastern rim.

The most conspicuous feature of the deep basin of the Sea of Japan is the Japan Basin, a rather featureless plain lying below the 9842 ft (3000 m) contour and occupying the north and northwestern half of the sea. To the south the basin is more complicated, being broken by the Yamato Bank and the Syunpu Bank, with a deep trough running between them. The seafloor in the south is interrupted locally by the Tsushima Islands in the Korea Strait, the Oki Islands off the west coast of southern Honshu, and the Ullung Islands off the coast of Korea.

The dominant current in the Sea of Japan is the warm, northward-flowing Tsushima Current which splits off from the KUROSHIO CURRENT to flow into the sea through the Korea Strait. A branch of the Tsushima, called the East Korean Warm Current, flows up the coast of Korea and then turns eastward to rejoin the Tsushima Current. The Tsushima eventually flows into the Pacific Ocean through the Tsugaru Strait (Tsugaru Warm Current) and into the Sea of Okhotsk through the Soya Strait (Soya Warm Current). Three distinctive cold currents (North Korea Cold Current, Maritime Province Cold Current, and the mid-Japan Sea Cold Current) flow down from the north to mix with the warmer water in the central part of the sea.

In the northern part of the Sea of Japan ice begins to form in November (particularly along the Siberian coast). By mid-February ice extends out into the sea and navigation is severely restricted. The waters are free of ice by May.

As might be expected for a sea that is nearly enclosed by land, the Sea of Japan has extremely low tides. The range along the coast of Japan is about 0.6 ft (0.2 m), and a range of 1.3–1.6 ft (0.4–0.5 m) is found along the Siberian coast. The range in the Korea Strait, because of its access to the Yellow and East China Seas, is about 6.6 ft (2 m).

It is in the Sea of Japan that many of the diving women of Japan (AMA) and Korea (hae-nyo) have plied their arduous trade for over 1000 years.

JAVA SEA, located in the western PACIFIC OCEAN, is bounded on the north by the southern limit of the SOUTH CHINA SEA (Gaspar and Karimata Straits), the southern coast of Borneo, and the southern limit of the Makassar Strait between Borneo and Sulawesi; on the east by the western limit of the FLORES SEA along a line running from Sulawesi to Flores; and on the south by the northern limit of the BALI SEA, Java, and Sumatra. The Java Sea covers an area of 167 138 mi² (433 000 km²), occupies a volume of 4789 mi³ (20 000 km³), and has a mean depth of 151 ft (46 m).

The Java Sea lies on the Sunda Shelf (CONTINENTAL SHELF) which extends southeastward from the Gulf of Thailand to the edge of the Flores Sea. The shallow floor of the sea is furrowed by drowned river channels that are thought to have been cut during the low stand of sea level that accompanied the last Ice Age.

The Java Sea is under the influence of the monsoons that blow from the northwest in winter and the southeast in summer. The surface WATER temperature ranges from 81.5° F (27.5° C) in winter to 84.2° F (29° C) in summer. Surface currents flow westward from September to May and are reversed during the remainder of the year.

JELLYFISH is the common name for any of the two-layered bell-shaped COELENTERATES (also called Cnidarians) belonging to the classes Hydrozoa and Scyphozoa of the phylum Coelenterata (which means hollow gut, referring to a baglike digestive cavity).

Jellyfishes, which are also called medusae, are found in all geographical parts of the world's oceans and at the surface, in the mid-depths as well as on the bottom. Their size can vary from microscopic to several feet in diameter.

These omnivorous animals belong to the same division of life which includes the corals, anemones, HYDROIDS, and gorgonians. (See CORAL; SEA ANEMONE.) Structurally all are similar, having an outside cell layer (ectoderm) and an inside layer (endoderm), with a jellylike substance between the two layers. These make up an open pulsating balloon formation with venomous tentacles surrounding the opening. The strength of the poison carried in the stinging cells (or nematocysts) attached to the tentacles depends upon the particular species of jellyfish. The most lethal of all the jellyfishes to human beings is the Australian sea wasp (a box jellyfish of the group Cubomedusae). This animal is perhaps the most dangerous of any to be found in the ocean. A sting from any one of its 15 tentacles has been known to kill a human being in a few minutes.

Many species of FISH and sea turtles eat jellyfish. Jellyfish, also being omnivorous, consume fishes which they capture either by stinging them or holding them with their sticky tentacles and then bringing the food to the mouth located on the underside of the bell-like body.

The jellyfish possess the ability to produce colonies of many individuals from a single fertilized egg. The PORTUGUESE MAN-OF-WAR, *Physalia,* a siphonophore, illustrates this point, for it actually represents a colony of thousands of animals developed from such a single egg.

JETTY See BEACH.

JEWFISH See GROUPER.

JOHNSON, MARTIN WIGGO (1893–), an American biologist, is best known for his work on marine life and its influence on sound in the sea and for his book *The Oceans* written with H.U. Sverdrup and R.H. Fleming in 1942. In recognition of his accomplishments, Johnson was awarded the Agassiz Gold Medal by the National Academy of Sciences in 1959.

During the early days of World War II the Navy was plagued by an unusual background noise which threatened to disrupt some of its more sensitive acoustic sensors. (See UNDERWATER SOUND.) The noise was a high-frequency crackle that resembled static and only appeared at certain times and localities. Particularly susceptible were harbor areas with muddy, littered bottoms and heavily fouled wharfs. Johnson, after considerable research, was able to show that the sound was produced by the "claw clicking" of many snapping SHRIMP. While nothing could be done about the shrimp, the location where this interference might occur could be predicted with some accuracy.

In 1942, C.F. Eyring, R.J. Christensen, and R.W.

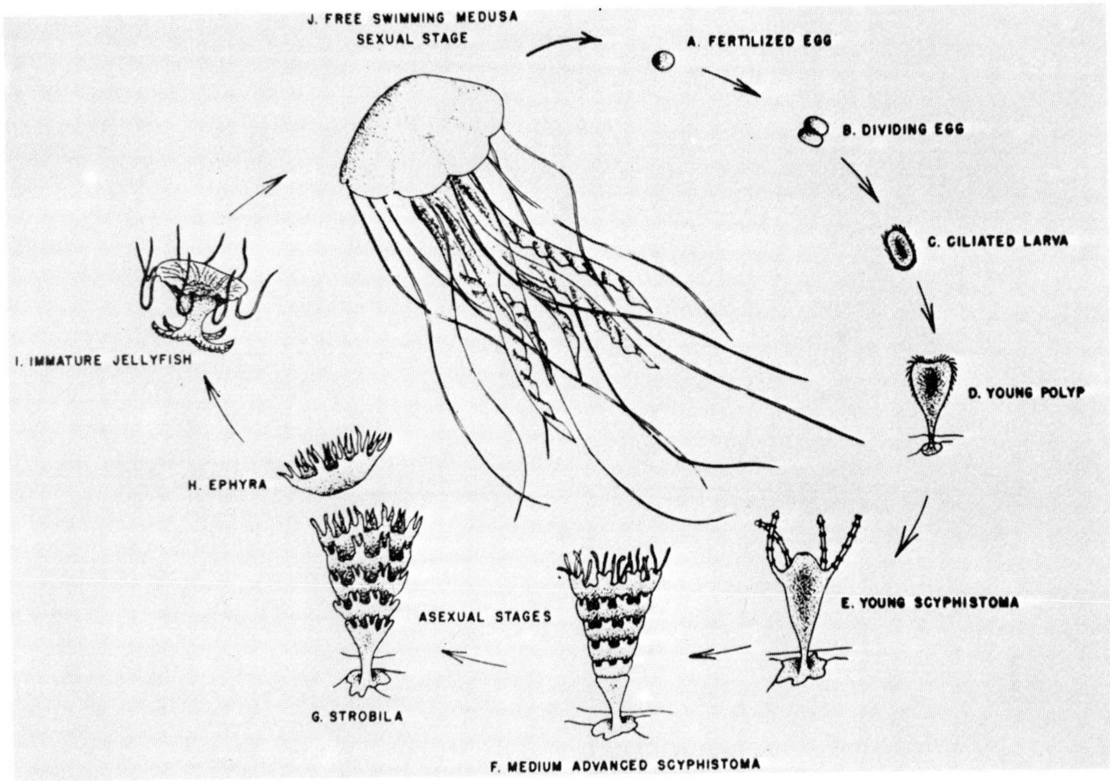

JELLYFISH. The diagram shows the stages of development in the reproductive cycle of the jellyfish *Aurelia.* The adult's long tentacles contain nematocytes, or stinging cells, used to disable enemies and prey. *(NOAA)*

Laitt observed that acoustic signals were often reflected from a mysterious scattering layer that prevailed at a depth of several hundred meters during the day but was observed to rise toward the surface during the late evening hours. The deep scattering layer, as it was called, was troublesome in that it was often interpreted as the bottom when, actually, the bottom lay far below. (See DEEP SCATTERING LAYERS.) Again, Johnson was able to unravel the mystery by showing that the sound energy was being scattered back to the surface by a layer of marine life made up of PLANKTON and the larger animals that feed on them.

Johnson was born in Chamber, South Dakota, on September 30, 1893. He received his B.S. (1924), M.S. (1930), and Ph.D. (1931) in zoology from the University of Washington (Seattle). He was curator of the University's Friday Harbor Laboratories from 1924 to 1929, biologist at the Passamaquoddy Bay International Fisheries Commission from 1932 to 1933, and associate at the University of Washington from 1933 to 1934. In 1934 he joined the faculty of SCRIPPS INSTITUTION OF OCEANOGRAPHY.

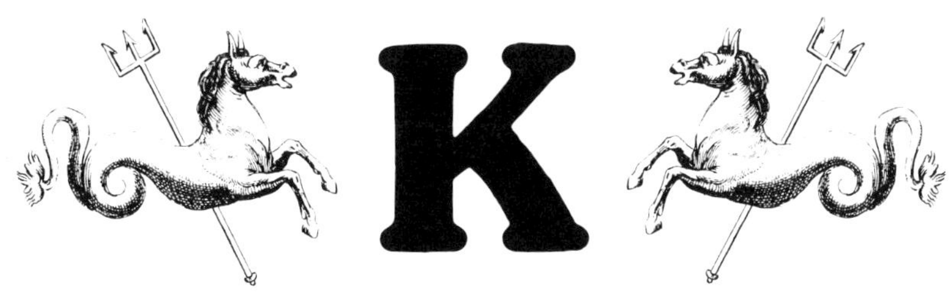

KARA SEA, one of the seas of the ARCTIC OCEAN, lies off the north coast of the West Siberian Lowlands (U.S.S.R.). The western boundary of the Kara is provided by the long island of Novaya Zemlya and a line from its northern tip to the easternmost point on Franz Josef Land on the edge of the permanent ice pack. The northern boundary is a line connecting Franz Josef Land to Komosomolets Island at the northern extreme of Severnaya Zemlya. The archipelago of Severnaya Zemlya and Cape Pronchishchev on the mainland serve as the eastern boundary. The Kara occupies an area of 340 838 mi^2 (883 000 km^2) and a volume of 24 950 mi^3 (104 000 km^3). Lying totally on the CONTINENTAL SHELF, the Kara has a mean depth of only 387 ft (118 m).

About 360 mi^3 (1500 km^3) of fresh water is carried into the Kara Sea each year, primarily by the Ob and Yenisei Rivers. This makes for surface salinities of 7–10 ppt wherever the lighter fresh water persists. By contrast, the Atlantic and Arctic waters flowing in under this layer have a salinity of 32–34 ppt. During the summer the temperature of the brackish surface water is typically 41–44.4° F (5–8° C), while the more saline bottom water is 32 to 29.3° F (0 to − 1.5° C). SEA ICE begins to form over the Kara in September, and melting usually does not begin until June. Icebergs from glaciers on boardering islands are most evident in August.

The first man of record to enter the Kara Sea by ship was the Dutch ship captain, seal hunter, and navigator, Willem Barents. In 1596, Barents, in a ship commanded by Jakob van Heemskerk (Barents had commanded two previous voyages), set out on his third attempt to find a Northeast Passage to China. In September 1596, after reaching the Kara Sea in July, his ship was frozen in the ice off the northeast coast of Novaya Zemlya. Barents became the first explorer to winter-over in the Arctic. He died on the return voyage the following year.

KELP is the name applied to the brown ALGAE belonging to the orders Laminariales and Fucales (phylum Phaeophyta). Kelp typically grow on rocky bottoms of the ocean and attain their greatest size in cold waters, with lengths as great as 100 ft (30.48 m) and blades slightly more than 4 ft (1.22+m) in width.

KILLER WHALE, *Orcinus orca,* is a predatory cetacean found only in cold ocean waters. See PORPOISE; WHALES.

KING WAVE is a wave of an unusual type which may suddenly rise 30–40 ft (9–12 m) up a cliff face to sweep unsuspecting fishermen and sightseers into the sea. Along many rocky coastal areas of Australia and South Africa there are signs that warn of the danger of king waves. In spite of the warning, a number of lives are lost to these waves each year.

King waves occur under a unique set of conditions involving deep water just offshore, a rocky coast, and large swell (or possibly, high waves) which approaches the coast at a shallow angle—usually 20° or less.

Waves and swell approaching a rocky coast at or near normal (90–75°) are reflected back to sea at an angle roughly perpendicular to the coastline. Waves which impinge on the coast at angles of 75–20° are not reflected back upon themselves but glance off at an angle similar to the incident angle, but in the opposite quadrant. Waves and swell which approach the coast at an angle of 20° or less, on the other hand, are not reflected. Instead, the water arriving with the waves flows along the coast, gain-

KELP. Kelp beds glisten along a pebble beach in California. Seaweeds are a good source of vitamins and minerals. Some have important medicinal applications, and some are used in food manufacture. (*U.S. Navy*)

ing velocity and increasing the height of the water surface, provided that enough new water is added to the system. This set of conditions often results in a king wave. Several such waves may arrive in succession, or the arrival of individual waves may be separated by several weeks.

In king wave territory one must be mindful of the danger up to as much as 40 ft (12 m) above sea level. The formation of a dangerous wave is usually accompanied by a hissing noise. As the wave gains in height and velocity, the hissing noise is replaced by a low rumble caused by the transport of shingles and blocks of rock many feet in diameter. The danger from a king wave stems primarily from the fact that it can arise very quickly and strike with great intensity.

KNUDSEN, MARTIN (1871–1949), the Danish scientist, who, in collaboration with another Dane, Jacob Jacobsen, evolved (in 1940) a workable definition of the CHLORINITY of SEAWATER. Knudsen also served as director of the Standard Seawater service of the Hydrographic Laboratories in Copenhagen. See DENSITY; NORMAL WATER.

KERMER, GERHARD, also known as Gerhardus Mercator, was a sixteenth century Flemish geographer who revolutionized the process of map- and chart-making. Mercator published a world map constructed on the projection that now bears his name. See NAUTICAL CHART.

KRILL is the common name for tiny floating (or planktonic) CRUSTACEANS resembling SHRIMP; these organisms are members of the Euphausiacea, a well-defined order closely related to the decapod crustaceans. All are pelagic and occur in all oceans. They are found in vast numbers in the polar seas. Characterized by the possession of well-developed, light-producing organs on the underside of their bodies and on the stalks which also support their eyes, Euphausiids emit a brilliant blue-green light from their photophores.

The word *krill* is of Norwegian origin and means "small fish." These tiny crustaceans of which there are about 80 species, serve as the diet for MARINE BIRDS, SEALS, PORPOISE, and certain FISH and WHALES (e.g., the baleen whales). Krill, which is one of the most important single elements of the marine biomass, is certainly the key organism in the whole ECOLOGY of the Antarctic, with entire communities of fish and animal life dependent upon it for food.

A typical example of krill is *Euphausia superba,* or lobster krill, a bright red, shrimplike crustacean that grows about 1 in (3 cm) a year, reaching a size of approximately 2 in (5–6 cm) in its 2-year life cycle. This plentiful animal spawns from November to March, with each female producing from 2000 to 30 000 eggs, depending on its size.

Euphausia superba and other krill species are herbivores and feed efficiently on the PLANKTON that is abundantly dispersed in waters of their environment. They occur in swarms or patches particularly in LATITUDE 60 and 75°S, such as the WEDDELL SEA, in waters of 0.3–2.0° C TEMPERATURE range. In the swarms [usually at depths of 33 ft (10 m) from the surface] the animals are so densely packed as to give a reddish hue to water areas ranging in size from a few yards to more than 6 mi (10 km) across. Whales congregate in such areas, and one whale consumes between 2 and 3 tons (1.8 and 2.7 metric tons) of this krill per day. The blue and fin whales of the southern waters feed almost exclusively on the *Euphausia superba,* although there are other members of the same genus (e.g., *E. frigida* and *E. triacantha*) present in the same waters. The whale's selectivity of *E. superba* appears to be due to the habit of this species to mass together in enormous swarms. The krill of the Arctic is primarily *Meganyctiphanes norvegica.*

Krill are rich in vitamin A and protein content (approximately 20 percent); a pound of krill yields at least 460 cal (about the same as other shellfish) when consumed by human beings. Undoubtedly, krill have a considerable potential for helping to feed the world population in the future. They represent a very large resource and one that can be systematically harvested and processed for human consumption.

In this regard, the Russians and Japanese are perhaps the most advanced in exploiting krill. Both utilize large stern trawlers, and catches up to 1000 tons (91 metric tons) in a 2-month (December to January) period have been reported. Several types of krill products have been made and sold for food and food flavoring.

KUENEN, PHILIP HENRY (1902–), a Dutch geologist, and one of the early experimental geologists, devoted a large part of his career to the study of marine geology with particular emphasis on coral reefs, submarine canyons, and turbidity currents. (See CORAL REEF; SUBMARINE CANYON; TURBIDITY CURRENTS.) Kuenen's interest in the ocean began when he was appointed to the scientific staff of the Snellius deep-sea expedition to the Moluccas in 1929–1930. The result of his research in those coral-studded waters strongly supported Charles Darwin's theory regarding the influence of subsidence on the formation of coral atolls and R.H. Daly's theory of glacial control to explain the upper part of such structures. (See DARWIN, CHARLES.) He was also able to show that the shape of the basins in the vicinity of the Moluccas had a systematic relationship to the gravity field set forth by F.A. Vening Meinesz.

Kuenen's next major contribution to oceanography was to bring his experimental talents to bear in support of Daly's 1936 theory that submarine canyons were formed by the eroding action of turbidity currents during the Ice Age. Using laboratory models, he showed that not only fine-grained clays but coarse sand as well could be transported by these currents. Kuenen went on to suggest that sand found on the deep seafloor was derived from the CONTINENTAL SHELF and had been transported there by turbidity currents. He believed, along with Daly, that the erosional power of these currents was sufficient to explain the existence of submarine canyons. This claim is still controversial, but most oceanographers agree that turbidity currents play a role in keeping canyon floors clear of sediment that slumps from the sides. Certainly the importance of these currents in transporting sediment to the ABYSSAL PLAIN cannot be overstated.

One of the very few early experimental geologists, Kuenen conducted laboratory studies of salt domes, volcanic cones, crustal folds, and other tectonic and sedimentary structures. These experiments have added much to our understanding of the process by which these structures are formed.

Kuenen was born July 22, 1902, in Dundee, Scotland. He was educated at the University of Leiden and held a number of staff positions there and at Gröningen. He became a full professor in 1946. His best-known book is *Marine Geology* published in 1950.

KURIL CURRENT See OYASHIO CURRENT.

KUROSHIO CURRENT, Japanese for "black stream," is the western Pacific counterpart of the GULF STREAM found off the east coast of the United States. The Kuroshio is one of the swiftest currents in the world [15.7 in/s (40 cm/s)] and can be identified as a relatively narrow stream of about 50 mi (80 km) in width running from the northeastern coast of the Philippines to the eastern coast of Japan. Flowing northward as a continuation of the North Equatorial Current (see EQUATORIAL CURRENT SYSTEM), the Kuroshio and the Kuroshio Extension eventually merge with the North Pacific Current that circles clockwise around the northern rim of the Pacific Basin. (See CURRENTS.)

As the Kuroshio Current reaches the southern Japanese island of Kyushu, some of its waters flow along the west coast of Kyushu and through the Tsushima Strait into the Sea of Japan to form the Tsushima Current. The major part of the WATER flows northeastward along the southeastern coast of Japan before turning eastward at about the northern extremity of Honshu. After leaving the coast of Japan, the Kuroshio splits, with one branch continuing east as the Kuroshio Extension and the other turning to the southwest to become the Kuroshio Countercurrent. The North Pacific Current is a further continuation of the Kuroshio Extension which gradually turns south, after crossing the Pacific, to feed the CALIFORNIA CURRENT.

The picture of the Kuroshio Current as a relatively narrow stream transporting some 1.4–1.8 billion $(1.4-1.8 \times 10^9)$ ft^3 [40–50 million $(40-50 \times 10^6)$ m^3] of water per second at velocities ranging up to 6 kn is, at times and places, misleading. Like its Gulf Stream counterpart, the Kuroshio fluctuates in speed with location, time, and season, and at times develops extreme meanders that may swing up to 310 mi (500 km) from the main axis. The greater understanding provided by oceanographic research over the past decade or so indicates that such behavior is characteristic of a major ocean current.

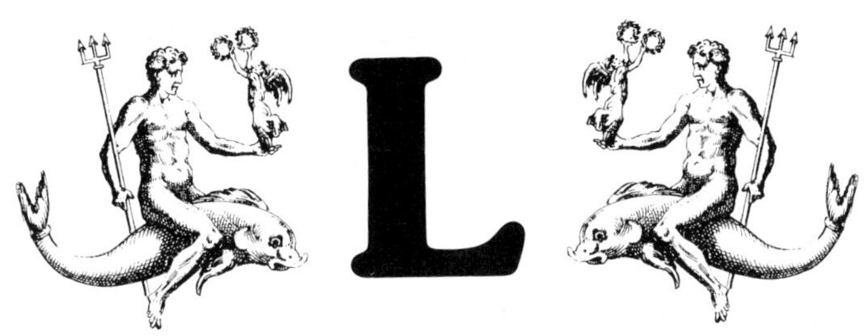

LABRADOR SEA, an arm of the North ATLANTIC OCEAN, is located between Greenland and Labrador. The southern boundary is a line from Cape Farewell on the southern tip of Greenland to Cape St. Charles on the coast of Labrador just north of Belle Isle Strait which leads into the Gulf of St. Lawrence. The northern boundary is the 66° north LATITUDE line joining Greenland and Baffin Island just north of the ARCTIC CIRCLE. The Labrador Sea is not officially recognized as a separate body of water, but scientists find it a useful distinction.

The CONTINENTAL SHELF of the Labrador Sea is about 31 mi (50 km) in width off the west Greenland coast and about twice that width along the coast of Labrador. The shelf frames the Labrador Basin in a horseshoe configuration which opens to the south. The northern limit, in Davis Strait, is a sill with a depth of about 2625 ft (800 m). The sill separates the Labrador Sea from Baffin Bay. Southward from the sill the depth continually increases until it reaches 13 123 ft (4000 m) in some locations along the southern limit of the sea. The bottom is made up of globigerina ooze mixed with clays and silts derived from the bordering continental shelf. The center of the basin is marked by the serpentine Mid-ocean Canyon which heads up on the sill in Davis Strait and runs southward to the Sohm ABYSSAL PLAIN off the tip of the Grand Banks. This canyon is thought to be the result of a TURBIDITY CURRENT which periodically carries large amounts of sediment down the basin to the deep ocean floor. (See SUBMARINE CANYON.)

Circulation in the Labrador Sea is provided by the West Greenland Current which flows northward along the west coast of Greenland and the Labrador Current which flows southward from Baffin Island. (See CURRENTS.) Between these two currents is a large body of WATER that results from the mixing of Atlantic water and Arctic water. In winter, when it is cooled to around 37° F (3° C), the density of this water becomes such that it sinks and flows southward. This cold, dense water can be traced to regions well south of the equator. (See UPWELLING.)

The Labrador Sea is only partly ice-covered during the winter months. The southeastern extremity is ice-free year-round except for the icebergs floating south from the Greenland coast. Ice begins to form along each flanking coast as early as October, and by December it has extended well out from the coast, particularly the Canadian coast.

The tidal range in the Labrador Sea is about 5.9 ft (1.8 m) along the western coast, and as much as 13 ft (4 m) along the eastern coast.

LACCADIVE SEA occupies a wedge-shaped area off the southern tip of India between the ARABIAN SEA on the west and the Bay of Bengal on the east. The official boundary is a line beginning at Sadashivgad Light on the west coast of India, to Corah Divh down the west side of the Chagos-Laccadive Plateau to Addu Atoll (Maldives), then northeast to Dondra Head on Sri Lanka (Ceylon), across Adams Bridge to India, and around the coast to Sadashivgad Light. Most oceanographers consider the Laccadive Sea to be a part of the Arabian Sea, and its major features have been discussed under that heading.

LAMPREY or **SEA LAMPREY,** is the common name for parasitic marine animals (cyclostomes) of the order Petromyzonida; some lampreys attain a length of 3 ft (approx 1 m).

There are some 24 species of lampreys living in North American, Eurasian, and sub-Antarctic waters of the world's oceans. Many lampreys are anadromous and leave the ocean waters to spawn in lakes and rivers. These parasites are not really eels since they lack paired FINS and jaws. (See EEL.)

LAMPREY. (*Top*) A lamprey attaches itself to a whitefish in a death lock. Once attached, the lamprey stays in this position until it or its victim dies. (*Bureau of Sport Fisheries & Wildlife*) (*Left*) The cavernous mouth of a typical mature sea lamprey yawns open, providing an astonishing view of its dentition. All the teeth present are hard and well equipped for rasping away fish flesh. (*NOAA*)

Equipped with a funnel-shaped mouth, the sea lamprey attaches itself to the side of a fish and drills through the scales with its teeth; it then lives off the host's blood and fluids. The animal is sometimes called a stone-sucker because of its ability to fasten itself to rocks and stones while awaiting its prey.

Perhaps the best-known and most infamous of the sea lampreys is the species *Petromyzon marinus.* In the 1920s, this fish migrated into the upper Great Lakes of the United States where it proceeded to destroy the lake trout population. This resulted in the collapse of the commercial fishing industry of that region. (See COMMERCIAL OCEAN FISHING.) In the 1950s and 1960s, a chemical known as TFM (or 3-trifluoromethyl-4-nitrophenol) was used effectively to kill the lamprey larvae in some 300 streams leading to the Great Lakes. This control, plus a program for restocking the waters with other fish species, such as SALMON, has largely restored the area.

LAPTEV SEA, occasionally referred to as the Nordenskjöld Sea, is one of the peripheral seas of the ARCTIC OCEAN and lies off the north coast of central

Siberia (U.S.S.R.) in the vicinity of the Siberian Lowlands. The Laptev is bounded on the west by the eastern coast of the Archipelago of Severnaya Zemlya (North Land), and on the east by the western coast of Kotelny Island and the Lyakhovskiye Islands. To the north the Laptev merges with the WATER of the Arctic Ocean, its northern boundary being marked by a loxodromic curve (rhumb line) drawn from Arktichesky Mys (Arctic Cape) at the northern tip of Severnaya Zemlya, to a point 80 mi (150 km) north of Kotelny Island where the meridian of 139°E crosses the edge of the CONTINENTAL SHELF. The area thus bounded is 208 40 mi² (540 000 km²). The Laptev being a shelf sea (i.e., a sea whose waters are confined largely to the continental shelf), some 64 percent of its area is less than 328 ft (100 m) deep. But, toward its northern boundary the bottom dips into the Nansen Basin with a depth of 9842 ft (3000 m) which results in a mean depth of 1720 ft (519 m) for the Laptev Sea.

Located over 400 mi (643 km) north of the ARCTIC CIRCLE and strongly influenced by the permanent ice pack of the Arctic Ocean along its northern boundary, the Laptev Sea lies in one of the most hostile environments on earth. The minimum winter temperature is − 58° F (− 50° C), with a summer high of 86° F (30° C), for a mean annual temperature of 9° F (− 13° C). The water temperature ranges from 32–52° F (0–11° C). Precipitation rarely exceeds 4 in (10 cm) per year.

The Laptev Sea is covered with ice, due to winter freezing, most of the year, and, in some years, remains ice-covered along its northern boundary throughout the short (2–3 month) summer. Numerous icebergs calved from the several glaciers on Severnaya Zemlya are added to this ice. Severe coastal erosion results from the grinding of this SEA ICE along the beaches and shallow offshore areas. This erosion, combined with the 17.5 million (17.5×10^6) tons (15.9 million metric tons) of sediment carried annually into the Laptev by the Lena and Yana Rivers, results in a complex and ever-changing coastal zone along the southern boundary. The coastline is irregular and deeply indented with numerous bays such as Khatangsky Bay, Anabarsky Bay, Oleneksky Bay, etc.

In spite of its harsh environment, the Laptev Sea contains and supports an abundance of life. Many forms of phytoplankton and zooplankton abound. (See PLANKTON.) Representatives of the FISH are Siberian crisco, Arctic char, and sturgeon. The WALRUS and PORPOISE are among the mammals. And, memorializing mammals long dead is the Mammoth Coast along the strait of Dmitri Laptev which contains abundant fossil remains of the wooly mammoth.

LARVA is the independent, immature, often wormlike form that develops from the fertilized egg of some ocean animals; it usually undergoes a series of form and size changes prior to taking on the characteristic features of the parent.

A larva (plural: larvae) hatches from the fertilized egg of the female of the species. For example, although some fishes are viviparous, or bring forth live young, in many of the bony fishes (or the class Osteichthyes) the eggs given off are small and do not contain sufficient food to allow the embryo to finish growing. Thus, the egg is hatched as a larva, or a small, partially developed fish which seeks food from the ocean environment. Most of the cartilaginous fishes (the class Chondrichthyes), or those that have a cartilage skeleton, such as the sharks and RAYS, produce large eggs containing sufficient food within them so that no larvae as such are formed. (See SHARK.)

Marine biologists have estimated that there are approximately 135 000 species of shallow-water, bottom-dwelling animals, and more than 100 000 produce free-swimming larvae. Thus, much of the PLANKTON of the upper layers of the oceans consist of these larvae. Representative of this is that produced by snails, MUSSELS, CRUSTACEANS, SEA STARS, and worms.

The several aspects of larval development can be illustrated by the sequence of developing stages found in the Brachyura, the true CRAB. These hatch as free-swimming zoeae from eggs carried by the female and normally pass through a consistent number of larval stages and molts. The final zoeal molt produces an intermediate larval form, the megalopa, and, at metamorphosis, the first crab stage appears.

A vast number of larvae, especially those in the pelagic or free-swimming stage are consumed by a wide variety of marine organisms, including other larvae.

See also OCEAN FOOD CHAIN.

LATERAL LINE is the name of a line along the sides of the body of most ocean fishes. (See FISH.) It is a line formed by a row of small sunken pits containing pressure-sensitive sense organs. Nerves run from the lateral line to the brain. PRESSURE changes in the water are picked up by fishes via their lateral-line organs. The knowledge of these changes assists fishes to navigate around solid objects and to avoid predators whose presence they sense by pressure fluctuations in the water caused by the predator's movements.

LATITUDE of any point is the angle between the local plumb line and the equatorial plane. Because

LARVA. This lobster larva has already developed the basic body shape of an adult lobster. Twenty to sixty thousand larvae are produced at a time by a single female lobster. *(NOAA)*

the earth can be considered as having the form of a spheroid, and because the plumb line, for all practical purposes, is perpendicular to the surface of the spheroid, any plane parallel to the equator cuts the surface of the spheroid in a circle, and all points on this circle have the same latitude. These circles are called parallels of latitude. Latitude is measured in degrees, minutes, and seconds north and south of the equator. Since the earth is not a sphere, the distance represented by a unit of latitude increases by about 1 percent between the equator and the poles. At the equator, 1° of latitude is equivalent to 362 749.9 ft (110 567.2 m), and at the poles, it is 366 463.1 ft (111 699.3 m).

LAW-OF-THE-SEA CONFERENCES are a series of international meetings having the objective of codifying agreements into an internationally accepted law (or body of laws) chiefly concerning the extent of a nation's sovereign rights of ownership to its bordering (territorial) seas; the rights of air space over the territorial sea, as well as its bed and subsoil; and the rights to the contiguous zone (or the intermediate area between the territorial sea and high seas) and the high or open sea.

Before the end of the Middle Ages, all parts of the world's oceans were regarded as the property of all mankind. When shipping and trading began to develop in the fifteenth century, nations started to claim rights not only to the seas bordering their coasts, but to a large portion of the oceans; as did Portugal for much of the PACIFIC OCEAN.

In 1609, Grotius in his book, *Mare Liberum*, expounded the principle of the freedom of the seas. The Dutch used this argument in their negotiations with Spain, Portugal, and Great Britain in 1613 and 1615 to allow them to trade in the East and West Indies and to hunt for WHALES in the areas around Spitsbergen. These were places that were considered as being out of bounds for all nations except those which had claimed them first.

Although such nations slowly and grudgingly accepted the fact that the open oceans were just too vast to possess, the principle that any state might claim and exert its rights to protect an area along its coast was generally recognized in the seventeenth and eighteenth centuries.

The criterion for how much of a belt could be claimed was very subjective. Sometimes the distance that could be covered in a 2-day ship voyage was taken, but usually the yardstick was the distance covered by a cannonball shot from a nation's shore-based fortress. This distance varied from 1 to 6 nautical miles in the opinion of seventeenth century seafarers, whose eyesight was apparently not too reliable if we keep in mind that the maximum range of a seventeenth century cannon was no more than 1 nautical mile.

In 1793, the United States Secretary of State, Thomas Jefferson, using the eighteenth century cannon capabilities as a criterion, interpreted the territorial sea of the United States to be a "limit of one sea league" (or 3 nautical miles) from its shore. Many countries adopted this concept. Other nations, over the years, made their own law-of-the-sea coastal rules to protect their national security, shipping, and fishing.

While an attempt was made in 1930 to codify the law of the sea on an international basis, no agreement among nations could be reached at that time. Morever, many nations began unilaterally to declare their jurisdiction not only over their territorial seas, but over areas contiguous to them. This was based primarily on the continual conflict that different countries have had in the competition for fish resources in particular areas of the high seas—for example, off the Grand Banks in the North ATLANTIC OCEAN.

The possibility of a conflict between countries led to the 1958 Conference on the Law of the Sea. Held in Geneva, this conference (attended by 86 nations) represented an effort to formulate new international laws governing all users of the sea and its resources. However, because of vagueness in some of the wording of agreements reached, plus the increase in knowledge of seabed resources and in the technology of how to exploit them, several nations still continued to act unilaterally in interpreting the extent of their ocean jurisdiction. Also in 1958, 25 countries, including the United Kingdom, the United States, and the Netherlands, fixed a width of their territorial seas at 3 mi and 11 countries decided on a width of 12 mi. A subsequent conference held by the United Nations in 1960 failed to evolve any substantial agreements, as did one in 1974 attended by 147 nations. By 1972, 51 countries had decided upon the 12-mi-limit zone and 16 others fixed on zones varying in width from 13 to 200 mi.

Law-of-the-Sea Conferences held in 1976 and 1977, tried, in essence, to adopt an international treaty of a 200-mi economic zone off the shores of coastal countries and a 12-mi offshore limit for total territorial control instead of the traditional 3 mi. Also, various provisions of the treaty stated the establishment of an international organization on deep-bed mining licensing rules. Such a treaty, *in toto,* was not adopted, and countries continued to make their own laws and adhere to their own codes of conduct regarding the oceans.

See also COMMERCIAL OCEAN FISHING; MINING.

LICHENS, or **LICHENES,** is the name applied to a group of some 18 000 species of plants made up of ALGAE and fungi growing together in such a way that the association involves advantages for each (mutualism). (See FUNGUS.)

Lichens are widely distributed throughout all parts of the world from the tropics to the polar zones. The marine lichens (e.g., the blackish-colored *Verrucaria maura*) are found encrusted on the shore environments of the world's oceans and are composed of single-celled algae and multicellular fungi living in a closely interdependent existence. In this relationship, the algae are photosynthetic (produce food by PHOTOSYNTHESIS) and the fungi provide water and mineral elements from the substratum to which they adhere.

See also SYMBIOSIS.

LIMPET is the name for several species of marine mollusks of the families Patellidae and Acmaeidae in the class Gastropoda that also includes the snails and slugs. (See GASTROPODS; MOLLUSK.) Limpets are widespread and found in most rocky shores free from winter ice. They are especially common in the Southern Hemisphere and in the PACIFIC OCEAN.

Most limpets are small, 2 in (5 cm) or less in length, and many resemble BARNACLES, although the latter are CRUSTACEANS rather than mollusks. Like barnacles, most species of limpets have adapted to the upper tidal regions of rocky shores where they feed on ALGAE and other vegetation. Unlike the barnacles which are firmly anchored for life, the conical, tentlike, shelled limpets have mobility. At night they forage for food, which they chiefly scrape off the rocks. In some species such as the common limpet, *Patella vulgata,* the animal returns at night to its own rocky resting place, and as it grows, its shell shapes to fit itself to the particular spot on the rock. Also, the shape of the shell is influenced by the limpet's position in the intertidal zone. The ones that remain underwater evolve a flatter shell than those that are exposed to the air each day.

Species of the *Patella* and *Acmaea* limpets have a relatively long life, and some live as long as 17 years. Reproduction, in most cases, is by separate sexes rather than asexual or hermaphroditic, and the eggs shed into the water are hatched into larvae. (See LARVA.)

The limpet is related to the ABALONE (genus *Ha-*

liotis), the periwinkle (*Littorina*), and the conch (*Strombus*), among other shelled snails. Interestingly, the ubiquitous periwinkles, or true marine snails, which, like the limpets, are characteristic of rocky shores the world over, seem to avoid SEAWATER as much as possible. They tend to live in tightly packed clusters well above the high-water mark and only on occasion dampen their gills with it, often surviving dry as long as a month or more.

LINGCOD is a bony fish of the family Hexagrammidae, one of a number of species called greenlings; the lingcod is not a true COD.

The lingcod (*Ophioden elongatus*) is a bottom-dwelling voracious fish that inhabits the intertidal zone, reefs, and kelp beds where there are strong CURRENTS. (See ALGAE; CORAL REEF.) The adults are found at depths from 330 to 600 ft (100 to 182 m) along the entire PACIFIC OCEAN coast of the United States. They range from the Baja Peninsula to Northwest Alaska and are especially abundant in the northern colder waters.

The color of the lingcod varies with its habitat. Basically, the fish has a coloration ranging from a mottled brown to bluish-green with cream-colored undersides. Its spots are brown, green, or tan, outlined in orange or light blue. Female markings are normally lighter in color, with orange tracings rather than blue. The lingcod has a large protruding mouth armed with caninelike teeth, two large, fleshy flaps (cirri) above the eyes, double nostrils, and a long deeply notched, dark dorsal fin. (See FINS.) The body and head are covered with smooth, small scales. Males, which are usually smaller than females, range up to 3 ft (0.9 m) in length, while females may attain a length of 5 ft (1.5 m) and an average weight of 5–20 lb (2.268–9.072 kg). However, fish up to 40 lb (18 kg) are not uncommon.

In one spawning, the female deposits more than half a million eggs in large adhesive pinkish-white masses on sheltered, rocky locations or in kelp beds below the lowest tidal levels. In ebb TIDES, they can be seen clinging to kelp or rocks. After fertilization, the male fans the eggs with his fins to provide good

LOBSTER. (*Left*) A spiny lobster treks across the sand of the sea floor, accompanied by a stone crab. (*Oceanographer of the Navy*) (*Above*) True lobsters differ from langoustes and spiny lobsters in having two large, strong, front-leg pincers, one of which is usually larger than the other. The lobster is a bottom dweller, but swims well and can use its strong tail for swimming backwards.

circulation of oxygen-rich waters surrounding the eggs and protects them from intruders until they are hatched. The principal spawning season is from December to March. Little is known about the early stages of lingcod development, but fingerlings from 3 to 5 in (7.6 to 12.7 cm) long are taken occasionally by seining in the EELGRASS during the summer. Several studies have indicated that lingcod do not travel widely from their natural habitat. The young feed on small CRUSTACEANS, while adults graduate to HERRING, FLOUNDER, COD, SQUID, and small lingcod.

Commercial fishermen obtain most of the lingcod catch with otter trawls. However, some fish are taken with set lines, hand lines, and troll lines. The vast majority of the catch is taken from Washington State waters, followed by catches from Oregon, Alaska, and California. Fishing for lingcod is year-round along the entire Pacific Coast but is best in California from April to October. Further north the fishing is best from October to May. Lingcod is a highly prized fish which gives anglers an exciting battle on light tackle. It is highly desirable as a fine eating fish and is marketed as dressed, fillets, and steaks. Smoked lingcod is another delicacy found in the markets.

LINNAEUS, CAROLUS (1707–1778) or Karl Linné, a Swedish botanist, is acknowledged as the originator of the first rigid and orderly system which has provided the basis of modern day scientific methods for the naming and classifying of plants and animals, including those found in the world's oceans.

Linnaeus, whose name became Karl von Linné in 1761 when the King of Sweden ennobled him, was born in the small Swedish village of Rashult. In his early student days, studying first for the ministry and later as a medical student at the Universities of Lund and Uppsala, he was most interested in the study of botany, a science that had been neglected for many centuries. Linnaeus' genius for identifying and classifying plant life was widely recognized while he was still a student. As a result of this recognition, he was commissioned in

1732 by the Academy of Sciences at Uppsala to conduct a scientific expedition to Lapland to explore and identify the resources of that area, which, at that time, were little known. The botanical results of this exploration were published in Linnaeus' *Flora Lapponica.* In 1740, after obtaining the degree of doctor of medicine from the medical school at Harderwijk, Holland (in 1739), he published another book called *Systema Naturae Fundamenta Botanica.* This work received wide international acclaim, and in 1741, he accepted the chair of medicine at the University of Uppsala where he remained until his death in 1778.

During his work at Uppsala, Linnaeus set forth in his *Species Plantarum,* published in 1753, what has become known as the binomial system of plant nomenclature. This naming of the plants by two names—the genus and species—was an accomplishment of major importance. It not only greatly simplified the previous practice that consisted of identifying plants with a single name followed by a set of descriptive nouns and adjectives, but the reform facilitated the identification of the great increase in newly discovered plants of the time. (See TAXONOMY.)

LOBSTER is the common name for several demersal (bottom-dwelling) decapod CRUSTACEANS which are members of the family Homaridae, most of which have commercial food value.

Lobsters of the family Homaridae are popularly called *homards* or lobsters with claws. The largest is the species *Homarus americanus,* found in the shallow northern waters around Maine and Nova Scotia. Many specimens attain a 40-lb (18 kg) weight and a 3-foot (1-m) length, although the average size is much smaller [about 3–4 lb (1.36–1.81 kg)]. Various conservation measures have been taken in the United States to protect the female lobsters and to limit the size of lobsters that can be caught. However, overfishing, water pollution, and other factors seriously threaten the continued commercial existence of this particular shellfish, one of the most highly prized seafood organisms in the world. Because of this, MARICULTURE (or systematic cultivation) of the American lobster is being attempted. The European homard *Homarus vulgaris,* or common lobster, found on the rocky shores of the Atlantic coasts of Europe, is smaller than the American species, as are the Norwegian lobster *Nephrops norvegicus* and the South African Cape lobster *Homarus capensis.*

The shell (external skeleton) of the lobster must be shed or molted periodically as the animal grows. After each molt, a lobster increases in carapace length (the distance from the rear of the eye socket to the rear end of the body shell measured along a line parallel to the center of the body). Concurrent with this increase, it starts to form a new shell casing. At about the fourth molt, bottom-dwelling habits take place. It is here, later on and after more molts, that reproduction of the species also takes place, but only after the female has molted. To avoid becoming an item on the menu of the other cannibalistic lobsters, including her mate, the female emits a substance (pheromone) that encourages sexual reproduction. After mating, the fertilized eggs are carried in a pouch in the female for 2–12 months. After that time, the eggs are discharged and are attached to the swimmerets, or pleopods, under the tail of the female. Another year is required for these eggs to hatch into pelagic larvae which are produced in numbers ranging from 20 000 to 60 000 at a time from a single female. (See LARVA.) The larval stages, of which there are a number, last for varying periods of time.

As in the case of most of the larger marine crustaceans, the lobster rarely moves away from the ocean floor. However, the animal swims well and when agitated will swim backward using its muscular tail. It feeds on live FISH and mollusks, although it is also a scavenger. (See MOLLUSK.) Prey are crushed by the strong pincers on the front legs of the lobster.

The true lobsters are distinguished especially by the possession of stout, well-developed pincers or claws. The langoustes, or spiny lobsters, differ from the true lobsters chiefly in that they either lack or have only weakly developed pincers on front legs. The langoustes are found in such southern waters as those off the coast of California, Florida, and southern Europe. The American species belong to the genus *Panulirus* and the European to the genus *Palinurus.*

The crayfish, or crawfish, (genera *Astacus* and *Cambarus*) comprise many widely distributed freshwater species. These are, in general, important in the natural web of food relationships. In the United States the Louisiana red swamp crawfish (*Procambarus clarkii*) is a real delicacy and is also relished throughout Europe. In 1976, it held the distinction of being the only crustacean grown commercially in the United States on a large-scale basis.

LONGITUDE is the angular distance in a great circle of reference measured from an accepted origin to the projection of any point on that circle. The longitude of a point on the surface of the earth is the angular distance measured along the equator from the point where the meridian (LATITUDE) through

Greenwich, England, cuts the equator to the point where the local meridian of the point of interest cuts the equator. The Greenwich Meridian cuts the equator in the vicinity of the Gulf of Guinea off the west coast of Africa. Longitude may be measured east or west of this point and may be expressed either in units of time (hours, minutes, and seconds) or in units of angle (degrees, minutes, and seconds). In navigation, west longitude is taken as positive (+) and east longitude is taken as negative (−). The maximum east or west longitude of any point is 12 h or 180°. This point lies on the equator between Howland Island and the Gilbert Islands in the PACIFIC OCEAN.

LONG-PERIOD WAVE See WAVES.

LONGSHORE CURRENT See BEACH.

LOOMING See FATA MORGANA.

LOWER HIGH TIDE See TIDES.

LOWER LOW TIDE See TIDES.

MACKEREL is the name for about 50 species of pelagic, carnivorous fish of the family Scombridae in the order Perciformes. Mackerel are characterized by a long, slender, streamlined body, a pointed head, a large mouth, a deeply forked tail, and the absence of a SWIM BLADDER in some species. The mackerel, such as king mackerel (*Scomber cavalla*), common mackerel (*S. scombrus*), and Spanish mackerel (*S. maculatus*) among others, are regarded as important and choice food fishes.

Most mackerel are found near the surface of the tropical and temperate parts of the world's oceans. While the king mackerel may reach 5 ft (1.5 m) in length and weigh up to 100 lb (45 kg), most other species are medium-sized or small fish [ordinarily 3–4 lb (1.35 to 1.81 kg) and not over 2 ft (61 cm) in length at a 2-year mature age]. The migration range of the North American or common mackeral is from the Canary Islands to Norway and from the Chesapeake Bay to the Gulf of Maine. They swim in close-ranked schools, often of great size, in the waters of these areas, especially in the early summer when they go to northern coastal areas to spawn. Here, from June to October, they feed on small fishes, especially the HERRING, which they hunt primarily by smell. After October, they inhabit deeper waters and exist chiefly on a diet of SHRIMP, small CRUSTACEANS, and ocean-floor life. In January, when they return to the surface, their eating habits change and their favorite food is the copepod, *Calanus.* (See COPEPODS.) Since they lack a swim bladder, the common mackerel can move from the depths to the surface quickly.

Like the common mackerel, the Pacific mackerel (*Pneumatophorous diego*) is also a valuable commercial fish. However, it possesses a swim bladder and lives in waters ranging from the GULF OF CALIFORNIA to Alaska and on the other side of the PACIFIC OCEAN.

The so-called horse mackerel is not a mackeral but a member of the family Carangidae (e.g., the scad, *Trachurus trachurus,* among others).

MALOCOSTRACANS See CRUSTACEANS.

MANATEES See SEA COWS.

MANGANESE NODULES (IRON-MANGANESE NODULES) are small [usually 0.39–6 in (1–15 cm) in diameter], dark brown, potato-shaped concretions consisting primarily of manganese salts and manganese oxide minerals (e.g., todorokite and birnessite) along with iron oxides; they are formed in the world's oceans as a result of complex precipitation or sedimentation processes.

First discovered by the CHALLENGER EXPEDITION (1873–1876), manganese nodules have been found on the floor of the ocean throughout many parts of the world. At the turn of the century Alexander Agassiz, on the *Albatross* expedition, dredged up such nodules from many parts of the eastern PACIFIC OCEAN and concluded that they covered an area larger than the United States. It is now estimated that the Pacific Ocean alone contains about $1\frac{1}{2}$ trillion ($1\frac{1}{2} \times 10^{12}$) tons of nodules, which are forming at the rate of approximately 10 million (10×10^6) tons per year. These nodules vary in grade or chemical composition, particularly in respect to their copper, nickel, manganese, and cobalt content. In addition, they vary widely in shape, surface texture, and internal structure from locality to locality. While manganese dioxide in deep-sea sediments takes on such forms as grains, slabs, coatings on rocks, and impregnations of porous materials, in general, the nodules are structurally formed in layered concentric rings of varying colors, around a small nucleus of rock, plant, or animal remains. The formation usually occurs in areas of low sedimentation rates where oxidizing conditions exist at the water-sediment interface. However, this complex formation process is poorly understood. Present

MANGANESE NODULE. (*Left*) A cross section magnified about five times shows the growth rings of a manganese nodule. Preserved tubes, built by marine organisms for shelter, seem to alternate with the nodule's concentric growth rings, which are believed to be related to the biological activity on the nodule's surface during its growth. (*Scripps Institution of Oceanography*) (*Right*) Manganese nodules have been found in the Atlantic, Pacific, Indian, and Southern Oceans. The Pacific alone is estimated to contain over 1600 trillion tons of these concretions. (*Smithsonian Institution*)

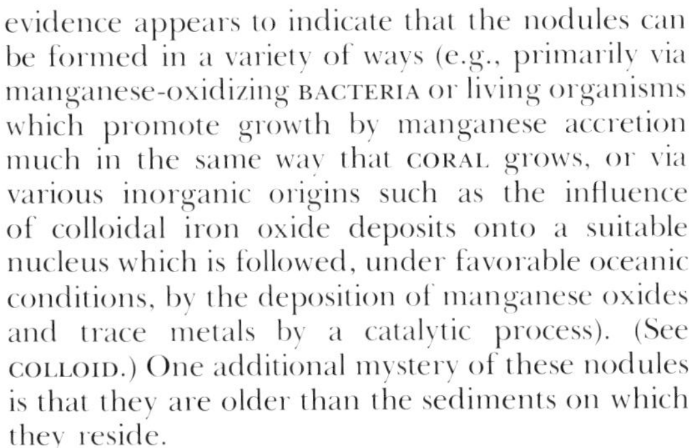

evidence appears to indicate that the nodules can be formed in a variety of ways (e.g., primarily via manganese-oxidizing BACTERIA or living organisms which promote growth by manganese accretion much in the same way that CORAL grows, or via various inorganic origins such as the influence of colloidal iron oxide deposits onto a suitable nucleus which is followed, under favorable oceanic conditions, by the deposition of manganese oxides and trace metals by a catalytic process). (See COLLOID.) One additional mystery of these nodules is that they are older than the sediments on which they reside.

Manganese nodules have been located, especially by means of bottom photography and deep-water television, in many parts of the oceans in potentially commercial quantities. They are usually found at 6600–13 200 ft (2000–4000 m) depths on the sediment surface and in areas of oxygen-rich water. Some of the most potentially attractive deposits, from a commercial standpoint, are to be found in the North Pacific Ocean, south of Hawaii. Here, in an east-west belt 125 m (200 km) wide (between 6° and 20°N LATITUDE 110 and 180°W LONGITUDE) in water of about 13 200 ft (4000 m) deep, the deposits are of sufficiently high abundance and metallic content. Other sites in the Pacific Ocean, ATLANTIC OCEAN (e.g., the Blake Plateau), and INDIAN OCEAN have been investigated by commercial interests, and the potential supply of the valuable elements that these manganese nodules represent has inspired several countries, notably Great Britain, France, the United States, West Germany, Japan, Canada, and the U.S.S.R. to have active interests in their recovery. This reflects the growing worldwide demand for metals such as nickel, copper, and cobalt which are concentrated in the deep-sea nodules.

Since the nodules occur on the ocean floor, usually in water deeper than 6000 ft (2000 m), reliable and economic recovery methods are difficult to devise. However, three basic systems have been designed. These are the airlift, the hydraulic lift without air, and the mechanical lift such as the continuous-line bucket system.

The significance of these efforts becomes apparent when one considers the fact that manganese is an essential ingredient in the large-scale manufacture of wrought steel and that there is no effective substitute for it. Other uses of the nodules themselves have been reported as effective converters of unburned hydrocarbons to carbon dioxide in automobile exhausts and efficient absorbers of sulfur from stack gases. Also, in addition to manganese, the nickel, copper, cobalt, and other minerals contained in these nodules on the ocean floor are conservatively estimated to be worth over 3 trillion dollars in value.

See also MINING.

MANGROVES is the name commonly applied to essentially three different trees—namely, the red, black, and white mangrove. These woody plants occur in muddy swamps at the mouths of rivers, in lagoons, and elsewhere in the tropics under tidal conditions. Such trees extract fresh water from ocean water, and the salt-secreting cells on the leaves of some species secrete a solution containing up to 7 percent sodium chloride under a layer of oil.

The mangrove family actually consists of some 15 genera and about 100 species of trees and shrubs, of which most are small in size. The Rhizophoraceae, or mangroves proper (genera *Rhizophora, Brugiera, Ceriops,* and *Kandelia*), grow in coastal mud flats and estuaries of saline or brackish water. Here, together with certain other genera (i.e., *Aegiceras* and *Avicennia*), they often constitute mangrove swamps.

The most prevalent of the trees growing in such areas is the red mangrove (*Rhizophora mangle*), which fringes both Atlantic and Pacific tropic and subtropic shores and thrives prolifically in estuaries and lagoons. The other species of the genus *Rhizophora* grow only in the Eastern Hemisphere. However, all these species are basically alike. They consist of evergreen shrubs or small to large trees with the trunk supported on an array of arched stiltlike roots (plus roots developed from lateral branches) that form thickets in the silt along low muddy seashores and fringe inland rivers and streams where the water is brackish. In many species, reproduction is viviparous; i.e., it takes place by means of seeds that germinate while still attached to the trees. These seeds develop an appendage with root beginnings, and when this radicle, about 1 ft (30.5 cm) long, falls from the tree into the mud, it is ready to form the roots and leaves of a new tree. Some of these falling root embryos are carried away by WAVES and tidal action and float until they are grounded. Borne on equatorial currents, seedlings have even crossed the Atlantic.

Studies in Florida have shown that almost 95 percent of the annual leaf production of the red mangrove falls into the warm shallow waters of the coast. Here, these simple leathery leaves are grazed, eroded, and dispersed by tides and currents over a very wide area. In this cycle, the leaves are first colonized by BACTERIA and FUNGI. The fungi (e.g., *Phytophthora vesicula*) grow filaments or hyphae in the pores of the leaf; these filaments penetrate the cell walls to obtain carbohydrates and other cellular nutrients. The PROTOZOA, bacteria, and several other species of fungus that combine to further degrade the leaf then attract such marine organisms as COPEPODS, worms, and small crabs. These organisms feed upon the leaf as well as break it into fragments. The leaf fragments are then consumed by detritivores, whose excreta are consumed by smaller detritivores. It has been found in the examination of the stomach contents of over 100 species of fish that these detritivores are eaten by small fish which, in turn, are eaten by larger ones, often well away from the mangroves.

The inner wood (heartwood) of the red mangrove is light red, deepening to dark red or reddish-brown, while the outer wood (sapwood) varies from a yellowish to a grayish or pink color. The wood is very hard and dense; weight is 60–70 lb ft^{-3} (960–1120 kg m^{-3}). The growth rings when present are usually indistinct. Red mangroves are found especially in the Orinoco Delta, Ecuador, along the Amazon estuary, in Baja California, and in Southern Florida.

Some common names for this tree are mangrove, red mangrove, (English); mangle, mangle colorado (Spanish); paletuvier, p. rouge (French); mangue, mangue vermelho (Portuguese); candelon, mangle dulce, tab-che (Mexican); mangle geli (Ecuadorean); ratimbo, m. verdadeiro, mangarabeira, mangue bravo, m. do brejo, m. preto (Brazil).

The genus *Avicennia* consists of some 14 species of small to medium-sized evergreens. The black mangrove (*Avicennia marina*) is an example of those species that especially grow in the tidal marshes and along the coasts. This distinct species of the Western Hemisphere (*A. marina*) occurs in Southern Florida, the West Indies, the north coast of South America, and both coasts of Mexico and Central America. The trees in these locations grow in association with *Rhizophora* and other plants of the mangrove formation. Although their heights sometimes reach 60–70 ft (18.2–21.3 m), with a trunk diameter of 2 ft (61 cm), the average size is around 20–25 ft (6.1–7.6 m) in height, with heavy aerial roots growing in and out of the mud in an arched and entangled fashion. The stringlike leafless roots (pneumatophores) of some species trap the silt from the rivers and the flotsam of the sea, and in doing so protect and extend the shoreline.

Some species of black mangrove trees, which also have aerial roots, have simple leaves that are downy in texture and about 2–4 in (5–10 cm) long. The bark is dark brown and scaly, and the flowers are white, fragrant, and a source of honey. The heartwood is brown to black, and the outer layers are white to gray in color. Like that of the red mangrove, the wood of the black mangrove is heavy and hard with about the same specific gravity (0.95).

Common names are black mangrove, blackwood (Florida); mangle negro, m. prieto (Cuba); manglier noir, paletuvier (Haiti); black mangrove, limewood, olive mangrove (Trinidad); palo de sal (Honduras); mangle blanco, m. negro, puyeque (Mexico); mangue bronco, mangue amarello, ceriuba, ciriuba (Brazil); mangle salado (Ecuador).

The white or button mangrove (*Conocarpus erecta L.*) also grows in swamps populated by red and black mangroves, especially along the shores of tropical America and West Africa. It occurs both as a shrub and a tree, usually small but sometimes 60 ft (18.2 m) high and 30 in (75 cm) in diameter. As a tree, it does not have aerial roots as do the red or black mangrove. The inner wood is olive brown often with a reddish tinge, and the outer wood is a lighter color. It is moderately heavy and hard.

Common names are button mangrove, b. wood, b. tree (some parts of the West Indies); botoncillo, mangle blanco (Venezuela); jele (Ecuador); mangue (Brazil); botoncahui, botoncillo, mangle negro, exkanche, extabche (Mexico); mangle boton, saragosa, yana (Cuba); mangle (Dominican Republic); mangle pinuelo (Panama).

Still another member of the mangrove formation along the shores of tropical America and West Africa is also termed the white mangrove or white buttonwood in Florida. This species (*Languncularia racemosa*) is quite similar to the button mangrove (*Conocarpus erecta L.*).

Of the aforementioned mangrove species, the red mangrove is not only the most prevalent but perhaps the most important from a commercial (usually localized) standpoint. For example, tannin is obtained from its bark; it is used as a fuel, for lumber in building houses, boats, posts, etc. However, the basic importance of all mangrove trees lies in the fact that they protect the land from erosion, and as they gradually advance in the shallow water, silt and debris are deposited around their roots and new land is formed. Moreover, the dense tangle of roots and foliage provides a habitat upon which many birds and animals depend for their life and livelihood.

MARICULTURE (sometimes called sea farming or marine aquaculture) is the systematic cultivation or husbandry of certain natural marine life for harvest and use by human beings.

The mariculture activities in various parts of the world are not new, dating back several centuries in the Indo-Pacific and European countries. However, in its present form it is a comparatively recent development that has evoked a worldwide upsurge of commercial interest. Accumulated scientific knowledge, the growing recognition that aquatic food sources represent an increasingly important reservoir of protein for the world's population, and better management practices in both capture and culture fisheries, among other factors and developments have been responsible for this interest. For instance, various studies and developments carried out in the United States under the SEA GRANT PROGRAM as well as the work of the World Mariculture Society (Seattle, Washington) have done much to advance this interest.

Mariculture systems are many and varied, but almost all have a fundamental requirement—the need to contain the animal or plant to be cultivated in a prescribed aquatic environment. The basic properties of this life-supporting environment must be considered in planning for cultivation. These include TEMPERATURE, oxygen availability, food availability, methods of waste disposal, and space availability. In the matter of space or containment many different methods, varying in degree of sophistication, are employed. These range from culturing (in cages) in controlled waters to using enclosed offshore areas, estuaries, and ponds, to producing young fish (juveniles) in hatcheries. (See ESTUARY.)

When cages or pens are used for culturing, the organisms are held in these enclosures (or feed lots) and fed a prepared diet until they attain a marketable size. Regulation of the water temperature is very essential, and this temperature must be cold enough to minimize the proliferation of disease but sufficiently warm to enhance growth of the organisms within an economic time frame. The water for the system must be supplied at a high flow rate in order to provide oxygen and to dispose of excreta and other wastes as well as bacteria.

In the so-called grazing systems, space is not a major problem since the culturing is done in areas such as diked tidal ponds. Feed consists of natural foods in these sites where small fish, SHRIMP, and various mollusks are grown. (See MOLLUSK.) New nutrient-rich water must be periodically introduced in order to ensure a proper oxygen supply and a sufficient amount of food for the system. SHELLFISH, such as MUSSELS, are grown in estuaries in great densities. They "graze" on PLANKTON and are grown on ropes suspended from floating platforms or rafts.

In the farming of the Atlantic SALMON (*Salmo salar*), several countries, notably the United States and Great Britain, employ two culture systems. In one, ocean ranching, advantage is taken of the salmon's instinctive homing behavior. The fishes are hatched and reared for a time under carefully controlled conditions in hatcheries and are released to grow in the wild. When they are ready to spawn, the fishes return to the hatchery where they were born and are then harvested. The second type of culture system is penrearing. Here, the fishes, such as king salmon (*Oncorhynchus tschawytscha*), are reared in floating pens and brackish water ponds.

MARICULTURE. (*Left*) Fish-cultivation techniques have greatly increased the supplies of many commercially important fishes. These bream fingerlings come from the Fish Cultural Station in Corning, Arkansas. (*NOAA*) (*Right*) Japanese oyster farmers harvest a string of oysters. Japan boasts an enormous fleet of boats used in oyster farming.

Only in Japan has mariculture developed to the stage where several species of other finfishes are commercially grown. The majority of these, however, are freshwater species, and the small number of cultured marine fishes includes, in addition to the salmon, yellowtail (a scombroid fish related to the amberjack), puffers, both black and red porgies, and a few others, including TUNA, now being reared experimentally.

Several other species of finfishes (e.g., milkfish in Southeast Asia, yellowtail and seabream in Japan, whitefishes and sturgeons in the Soviet Union) are also being systematically raised (or the methods of commercially growing them are being studied) in various parts of the world.

The cultivation of mollusks (CLAMS, oysters, and mussels) is one of the world's oldest and probably the most successful form of mariculture. In Spain, mussels are cultured by the use of rafts which are also employed in Japan for growing oysters. In the United States (e.g., Puget Sound) and Europe the seed oysters, imported from Japan and Korea, are hatchery-raised and cultured in specially selected growing areas. In all cases, the mollusks feed on SEAWATER plankton, and the yields from shellfish culture range from 10 to more than 1000 tons per acre per year in the more successful raft-culture practices.

CRUSTACEANS (e.g., CRAB, SHRIMP, LOBSTER) are good protein sources, and shrimp culture of the Kruma shrimp (*Penaeus japonicus*) has been practiced in Japan since 1934. Here, large numbers of larvae are reared in controlled conditions through to juveniles which are used for stocking large saltwater lagoons. (See LARVA.)

The controlled culture of Penaeid shrimp has been subsequently tried in a number of countries with varying degrees of success. In the Philippines, government-operated hatcheries are used for the culture of the shrimp larvae (e.g., *Penaeus monodon*) and the juveniles produced are supplied to the owners of coastal lagoons. Considerable experimental work on the culture of shrimp has taken place in the United States.

In what is termed *lobster ranching,* the larvae and juveniles are cultured for a period in a hatchery, and then the juveniles are released to the controlled areas of the ocean. However, from a number of biological aspects, the lobster is much more difficult to cultivate than Penaeid shrimp. It has a low reproducibility rate or fecundity and is more ferocious and territorial in behavior. Because lobsters enjoy a good market at a very high price in northern Europe and the United States, lobster cultivation has been the subject of much research in these areas.

With a few exceptions, the plants of importance as

food in the ocean are ALGAE. The earliest records of the use of marine algae are in Chinese writings of the first century. About 1660, the Chinese introduced dried algae into Japan. The Chinese and Japanese continue to eat more algae than do any other peoples. The Japanese eat certain seaweeds much as people in the United States eat tomatoes and lettuce. However, although there are about 9 times as many plants in the world's oceans as on the mainlands, only a very few of these (e.g., KELP) are now being adapted as food products. Some work in the United States has been carried out in seaweed harvesting to supplement shellfish and salmon mariculture. Many of these seaweeds as well as the variety of most large algae are not rich in protein but are still valuable to humans as condiments, as sources of essential minerals, or as sources of industrial colloids. On the other hand, microalgae have extremely high protein contents, and, while their very dilute presence in the oceans precludes their being captured for food, culturing them may conceivably make a significant contribution to the protein nutrition of husbanded animals and possibly humans.

In summary, the success of all mariculture activities is dependent on many ecological factors such as CLIMATE, TIDES, SALINITY, etc. Therefore, considerable differences exist in application and operation, depending on region and LATITUDE. Moreover, traditional food preferences and eating habits differ by country. The final success of most mariculture projects depends on an intelligent research and development plan. Intensive research at the laboratory and pilot-plant level, conducted in close cooperation with engineers and economists, is essential for providing industrial mariculture with a sound and viable foundation. Such research includes not only a complete biological study of the species in culture—their life history, genetics, environmental tolerances (temperature, salinity, etc.), nutrition, and use to human beings as a food, fertilizer, feed, etc.—but also a study of the types of cultivation installations required, yields per unit acre, and the marketing practices and economics involved.

Implicit in all this is the need to control domestic and industrial POLLUTION since such pollution renders many desirable areas worthless for any use, especially mariculture. Pollutants (e.g., pesticides) are accumulated in food organisms and passed on to animals higher in the OCEAN FOOD CHAIN.

MARINE See OCEAN.

MARINE AIR is the atmosphere above the surface of the world's oceans.

The atmosphere, or the envelope of air surrounding the earth, is divided into two principal regions: The troposphere and the stratosphere. The former, closer to the earth's surface [sea level to about 35 000 feet (10.5 km) elevation], is a region in which the infrared radiation is absorbed mainly by water vapor to raise the surface temperature. The stratosphere, the region above the troposphere, contains little water vapor, and the ozone (O_3) absorbs ultraviolet light. The top of the stratosphere is at 164 000 ft (50 km) elevation.

In the troposphere the atmospheric gases (e.g., nitrogen, oxygen, carbon dioxide, argon) present are mixed by convection so that there is no separation of these different gases. The result is a fairly homogeneous mixture except for water content which varies with location, elevation, and season.

In the stratosphere, ozone is produced by the photochemical dissociation of molecular oxygen (O_2) and the subsequent reaction of atomic oxygen (O) with O_2. This ozone participates in a number of reactions with nitrogen oxides and other species present in the upper atmosphere.

There is a lack of detailed understanding of the several reactions and cycles of natural and artificial substances in both the stratosphere and troposphere. However, such an understanding is essential to assess human impact, as a result of technological achievements, on the possible inadvertent modification of the atmosphere. In addition, the importance of atmospheric chemistry in understanding cloud and rain formations, atmospheric mixing processes, and changes in the global radiation budget cannot be overstated.

Fundamental to an understanding of atmospheric behavior is an understanding of processes taking place at the air-sea boundary. Here, nearly one-half the heat energy, which the atmosphere receives for maintaining its circulation, is derived from the condensation of water vapor, principally from ocean water evaporation.

More than 70 percent of the earth's surface is covered by the oceans, and all the atmospheric gases exist in these waters. The concentrations of such atmospheric gases depend upon SALINITY, pressure, and temperature conditions as well as reactions within the oceans.

In the atmosphere, both nitrogen and oxygen are maintained at their levels by biological processes. (See NITROGEN CYCLE; OXYGEN CYCLE; PHOTOSYNTHESIS.) Carbon dioxide in the atmosphere is largely controlled by mineral deposition in the ocean, by the burning of fossil fuels, and by respiration and decay processes. Carbon dioxide passes from the atmosphere to the ocean at the sea surface more slowly than does oxygen. Carbon monoxide in the air is thought to emanate from the oceans due to the incomplete oxidation of organic matter in SEAWATER or as a metabolic product of organisms. Ammonia occurs only in near-surface or surface ocean waters and is rapidly oxidized to nitrate or utilized directly by phytoplankton. (See PLANKTON.) Three cosmic-ray produced radioisotopes, hydrogen 3, carbon 14, and argon 39, are present in the atmosphere, and these enter the ocean and are returned to the atmosphere as gaseous molecules, $^{3}H^{1}HO$, $^{14}CO_2$, and ^{39}Ar.

Among the other elements and their compounds found in the atmosphere above the world's oceans are gaseous chlorine or chloride, gaseous bromine or bromide, and iodine. Two sources are believed to be responsible for these substances in uncontaminated marine air. They are sea salt particles and volcanism, with volcanoes being the source of gaseous chlorine.

The salt particles are present in the atmosphere at relative humidities over 70 percent. Over most of the oceans, the relative humidity varies between 70 and 80 percent in the atmospheric layers immediately above the surface water. Some believe that the chloride in escaping aerosol particles from the ocean reacts with ozone to form gaseous chlorine. Others have postulated that the initial form of this chlorine is either hydrogen chloride or nitrosylchloride (NOCl).

The average particulate mass of the marine air salt is 0.01–0.001 picograms (pg) at 90 percent relative humidity. (See PARTICULATE MATTER.) In rain at sea, the chloride/sodium ratio is usually equal to or slightly below the seawater ratio of 1.80 : 1.

Bromine in marine air in relation to chloride in airborne particulates in ocean and coastal areas is lower than the seawater ratio of 0.0034 : 1. In the gaseous form, bromine in some areas constitutes 80 to 90 percent of the bromine content as compared with about 35 to 40 percent of the total chloride. Gaseous residence times for each are much longer (e.g., that of gaseous bromine is 7 times longer than that of the particulate form).

Iodine in marine air (in particles and rain) has significantly higher ratios than chloride and bromine. These ratios of iodine to chloride vary from 100 to 1000 times the ratios present in seawater. This is thought to be the result of either iodine buildup on the saltwater particles when they are produced at the surface (because of the association of iodine with surface-active organic material present at the surface) or the release of iodine gas from the surface and its subsequent deposition on the particle surface.

In the case of fluorine, some rain analysis conducted over land has shown that the fluoride/chloride ratios were 10–1000 times the seawater ratios. At various mid-ocean sites, this ratio was approximately the same as that of seawater, while near the coast of the United States, the fluoride/chloride ratios were 2–25 times higher.

MARINE BIRDS or **OCEANIC BIRDS**, is a collective term applied to those various species of warm-blooded vertebrates of the class Aves which have feathers, wings, and beaks and which live along the shores of the world's oceans or spend their lives on and over the oceans, returning to land only to nest; both the shore birds and the sea birds obtain a substantial part or all of their food from the marine environment.

Most authorities differentiate between the categories of ocean shore birds and sea birds. That is, not every bird that is somehow associated with the ocean can be correctly termed a sea or ocean bird. Rather, the true sea birds (such as the pelagic birds: the fulmars, petrels, and shearwaters) are those that live on and over the sea during the major portion of their lives. Accordingly, such birds with this particular life cycle have various adapted mechanisms and characteristics that other birds, including the ocean shore birds (e.g., the sandpipers), do not possess. Among these adaptations are the following: Most sea birds have a salt gland which secretes excess salt in order to prevent dehydration; most (except the albatross) have the ability to change the focal length of the lenses in their eyes and thus are able to see underwater; their bodies are especially configured for flying and/or swimming rather than walking on land, where they are quite clumsy due to the placement of their legs—far back in relation to their bodies.

In all, there are approximately 8000 different species of birds, belonging to 28 different orders and many families. Of these, the avian order Procellariiformes contains the largest number and variety of true ocean birds, all of which live beyond the continental edge. Included in this group are the following pelagic birds:

- *Albatrosses, shearwaters, fulmars, and petrels* (*order Procellariiformes*). The albatrosses are chiefly Southern Hemisphere inhabitants, although a few breed in the North PACIFIC OCEAN. There are two genera (*Diomedea* and *Phoebetria* in the family Diomedeidae) of these large, long-winged birds; all species are capable of long flights. The wandering albatross, *Diomedea exulans*, has a wingspan of over 12 ft (3.6 m). A variety of ocean life, especially fishes, makes up their diet. The famous gooney

Two gooney birds (black-footed albatrosses) "dance" on the beach at Sand Island, Midway. *(Bureau of Sport Fisheries & Wildlife)*

The Laysan albatross possesses long wings, a characteristic hooked bill, and long, tubular nostrils. *(Bureau of Sport Fisheries & Wildlife)*

A magnificent male frigate bird displays his inflatable crimson throat pouch, which develops at courtship time. The frigate bird is also known as the man-o'-war bird. *(Bureau of Sport Fisheries & Wildlife)*

A booby watches over her offspring. *(Bureau of Sport Fisheries & Wildlife)*

Emperor penguins walk along the ice edge in Antarctica. *(U.S. Navy)*

A brown pelican soars over a nesting island near Ocracoke, North Carolina. *(U.S. Fish and Wildlife Service)*

A cormorant dries its watersoaked feathers in the sun after diving for the small fish on which it feeds. *(U.S. Fish and Wildlife Service)*

Cormorants, anhingas, and pelicans rest in the trees and waters of the Pelican Island National Wildlife Refuge in Florida. *(Bureau of Sport Fisheries & Wildlife)*

A colony of wood storks, also called wood ibises, rest on mangroves in Florida. *(U.S. Fish and Wildlife Service)*

An osprey peers down into the night. *(Bureau of Sport Fisheries & Wildlife)*

A female anhinga, perches on a branch. Its neck reveals its relation to the cormorant. *(Bureau of Sport Fisheries & Wildlife)*

Wind blows the long neck feathers worn by the great blue heron during the breeding season. *(Bureau of Sport Fisheries & Wildlife)*

A California gull lectures her young. *(Bureau of Sport Fisheries & Wildlife)*

The least tern, like almost all terns, has a forked tail and a black cap. *(Bureau of Sport Fisheries & Wildlife)*

An eastern green heron in mangrove thicket stares into the distance. *(Bureau of Sport Fisheries & Wildlife)*

The tufted puffin shown here is native to the banks of the Bering Sea, where it is an effective hunter of small fish in the open sea. *(U.S. Fish and Wildlife Service)*

A reddish egret feeds in the surf on Sanibel Island. (Bureau of Sport Fisheries & Wildlife)

birds of Midway Island are members of the albatross family.

The shearwaters and fulmars are also migratory and wholly ocean birds belonging to the family Procellariidae. The shearwaters, typified by the Manx shearwater (*Puffinus puffinus*), an inhabitant of the MEDITERRANEAN SEA and North American shore areas, nest in very large groups. Large flocks of shearwaters cruise both coasts of the United States. There are four Atlantic shearwaters and six Pacific species. Some, like the sooty shearwater (both U.S. coasts), are dusky all over; others, like the greater shearwater (Atlantic) are white below. Shearwaters look like small gulls but have "tubed noses" and their stiff-winged flight makes them look quite unlike the gulls. The fulmars, which nest on ledges along ocean cliffs, can be represented by the northern fulmar, *Fulmar glacialis,* while the small, swallow-sized stormy petrels (of the family Hydrobatidae) are perhaps best typified by Leach's stormy petrel. *Oceanodroma leucorhoa,* of the northern oceans and the Peruvian stormy petrel, *Oceanodroma tethys.* The five species of diving petrels belonging to the single genus *Pelecanoides,* family Pelecanoididae, are found in Antarctic waters but migrate to the shores of Peru, New Zealand, Australia, and other parts of the South Atlantic Ocean and Pacific Ocean. Some diving petrels breed along the western coast of South America from Cape Horn to the northern islands off Peru (latitude 6°25′S). Diving petrels do not breed north of about latitude 37°S in any other part of the world.

Some other marine birds, including the shore birds and those inhabiting near-in ocean waters, belong to the following orders:

- *Penguins (order Sphenisciformes).* The six families of flightless birds of the Southern Hemisphere are truly the most marine of all the birds. They lack any ability to fly but swim under water very efficiently. Unlike hollow-boned airborne birds, penguins possess solid bones for ballast when they dive, often to 800 ft (240 m) or more. Penguins are the only birds that have developed the art of porpoising as they swim—leaping clear of the water and then slipping in again. Their diet consists of FISH, CRUSTACEANS, and small SQUID. Two families, Aptenodytes (e.g., the emperor penguin, *A. fosteri,* and the king penguin, *A. patagonica*) and Pygoscelis (e.g., the Adélie penguin, *Pygoscelis adeliae,* and the gentoo penguin, *P. papua*) are found in the Antarctic. Penguins of genus *Spheniscus* are found along the west coast of South America southward of LATTITUDE 6°S, and also an endemic species (*S. mendiculus*) lives on the Galapagos Islands at the equator.
- *Divers and loons (order Gaviiformes).* Four species are found in the Northern Hemisphere chiefly along the ocean coasts of Northern Europe and North America. Representative of these birds is the common loon (*Gavia immer*), an inhabitant of North America, Iceland, and Greenland during the breeding season, moving south during the winter. Such birds are built like torpedoes for submarine diving. Their feet are placed so close to

the tail that they must run for a distance before getting into the air.

• *Pelicans* (*order Pelecaniformes*). The distribution of this order of large and huge-billed birds is worldwide. There are six living families of the order. The best-known of these full-webbed swimmers in the United States is the brown pelican which is found on both southern seacoasts. The man-o'-war bird, *Fregata magnificens* (also called the frigate pelican), has a wingspan of $7\frac{1}{2}$ ft (2.2 m) and is scissor-tailed. It lives throughout tropic oceans. One species of the family Pelicanidae, the Peruvian pelican (*Pelicanus occidentalis*), is a large ungainly bird weighing about 16 lb ($7\frac{1}{4}$ kg) with pearly gray plumage and a white head, black or orange-yellow pouch with greenish bill, and lavender-gray legs. It may hurtle into the water with a resounding splash or may swim and scoop while fishing. This bird is the most conspicuous among the flock birds on the Peruvian Coast because of its large size. It does not stray out of sight of the coast and has never been sighted off the outlying islands.

In addition, included in this Pelican order are the tropic birds (family Phaethontidae), the gannets and boobies (family Sulidae), the cormorants (family Phalacrocoracidae), and the frigate birds (family Fregatidae). Most species of these families, as a group, are inhabitants of coastal and estuarial waters in various parts of the world (especially the tropics and temperate zones). They fly quite well and are voracious eaters of ocean fish.

• *The herons and storks* (*order Ciconiiformes*). Five families, the herons (family Ardeidae), storks (family Ciconiidae), spoonbills (family Threskiornithidae), shoebills (family Balaenicipitidae), and hammerheads (family Scopidae), of these widely distributed birds are included within this order. In the main, all are distinguished by long featherless legs adapted for wading, especially in marshy areas, rivers, ponds, and on the seashore where their diet consists of frogs, reptiles, CRUSTACEANS, mollusks, and insects. (See MOLLUSK.) The gangly herons wade in the shallows and stand motionless for minutes at a time, waiting for food which they will spear with a lightning thrust of their sharp bills. They can stretch their necks so that they are long as their legs, or they can fold them in so that their heads rest on their shoulders. When they fly, they tuck their collapsible necks into an S curve and their long legs trail behind.

• *Ducks, geese, and swans* (*order Anseriformes*). Some ducks, geese, and swans can be classified as oceanic birds since they live supported by EELGRASS and SEAGRASS ecosystems in estuarine and coastal waters, although for the most part the many and widely distributed species of this order of birds are freshwater inhabitants. Some notable exceptions are ocean shore dwellers such as the *Anser* and *Branta* true geese, migratory birds that breed in the cold tundra regions and in winter move to the African, Mediterranean, and United States coasts: the wild ducks, or mallards (e.g., *Anas platyrhynchos*) that feed on ocean life at certain times of the year; and the common eider duck (*Somateria mollissima*), a bird found throughout the ARCTIC OCEAN and in northern temperate waters.

• *Sea eagles* (*order Falconiformes*). The sea eagles, or fish eagles, include birds such as the European or white-tailed eagle (*Haliaetus albicilla*), a very large migrating bird that appears at certain times of the year off the coasts of North America, Europe, and Northern Asia where its food consists almost exclusively of marine fishes, young SEALS, and porpoises. (See PORPOISE.) In this order, the widely distributed osprey or fish hawk (*Pandion haliaetis*) resembles a small eagle but has clear white underparts. In addition, among other anatomical features it has a reversible hind toe, especially well-developed claws, and spicules on the scales of its feet.

• *Gulls* (*order Charadriiformes*). Almost all species contained within this order are shore birds found on or near water. Typical examples are the American oyster catcher (*Haematopus ostralegus*), a bird of the tropical and temperate coasts of North and South America that feeds on MUSSELS and oysters, and the herring gull, *Larus argentatus*, a common species of northern North America and Northwestern Europe. The herring gull is the only larine gull that nests in the continental United States. Others of circumpolar distribution, such as the glaucous gull (*Larus hyperboreus*), Thayer's gull (*L. thayeri*), and Kumlien gull (*L. glaucoides*), look remarkably alike and live together, yet they do not interbreed. Cousins of these birds are the migratory sand larks or beach plovers (e.g., the Northern European little ringed plover, *Charadrius dubius*); the crab plover, *Dromas ardeola*, that lives on the coasts of East Africa and the western shores of the INDIAN OCEAN where it subsists on CRAB; and the Antarctic sheathbills (*Chionis alba* and *C. minor*) whose diet consists of seaweed, mussels, small crustaceans, and the eggs and young of other birds.

The skuas (or jaegers), abundant in both the polar regions (e.g., the Arctic skua. *Stercorarius parasiticus*), are also members of the gull order (family Stercorariidae) that subsist on marine fish as well as on the eggs of penguins. Skuas possess amazing homing abilities, have hawklike habits, and are capable of plunge-diving to catch ocean fish.

The gulls of the Laridae family are excellent flyers which swim well in the ocean but never dive. Combining the gifts of swimming and soaring, these gulls float buoyantly on the waves or glide on the breeze like a sloop before the wind. Most gulls are larger than pigeons, with wings that are much longer in proportion. They are largely white with pearly gray backs. The young ones run to dappled browns. The dozen species of gulls belonging to this family are distributed widely around the ocean waters of the United States, Europe, Northern Africa, and Australia. Some species are sedentary, while others are migratory. One gull, the southern kelp gull (*Larus dominicanus*) is a sub-Antarctic ocean bird of circumpolar range which breeds in numbers on Tierra del Fuego and upon other heavily glaciated, ice-covered islands. This bird also extends its breeding range along the tropical western coast of South America practically as far as the PERU CURRENT is in contact with the land.

Gulls are known as the symbols of grace and maneuverability, yet they cannot compare in these attributes with their close relatives, the terns. Most terns can be immediately recognized as terns by two features—a forked tail and a black cap. In fall and winter the cap is back over the ears and a white forehead shows.

Terns are more streamlined than gulls. Their wings are narrower, their tails longer, and their bills sharper.

A dozen terns nest in the United States. Some are the common tern (Atlantic Coast), least tern (both coasts), royal tern (only coasts of southern states), and the black tern (inland and both coasts). These larine terns, or sea swallows, are all poor swimmers but are extremely active birds which dive into the ocean waters at great speeds to capture prey. The little tern (*Sterna hirundo*) is found worldwide on all coasts of the world's oceans.

Larger than the terns, but smaller than the gulls, are the three species of tropical skimmers (family Rhynchopidae) found in the estuarine regions of tropical Africa, India, and the United States and South America where they feed primarily on fish. In the United States, the black skimmer, a bizarre bird, is often found with the terns on the Atlantic coast. It is black above, with a long unequal red bill.

Auks, puffins, and guillemots belong to the gull order and family Alcidae. (The diving petrels are similar in anatomy and habits to the auks.) These short-winged birds are inhabitants of the northern seas and also somewhat resemble the penguins of the Southern Hemisphere. They feed on the fishes and PLANKTON of the areas they inhabit: near Iceland, Greenland, and Labrador.

In summary, birds that live away from land for long spans of time and secure their food from the open oceans are called *pelagic birds*. The largest number of these belong to the order Procellariiformes, which includes the albatrosses, fulmars, petrels, and shearwaters. Such birds possess large nasal salt glands which control salt secretion.

Perhaps the most interesting of the migratory birds are the wandering albatross (*Diomedea exulans*) and the arctic tern. The Southern Hemisphere albatross is a large [some have a wingspan up to 12 ft (3.7 m)], strong, ocean wanderer that has been known to circumnavigate the globe. These birds are white with brown wing tips, a buff-colored beak, and blue-gray legs and feet. They nest mostly on glacial Antarctic islands such as South Georgia. The arctic tern (*Sterna paradisea*) is about the size of a small seagull and makes a round trip of some 22 000 mi (35 000 km) between the Arctic and Antarctic regions each year.

Certain other migratory birds spend the summer in Europe and winter in South America; others, such as the great cormorant (*Phalacrocorax carbo*) and the arctic skua (*Stercorarius parasiticus*), fly from Northern Europe to the African coasts. The white-breasted cormorant, or guanay (*P. bougainvillei*), is a member of the Antarctic branch of the cormorant family. Its closest kin are the cormorants of the Strait of Magellan, New Zealand, sub-Antarctic islands, and the shores of the Antarctic Continent, but its relationship with the other cormorants of South America or with those of the Northern Hemisphere is distant. The range of the guanay extends along the west coast of South America from central Chile to within 6° of the equator. About 20 in (50 cm) high, weighing about $4\frac{1}{2}$ lb (2 kg), this long-winged, long-billed cormorant, with glossy green and blue-black neck and back and glossy white breast and pink feet, has a conspicuous naked skin area around the eye. A very capable and strong flyer, feeding exclusively upon the sea-surface organisms, the guanay is the most famous of the guano birds of Peru.

The navigational mechanisms of these birds are not well understood. Neither are those which give the small flightless Adélie penguin its ability to home from its winter quarters in the ice pack to its breeding site of the previous year. Also, little is known about the migrations of many of the diving birds (e.g., auks, guillemots, puffins, and other of the Alcidae) of the polar regions.

The influence of all birds on oceanic surface life is quite substantial. For example, millions of tons of anchovies are consumed yearly by guano-producing birds off Peru; and without a doubt many millions of tons of other forms of ocean life are removed yearly from the world's oceans by birds of less discriminating diets.

MARINE MAMMALS are those warm-blooded ocean-inhabiting vertebrates, or some 117 species that are members of the class Mammalia and the present orders of Carnivora, Pinnipedia, Sirenia, and Cetacea (suborders Mysticeti and Odontoceti). Primarily the marine mammals are characterized by the fact that they possess mammary glands and the young are nourished by milk. Hair is never entirely lacking in these aquatic animals, and the brain is well developed. They are viviparous, and the fetus is enclosed in the allantois (a fluid-filled sac) and amnion (membrane enclosure of the sac).

Among the several marine mammals are the following:

- SEA OTTER, *Enhydra lutris*, order Carnivora, family Mustelidae.
- POLAR BEARS, *Thalarctos maritimus*, order Carnivora, family Ursidae.
- SEALS, order Pinnipedia, including the sea lions of genera *Otaria, Phocartos, Neophoca, Zalophus*, and *Callorhinus;* and walruses, *Odobenus rosmarus* and *O. divergens*, genus *Odobenus;* the family Phocidae, including seals of the

genera *Phoca, Pusa, Halichoerus, Histriophoca, Pagophilus, Erignathus, Monachus, Lobodon, Omnatophoca, Hydrurga, Leptonychotes, Cystophora,* and *Mirounga.*

- SEA COWS and Dugongs, order Sirenia, including the genera *Dugong, Hydrodamalis,* and *Trichecus* (manatees).
- CETACEANS, order Cetacea with suborder Mysticeti, the baleen WHALES, and Odontoceti, the toothed whales, including the DOLPHIN and the PORPOISE. In the suborder Mysticeti there are three families (Balaenidae, Eschrichtidae, and Balaenopteridae) and a total of five genera (*Balaena, Caperea, Eschrichtius, Balaenoptera,* and *Megaptera.* In the suborder Odontoceti there are five families [(1) Platanistidae, (2) Delphinidae, (3) Monodontidae, (4) Physeteridae, and (5) Ziphiidae] and a total of some 31 genera [(1): *Platanista, Inia, Lipotes,* and *Pontoporia;* (2): *Steno, Sousa, Sotalia, Tursiops, Grampus, Lagenorhynchus, Stenella, Delphinus, Lissodelphis, Cephalorhynchus, Peponocephala, Feresa, Pseudorca, Globicephala, Orcinus, Orcaella, Phocoena,* and *Phocoenoides;* (3): *Delphinapterus,* and *Monodon;* (4): *Physeter* and *Kogia;* (5): *Tasmacetus, Mesoplodon, Ziphius, Berardius,* and *Hyperoodon.*

MARINE MEDICINE See PHARMACEUTICALS FROM THE OCEAN.

MARINE OPTICS is the study of the science of light and vision in the marine environment.

Optical or light transmission can be discussed from the standpoint of refraction, reflection, absorption, scattering, and color.

The principal physical factor affecting visibility in the underwater marine environment is that light behaves differently in the denser water medium than it does in air. Parallel rays of light hitting the water surface at an angle of incidence will be in part reflected and in part refracted. In general, the angle of reflection is equal to the angle of incidence. The light rays that pass from air into the water are bent, and the effect known as *refraction* occurs. The refraction that takes place between air and water is directly proportional to the relative speed of light in the two media. The light waves decrease in speed as they pass from the air into the water, and the relationship between the angle of incidence I and the angle of refraction R is given by SNELL'S LAW:

$$\frac{\sin I}{\sin R} = \text{refractive index, which for air and water is about } \tfrac{4}{3}$$

Refraction can cause an object to appear smaller or larger than its true size, and to appear in a position other than its true position. In water, with the human eye functioning normally, refraction causes nearby objects to appear larger and closer than they actually are.

Incident light radiation in SEAWATER can be extinguished by absorption and scattering caused by the seawater itself or diffused by scattering, diffraction, and reflection caused by suspended PARTICULATE MATTER in the water. The depth to which light rays will penetrate is directly dependent on this suspended and dissolved particulate matter. The rate at which the downward-traveling radiation decreases in water is determined by the extinction or absorption coefficient. The values of these coefficients have been measured by oceanographers. At times the diffusion that takes place scatters light into areas that would otherwise be in shadow or have no illumination. Hence, it is helpful to the diver and underwater photographer. At other times it is detrimental because it produces backscatter which interferes with vision and underwater photography when artificial lighting is involved.

The color of ocean water is due to a variety of factors: e.g., sky cover, water depth, and the particle content of the water. The particles may consist of nutrients, sediments, or plant and animal matter, both alive and dead. Open ocean water of high transparency (e.g., the water of the GULF STREAM off the eastern coast of the United States), when observed from a surface platform, is usually deep blue owing to light scattering on the water molecules or finely suspended particles. Blue light, being of short wavelength, is scattered more effectively than light of longer wavelengths. Ocean water of a variety of shades of green also occurs, and these shades range from yellow-green to blue-green, depending upon the amount of suspended material present.

As sunlight enters the ocean water, not only is the intensity reduced, but also certain light rays are absorbed more than others. For example, as the light travels to deeper water, red light is filtered out at relatively shallow depths in clear water. Orange is filtered out next, then yellow, green, and blue. This filtering of the various colors is dependent on depth and other factors such as SALINITY and turbidity. Seawater tends to readily absorb the long-wave and short-wave parts of the visible spectrum so that most of the energy utilized for PHOTOSYNTHESIS in the oceans comes from the blue, green, and yellow regions of the spectrum. In certain areas the presence of dissolved yellowish matter related to humus and regarded as the product of phytoplankton (microscopic floating plants) metabolism is thought to be the cause of the transition from the typically blue color of the ocean to green. The presence of colored suspended forms of algae explains the brown or brown-red color often observed in coastal areas. Also, the color variations that frequently occur in these regions are caused by the addition of sediments and nutrients from river effluent and by disturbance of bottom sediments as a result of wave action in shallow waters.

The transmission of light underwater can alter the color appearance of an underwater object radi-

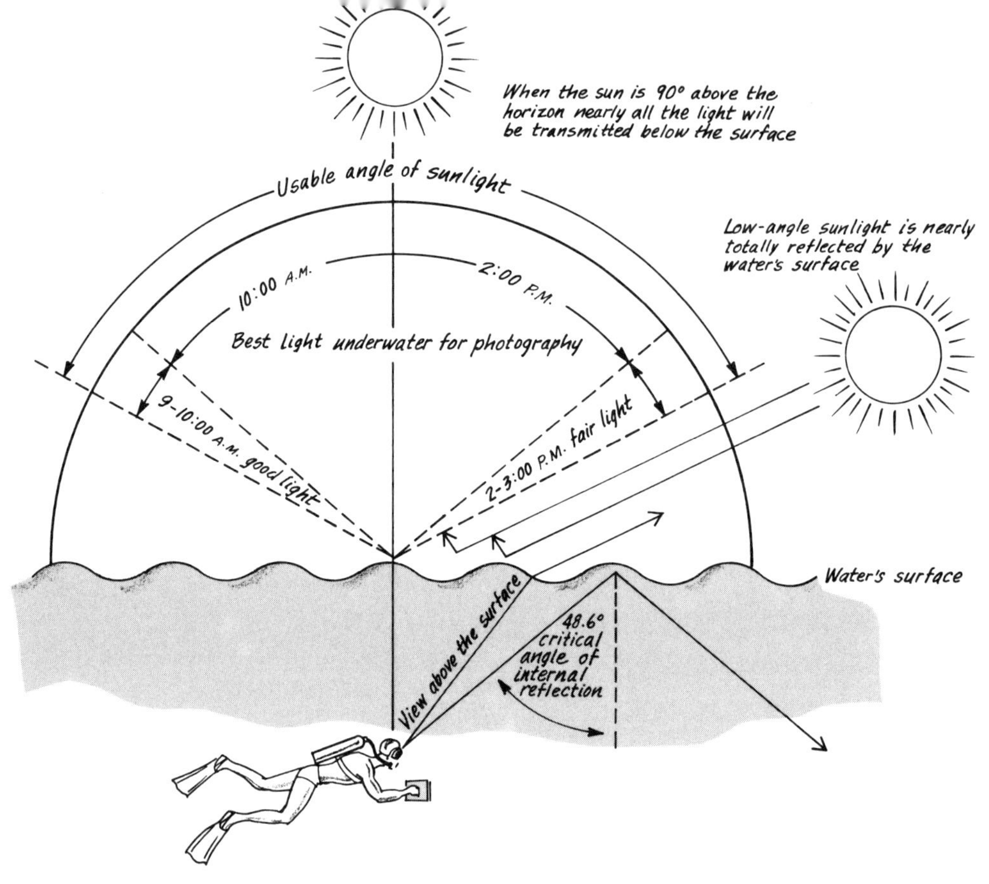

MARINE OPTICS. Diurnal variation of light under water. (*NOAA*)

cally. For example, an object that is red on the surface will appear black when placed underwater, since seawater absorbs red. In the same way a blue object placed in green water will also appear black, and a yellow object in green water will resemble an object that was white or pale gray on the surface. Red and blue-green objects in green water and yellow and dark-blue objects in clear ocean water will retain their color to considerable depths.

MARINE SEDIMENTS include all that material which, as a result of several transportation and formative processes, has accumulated on the basaltic plate which forms the crust of the ocean basins. The most important source of marine sediments is the land. For instance, it has been estimated that the rivers of the world carry some 2.8 mi³ (12 km³) of sediment and dissolved minerals into the sea each year. In addition, wind, ice, and volcanoes play an active role in moving material from the land to the sea. The amount of fine dust, sand, and volcanic ash blown into the sea by the wind is not known, but analysis of the bottom sediments indicates that it has been significant. Marine aeolian (wind-blown) sands on the ocean floor around the Cape Verde Islands were transported from the Sahara Desert by the wind. Volcanic ash is particularly evident in the marine sediments around the volcanic islands of the PACIFIC OCEAN. Glaciers incorporate large quantities of sediment of various sizes (boulder to dust size) into their mass as they grind down to the sea. As icebergs (see SEA ICE) calve off and float away, they slowly release their sediment burden as they melt. Today glacial sediment is deposited as far toward the equator as 45° north and south. However, during the last Ice Age such sediments were carried as far as 25° north and south.

Terrigenous sediment, consisting of gravel, sand, clay, and mud mixed with varying amounts of organic debris, is largely confined to the CONTINENTAL SHELF, CONTINENTAL SLOPE, CONTINENTAL RISE, and ABYSSAL PLAIN of the world's oceans. Once the sediment has reached the seafloor, it is dependent on wave action, ocean currents, avalanches, and turbidity currents for any further movement. A TURBIDITY CURRENT, activated by earthquakes and avalanches, is particularly active in transporting sediment down the continental slope and submarine canyons and depositing it on the continental rise and the smooth abyssal plain. (See SUBMARINE CANYON.) It has been estimated that the continental rise contains about half the sediment in the ocean basins.

The deep ocean floor near the continents often contains mud deposits that are blue, green, red, white, or black in color. The blue color is due to organic matter and to ferrous oxide or ferrous hydrate. Red muds get their color from an abundance of ferric iron oxide which indicates an oxidizing condition. Green muds result from the inclusion of significant amounts of glauconite. The green of glauconite, in turn, comes from ferrous iron and organic matter. The white muds are derived from organic reefs (see CORAL REEF) and consist of calcium carbonate. The black muds result from the inclusion of organic matter and iron sulfide. These muds usually smell strongly of hydrogen sulfide.

The next largest category of marine sediments is the pelagic sediments, or those which accumulate in the deep-sea basin far from land. The more prevalent pelagic sediments are red clay, DIATOM OOZE, and radiolarian, globigerina, and pteropod oozes.

Red clay covers an estimated 6.2 million (6.2×10^6) mi² (15.9 million km²) of the floor of the Atlantic, 27 million mi² (70.3 million km²) of the Pacific, and 6.2 million mi² (16 million km²) of the Indian Ocean. It gets its name from the brick-red samples taken from the Atlantic, but its color is chocolate-brown in the Pacific and Indian oceans. The color tends to change to blue as it approaches the continental border. Red clay contains very little calcium carbonate; it is plastic and greasy to the touch. The origin of red clay has been debated for many years. There is some evidence that the clay is actually a colloidal clay derived from land and distributed to the deep-ocean basins by currents.

Diatom ooze is made up of tiny siliceous skeletons of the microscopic marine plant known as the diatom. Deposits of this yellowish-to-cream colored material covers about 1.6 million mi² (4.1 million km²) in the Atlantic, 5.5 million mi² (14.4 million km²) in the Pacific, and 4.9 million mi² (12.6 million km²) in the Indian Ocean. Diatom ooze is most likely to form under those areas where significant UPWELLING brings nutrient-rich WATER to the surface. Therefore, the largest deposits of diatom ooze lie in a belt surrounding the continent of Antarctica and along the rim of the North Pacific.

Radiolarian ooze is red to chocolate-brown in color and is composed of the siliceous skeletons of tiny planktonic animals. This type of ooze covers about 2.5 million mi² (6.6 million km²) in the Pacific and 2.4 million mi² (6.3 million km²) in the Indian Ocean. The radiolarian ooze of the Indian Ocean occurs as patches, but in the Pacific it forms a belt beneath the North Equatorial Current (see EQUATORIAL CURRENT SYSTEM) that extends from about 170°W LONGITUDE to near the west coast of South America. Significant deposits appear to be absent in the Atlantic.

MARINE SEDIMENTS. Sediment pockmarks and trailing patterns at 13 200 ft (396 m) in the Indian Ocean suggest ocean bottom currents moving from the upper left to the lower right in the photo, which encompasses a 6 × 8-ft area. (*NOAA*)

Globigerina ooze is milky white, rose, yellow, or brown and is made up of the calcareous remains of the foraminifera known as globigerina. It is the most widespread of the organic oozes, covering 15.5 million mi² (40.1 million km²) in the Atlantic and 13.3 million mi² (34.4 million km²) in the Indian Ocean. As a general rule globigerina ooze is found only in depths ranging from 6000 to 15 000 ft (1829 to 4572 m). This is probably due to the presence of carbonic acid in the greater depths of the ocean. The CO_2 content of seawater increases with increasing pressure and with decreasing temperature. Therefore, at great depths calcium carbonate is taken back into solution and the calcareous skeletons do not accumulate.

Pteropod oozes are white to light brown with a reddish pink or yellow tinge and are composed of the calcareous skeletons of the pteropod. Pteropod ooze covers about 0.6 million mi² (1.5 million km²) in the Atlantic but is not found in significant accumulation in the other oceans. Being calcareous, the fragile skeleton dissolves at great depth.

Over a century ago the globe-circling CHALLENGER EXPEDITION discovered the existence of MANGANESE NODULES—concretions of ferromanganese oxide—lying abundantly on some areas of the deep seafloor. Subsequent investigations have indicated the existence of these nodules in the Pacific, Atlantic, Indian, and Southern oceans. The Pacific alone is estimated to contain about 1.5 trillion (1.5×10^{12}) tons (1.4 trillion metric tons) of nodules, and to be adding to that quantity at the rate of 10 million tons (9 million metric tons) per year. The nodules, which contain valuable amounts of manganese, copper, and nickel, are only one nodule thick in the surface layer, but other layers exist at deeper levels in the sediment. Many nations have an interest in mining manganese nodules, and techniques for their recovery have been developed. However, exploitation will have to wait until ownership among the nations of the world can be negotiated. In the meantime two perplexing questions remain to be worked out: how are the nodules formed and how do they remain exposed at the water-bottom interface while they are forming when the rate of sediment accumulation on the bottom exceeds their growth rate?

MARINE SNAILS. See GASTROPODS.

MARINE SNOW is a coagulate of suspended living and dead organic matter and inorganic debris of the ocean which is present in concentrated amounts at density boundaries such as the THERMOCLINE. See PARTICULATE MATTER.

MATA-MALU is the Samoan name for the SEA ANEMONE, *Rhodactis howesii,* which contains a lethal poison.

MAURY, MATTHEW FONTAINE (1806–1873), a U.S. naval officer, is generally accepted as the man who gave OCEANOGRAPHY its first real impetus. Maury was appointed a midshipman in the U.S. Navy in 1825 and was immediately assigned to sea duty. He spent the next 14 years on various Navy ships, including the *Vincennes* of the U.S. Exploring Expedition which circumnavigated the earth. A man of great ambition and vision, Maury directly applied his self-taught knowledge of mathematics, astronomy, and navigation, plus his observations and deductions to his work at sea. His first book, *A New Theoretical Treatise on Navigation,* was published as a navigation textbook for naval officers in 1836. An injury in 1839 caused him to be permanently lame and thus prevented him from continuing his active naval career at sea.

However, Maury, as supervisor of the Depot of Charts and Instruments in Washington, D.C., began collecting information systematically from naval vessels and merchant ships on CURRENTS, weather, winds, and other useful data recorded in Maury's specially designed ships' logs. The 76 charts classified into six distinct series and classes (track charts, trade winds, pilot charts, thermal charts, storm and rain charts, and whale charts) and the notices to mariners that evolved from this information were published and widely distributed by the U.S. government from 1847 until 1860. They gained immediate popularity and usefulness with mariners throughout the world. In 1855, Maury published *The Physical Geography of the Sea.* This pioneering work immediately won him international acclaim. While later discoveries and studies changed some of the conclusions Maury had made in his book, his initiative and the systematic methods of compiling and utilizing ocean data he employed formed the basis for oceanography.

Always a controversial figure during his lifetime, admired and respected by many of his contemporaries and scorned and criticized by others, Maury, at the outset of the U.S. Civil War, left his post as Director of the Naval Observatory and threw his fortunes in with the Southern Confederacy. He resided in England and Mexico during the 1860s, but later returned to America where he spent his last years as professor of physics at the Virginia Military Insititute (Lexington, Virginia) from 1868 to 1872. He died in Lexington on February 2, 1873.

MEDITERRANEAN SEA is a near-landlocked sea bounded by the coasts of Europe, Africa, and Asia, and extends from the Strait of Gibraltar [as little as 13 mi (21 km) wide] in the west to the Dardanelles [as little as 1 mi (1.6 km) wide] in the east. The Suez Canal [about 300 ft (91 m) wide] also provides a small outlet on the southeast. The Mediterranean is conveniently divided into an eastern and a western basin joined by the Strait of Sicily [77 mi (124 km) wide] and the Strait of Messina [1.7 mi (2.7 km) wide]. Each is further divided into a number of seas. The western basin consists of the Alboran Sea, lying west of the meridian of the Alboran Islands (3°00′W); the Balearic or Iberian Sea, lying between the Balearic Islands and the coast of Spain; the Ligurian Sea, lying between Corsica and the Ligurian coast of Italy; and the Tyrrhenian Sea, bounded by the coast of Italy, Sicily, Sardinia, and Corsica. The eastern basin consists of the Ionian Sea, lying in the pyramid formed by the southern extremity of Greece and Sicily and the heel of the Italian boot; the Adriatic Sea, lying north of the Ionian Sea and flanked by Italy and Yugoslavia; the Aegean Sea, lying between Greece and Turkey; and the Sea of

MEDITERRANEAN SEA. It is easy to distinguish the Strait of Gibraltar in the center of this satellite photo, and the Mediterranean Sea lies to the right under light cloud cover. Most of the cloud activity is concentrated over the Costa del Sol. (*NASA*)

Marmara, lying between the Dardanelles and the Bosporus. The major islands in the two basins are Cyprus, Crete, Sicily, Sardinia, Corsica, and the Balearics. The Sea, thus defined, covers an area of 971 176 mi² (2 516 000 km²), occupies a volume of 901 582 mi³ (3 758 000 km³), and has a mean depth of 4902 ft (1494 m). The greatest depth in the western basin is 12 201 ft (3719 m) and occurs in the Tyrrhenian Sea, while the greatest depth in the eastern basin is 18 143 ft (5530 m) and occurs in the Ionian Sea. The Mediterranean gets its name from the Romans and means "middle of the earth." The Romans also referred to it as *Mare nostrum* which means "our sea," a name that the Italians revived during World War II.

The Mediterranean Sea occupies an arid region in which evaporation exceeds both precipitation and runoff from rivers. The sea loses about 1000 mi³ (4168 km³) of WATER each year to evaporation, and runoff repays only about one-tenth of this loss. Were it not for the inflow of Atlantic water through the Strait of Gibraltar, the Mediterranean would dry up in about 1000 years. Such an event would leave a vast, bone-dry chasm over 2000 mi (3218 km) long, up to 1080 mi (1738 km) wide, and as much as 18 143 ft (5530 m) deep. As unlikely as it sounds, this is what happened in the comparatively recent geologic past. Some 15–20 million ($1.5-2.0 \times 10^7$) years ago the Mediterranean was part of an enormous waterway that connected the ATLANTIC OCEAN, PACIFIC OCEAN, and INDIAN OCEAN and included the Black Sea and Caspian Sea. Some 7–8 million years ago this pattern was disrupted by mountain building (the Carpathian Mountains) and other structural adjustments between the continents of Europe, Africa, and Asia. As a result, the basin was cut off from the area and became a desert. Water still poured in from such rivers as the Rhone of France and the Nile of Egypt, but evaporation prevented any accumulation.

Today the evidence for the desiccation of the Mediterranean lies in a hard evaporite layer which runs throughout the basin beneath 1000 feet (305 m) of recent, unconsolidated sediment and in the existence of numerous salt domes in the subfloor structure. These domes have been squeezed up into the overlying sediment from a bed of salt several hundred feet (1 ft = 0.3 m) thick that precipitated from the Mediterranean water as it became more and more concentrated through evaporation. Additional evidence comes from rivers such as the Rhone and Nile that for millions of years have drained into the Mediterranean basin. As the water level in the basin dropped, these rivers were forced to cut their valleys deeper and deeper in an effort to keep pace with an ever-lowering sea level. Drilling and seismic profiling have been able to trace the Rhone valley to 3000 ft (915 m) below the level of the sediment which subsequently covered it. Similar evidence indicates that the Nile was induced to cut its valley in solid granite 700 ft (213 m) below present sea level.

About 5.5 million years ago a structural adjustment opened the Strait of Gibraltar to produce a waterfall that must have been roughly 1000 times the size of Niagara Falls. At this rate the vast Mediterranean basin would have filled to its present level in about 100 years.

The CONTINENTAL SHELF of the Mediterranean Sea is only moderately well developed—averaging about 25 mi (40 km) in width. The shelf slopes steeply to the deep basin and is frequently cut by submarine canyons. Several large alluvial fans—the Rhone and Nile being the more prominent—debouch onto the deep floor, their surfaces cut by fan valleys (see SUBMARINE CANYON) down which sediment is distributed to the ABYSSAL PLAINS. The sediment cover consists of calcareous debris from plant and animal life, volcanic debris, and both clay and sand transported to the basin by wind and rivers. In the eastern basin there are black muds indicating that the bottom waters may have become stagnant (see BLACK SEA) during the last Ice Age when the flow into the basin was restricted by a low stand of sea level. The western basin is more oceanlike and contains several abyssal plains, with the Balearic being the largest. The eastern basin, on the other hand, is very different in that it is in a state of continual adjustment to stresses. The basin is partitioned by the Medterranean Ridge which runs eastward from the sole of the Italian boot.

Since evaporation exceeds precipitation and river runoff throughout the Mediterranean, the surface water is more saline (38–39.9 ppt) than Atlantic water. This heavier, though warmer [70–77°F (21–25°C)], water flows out through the deep 1000-ft (305 m) Straits of Gibraltar and sinks to an equilibrium depth of about 3280 ft (1000 m). This Mediterranean water can be detected for thousands of miles into the Atlantic. To compensate for this outflow of surface water, Atlantic water, in far greater quantities [61 800 000 ft^3/s (1 750 000 m^3/s)], flows into the sea as a surface current. This lighter water is eventually cooled by evaporation and sinks to lower levels in the water column. This process is sufficient to renew the water in the entire Mediterranean every 70 years. But, the sill depth between the sea and the Atlantic is such that only the relatively depleted surface water in the two bodies is exchanged. Therefore, the Mediterranean Sea is the most nutrient-poor large body of water in the world.

MEDUSAE See JELLYFISH.

MENHADEN (also called pogy, mossbunker, or fatback) is the name for a fish of the order Clupeiformes, family Clupeidae—the HERRING family. Menhadens are one of the least known but most abundant of fishes in the world's oceans. Although seldom used for human consumption, they are of great industrial importance, especially as fertilizer. In fact, the name *menhaden* comes from *munnawhatteaug*, an American Indian word, meaning "that which fertilizes."

Menhaden are found in temperate waters along the coast of North America. Atlantic menhadens (*Brevoortia tyrannus*) are found from Nova Scotia to central Florida. Gulf menhaden (*B. patronus*) occur from southern Florida to Veracruz, Mexico. They are not equally abundant throughout their range but are concentrated in certain localities during certain periods of the year. They live in near-surface waters overlying the inner half of the CONTINENTAL SHELF during the warmer months. They are rarely seen in surface waters during the colder months, and there is evidence that during this period they live in deeper waters over the continental shelf.

These fishes have a tendency to school according to size and age. This habit, varying from hundreds to thousands of individuals in a school, makes netting (purse seine) them relatively easy for fishermen.

In calm weather, vast schools of menhaden may be seen lifting their snouts up out of the water as they feed, and they occasionally break through with their top fins and tails. The schools are often so dense they seem to darken the surface like great cloud shadows on the waters of the ATLANTIC OCEAN coast and the Gulf Coast.

Menhaden vary in color from dark blue to green, blue-gray, or blue-brown above and have silvery sides, belly, and fins with a yellow luster. There is a dusky spot on each side of the fish behind the gill openings, followed by a varying number of smaller dark spots. The body is somewhat flattened sideways like other members of the herring family. The scaleless head is very large in proportion to the body. The body scales are nearly vertical (not rounded) and are edged with long, comblike teeth instead of being smooth. The large gaping mouth is toothless, and the lower jaw projects beyond the upper. The tail is deeply forked. Menhaden feed on microscopic plants and small CRUSTACEANS which are screened out of the water with highly specialized, sievelike gill rakers. Adults of the Atlantic species average about 1 lb (0.4536 kg) in weight and from 12 to 15 in (30 to 37.5 cm) in length. They live for 7 to 8 years. The Gulf menhaden is a smaller fish which usually weighs less than $\frac{1}{2}$ lb (0.227 kg) and lives less than 3 years. The menhaden has a large ratio of gill surface area to body weight (1770 mm^2 or 2.74 in^2 of gill per gram of body weight). The gill-surface-area measurement indicates how much oxygen a fish can take in. By comparison, the MACKERAL has 1160 mm^2 of gill per gram of body weight.

The menhaden is an extremely oily fish with many small bones. These qualities have largely prevented the species from gaining popularity as a table fish. However, the menhaden is valuable for the industrial products (oil and meal) made from it; practically none is used directly for human food. About $\frac{1}{3}$ of all the fish meal prepared in the United States and $\frac{1}{4}$ of the oils processed from ocean animals comes from menhaden. Most of the meal is used as a supplement in feed for hogs and chickens; the oil has been used in a variety of commercial products such as margarine, paints, insect sprays, printer's ink, soaps, and lubricating oils as well as in leather tanning and aluminum casting.

MERMAID'S PURSES is the name commonly applied to the egg cases of the SKATE and SHARK. Skates and some types of sharks eject their fertilized eggs one or two at a time in flattened capsules which are formed around the egg as it passes down the oviduct. The outside layers of these capsules harden in the SEAWATER, producing the leathery mermaid's purses which are often washed up on sea shores or found in seaweeds along the high water mark. (See ALGAE.) Each purse has a short, hollow horn at each corner through which water can freely pass to aerate the eggs. Each case contains a single embryo which takes up to 15 months to develop, depending on the species.

METEOR was the German oceanographic research expedition of 1925 in the ATLANTIC OCEAN. In the quest for a useful method of extracting gold from the oceans, Germany assigned, in 1925, its most famous chemist Fritz Haber and the ship *Meteor* to the task. The ship cruised the South Atlantic Ocean in this fruitless effort, but, in doing so, it also accumulated a considerable amount of oceanographic information, especially relative to the distribution of nutrient substances found in PLANKTON and other ocean organisms.

MINING (OCEAN) is the process of recovering ENERGY sources and mineral resources from deposits on and under the ocean bottom as well as from SEAWATER.

It has been estimated that the ocean floor contains more than 30 percent of the world's remaining petroleum and natural gas and about 50 percent of its hard minerals. In the face of the world's increasing demand for such sources of energy and for the mineral commodities, ocean mining capabilities have been advanced and exploitation of such resources, in some cases, has taken place.

Large-scale development of the technology for mining the offshore undersea resources of petroleum started in 1947 off the coast of Louisiana. The experience that accrued from this venture furnished much of the basis for the expertise that went into the exploration and production used in other areas of the world (such as the NORTH SEA and offshore Indonesia).

In 1977 the world's active drilling rig fleet consisted of 439 units. Of these, 24 were submersibles, 93 were drillships and drill barges, 197 were jack-ups, and 125 were semisubmersibles.

These rigs for drilling or mining consist essentially of three types: (1) permanent platforms which are a logical extension of the oil- and gas-production technology used on land; (2) the mobile rig of which the jack-up is a common type—this appropriately designated platform is floated to the candidate drilling site; then its long legs are lowered and securely placed on the ocean floor, and the drilling deck is jacked up above the ocean surface; and (3) the floating rigs—a class that includes the towed-to-site semisubmersible rig, which is more stable than the drillship because its latticed structure offers less resistance to WAVES. When one of these rigs has been towed into position, the large, buoyant tanks and columns on which it floats are ballasted with seawater until they sink far enough to stabilize the rig. The structure, with its large drilling deck raised high above the surface, is stable in all but the highest SEA STATE conditions. When in position, the rig is secured in place by several cables, and each of these is heavily anchored. Another method of holding the drilling vessel to the well is called *dynamic positioning*. Here propellers beneath the hull of either a drillship or a semisubmersible can be positioned in different directions in order to provide variable thrust to counter winds and CURRENTS that push the vessel away from the drill hole. The propellers are controlled by an on-board console, which takes its position from an acoustic signal beacon placed on the seafloor wellhead.

Such types of platforms accounted for approximately 20 percent of the world's daily petroleum production, or about 10 million (10×10^6) barrels. In 1977 this offshore drilling was done chiefly at water depths less than 660 ft (200 m). However, potentially important oil fields are being discovered at varying ocean depths every year, with the nations of the world investing considerable capital in offshore ventures. More than 85 countries were engaged in offshore activities, and discoveries have been reported from the shelves of North and South America, Australia, Japan, and the countries around the MEDITERRIAN SEA, the RED SEA, and the ARABIAN SEA; the Union of Soviet Socialist Republics; and, in the NORTH SEA, and the SOUTH CHINA SEA. In 1977, 32 of these countries produced petroleum from their continental shelves. (See CONTINENTAL SHELF.)

Most sand and gravel is mined at or near the beach. However, various near-offshore areas are also exploited for these materials, as are calcium carbonate shells and sands fragmented from them for use in the manufacture of portland cement and lime. OYSTER shells and other calcareous shells are mined from several locations in the United States and other parts of the world.

Consolidated deposits of iron ore, coal, and sulfur have been mined from the ocean floor in the past by many nations (e.g., Finland: iron ore; Japan, United Kingdom, Turkey, and Taiwan: coal; United States: sulfur).

Several minerals occur in association with, or under, the beach sands overlying the bedrock. These so-called heavy minerals are mainly derived from sources on land. As the rocks on land are subjected to the action of wind, rain, ice, and other destructive agents, the rocks yield to the processes of chemical and mechanical weathering. The weathered fragments are transported by streams and wind to the oceans. There, the winnowing action of the WAVES serves to concentrate the heavier minerals and metals into profitably minable deposits. The heavier the mineral, the closer to shore it is deposited. Consequently, these placer deposits are located on present beaches, on submerged beaches, and a few miles offshore. They usually include metals like gold, tin, and platinum; and the relatively lighter minerals like diamond, the titanium minerals ilmenite and rutile, the tungsten minerals scheelite and wolframite, the iron ore magnetite, chromite, and zircon.

Most authorities agree that a substantial number of economically feasible deep-ocean mine sites will be available for commercial development through the next century. Such ocean mining could make a significant contribution toward meeting the world's mineral resource needs of the future, particularly as a source of nickel, copper, cobalt, and possibly manganese. It is estimated that between 180 and 460 commercially viable first-generation mine sites exist on a global basis. Even if only 15 to 20 of these mine sites can be operational by the year 2010, the deep ocean floor could be a source of raw materials for many decades to come. It is only through more extensive prospecting in all ocean areas that the full potential can be defined, although studies show that the area covered by, and the potential tonnage of, MANGANESE NODULES are both large, with nickel potential substantially exceeding the 50 million tons (45 million metric tons) of land-based nickel reserves. Hence, mining techniques to obtain them have been proposed by several commercial firms, and mining operations are expected to start by the early 1980's. However, the capital costs will be very high and the law-of-the-sea ramifications relative to this type of venture are as yet unclear. (See LAW-OF-THE-SEA CONFERENCES.) Equally unclear are the legal and technological possibilities for exploiting gas, oil, and potentially massive deposits of coal, platinum, gold, copper, lead, zinc, iron, silver, nickel, tin, uranium, and molybdenum that are believed to exist in the Antarctic area.

However, in addition to the potential of new mining development, the possibilities for expanding past mining underseas operations also appear likely. In addition, although seabed deposits are generally not as abundant or as valuable as their counterparts on land, several exceptions are found around the world, such as the tin deposits off the coast of Cornwall in Great Britain and off the coast of Indochina, the diamond deposits off the coast of southwest Africa, the zircon sands of New South Wales, Australia, and phosphate ore off the west coast of the United States.

Although they were not mined in the strict sense of the word, dissolved minerals in seawater that have been recovered in the past are sodium, magnesium, calcium, and bromine. Practically all of the U.S. magnesium and bromine is chemically extracted from seawater. These operations, plus those involving minerals that can be recovered as a side benefit to the DESALINATION of seawater, may well be increased in scope and size. Also, uranium in seawater [estimated at three parts per billion (10^9)] can be recovered by extracting it by absorption techniques from warm wastewater or the seawater discharged after being used as a coolant in thermal or nuclear power stations.

The recovery of hydrogen gas from seawater for use in future energy conversion systems also holds fascinating possibilities.

All such mining and recovery operations of the future pose not only technical problems but legal and social ones as well. If these can somehow be solved, the potential yield is exceedingly large, for the world's ocean resources are as yet virtually untapped.

MIRAGE See FATA MORGANA.

MIXED TIDE See TIDES.

MODE (**MID-OCEAN DYNAMICS EXPERIMENT**) was a large, intensive program conducted by U.S. and U.K. physical oceanographers in the North ATLANTIC OCEAN in mid-1973.

The MODE program was designed to provide a more detailed understanding of fluctuations of ocean circulation by combining oceanographic observations with advanced numerical theory to establish the dynamics and statistics of medium-sized eddies, their ENERGY sources, and their role in the general circulation of the ocean. MODE I was an experiment in the organizational aspects of OCEANOGRAPHY nearly as much as it was a scientific project, and its organization has become a model for several large oceanographic projects.

Among the findings of MODE were that open-ocean eddies, which are dynamic features, have some profound effects on mean ocean circulation. These eddies were measured by some well-tried INSTRUMENTATION and techniques as well as by others that were entirely new. Thus, in many cases, MODE I was a proving ground for new oceanographic instruments.

MOHR TITRATION (MOHR-KNUDSEN METHOD) is a standardized yet empirical analytical chemistry method used in determining the SALINITY and CHLORINITY of SEAWATER. Basically, as in all titration methods, the composition of the solution is analyzed by adding known amounts of a standardized solution until a given reaction (color change, precipitation, or conductivity change) is observed. In the Mohr titration, the halide ions in seawater react with standardized silver nitrate solution which contains a potassium chromate indicator (see HALOGENS). (See also INSTRUMENTATION.)

MOLLUSK or **MOLLUSC** is the name applied to any member of the phylum Mollusca, the unsegmented bilaterally symmetrical invertebrates that include oysters, CLAMS, mussels, SQUID, OCTOPUS, etc. The bodies of these animals are soft, a fact from which the name mollusk is derived (*L. Mollusca,* a soft-shelled nut). Divided into six classes: Monoplacophora, Amphineura, Scaphopoda, Gastropoda, BIVALVIA or Pelecypoda, and Cephalopoda, comprising over 100 000 described species, the phylum is characterized by a shell-secreting organ, the mantle, and a food-gathering organ, the radula, located in the mouth.

Mollusks are the most adaptive of all the animal groups. They can be found at great ocean depths as well as in shallow waters; at high altitudes on land as well as in the lowlands; in swift flowing rivers and quiet ponds. Moreover, all mollusks, except for some types of clams and snails, can be found in all the HABITABLE ZONES of the world's oceans.

In the ocean's littoral zone, or along the coasts, mollusks are usually numerous, especially clams and snails; lesser numbers of squids and chitons can be found. Mollusks in the pelagic zone, or the surface waters of the open sea, are fewer in number than in the littoral zone and more specialized, e.g., a few of the cephalopods (squids, octopuses and nautiluses) and some shell-less snails and shell-less clams. In the abyssal zone, the darker ocean depths, are found the bivalves, octopuses, the giant squid and, in general, animal life that resembles ancient forms found in previous geologic ages.

MOLUCCA SEA is located in the western PACIFIC OCEAN and lies within the northern tier of the Spice Islands (Moluccas). The boundary is a line running clockwise from the north point of Sulawesi (Celebes), the Sangihe Islands, Talaud Island, the northern and southern point of Halmahera, Obi

Major, the Sulu Islands, Banggai, and around the Gulf of Tomini back to the north point of Sulawesi. The Molucca Sea covers an area of 118 502 mi^2 (307 000 km^2), occupies a volume of 138 668 mi^3 (578 000 km^3), and has a mean depth of 6168 ft (1880 m). The maximun depth is 15 780 ft (1880 m).

The floor of the Molucca Sea is a complex of ridges, deep basins, and somewhat shallower depressions. Notable among these is the Sangir Trough [12 532 ft (3820 m)] in the northwest, the Talaud and Morotai Basins [11 318–12 762 ft (3450–3890 m)] on the northeast, and the Mangole and Batjan Basins [11 515–15 780 ft (3510–4810 m)] in the south near the island of Obi Major. The Sangir Trough is separated from the Talaud and Morotai Basins by a ridge that cuts through the center of the sea. The deeper portions of the seafloor are covered with terrigenous and volcanic muds and globigerina ooze. The Molucca Sea lies in an explosive area of the western Pacific. It is ringed by active and dormant volcanoes and is the site of many deep-focus earthquakes.

Pacific WATER enters the Molucca Sea in the north and flows southward into the CERAM SEA, BANDA SEA, and FLORES SEA. Surface currents, under the influence of prevailing winds, generally flow southward along the eastern edge of the sea and northward along the western edge. In the north the flow is generally eastward.

MONSTER BUOY is a large instrumented ocean platform capable of being moored in deep waters and operating unattended for a year at a time; the collected information on several ocean properties can be either stored on tape in the buoy or relayed upon demand to a shore station. See INSTRUMENTATION.

MORAY, or **MORAY EEL,** is the name for about 120 species of elongate, serpentine marine fishes of the EEL family Muraenidae (order Anguilliformes); all have thick, scaleless skins, small rounded gill openings, large snakelike mouths with strong teeth, and lack pectoral or pelvic fins. In some types, the dorsal, anal, and caudal fins are very small or concealed under the skin so that the eels appear finless.

The morays live in the tropical and subtropical waters of the world's oceans. In these warmer seas they inhabit reefs and rocky areas from inshore waters to a depth of about 150 ft (45.4 m) and, being poor swimmers, only rarely travel into open water. (See CORAL REEF.) While morays have been found living in the deeper rocky parts of the Atlantic, these animals are most common in the coastal waters off Florida, California, Hawaii, Java, and atolls in the South PACIFIC OCEAN as well as in the MEDITERRANEAN SEA. They range in size from 6 in (15 cm) to over 10 ft (3 m) for the Indo-Pacific species. Many display varied color patterns, and others are a uniform brown or olive or yellow with a brown network. Often this coloration is affected by a skin coating of ALGAE, which can be yellow, brown, or green.

The diet of the morays includes CRUSTACEANS, mollusks, and fish, all of which they must swallow whole and quickly since they require a continuous flow of water through their mouths in order to breathe. Nocturnal feeders by nature, they come out of the crevices among the coral or rocky reefs during the day only if disturbed.

When caught, morays are capable of twisting into a knot in an attempt to free themselves. They also have the ability of moving the forward part of the body extremely fast and striking out from their shelter to bite any object, including the probing fingers of a LOBSTER- or ABALONE-seeking skin diver. Such wounds can be quite severe.

Very little is known about the breeding habits of the morays. Most authorities believe that there is no migration involved, as is the case with the eels (e.g., *Anguilla anguilla,* the European eel, and *A. rostrata,* the American eel) that travel long distances from rivers, lakes, and ponds to spawn in the ocean. When the moray spawns, swarms of females lay large quantities of yolked eggs from which hatch long, ribbon-shaped leptocephalus larvae that are quite similar to the larvae of such fish as the tarpon.

Representative species of morays are the Indian Ocean spotted reef eel (*Muraena pseudothyrsoidae*), the Indo-Pacific zebra moray (*Echidna zebra*), the Florida and West Indies green moray (*Gymnathorox funebris*), the California moray (*G. mordax*), the Mediterranean moray (*Muraena helena*), and the Hawaiian dragon eel (*M. pardalis*).

Some species are edible and are considered by some people to be quite a delicacy. On the other hand, the flesh of some morays can be highly poisonous to humans when eaten.

MUNK, WALTER HEINRICH (1917–), an American geophysicist and oceanographer, is best known for his research on wave motion in the ocean and the irregularities in the rotation of the earth. Born in Vienna, Austria, on October 19, 1917, Munk came to the United States in 1932. He received a bachelor's degree in applied physics from the California Institute of Technology in 1939. A master's degree in geophysics was earned at Caltech in 1941, and a doctorate in oceanography from the

University of California in 1947. He served as assistant professor (1947), associate professor (1949), and professor of geophysics (1954) at the SCRIPPS INSTITUTION OF OCEANOGRAPHY. From 1959 he served as associate director of the University of California's Institute of Physics and Planetary Physics and as director of the La Jolla Laboratories.

Summer work with Harald Sverdrup and Roger Revelle at Scripps turned Munk's interest to oceanography toward the end of his undergraduate training. (see REVELLE, ROGER RANDALL; SVERDRUP, HARALD ULRIK.) Over the years, that interest has centered around the study of wave generation, progression, and decay. Early interest dwelled on the microscale waves—capillary waves and high-frequency gravity waves—that add much to the observed roughness of the sea surface. Over the years Munk's research moved increasingly to the lower-frequency waves characterized by the long swell that strike the California coast, and tides. In collaboration with Frank Snodgrass, Munk detected waves in the PACIFIC OCEAN that had been generated 7456 mi (12 000 km) away in the INDIAN OCEAN. It was found that these low-frequency waves traveled a great circle route which entered the Pacific through the TASMAN SEA. Recorders stretching from Australia to Alaska monitored their progress through the Pacific. Similarly, Munk and his colleagues placed sensors on the deep ocean floor to measure tides as a means of improving tidal predictions.

For his many contributions to a better understanding of the dynamics of the oceans, to the effects of perturbations in the earth's rotation, and to astronomy, Munk was awarded the Arthur R. Day Medal of the Geophysical Society of America (1965), the Sverdrup Gold Medal of the American Meteorological Society (1966), the Gold Medal of the Royal Astronomical Society of London (1968), the Marine Technology Society's first award for Ocean Science and Engineering (1969), the Maurice Ewing Medal of the American Geophysical Union (1976), and the Alexander Agassiz Gold Medal from the National Academy of Sciences (1976).

MUSSELS is the name for certain of the bivalve mollusks of the order Anisomyaria and family Mytilidae. See BIVALVIA; CLAMS; MARICULTURE.

MUTUALISM See SYMBIOSIS.

MYTHS AND LEGENDS (related to the OCEAN) are stories, some of which are popularly taken as historical though not verifiable fact, and others (especially myths) which are ostensibly historical but invariably perpetuated to explain some particular belief, institution, or natural phenomenon. The myths and legends about the ocean itself, life in the ocean, and the ocean's many fascinating interactions with human beings are as old as recorded history. While some of these have been categorically disproven over the years as having no factual basis, still others persist to this day. Perhaps, as it has been said, mythology is history seen through the eyes of the intellectually immature.

In this brief discussion a sampling is presented of a few such stories such as the lost land of Atlantis, the legend of the PORPOISE, and the Bermuda Triangle.

Atlantis The story of the sunken continent of Atlantis is one of the oldest and most interesting, as well as the most persistent, of all legends. More than 5000 existing accounts tell about an advanced Atlantean civilization that was destroyed by a sudden cataclysm.

The Greek philosopher Plato, in the third century B.C., following, in general, Solon, his source of information, first described this lost land in his story *Timaios.* Although the account was invented perhaps solely for literary purposes, the embellishments and plausible details that Plato added in his fascinating work have provided the basis for most subsequent speculation on the existence and whereabouts of such a place.

Plato thought Atlantis to be somewhere west of the Pillars of Hercules (a place now assumed to be in either the Strait of Gibraltar or the Strait of Messina). Beyond these Pillars, the Greeks of Plato's time believed that a mysterious river, which they called Oceanus, circled the earth while the MEDITERRANEAN SEA separated their two known continents. Other continents, such as Atlantis, were said to exist in the river Oceanus; and, according to Plato, Atlantis was a continent inhabited by a people who not only built temples, palaces, harbors, naval storehouses, ships, and a system of canals covered by bridges, but also had developed an advanced form of agriculture and commerce. By the year 9600 B.C., as the story goes, this superior and powerful civilization had also conquered all the then known world except Greece. Greece was saved only because Atlantis was the victim of a massive earthquake and was engulfed by the river Oceanus. Thus, the continent disappeared overnight without leaving a trace.

Several articles and books by scientists and scholars, among others, have attempted to substantiate the existence and actual geographic location of Atlantis. Therefore, various theories concerning the place—along with possible oceanographic, geologic, biological, and anthropological evidence—have been put forth. For example, Atlantis has been said by some researchers to have been located in the Cretan empire around the island of Thera; others believe that Lanzarate, with its ten sister Canary Islands in the Azores, represents the remains of the lost continent. A case has been made for Helgoland in the NORTH SEA, and some even think Atlantis was located beneath the Sahara. However, the most accepted location is somewhere near the middle of the North ATLANTIC OCEAN, although the evidence even on this is quite inconclusive.

While the existence of Atlantis as an historic fact is as yet unproven, indications are that the hard core of the story seems to have some substance. Perhaps future oceanographic expeditions will uncover the convincing missing links that will establish Atlantis as a real entity rather than a myth. In this regard it is well to recall that for thousands of years the cities of Troy, Pompeii, and Herculaneum were also considered to be mythical.

Legend of the Porpoise Porpoises belong to the marine mammal family known as Phocaenidae. These social animals are found in tropical and temperate parts of the world's oceans as well as in several large rivers of South America and Asia.

Porpoises have a well-developed brain (just as convoluted and complex as a human brain although a great deal larger), and they use an elaborate system of behavior and sounds to communicate between individuals. Moreover, via a highly developed SONAR system these animals are able to echolocate. Most porpoises are excellent swimmers and can propel themselves through the water at rates in excess of 20 kn for short periods, and there are records of observations of porpoises swimming before a ship at over 40 kn, using the bow wave for extra speed.

For as many years as there have been oceangoing sailors, human beings have been fascinated by such qualities as these and have been strangely drawn to porpoises. Hence, a rather large body of half-truths, legends, and half-legends have grown up about these playful and intelligent marine animals.

The oldest myth, in this regard, concerns the origin of porpoises. It seems that the Greek god Dionysus, the god of wine, was once attacked by a pirate crew. Dionysus used his magic powers to change the oars of the ship into snakes and to cause the pirate sailors to dive into the sea where they turned into the world's first porpoises. From that day on, as the story goes, porpoises have befriended sailors and especially those shipwrecked in the sea.

Actually, there are many cases on record where porpoises have done just that—befriended sailors in trouble. In Australia, for example, a purposeful porpoise, called Pelorus Jack, met early sailing ships carrying colonists to that continent and guided the ships into sheltered anchorage. Also, there are documented accounts of the spectacular feats, over a period of years, of a porpoise in U.S. waters called Hatteras Jack. This animal, upon hearing foghorn signals from sailing ships of the late 1700s, would surface and then act as a pilot to guide the vessels past dangerous shoals and reefs into the relatively safe waters of North Carolina's outer-bank harbors. In addition, there has been a multitude of stories of porpoises protecting shipwrecked humans in the water by engaging in fights with sharks and by nudging and helping keep the human beings afloat and directing them toward shore.

All such tales, some quite true and others perhaps embellished and exaggerated in detail, have helped to support the idea, in some people's minds, that porpoises may once have been humans or land animals like humans. Their arguments essentially state that human beings or creatures who were like human beings, after being mysteriously drawn back to live in the sea, gradually changed their shapes into those of fishes. The body was transformed for adaptation to ocean life, and the front limbs or arms became flippers. Porpoises do have finger bones in their flippers which are used only for balancing or steering. The legs disappeared, and a tail developed so that it could be used for motion or propulsion. The tail is quite unlike that of a fish since it can be moved up and down rather than from side to side. The nose turned into a blowhole—one large nostril about an inch (2.54 cm) in diameter, with inner and outer valves to seal out the water. The lung system is like that of human beings, and when a porpoise surfaces for air, it opens the blowhole wide and inhales 4–10 qt (3.8–9.5 L) of air in less than a half a second. Also, like some land animals, porpoises produce their young (gestation period is 1 year) as babies requiring milk from their mothers.

Some believe that rather than porpoises having once been human, human beings may, at one time, have lived in the waters of the warmer oceans of the world. It is more than mere coincidence, they state, that human beings and marine animals are sleek, whereas most land animals, other than humans, are hairy. Moreover, the subcutaneous layer of fat that human beings tend to accumulate in contrast to most land mammals, the ease and naturalness with which human beings swim, the supporting buoyancy of the water which may have provided human beings with the means of evolving their erect position, and other such evidence all seem to corroborate the theory that human beings swam in the ocean before they walked on land.

This myth of the origin of the gentle and remarkable porpoise, known as the most intelligent of all sea life, makes a good story. Since there has been very strong experimental evidence that this animal can mimic the human voice, perhaps the porpoise may even be able to communicate with humans some day. The story of the porpoises promises to be an even better one when and if such a day arrives.

The Bermuda Triangle A triangular-shaped zone in the ATLANTIC OCEAN, bounded on its apexes by Bermuda, Miami (Florida), and San Juan (Puerto Rico), the "Bermuda or Devil's Triangle" has been considered for many years by more than a few people to be an area possessing supernatural qualities.

This belief is based upon the unusually high incidence of mysterious disappearances of commercial ships and pleasure craft, as well as aircraft, that have traversed this region.

Among such publicized losses was the *U.S.S. Cyclops,* a 19 360-ton-displacement fuel ship of the U.S. Naval Overseas Transportation Service. This well-equipped vessel, with a crew of 236 and 73 passengers, departed from Barbados in the CARIBBEAN SEA on March 4, 1918 bound for Baltimore, Maryland. Since that departure there has never been any trace of this ship, and the disappearance represents one of the most baffling mysteries in the annals of the U.S. Navy. All attempts to locate the *Cyclops* proved unsuccessful. Many theories have been advanced, but none reliably accounts for her disappearance. Incidentally, there were no enemy submarines in the area at the time.

Another large [590-foot (17 meter), 15.020-gross ton] Panamanian ship, the *Sylvia L. Ossa,* bound from Brazil to Philadelphia, Pennsylvania, with a cargo of iron ore and a crew of 37 disappeared in the Triangle on October 17, 1976. This vessel, two days before its disappearance, had reported that it had encountered sudden gale-force winds and high seas some 600 nautical miles (1111 km) from Philadelphia, Pennsylvania, and that it would be late in its scheduled arrival because of reduced speed. No other communication was ever received, and, when the ship failed to arrive, an 8200 mi² (21 238 km²) air-sea search, beginning at the point where it was last heard from, was immediately made of the area. The weather conditions during this search were of a visibility of more than 40 mi (64 km) and calm seas. Some debris, identified as being from the *Sylvia L. Ossa,* was found. This included a life preserver from the missing vessel with burn marks on it.

Another well-publicized loss in the same area was that of Flight 19. This flight, consisting of five U.S. military aircraft (torpedo bombers), departed from Fort Lauderdale, Florida, on the afternoon of December 5, 1945, for an authorized advanced over-water navigational training flight. All five airplanes were manned by qualified aviators having several hundred hours flight time, of which at least 55 was in the type of aircraft they were flying at this time. The prevailing weather conditions in the area were considered average for such flights.

At 4 P.M. on December 5, 1945, a radio message between two planes on the flight was intercepted back at the Fort Lauderdale Naval Air Station. This message was the first indication that Flight 19 was lost, and the pilots stated that they were experiencing malfunction of the compasses on the planes. Communications to the flight were impossible to establish, although indications were that it became lost somewhere in the Bermuda Triangle. The flight was never heard from again, and no trace of the planes was ever found even though an extensive 5-day search operation, employing many aircraft and surface vessels, was carried out. Ironically, one of the search aircraft, a Navy patrol plane, was also lost in the first evening (December 5) of the search operation, and it too vanished without any part of the plane or its crew ever found.

Traceless disappearances of several other surface ships and aircraft have lent credence to the popular belief in the mysterious character of the Bermuda Triangle.

A great many theories have been offered attempting to explain these many disappearances throughout the history of the region. The most feasible of these seem to be environmental in nature and those which indicate human error. However, even in this explanation, there exists an enigma of sorts.

The area, from an oceanographic research standpoint, is perhaps one of the best-known ocean parts of the world, including bottom and sub-bottom. U.S. Navy and civilian oceanographic ships very frequently carry out research there and have found it to have no unusual undersea volcanoes and no seismic disturbances. However, it has also been determined that some unique environmental features exist in the region; it is one of two places on the earth that a magnetic compass *does* point to true north. Normally it points toward magnetic north. The difference between the two is known as compass variation. The amount of variation changes by as much as

20° as one circumnavigates the earth. If this compass variation or error is not compensated for, a navigator could find himself far off course and confused. (An area called the "Devil's Sea" by Japanese and Filipino seamen, located off the east coast of Japan, also exhibits the same magnetic characteristics. Like the Bermuda Triangle, it is known for its mysterious disappearances.)

Still, the other environmental factor is the nature of the GULF STREAM, an extremely swift and turbulent ocean current that can quickly erase any evidence of a disaster. (see CURRENTS.) The unpredictable Caribbean-Atlantic weather pattern also plays its role. Sudden local storms often spell disaster for pilots and mariners.

Finally, the topography of the ocean floor varies from extensive shoals around the islands to one of the deepest marine trenches in the world. With the interaction of the strong currents over the many reefs, new navigational hazards are often very rapidly developed.

Not to be underestimated in all of this is the human error factor. A large number of pleasure boats travel the waters between Florida's Gold Coast and the Bahamas. Very often, crossings are attempted with too small a boat, insufficient knowledge of the area's hazards, and a lack of good seamanship.

The U.S. Coast Guard, an agency that has had to respond to many public inquiries about the region, has stated that it is not impressed with supernatural explanations of disasters at sea. It is the U.S.C.G. position that experience shows that the combined forces of nature and the unpredictability of human beings outdo even the most far-fetched science fiction many times each year.

NANSEN, FRIDTJOF (1861–1930), Norwegian Arctic explorer, zoologist, statesman, and humanitarian, is noted in scientific circles for his trek across the Greenland ice cap in 1888 and his drift across the ARCTIC OCEAN aboard the *Fram* in 1893–1896. Nansen also won the Nobel Peace Prize (1921–1922) for his work in the repatriation of 500 000 prisoners of war from Siberia, China, and other parts of the world, and for directing the relief of families suffering from famine in Russia (1921–1923).

The son of a lawyer, Nansen was born on an estate near Oslo (Christiania) Norway, on October 10, 1861. Entering Christiania University in 1880, Nansen concentrated on zoology. In 1882 he sailed to Greenland waters to do research which inspired his long fascination with the Arctic. Returning to Norway, he became the curator of the zoological collection at the Bergen Museum. In 1888 he received his doctorate from Christiania University.

In 1887, after being refused funds by the Norwegian government, Nansen won financial support from a Danish citizen to carry out his plan to cross the Greenland ice cap on skis. Accompanied by five companions, he left Norway in May 1888. After considerable difficulty in landing because of ice conditions, the party began their east-to-west trek on August 16. Nansen reached the village of Godthaab on the west coast of Greenland in early October but was forced to winter over because the last ship had already departed. The winter gave Nansen the opportunity to study the Eskimo and eventually to write a book entitled *Eskimo Life* (1891).

After the Greenland success Nansen had less difficulty in raising funds for this next venture—a drift across the Arctic ocean with the ice pack. With funds supplied mostly by private contributions, Nansen built a ship which he named *Fram* (forward). The most distinctive feature of the ship was its rounded hull which would allow the ship to ride up as the ice pack began to squeeze in on it. With 12 companions Nansen departed for the Arctic on June 24, 1893. On September 22 the *Fram* was made fast to the ice at 77°43′N, 134°E, just northeast of Cape Chulyskin. As the drift progressed, Nansen realized that the path would not take the ship across the North Pole as calculated. Therefore, in the spring of 1895 Nansen, with a single companion, set out for the Pole on skis. Ice conditions made the march impossibly difficult, and they turned back to Franz Joseph Land, 700 mi (1126.5 km) away, on April 8. They had reached 86°14′N—a record. The *Fram* arrived back in Norway 8 days after Nansen.

On his return to Norway, Nansen was made a professor of zoology at Christiania University. However, his interest shifted to physical oceanography, and his status was changed to professor of oceanography in 1908. From 1896 to 1917 Nansen's energy was directed to scientific study. He participated in the creation of the International Council for the Exploration of the Sea, accompanied the cruise of the *Michael Sars* to the NORWEGIAN SEA (1900), the *Fridtjof* through the northeastern North Atlantic (1910), the *Veslemoy* to Spitsbergen (1912), and the *Armauer Hansen* to the Azores with B. Helland-Hansen (1914). In 1939 Nansen traveled through the BARENTS SEA and KARA SEA, to the Yenisei River, and back through Siberia. These studies resulted in numerous publications, many of which Nansen illustrated himself. He contributed to the design of oceanographic instruments, to the explanation of the wind-driven ocean currents, and to the manner in which the deep water layer of northern waters is formed.

With time Nansen became increasingly interested in the relationship between the individual and the state. In 1905 he involved himself in the debate over the dissolution of the union between Norway and

Sweden and became the first minister to London (1906–1908) upon establishment of the Norwegian monarchy. In 1917 he headed the Norwegian Commission to the United States to negotiate the release of essential supplies to Norway. In 1920 the League of Nations appointed him High Commissioner responsible for the repatriation of 500 000 prisoners of war. Upon completing that task, Nansen was asked by the International Committee of the Red Cross to direct an effort to bring relief to the famine victims of Russia. For these services and his efforts on behalf of Russian, Armenian, and Greek refugees, Nansen was awarded the Nobel Peace Prize for 1922. He used the prize money to support international relief efforts.

NAUTICAL CHART is a special type of map designed to inform marine navigators of the nature and form of the ocean bottom, the location of channels, reefs and shoals, and navigational aids; it provides an accurate graphical depiction of hidden hazards and safe passageways to the mariner. Nautical charts are usually augmented by official publications for coastal piloting and other types of sailing directions which furnish pertinent and timely information not shown on the charts.

The first nautical charts date back to the fourteenth century. At that time they consisted of maps with descriptive notes and loxodromic (rhumb) lines plus a network of lines corresponding to the principal points of the compass. On such early charts, various coastal areas of countries were shown in their correct relationship, based on bearings, and the distances between ports were given.

In 1569 Gerhard Kremer published a world map using a new and improved navigation system. (See KREMER, GERHARD.) This Mercator projection, as it is now known, is a conformal cylindrical map projection in which the surface of a sphere or spheroid, such as Earth, is conceived as developed on a cylinder tangent along the equator. On this type of map, meridians appear as equally spaced vertical lines, and parallels, are shown as horizontal lines drawn farther apart as the LATITUDE increases, so that the correct relationship between latitude and LONGITUDE scales at any point is maintained. Nearly all nautical charts today, except those for high latitudes, are on the Mercator projection. The principal reason for the almost universal use of the Mercator projection is that the rhumb line, or that line which specifies the path of a ship on a constant compass course, is a straight line on the projection.

A considerable amount of the information shown on charts is given by internationally standardized symbols and abbreviations. The charts are also published in various colors to aid in the selective reading of the continents. In this and in several other respects nautical charts are similar to aeronautical charts. However, while nautical charts are concerned essentially with water areas and information for mariners, the aeronautical charts deal primarily with land areas and other information aimed toward aviators.

In the United States both nautical and aeronautical charts, as well as other publications to assist the navigator, are available from such sources as the U.S. Naval Oceanographic Office, the Coast and Geodetic Survey, and the Aeronautical Chart and Information Center of the U.S. Air Force.

Charts are indispensable in marine navigation. They delineate nautical and geographical features, including latitude and longitude scales, useful topographic features and aids to navigation, depth of water, magnetic information, warning notes, and information on electronic aids. These published charts are the product of precise geodetic, oceanographic, and cartographic work and are updated at frequent intervals to ensure maximum safety in navigation.

NAUTILUS is the common name for an unusual CEPHALOPOD belonging to the monogeneric family Nautilidae in the order Nautiloidea; *Nautilus pompilius* is the only well-known species. See CUTTLEFISH; OCTOPUS; SEASHELLS; SQUID.

NEAP TIDE See TIDES.

NEKTON is the term applied to free-swimming aquatic animals whose locomotion is essentially independent of water movement. See FISH.

NEMATOCYST is a fluid-filled, flask-shaped cell—often called a stinging cell or nettle cell—that is found in the tentacles of such COELENTERATES as the common JELLYFISH and the PORTUGUESE MAN-OF-WAR. A short trigger-hair projects outward from the cell opening. The fluid-filled cell contains a tiny "harpoon" attached to the cell wall by a slender, coiled filament. When the hair is touched, it triggers a compression of the cell which fires the harpoon under rapidly rising hydrostatic pressure. It is a combination of the harpoon's penetration and a venom it carries that causes the stinging, burning sensation after one brushes against a jellyfish. The nematocyst is used both in food gathering and in self defense.

See also CORAL; SEA ANEMOME; VENOMOUS MARINE LIFE.

NEMICHTHYIDAE is a family of bathypelagic eel-like fish in the order Apodes.

The Nemichthyidae are typified by the snipe EEL. Unlike the Anguillidae, they are nonmigratory and live in very deepwater.

NEUTRON ACTIVATION ANALYSIS (NAA) is an analytical technique which consists of bombarding a sample such as SEAWATER with neutrons. In this way radioisotopes are produced from the resident elements. The irradiation source can be either a nuclear reactor, an accelerator-type of fast neutron generator, or a neutron-emitter radioisotope such as californium 252. The energy spectrum of the resulting x-rays from the decaying radioisotopes is recorded and, analogous to optical spectroscopy, the energies of the spectral peaks and the areas under the peaks are identified and related to the amount of the elements present.

Neutron activation analysis is a highly valuable technique in certain oceanographic work. For example, trace elements present in the ocean environment play an important role in geochemical and biological processes—a role disporportionate to their extremely low concentrations. Trace elements also serve as excellent tags for the study of deep ocean circulation and as stand-ins for chemically related radionuclides. Neutron activation analysis has made it possible to perform direct instrumental and simultaneous analysis of numerous complex mixtures which previously required both extensive radiochemical separations and very sensitive counting techniques. The ability to make direct instrumental determinations is especially important in that it avoids many of the problems of inadvertent sample contamination that have long plagued trace element analysis and permits determinations on very small samples. See INSTRUMENTATION.

NITROGEN CYCLE is the system in nature whereby molecular gaseous nitrogen (N_2) in the atmosphere is entrapped and converted (fixed) by specialized organisms (and also by ionizing phenomena, e.g., cosmic radiation, meteor trails, and lightning, as well as by industrial processes) to nitrogenous compounds. Usable forms of such compounds are then made available to higher plants and ultimately to animals. These plants and animals die and, by a series of processes, return the fixed nitrogen to soil and water. Here, the nitrogen may be recycled through a new generation of plants and animals, or it may be broken down to gaseous nitrogen and returned to the atmosphere.

Approximately 65–70 percent of all the nitrogen in the world's oceans occurs as nitrate. The average nitrate concentration is approximately 450 mg of nitrogen per cubic meter of SEAWATER. Nitrogen occurs in a number of forms in seawater, among them:

- Molecular nitrogen, which originates in the atmosphere.
- Various complex organic forms, either the excretion from living organisms or the decomposition products of dead ones.
- Ammonium ION, principally from decomposition of organic matter.
- Nitrate, nitrite, and other oxides of nitrogen, from the stepwise oxidation of ammonium ion and possibly significantly from atmospheric nitrogen fixed by lightning.

Studies of the biological and chemical processes in the ocean have accounted for nine different states of oxidation of nitrogen (from N^{-3} to N^{+5}).

In its molecular form, nitrogen (N_2) makes up 79 percent of the atmosphere and is required by the entire biotic world for the construction of proteins, nucleic acids and other nitrogenous compounds. However, it cannot be utilized directly by the vast majority of living organisms. Nitrogen first is biologically fixed by a few species of microorganisms that oxidize (a process basically involving the removal of electrons from a substance) the NH_4^+ (or ammonium ion) to NO_2^- (nitrite) or NO_2^- to NO_3^-(nitrate).

As this nitrogen cycle relates to the ocean, both the NH_4^+ and NO_3^- are brought into it by the infusions of rivers and rains. Also, nitrogen fixed by atmospheric ionizing phenomena also enters ocean waters. In addition, dissolved gaseous nitrogen present in seawater is converted (or fixed) by biological or bacterial action. This nitrogenous material is then taken up and used by ocean plants and animals. As these plants and animals die, some of the dead organic matter becomes a part of the sediments and some is also utilized by marine ALGAE and subsequently by ocean animals. Most of the plants can absorb nitrogen in all three forms—ammonium, nitrites, and nitrates. However, most soluble and particulate nitrogen compounds in animal wastes are broken down first by bacterial action. (See PARTICULATE MATTER.) In this process, inorganic nitrogen (nitrites and nitrates) is reduced (basically, an addition of electrons) to gaseous compounds, such as molecular nitrogen or nitrous oxide, and these are released to the atmosphere.

Biological nitrogen fixation processes occurring

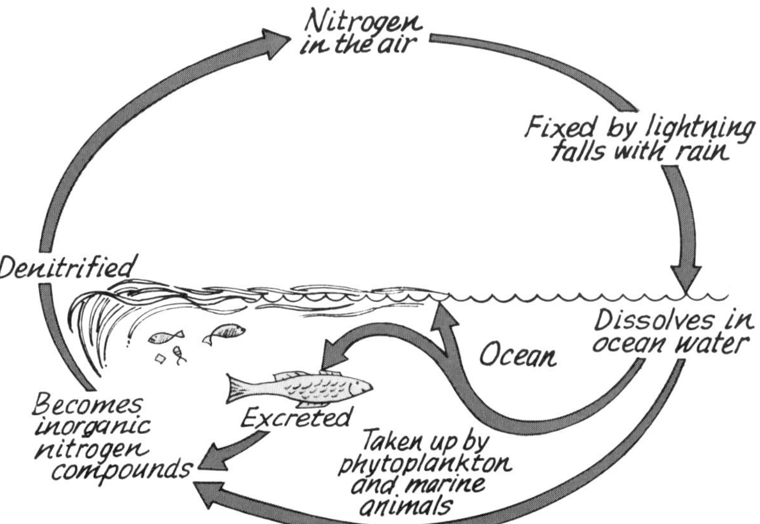

NITROGEN CYCLE. Atmospheric nitrogen is fixed (forms compounds) by the energy from lightning. These compounds fall with rain, are dissolved in the ocean, and are taken up by marine life forms. Excreted organic matter breaks down into inorganic nitrogen compounds which are denitrified and return as nitrogen to the air.

in the ocean environment are relatively small as compared with those occurring on the land, and nitrogen concentrations (NH_4^+, NO_2^-, and NO_3^-) in seawater are highly variable in depth, location, and season, as are the biota responsible for them. Although many of the intricacies of the nitrogen cycle in the oceans are largely unknown, several authorities deem it fortunate that nitrogen in its gaseous form is released from seawater back to the atmosphere. While the amounts removed as a product of biochemical oxidation processes of organic material are relatively small, without this denitrification, most of the atmospheric nitrogen probably would be contained in ocean sediments. Moreover, in the absence of denitrifying BACTERIA, seawater would be more acidic and initiate the release of carbon dioxide from ocean carbonates. Much of this denitrification takes place in anoxic waters of the ocean—waters essentially devoid of dissolved oxygen—where circulation is restricted and where the oxygen consumption exceeds the removal.

NOAA is the abbreviation for the National Oceanic and Atmospheric Administration, an agency of the U.S. Department of Commerce. NOAA was established in October 1970 to create a center for expanding effective and rational use of ocean resources, for monitoring and predicting conditions in the atmosphere, oceans, and space, and for exploiting the feasibility and consequences of environmental modifications.

NOISE See UNDERWATER SOUND.

NORDENSKJÖLD SEA is the name occasionally applied to the LAPTEV SEA, a 208 440 mi² (540 000 km²) body of WATER located off the north coast of central Siberia between the Taimyr Peninusla in the west, and the Strait of Dmitri Laptev in the east. The unofficial name honors Nils Adolf Erik Nordenskjöld, a Swedish polar explorer, mineralogist, and map authority. Aboard the *Vega* in 1878–1879, Baron Nordenskjöld became the first man to negotiate the Northwest Passage between the ATLANTIC OCEAN and PACIFIC OCEAN by sailing along the northern coast of Europe and Asia. On this voyage he also became the first scientist to explore the Laptev Sea.

By the decision of the Central Executive of the U.S.S.R. as reported in the International Hydrographic Review of 1936, the name Laptev Sea is the official name.

NORMAL WATER, also called STANDARD WATER, COPENHAGEN WATER, and EAU DE MER NORMAL, is water whose CHLORINITY lies between 19.30 and 19.50 (usually 19 372) parts per thousand (ppt) and has been determined to within ±0.001 ppt. In 1902, this type of solution was first prepared by the Hydrographical Laboratories in Copenhagen, Denmark. Such water is now prepared and distributed worldwide by the Standard Sea-Water Service located at the Institute of Oceanographic Sciences (ex-NIO—National Institute of Oceanography) at Wormly, Godalming, Surrey, England. This solution provides oceanographers throughout the world with standard seawater samples of a precise and certified chlorinity which also can be related to conductivity values. Thus, these samples are especially useful for calibration purposes of salinometers, are the basis for all chlorinity titrations (see MOHR TITRATION), and are used to standardize the silver nitrate that is used before and during a set of titrations.

Although it is sometimes referred to as Copenhagen Water, the source of normal water is in the North ATLANTIC OCEAN about 200 nautical miles (370 km) south of Iceland, in the open ocean away from major sources of pollution. Usually about 5300 qt (5000 L) are brought back at a time, very carefully filtered, irradiated with ultraviolet light to kill organic materials, and stored. The water is transferred into individual glass ampules in large batches, after adjustment by dilution to a SALINITY as close as possible to 35 ppt—the reference value used in most oceanographic tables. Each batch is carefully analyzed at the time of bottling. Since the motivation is to provide a standard for only one type of measurement, no particular care is taken to prevent entry or removal of other components present in small concentrations; thus, these samples cannot be used, for example, to provide a 75-year time series on the presence of minor constituents.

NORTH ATLANTIC CURRENT is a continuation of the GULF STREAM, originating at about 40°N LATITUDE and 50°W LONGITUDE in the vicinity of the Grand Banks off Newfoundland. The branches of the North Atlantic Current are often masked by shallow and variable wind-driven surface movements so that they are sometimes described as the North Atlantic Drift.

In the vicinity of the Mid-Atlantic Ridge (see CONTINENTAL DRIFT) the North Atlantic Current divides into two major branches; the northern branch flows between latitudes 50°N and 52°N with the Labrador Current on its northern flank. (See CURRENTS.) The southern branch flows along latitude 45°N and appears to carry undiluted Gulf Stream WATER. The northern branch continues to the east-northeast into the NORWEGIAN SEA, and part turns continually northward until it forms the IRMINGER CURRENT which curls west and south below Iceland until it finally rejoins the North Atlantic Current. The remainder of the current continues on to become the CANARY CURRENT which flows southward along the coast of Europe.

NORTH EQUATORIAL CURRENT See EQUATORIAL CURRENT SYSTEM.

NORTH SEA, an arm of the eastern ATLANTIC OCEAN, is partially confined by England and Scotland on the west, and Norway, Denmark, Germany, Holland, Belgium, and France on the east. The southern boundary is a line running from Walde Lighthouse on the French coast, across the Strait of Dover, to Leathercoat Point on the English coast. In the north the boundary is a line running from Dunnet Head in Scotland, through the Orkney and Shetland Islands, and then up the meridian of 0°53′W to the parallel of 61°N, and eastward along this parallel to the coast of Norway. The North Sea communicates with the Atlantic Ocean through the Strait of Dover in the south, and with both the Atlantic and NORWEGIAN SEA through the gap between Scotland and Norway in the north. Communication with the BALTIC SEA in the east is through the Skagerrak and Kattegat between Norway, Sweden, and Denmark. The North Sea approaches 600 mi (965 km) in length and is about 360 mi (580 km) wide in the northern sector. It covers an area of 231 600 mi² (600 000 km²), occupies a volume of 37 186 mi³ (155 000 km³), and has a mean depth of 299 ft (91 m). The several islands in the sea cover a combined area of 2500 mi² (6473 km³).

The North Sea is known as an epicontinental sea; i.e., it occupies a structural basin which lies totally on continental (as opposed to oceanic) crust. The basin is, in part, a geosyncline (site of prolonged sediment deposition) which has been folded into mountains at least twice in the past. On each occasion these mountains have been eroded away to leave a shallow basin between England and the continent. Roughly 230 million (230×10^6) years ago the land surrounding the North Sea was a desert, and due to high evaporation and a restriction in water flow from the north, huge evaporite deposits were formed. Today these evaporites are represented by the salt domes and structures found both in the seafloor and in Germany and Denmark. This history and the resulting structure of the floor of the North Sea are directly connected with the vast oil deposits now being exploited by the contiguous nations.

The floor of the North Sea is, of course, all CONTINENTAL SHELF. The southern half of the sea is a plateau with depths of around 131 ft (40 m). The floor slopes gradually to the north until it reaches a depth of around 600 ft (183 m) where the shelf breaks west of the Shetland Islands. Angling around the southern tip of Norway and out to the edge of the continental shelf is an unusual channel (Norwegian Channel) which ranges in depth to 1968 ft (600 m). The valley is thought by some oceanographers to be caused by the scouring of a continental glacier. (See SUBMARINE CANYON.) Other remnants of the last Ice Age (11 000–8000 years ago) take the form of stream valleys cut during the low stand of sea level and glacial moraines (accumulations of sediment at the leading edge of a melting glacier). Dogger Bank between England and Denmark is an example. It has a depth of only 43 ft (13 m). The sediment covering the seafloor is largely glacial gravel, sands, and silts that have, in places, been reworked by currents and wave action.

WATER circulation in the North Sea is influenced both by Atlantic water flowing in from the north and Baltic Sea water flowing in from the east. Very little water flows in through the Strait of Dover in the south. Most of the water is diverted from the NORTH ATLANTIC CURRENT which flows south in the vicinity of the Orkney and Shetland Islands. The current turns counterclockwise in the southern sector and flows out along the coast of Norway as the Baltic Current. Surface water temperatures range from 35.6–45.6° F (2–7° C) in February to 51.8–62.6° F (11–17° C) in August. SALINITY ranges from 31 to 35.25 ppt, depending upon position and season. Due to the large volume of fresh water flowing in from the continental rivers (Rhine, Elbe, Weser, Ems, and Scheldt), the water off the coast of Norway, Denmark, Holland, and Germany freezes over during even a mild winter. The western portion, however, is ice-free during even the severest winter due to the lesser inflow of fresh water and the influence of the North Atlantic Current.

TIDES in the North Sea are complex due to a minor component which moves north through the Strait of Dover and a major component that moves south through the Scotland-Norway gap. Tides range from 23 ft (7 m) along the coast of England to 5 ft (1.5 m) in most of the open sea. Storms sweep into the sea from the northwest and may do considerable damage along the 4000 mi (6436 km) of coastline. When the storm surge (see WAVES) associated with these disturbances coincides with both high waves and high tides, severe damage is done to the coastal areas, particularly the dyke-protected Dutch coast.

The North Sea is one of the great fishing areas of the world and provides half the wold's HERRING catch. Fishing is one of the major industries for the coastal peoples of all the contiguous nations. In 1958 the floor of the sea was divided into the British, Dutch, German, Danish, and Norwegian sectors for purposes of oil and gas exploration and recovery. The first gas well began producing in 1959, and the first oil well in 1969. Today the North Sea oil reserves are put at 23 billion (23×10^9) barrels—the ninth largest oil deposit in the world.

NORWEGIAN SEA is located between the coasts of Norway and Iceland and is surrounded by the NORTH SEA and the ATLANTIC OCEAN to the south, the GREENLAND SEA to the west, and the ARCTIC OCEAN and the BARENTS SEA to the north. More specifically, the Norwegian Sea is bounded on the northeast by a line joining the southernmost point of Spitsbergen to the North Cape on the coast of Norway. On the southeast the sea is bounded by the west coast of Norway between North Cape and Cape Stadt. On the south it is bounded by a line joining Cape Stadt (Norway) to Muckle Flugga, thence to the northwest extreme of Fugloe, and on to the east extreme of Gerpin in Iceland. On the west it is bounded by a line joining the eastern extreme of Gerpin (Iceland) and the southernmost point of Spitsbergen. The Norwegian Sea covers an area of 1 383 000 km² (533 838 mi²), has a volume of 2 408 000 km³ (577 703 mi³), and a mean depth of 1742 m (5715 ft).

The first explorer of record to reach the Norwegian Sea was probably Pytheas of Massalia. By the accounts passed down by such classical writers as Polybius and the learned geographer Strabo, Pytheas sailed out through the Strait of Gibraltar in the year 320 B.C. Turning north along the coasts of Spain and France, Pytheas circumnavigated the British Isles via Land's End, Kent, and the Orkneys. From Britain, the accounts indicate, Pytheas sailed north for 6 days to discover a land that lay 1 day's sail from the frozen sea and where, during the summer, a day is only 2–3 h long. Scholars disagree as to whether Pytheas was describing Iceland or the coast of Norway. In either case he probably crossed into the southern Norwegian Sea. If Pytheas' voyage can be taken as fact, then he must be counted as one of the greatest explorers of all time since his voyage was not surpassed for a thousand years.

According to an account handed down by King Alfred of Wessex, a Norse chieftain by the name of Ottar (or Othere), in about 870, sailed up the coast of Norway, through the Norwegian Sea, the Barents Sea, and into the WHITE SEA on the Murman Coast. On the basis of Ottar's discoveries King Harald annexed all territories as far north as the White Sea. His successors sponsored numerous expeditions to this region for both trade and conquest.

The Norwegian Sea approximates the area covered by the Norwegian Basin. The basin has a maximum depth of 12 992 ft (3960 m) and is bounded by the coast of Norway on the east, the Mid-Oceanic Ridge on the west, and the Wyville Thomson Ridge and Iceland-Faeroe Ridge on the south. The sediments in the basin are largely globigerina ooze and DIATOM OOZE in the deeper sections, clays and gravels near the coastlines, and volcanic ash admixtures around Iceland. (See MARINE SEDIMENTS.)

The tidal range in the Norwegian Sea ranges from a low value of 0.9 ft (0.3 m) in the Faeroes to 7.5 ft (2.3 m) at North Cape (Norway). Two water masses, the North Atlantic WATER and the Norwegian Deep Water, are identifiable in the sea. The latter is formed around Jan Mayen Island by the mixing of Atlantic and Arctic waters. Cooled to about 31.8° F (−1.0° C) the Norwegian Deep Water sinks beneath the North Atlantic Water. The resulting water layers are relatively stable in temperature; the layer from 328 to 1968 ft (100 to 600 m) ranging from 48.2° F (9° C) in the south to 42.8° F (6° C) in the north, and the layer below 1968 ft (600 m) is nearly uniform at 31.8° F (−1° C)

Water movement in the Norwegian Sea is dominated by the Norwegian Current, a branch of the NORTH ATLANTIC CURRENT which flows into the sea north of Scotland at a surface velocity of around 0.5 kn, and with a volume ranging from 3 to 6 million (6×10^6) m^3/s In the north the current branches, with one branch—the North Cape Current—flowing into the Barents Sea, and the main branch continuing on the merge around Spitsbergen with the Spitsbergen Current.

NUDIBRANCH is the term, meaning naked gill, applied to mollusks that lack a shell and mantle cavity and whose gills vary in size and shape. The sea slugs are nudibranchs (see MOLLUSK).

OCEAN comes from Oceanus (Greek—*Okeanos*) of Greek mythology. Oceanus was the eldest of the Titans and the son of Uranus (Heaven) and Gaea (Earth). He was the father, by his wife and sister Tethys, of the 3000 Oceanids (i.e., Doris, Electra, Amphitrite, etc.) who protected the ocean. Oceanus, often depicted as an old man with a long beard wearing a headdress made of bull's horns, was more of a spectator than an active participant in the activities of the gods. Homer portrayed Oceanus as the ever-flowing stream that surrounded the circular plain of the earth. (To the ancients the Atlantic must have appeared as the watery rim of a flat earth.) Thus, the world by which we denote the watery seven-tenths of the globe comes from the mythological god of all WATER, himself the product of heaven and earth, and the father of the nymphs who protect the hydrosphere.

The term *Oceania* (Oceana or Oceanica) was originally used to designate that portion of the Earth's land surface which remained after apportioning the continents of Eurasia, Africa, and the Americas. When first used, Oceania included Australia, the Indonesian (Indian) Archipelago, and the Pacific Islands. Today the term refers only to those lesser Pacific Islands which fall into the geographical divisions of Polynesia, Micronesia, and Melanesia.

Marine is frequently used to denote anything related to the oceans or the seas, from equipment (marine pumps) to naval infantry (Marines) to the science of oceanography itself (marine geology, marine sediments, marine climate, etc.).

The words *ocean* and *sea* are frequently used interchangeably. For instance, we speak of seafloor, bottom of the sea, and sea life to describe features that can be discussed with equal accuracy by using the word *ocean*. The term *high seas* is used to denote that part of the world's oceans that has not been politically allocated. *Territorial waters* denote that part which has been so allocated. To the ancients the Seven Seas were the INDIAN OCEAN (Erythraean), the RED SEA (Sinus Arabicus), the PERSIAN GULF, the BLACK SEA (Pontus Euxinus), the Sea of Azov (Polos Macotis), the Adriatic Sea, and the Caspian Sea. These represented the known water bodies of any magnitude at the time. Today, the term *Seven Seas*, when it is used at all, refers to the ARCTIC OCEAN, SOUTHERN OCEAN, Indian ocean, North ATLANTIC OCEAN, South Atlantic Ocean, North PACIFIC OCEAN, and South Pacific Ocean.

The word *sea* does not have a precise scientific definition, nor is it easy to arrive at one that accurately defines all those bodies of water that are now so designated. Perhaps a large body of oceanic water with distinctive physical and chemical properties which is partially surrounded by land comes as close as any. But it fails to include such seas as the Sea of Azov, Caspian Sea, Dead Sea, Salton Sea, etc., which are not connected to the ocean. And, what do you do with the SARGASSO SEA which lies in the middle of the North Atlantic?

The problem is further complicated by the words *gulf, bay,* and *bight,* which are used with equal casualness and interchangeability. A gulf is often defined as a body of water whose length is great compared with its width. The Gulfs (see GULF OF CALIFORNIA.) of Bothnia, Finland, and California are examples. Of course, by this definition the Red Sea should be called the Red Gulf. A bay can be defined as a body of water that lies between two widely spaced headlands, and a bight can be defined with the same words. The GREAT AUSTRALIAN BIGHT, Walvis Bay, and the Bay of Naples are examples. The margin of the world's oceans is populated by bodies of water which are, by these loose definitions, misnamed. One has but to look at the ARABIAN SEA and the BAY OF BENGAL which lie to either side of the Indian subcontinent to find an excellent example.

Seas may be classified as enclosed or partly enclosed. Those which are enclosed are also referred to as intercontinental seas, or mediterranean seas. The MEDITERRANEAN SEA, the GULF OF MEXICO, and the CARIBBEAN SEA are examples. The Arctic Ocean was once referred to as the Arctic Sea and qualified as a mediterranean sea. The partly enclosed sea is exposed to the ocean by a wide entrance such as the NORTH SEA and the WEDDELL SEA. Seas with such open communication with the ocean are often referred to as marginal or tributary seas. They may also be classified as shallow or deep pelagic seas, or, where they lie totally on the CONTINENTAL SHELF, as shelf seas.

Seas, gulfs, and bays are often connected to each other and to the ocean by relatively narrow bodies of water called passages, channels, or straits. These terms, too, are used interchangeably with no clear distinction between them. The Windward Passage, the Strait of Gibraltar, and the English Channel are examples.

OCEAN CIRCULATION See CURRENTS.

OCEAN CURRENTS See CURRENTS.

OCEAN ENGINEERING is the branch of engineering concerned with the design, construction, maintenance, and operation of structures, equipment, tools, and devices for doing work on and in the OCEAN. The following is one modern definition of engineering itself: those efforts of applied science that draw and rely upon the discoveries of all science and make especial use of the philosophy of mathematics as its language for the creation, design and execution of ideas, things, and processes to serve human beings in the most useful, economic, and safe manner possible.

A comparatively new term whose usage first became popular during the 1960s, ocean engineering has come to mean the employment of several branches of engineering coupled with an understanding of the behavior of the ocean environment in order to most effectively conduct engineering work in and on the oceans. Thus, the broad category of ocean engineering includes the application of marine engineering, naval architecture, and coastal engineering, as well as other forms of engineering for performing useful work in the oceans, especially at great depths. Strictly speaking, the marine engineer is a specialized ocean engineer whose efforts are devoted to the design, installation, and operation of both nuclear and conventional propulsion systems for surface ships and submarines. The naval architect is concerned with the other design aspects of surface ships, submarines, boats, and craft. As the name implies, the coastal engineer is concerned with work primarily in the boundaries of the ocean, the construct of seawalls, piers, harbors, breakwaters, groins, ramps, etc. (See BEACH.)

Historical Background Some of the earliest known peoples who applied their knowledge of engineering to the oceans were the Egyptians and Phoenicians. These ancient peoples built and sailed ships on the ocean waters encompassing their world. Having ships, they naturally constructed harbors. One notable example is the Egyptian Port of A-ur (on the Canopic branch of the Nile River) that dates back to 3000 B.C. Also, the harbor of Pharos was built about 2000 B.C., probably by the Minoan Cretans, nearby on the coast of Egypt. The most famous harbor of antiquity, Tyre, was built around 1000 B.C. Great harbors were also constructed by the Greeks and Romans, and by the third century B.C., the latter had developed a method of driving piles for foundations and cofferdams to construct concrete seawalls using a hydraulic type of cement (pozzolana).

With the decline of ancient civilization and the pall of the Dark and Middle Ages that enveloped Western Europe until the sixteenth century, recorded history (with the possible exception of the accomplishments of the ninth century A.D. Viking seafarers and the voyage of Columbus in 1492) does not delineate any significant events of those times that may be clearly classified as engineering in or on the oceans.

However, the challenge of the ocean led some sixteenth, seventeenth, and eighteenth century men of vision and intellect to direct their abilities to the design and construction of things as well as to the creation of ideas that could be of use in and on the ocean. In naming but a few representative individuals of this type, the first deserving of such recognition is Leonardo da Vinci, a man of varied and remarkable talents. In his famous notebooks, he discussed the nature of water, its incompressibility, and how fluids, in general, flow. In about 1500, he designed primitive DIVING equipment (including fins) that enabled individuals to enter and spend useful time under water. He also sketched out a number of designs for leather diving lungs and conceived the design of a submarine. In this latter regard, he may well have taken some of his ideas on the subject from the work of Roberto Valturio, a Venetian, who had made a submarine some 25 years previously.

Cornelius Van Drebbel, a Dutchman who lived in England during the time of James I (1603–1625), built the first workable submarine. Drebbel's "sub," equipped with "twelve-man rower-power," operated on the Thames River for almost 10 years.

The modern father of diving is considered by some to be Edmund Halley, the Englishman of Halley's Comet fame. He constructed, in 1690, a diving bell in which a man stayed at a depth of approximately 60 ft (18 m) for 90 min. Previous to this, in 1686, Halley had tried to make a systematic study of the winds and their relation to the main ocean CURRENTS.

Inspired by this work of Drebbel and Halley, André Fréminent, a Frenchman, made a 1-h descent, in 1774, to a depth of around 50 ft (15 m) at Le Horne, France, in a diving bell of his own design.

When the American colonies rebelled against the British Government in 1775, certainly a segment of what may be classified as modern ocean engineering also came into being. This was the creation of the first war submarine. Designed by David Bushnell, the submersible was called the *Turtle*. This craft, the first to use modern submarine principles, was well ahead of its time. Larger than most present-day underwater research vehicles, the *Turtle* had water-ballast tanks and pumps, two hand-operated screw propellers, and carried a mine containing 150 lb (67.5 kg) of gunpowder.

At the beginning of the nineteenth century, Robert Fulton, another American pioneer of ocean engineering, took on where Bushnell had left off and designed a somewhat similar (hand-operated through gearing), improved propeller-driven underwater craft, the *Nautilus*. This ship also had diving planes and a folding sail. Its copper hull was 21 ft 4 in (6.4 m) long and 7 ft (2 m) in diameter, with a heavy keel that could be released in an emergency. At the fore-end was a hemispherical conning tower with glass look-out scuttles. All of Fulton's tests with the submarine were successful. Discouraged, however, because his efforts did not receive the recognition he felt they deserved, and having reached the conclusion that submarine navigation could not become practical until the boat could be power-driven. Fulton went on to design the famous steamship *Clermont*.

Taking up where Fulton had stopped in his work on the submarine, many submersible vehicles were built and used in Europe and the United States during the 1800s. In 1848, a German, Wilhelm Bauer, designed the *Sea-Diver*, and the ship was used in a war against the Danes. Also, during the 1870s the U.S. Navy purchased a submarine called the *Intelligent Whale*. Prior to that, the English designer Winan built a side-wheel powered submarine. The French, throughout all this period, designed and launched the greatest number of submersibles, with the most sophisticated engineering of all being their 1886 electric-powered, steel-hulled *Gymnote*.

In mentioning a few other pioneers who helped lay the foundations of modern ocean engineering, one may begin with the work of Nathaniel Bowditch. In 1802, Bowditch published his first edition of the *New American Practical Navigator* (see BOWDITCH, NATHANIEL). This manuel paved the way for modern-day navigation and charting. The same year, Franz Joseph von Gerstner published the first theory of surface WAVES in deep water. In 1818, Sir John Ross invented a "deep sea clamm" for collecting ocean life from the deep sea. Michael Faraday discovered magnetically induced electric currents, and in 1832 he suggested that electromagnetic induction and SEAWATER action through the Earth's magnetic field might yield measurable signals of value. Faraday, a great thinker and experimenter, was not a mathematician. However, his ideas found their mathematical expression and implementation in the work of James Clerk Maxwell (1831–1879), a very great pioneer in several fields of physical science. Charles Wilkes, an American, in his 1839–1842 voyage, used copper wire for the first time to make ocean depth measurements. Edward Forbes, an Englishman, improved upon the design of Ross' naturalist dredge and made extensive collections of sea life off the coast of the Isle of Man and in the Aegean Sea (1841–1854).

With a knowledge of the work of these men and others of his time, the man who is generally recognized as having given OCEANOGRAPHY and ocean engineering their first real impetus was Lieutenant Matthew Fontaine Maury. (See MAURY, MATTHEW FONTAINE.) His charts and notices to mariners gained immediate popularity and usefulness with both naval and merchant marine people throughout the world. Maury published *The Physical Geography of the Sea* in 1855 and won world acclaim as "the pathfinder of the seas." In addition, because of his studies and investigations of the ocean floor in support of Cyrus W. Fields' North Atlantic telegraph cable project, Maury is considered to be not only the father of modern-day oceanography, but a significant pioneer in ocean engineering as well.

Sir Charles Wheatstone had proposed the first telegraphic cable connection between England and

OCEAN ENGINEERING

Drilling for oil began as early as 1891 on the coast of California. The drilling ship *Wodeco V* carries on the tradition near Africa's Ivory Coast. *(An EXXON photo)*

This is a view down to the deck from the vertiginous heights of the derrick on a drilling ship. *(An EXXON photo)*

The research submersible *Alvin*, operated by the Woods Hole Oceanographic Institution and the Office of Naval Research, heads down on an exploratory mission. *(Woods Hole Oceanographic Institution)*

Four marine scientists lived for two months, beginning on February 15, 1969, in the TEKTITE I habitat on the ocean floor near St. John, Virgin Islands. This experiment was conducted for long-term marine investigation and for research into human psychological and physiological reactions to extended stays in an isolated and hostile environment. *(General Electric Company)*

An underwater manipulator arm utilizes a simple U-shaped tool to turn a typical valve installation. The arm can be equipped with a number of different tools to accomplish various underwater engineering tasks. *(North American Rockwell Corporation)*

The Alaskan Star, a semisubmersible drilling platform, was used in the first exploratory drilling operations in the Gulf of Alaska. *(An EXXON photo)*

Tugs and their tow, the *Brent B* producing platform, arrive at the narrows of Stavanger Fjord on the way to the North Sea. *(An EXXON photo)*

A production platform squats in the Gulf of Mexico off the coast of Louisiana. *(An EXXON photo)*

France across the Strait of Dover, and the first cable was laid 10 years later in 1850 by John Watkins Brett and Jacob Brett. By the 1860s, the engineering profession in the United States and in many parts of Europe was making itself felt by various contributions in weapons, submarine technology, COMMERCIAL OCEAN FISHING, shipbuilding, ship propulsion, and navigation improvement—all ocean engineering–related pioneering endeavors. For instance, in 1866, Cyrus Fields, with the assistance of Maury, laid the first successful telegraph (or permanent connection) cable from England to Newfoundland. By 1876 submarine telegraph cables connected France and America, India and England, Australia and Singapore, and Brazil and Portugal; in 1921, a submarine telephone cable between Key West, Florida, and Havana, Cuba, was introduced.

During the late nineteenth century, innovations and improvements were made in oceanographic instrumentation and techniques. The 1872–1876 expedition of the *H.M.S. Challenger,* commissioned by the Royal Society of London to investigate the physical, chemical, and biological properties of the world's oceans, used such techniques in this survey, the results of which were published in a 50-volume work by Sir John Murray. Another example of novel instrumentation was the conception, in 1872, of the tide predictor by Sir William Thomson (Lord Kelvin). Heinrich Rudolph Hertz, using Maxwell's equations of Faraday's electromagnetic theory, confirmed that theory by experiment in 1887 and so established the foundations for radio communication and radio navigation.

Another manifestation of ocean engineering in those times was the experimental drilling of oil as early as 1891 in the shallow waters off California. (Oil companies began drilling on a small scale in the Louisiana bayous during the 1930s.) Also, through the late nineteenth and early twentieth centuries, the technology of ship construction improved so that ships' drafts exceeded the water depths of most ports. The fact necessitated extensive dredging operations in order to deepen the channels.

Regarding ships, contrary to most popular opinion, the concept of the hydrofoil ship is not new. Actually, such a fast surface craft that, by means of struts of foils, could have its entire hull lifted out of the water, thus greatly lessening friction during propulsion, was invented in 1887. The first patent was granted to a Frenchman, the Comte de Lambert. Eighteen years later, in 1905, Enrico Frolanini, working in Crocco, Italy, also built and demonstrated a 45-mph (72 km/h) hydrofoil ship. Then, in 1919, the famous inventor of the telephone, Alexander Graham Bell, designed and built an aircraft engine–powered hydrofoil that attained a pnenomenal speed of 70.85 mph (114 km/h).

By 1905 submarines had become commonplace in Europe. Because of this, the English deemed it necessary to develop some emergency means of allowing human beings to escape from a trapped submarine. Sir Robert Henry Davis perfected such a device, which he called the D.S.E.A. (Davis submarine escape apparatus).

A few of the more outstanding developments that followed were the invention, in 1921, of the sonic depth finder (this provided a means for an unprecedented increase in an ability to determine both the size and shape of the ocean); the building of underwater robot cameras in 1916 by Harold Hartman, an American electrical engineer, and in 1923 by W.H. Longley, an ichthyologist. This provided the basis for the development of the underwater camera which, for the period 1950–1965, was the primary method of actually obtaining any view of the seafloor); the development of SONAR during the 1930s; the doubling of the world catch of fish and shellfish in the 1920s [4.4 million tons (.4 million metric tons) were caught in 1900] and drilling of some offshore oil in California and Louisiana with considerable ingenuity demonstrated by adapting land rigs to barges, piers, and platforms.

Prior to World War II, most ocean-connected engineering developments were involved primarily with such classical work as shipbuilding, laying of ocean cables, building of piers and breakwaters, construction of bridges, and dredging of harbors. To a somewhat more limited extent, the work also involved the building of submersibles, diving vehicles, and ocean salvage equipments. However, all these applications of engineering and technology to the oceans, while very significant, represented only the beginnings of modern ocean engineering, especially since this work was carried out by engineers who usually had a rather meager knowledge of the complexities and unique behavior of the ocean.

During World War II, almost all the nations involved in the conflict were forced to design and build various ocean structures and equipments, as well as to employ techniques and concepts for performing warfare at sea. It was then that the knowledge gap that existed between engineering and oceanography or ocean science became most apparent. In the United States in 1943, in an effort to close this gap, the U.S. Navy was assigned the responsibility for supplying oceanographic data to all the military services. The Navy's Hydrographic Office conducted the major part of such work, although a considerable amount was done by various universities under contract and by other government organizations. Moreover, at that time many cooperative ocean engineering developments were made possible.

In the 30-year period that followed the end of the war, there was a growing awareness of the urgent need to continue to understand the behavior of the world's oceans and how to best apply engineering principles in order to realize the many and diversified potential benefits that more mature approaches for performing ocean work clearly promised.

In 1960, two events opened a decade of rapid expansion in regard to the modern-day foundations of ocean engineering work. The nuclear submarine *Triton* circumnavigated the world while submerged, and Jacques Piccard of Switzerland and Don Walsh of the U.S. Navy descended 35 800 feet (10 848 m) to the bottom of the Marianas Trench aboard the bathyscaphe *Trieste.* In both cases, it was demonstrated that with good engineering it was possible for humans to live underwater for prolonged periods and to use instruments for observing the extremes of the ocean.

More future opportunities were clearly pinpointed when the nuclear attack submarine, the *U.S.S. Thresher,* tragically sank in 8400 feet of water. The recovery events that followed this disaster proved human inabilities, as of 1963, to cope with submarine rescue in deep water. As a result, many developments in ocean engineering work were started and carried out.

However, two tragic events that occurred in 1966 and 1968 again brought such ocean engineering needs and opportunities into renewed perspective. The first involved the 1966 mid-air collision of two U.S. Air Force aircraft. The problem here was to locate and recover from the waters off Palomares, Spain, an unarmed nuclear weapon that had sunk in these waters as a result of the accident.

Coping with this problem led to many improvements and developments in ocean engineering.

While some of these improvements were being initiated, the U.S. nuclear submarine *Scorpion* sank in the ocean south of the Azores in May 1968. Here, the problems of search were formidable. However, the body of accumulated experience gained from both the *Thresher* and the Palomares searches did prove valuable. The *Scorpion* was discovered and photographed at a depth of 2 mi (3.2 km) by the newly designed ocean research ship *Mizar.*

Subsequent accomplishments include the improvement of offshore oil platforms. The standard drilling rigs of the 1960s were largely those fixed platforms of the type used during World War II as radar stations. Many of these were not substantial enough to survive the storm conditions to which they were exposed. Thus, for the offshore oil industry, a number of mobile so-called jack-up (self-elevating) rigs were designed and used on a worldwide basis. Drillships, unmoored and dynamically positioned, were also developed as were the semisubmerged, dynamically positioned platform structures. As of 1977, about 400 such platforms had been built. In addition, both the submersible and the SCUBA diver have been employed in a wider spectrum of these efforts—as well as in other ocean engineering work.

This list of some historical ocean engineering endeavors can perhaps best be summarized by stating that the field encompasses all the world's oceans and all engineering work that can be related to them and that in the past all ocean engineering work has been relatively minor compared with its tremendous future potential.

Some Basic Technological Problems in Ocean Engineering All ocean equipment or hardware consists of but two basic components—composition (or material) and configuration (or design)—thus, it is important to understand the strong relationship that exists between materials science, ocean science, and engineering. This understanding is crucial to the success of all oceanic engineering systems.

Such systems require reliable hardware. They require component materials that are able to successfully withstand pressures from 1 to 1000 atm, water temperatures from −2 to 38° C, plus a host of corrosive and mechanical forces, as well as effects of biological attack. Materials have to be put together in designs that will have to be based on an everyday familiarity with the conditions of the oceans and the high degree of engineering know-how required to cope with such conditions.

Marine hardware materials are seriously vulnerable to several forms of CORROSION. For instance, stress corrosion is a deterioration that unexplainably takes place in many materials under the influence of a tensile stress. This type of breakdown has long been a problem with some brass and steel alloys used in ordinary SEAWATER applications.

Moreover, the failure of materials in structural members for underwater hardware can occur in various ways that include fatigue failure, buckling, stress corrosion, and ductile or brittle fracture. All materials have a breaking point; they resist all stresses up to that point by either yielding or stretching a bit, but then they are suddenly broken by fracture: a cable snaps; a pipeline breaks in half; a submerged vessel bursts. Corrosion which can attack the internal structure of a material abets this ultimate fracture. In spite of continued studies, the degree of success in developing materials with greater fracture toughness, or the ability to withstand stress without breaking, is still a rather open question. Moreover, there is still a growing need for a greater understanding of how the ocean environment affects both fracture and fatigue in most materials, and how fatigue is affected by weldments and laminates.

Anything placed within the total marine environment must be capable of withstanding pressures which increase nearly 2.7 psi (1.9 N cm^{-2}) for every 6 ft (2 m) of depth. Interestingly, human beings can theoretically withstand enormous hydrostatic pressures on their outer bodies. However, the effects both of pressures and of "inert" breathing gases (nitrogen and helium) that must be used at high underwater pressures are not yet completely understood.

For a great many ocean hardware designs, high strength-to-density materials are needed. (The DENSITY of a material is its mass per unit volume, or density is equal to mass measured in grams divided by volume expressed in cubic centimeters.) Both operational and research submarine vehicles, for example, require hull materials to be tough and strong. (Strength, as used here, carries two meanings: toughness or the maximum energy that a material will withstand without breaking; the maximum force that a material will sustain without yielding or undergoing plastic deformation to more than a nominal design-allowable and acceptable extent.) Yet the vehicle, with all its machinery and habitability requirements, plus those of speed and endurance, must have a proper degree of buoyancy. Thus, the selection of the hull material must take into consideration strength and buoyancy requirements.

Hydrostatic pressure primarily represents a squeezing force or compressive loading on underwater structures. This fact gives rise to one of the primary reasons for considering the use of brittle materials (such as ceramics and glassy materials) for many types of deep submergence work. These materials can withstand high compressive loads. For example, glass, by its nature, collapses from tensile forces (or longitudinal stresses) when loaded in compression.

In the design of underwater structures, it is necessary to guard against such tensile forces. These types of stresses may even occur with materials of strong compressive strength because of local bending near structural stiffeners of vessels, or they may result from an elastic deformation in cavities within the materials themselves. In addition, tensile stresses may follow from local plastic upset in compression. It is also possible that hydrostatic pressure, combined with dynamic factors that induce motion in the ocean, may at times be translated into tension. Accordingly, various atmospheric forces above the ocean, underwater phenomena, and turbulences must somehow be dealt with by the ocean engineer. Among other things of this nature are underwater earthquakes and erosions that cause tons of sediment to periodically fall from SUBMARINE CANYON walls. These, in turn, cause the formation of turbidity currents which travel seaward with great speeds and which are capable of exerting strong mechanical forces. (See TURBIDITY CURRENT.)

In addition to pressure, the TEMPERATURE in the ocean may range from 28 to 90° F (−2–32° C). The low temperatures that prevail in many areas of the ocean can create serious diving problems, which illustrate that really efficient insulating materials must be used to fabricate the needed wet suits and/or dry suits required by free swimmers in their exploration and exploitation of the continental shelves and the sea beyond.

Ocean temperatures decrease with increasing depth [at 600 ft (183 m) the temperatures are about 68° F (20° C); at 4000 ft (1219 m), about 41° F (5° C); and below this depth, about 28° F (2.2° C) in the abyssal regions]. Over the temperature range of 28–68° F, the conductivity of seawater, an important consideration in the corrosion of materials, almost doubles. With higher water temperatures [above 59° F (15° C)], corrosion damage, FOULING, and attack by marine life are of great concern to ocean materials.

The fouling organisms of the sea that attach themselves to either mobile or fixed offshore platforms and structures are especially important from the standpoint of increased fuel consumption in the case of ships used for transporting personnel and equipment, or of increased lateral resistances to currents and waves in the case of fixed or moored ocean structures. They are also important because they can damage the protective coating that prevents corrosion.

Most typical of all the fouling organisms are the BARNACLES and the MUSSELS which, as a group, have a tolerance to a wide range of temperatures (some varieties of barnacles grow more abundantly at greater depths than others). Marine borers also have permanently damaged lead-sheathed cables at depths of 7000 ft (2133 m) in the INDIAN OCEAN. The variations in currents, primary food supply, temperature, salinity, dissolved oxygen concentration, pH, and turbidity determine this depth as well as the population densities of other species of fouling organisms—hydroids, ALGAE, calcareous worms, sea squirts, etc., found in the ocean.

A respectable body of engineering and scientific knowledge, not entirely coordinated, however, does exist concerning the materials and structures for surface-water applications, and to a more limited degree, for the deep parts of the ocean world environment. However, it is still somewhat problematic as to whether the major portion of such materials data can be reliably used in the planning and design of the needed wide gamut of structures, vehicles, and equipment destined to operate in all parts of the ocean.

Universities—Ocean Engineering There are many academic institutions, both in the United States and abroad, that include in their curricula course work directly applicable to ocean engineering. In the United States, some institutions awarding academic degrees in ocean engineering are as follows:

- Florida Atlantic University, Boca Raton, Florida (bachelor of science and master of science in ocean engineering)
- California State University, Long Beach, California (bachelor of science in engineering with ocean engineering option and master of science in engineering with specialization in ocean engineering)
- The U.S. Coast Guard Academy, New London, Connecticut (bachelor of science in ocean engineering and marine engineering)
- U.S. Naval Academy, Annapolis, Maryland (bachelor of science in naval architecture and bachelor of science in marine engineering)
- Massachusetts Institute of Technology, Cambridge, Massachusetts (bachelor of science in ocean engineering, naval architecture, and marine engineering; master of science in ocean engineering, naval architecture, marine engineering, and shipping and shipbuilding management; Ph.D. or Sc.D. in ocean engineering)
- Woods Hole Oceanographic Institution, Woods Hole, Massachusetts (Ph.D. in ocean engineering)
- The University of Rhode Island, Providence, Rhode Island (master of science in ocean engineering)
- The University of Connecticut, Storrs, Connecticut (master of science in ocean engineering)

These U.S. institutions are the primary ones awarding degrees in ocean engineering. However, many colleges and universities also grant academic degrees in naval architecture, marine engineering, and marine sciences.

Research Centers in Ocean Engineering Modern ocean engineering has developed rapidly within the past few years. Moreover, it is a discipline that includes under its purview a myriad of diverse seaward extensions of land engineering endeavors, i.e., the design and construction of piers, bridges, causeways, breakwaters, groins, ramps, structures, power and communication systems, vehicles, equipment and devices, etc. Because of these facts, new ocean engineering research centers are being established, and some of the older oceanographic centers are tending to embrace aspects of ocean engineering in their principal fields of ocean research. Thus, it is extremely difficult to name any more than a representative few of the more recognized ocean engineering research centers of today. These, in the United States, are

- The Coastal Engineering Department and the Engineering and Industrial Experiment Station, both of the University of Florida, Gainesville, Florida
- The Center for Engineering Research of the University of Hawaii, Honolulu, Hawaii
- The Civil Engineering Research Laboratories of the University of Illinois, Urbana, Illinois
- The Engineering Design and Analysis Laboratory of the University of New Hampshire, Durham, New Hampshire
- The Division of Engineering Research and Development of the University of Rhode Island, Kingston, Rhode Island
- The Office of Engineering Research of the University of Washington, Seattle, Washington
- The Engineering Experiment Station of the University of Wisconsin, Madison, Wisconsin
- The WOODS HOLE OCEANOGRAPHIC INSTITUTION, Woods Hole, Massachusetts
- The SCRIPPS INSTITUTION OF OCEANOGRAPHY, La Jolla, California
- The IIT Research Institute, Chicago, Illinois
- The Battelle Memorial Institute, Columbus, Ohio

The interested reader who wishes to dig deeper into the subject of ocean engineering research center listings, both on the national and international level, is referred, respectively, to *The Research Centers Directory,* edited by Archie M. Palmer, Gale Research Company, Book Tower, Detroit, Michigan (1975), and *The Ocean Research Index,* Francis Hodgson Ltd., P.O. Box 4, Guernsey, British Isles.

OCEAN FOOD CHAIN (food pyramid), refers to the transfer of food energy from the source in PLANKTON or micro-ALGAE through a series of marine organisms that feed on other marine organisms. At each transfer a large proportion of the potential energy is lost as heat; therefore, the number of steps in each sequence is usually limited to four or five.

Ocean food chains are of two basic types: the grazing food chain, which originates with the green plants (phytoplankton) or primary producers, goes to grazing herbivores (organisms that eat living plants), and ends with carnivores (animal eaters); and the decay or DETRITUS food chain, which starts from dead organic matter, goes to microorganisms, and ends with detritus-feeding organisms (detritivores) and their predators. These types of chains are interconnected with one another and may not always follow this particular sequence. For example, the main diet of some types of WHALES (such as the baleen whale) consists of plankton which is consumed daily in massive amounts. Moreover, the classification of a chain by successive categories (trophic levels) is often misleading because a species may occupy more than one trophic level during its life cycle or in response to changes in food availability. A modified food chain that expresses feeding relationships at various changing levels is called a food web. (See OCEAN FOOD WEB.)

A typical example of the ocean food chain, or food pyramid, is that beginning with 1000 lb (454 kg) of DIATOMS (single-celled plants) at the bottom of the pyramid. This amount provides food for 100 lb (45.4 kg) of tiny oceanic herbivores or zooplankton. These produce 10 lb (4.5 kg) of food for HERRING, which in turn, furnish 1 lb (454 g) for the COD. Thus, as one nears the top of the ocean food pyramid, the number of creatures becomes far less than at the bottom.

Compared with food chains on land, ocean food chains appear to be quite efficient. However, most ocean food chains have more links than the majority of terrestrial ones, and many in fact represent fairly complex webs.

OCEAN FOOD WEB is a modified OCEAN FOOD CHAIN that expresses feeding relationships at various changing trophic levels. Since there are many reversals and offshoots of the ocean food chain, the term *food web* provides a more accurate word picture of the varied and complex feeding patterns of marine organisms.

OCEAN GASES, See SEAWATER.

OCEAN GEOTHERMAL DEPOSITS, such as those found in the RED SEA, are deposits of mineral-rich mud resulting from the interaction with hot brines circulating up through the fissured seafloor.

Peculiarities in the SALINITY of the waters of the Red Sea have been known since the Russian Expedition of the *Vityaz* in the 1880s (see VITYAZ). In the mid-1960s, however, additional data were obtained by the British *R.R.S. Discovery,* followed by the U.S. vessel *Atlantis II,* and several others. The *Discovery* sampled water with a TEMPERATURE of 111° F (44° C) and a salinity of 256 ppt. *Atlantis II* measured a brine temperature of 133° F (56° C) and obtained bottom sediment samples having a temperature of 144° F (62° C), and containing a mixture of metal compounds, principally oxides and sulfides of iron, manganese, zinc, and copper.

Geothermal heat from the molten interior of Earth is transmitted to the SEAWATER through the fissures along the rift of the Red Sea floor. The heated waters dissolve salts from sedimentary rock formations and leach heavy metals out of crustal volcanic rocks, creating metal-saturated brine. As this metalliferous brine cools, it releases the sulfides of lead and zinc and the carbonates of iron contained in the WATER.

The submersible *Alvin* has also directly observed mineral-rich hot springs on the Galapagos spreading center as well as hydrothermal precipitates, including the sulfides of copper, lead and zinc which have been drilled in many parts of the ocean (such as the spreading center south of Baja California) during the GLOMAR CHALLENGER Deep Sea Drilling Project. Although existence of the hydrothermal process is well known, its mode of operation—How does the water react and interact with the hot rock and circulate through newly formed crust on the ocean floor?—is still far from being understood.
See also ENERGY.

OCEAN POLLUTION is the adverse alteration of the ocean environment, largely as a by-product of the actions of human beings, through direct or indirect effects of changes in energy patterns, radiation levels, chemical and physical constitution, and distribution, abundance, and quality of organisms.

The popular notion that the world's oceans are polluted solely by human beings ignores the fact that the oceans are the natural receptacle for all the waste matter produced by its many and varied forms of living ocean inhabitants. These same oceans are also the sump for virtually all the geological transportation processes that occur on land.

Such processes cause changes in the marine environment by the introduction of substances from outside the natural oceanic system.

Human beings, of course, have to live with these natural sources of waste. However, the dispersal of artificial substances to the oceans and the ability of these waters to accept them is quite another matter. As a result of an increasing world population and an increasing per capita use of energy, many materials in degraded forms have been identified throughout the world's oceans. While such sensational statements as "the oceans are dying" are scientifically untrue, it is true that some major sources of pollution in the world's oceans are

- Sewage outfalls
- Ocean disposal of sewage sludge and dredge spoil
- River discharge and land runoff
- Wastes from vessels
- Accidental spills of oil and chemicals
- Harbor debris
- Industrial wastes
- Thermal energy outfalls from DESALINATION plants, etc.
- Chlorinated hydrocarbons, such as DDT, and the production of the aldrin-toxaphene group: aldrin, chlordane, dieldrin, endrin, heptachlor, and toxaphene

Contaminants from these sources have been identified throughout the world's oceans. This is because human waste products ultimately can only be discharged into the atmosphere, onto land, or in water. Wastes discharged into the atmosphere are dispersed over Earth's surface laterally and vertically in amounts that change with time. Most of these kinds of contaminants do not remain airborne for long and eventually settle onto the land or water or are converted into a natural form. However, wastes distributed over the land are not dispersed as easily as those in the air. Water flowing as surface runoff or ground water can transport these wastes or their degraded products to the ocean. Some pollutants may be converted to natural forms, but the rates of conversion are slow; others discharged to surface or ground waters may find their way to the oceans after being retained in sedimentary deposits. The world's oceans, therefore, are the eventual sink for contaminants that are not retained in the air and on the land. The situation is made more difficult because the concentration of pollutants is highest in the coastal and inshore waters where the impact is greatest on human activities.

Tropical and temperate ocean communities may vary in their reaction to this pollution, and although some present evidence is conflicting, TEMPERATURE effects caused by hot-water effluents from power-plant stations affect shallow-water tropical communities the most. Such communities normally have small thermal tolerances, with the thermal maximum being close to the ambient temperature. The annual temperature range is also less than in temperate waters. Although biodegradation of chemical pollutants may be more rapid at higher temperatures, this effect may be nullified by lower tidal amplitudes in tropical waters.

In summary, aside from the physical and aesthetic aspects of pollution, most other major deleterious effects are toxicological. These present an array of complex environmental problems affecting human beings and marine organisms. All such adverse changes affect humans directly or indirectly through their supplies of food and other products, their physical objects or possessions, and their opportunities for recreation and appreciation of nature. Thus, important avenues of research for the oceanographer are defined by such phenomena. This research will provide the needed environmental information on which sound policy decisions can be made.

See also ECOLOGY.

OCEAN SUNFISH is the common name for ocean fish in the family Molidae, order Tetrodontiformes (an order that also includes the porcupine fish, filefish, PUFFERFISH, and trunkfish). The ocean sunfish is especially distinguished by its distinctive oval-shaped body. The mouth is small, and the teeth in both the upper and lower jaws are fused to form a single sharp-edged beak. The dorsal and anal fins are large and high, and the body ends abruptly in a low tail fin. The fish lacks pelvic fins and has a small spinal cord and brain.

The ocean sunfish, family Molidae, include three species: *Mola mola, M. lanceolata,* and *Ranzania truncata.* Of these, the first, *M. mola* (*mola* is the Latin word for "millstone"), is the largest and attains a length as great as 10 ft (3 m) and a weight exceeding a ton (907 kg). The other two are smaller, although all have compressed short bodies of a rounded outline. These bodies are gray to olive brown in color and have a very tough, wrinkled skin which overlies a coat of gristle. All three species are poor swimmers and are found in the warmer and more temperate parts of the world's oceans. Ocean sunfish are usually observed drifting on the surface of the ocean, especially during calm weather. They feed mainly on PLANKTON, JELLYFISH, and fish LARVA.

Among the many peculiarities characteristic of these fish is their method of propulsion. This is ac-

OCEAN POLLUTION. (*Left*) Dead fish bestrew the beach as a result of massive discharge of wastes into a biologically active marine area. (*Above*) A satellite photo highlights an irregular dark shape in the waters adjacent to Long Island and the New Jersey coast. The phenomenon marks disposal of waste chemical products considered harmless to local ecology. (*NOAA*)

complished by moving the large dorsal and anal fins from side to side, with the small pectoral fins acting as stabilizers and the rounded tail as a rudder. The gills are used for guiding its direction by ejecting a jet of water from them or out of the fish's mouth. The brain is unusually small and less developed in size than the two kidneys which are located just behind it, rather than farther back in the body as is more usual in fish. The ocean sunfish larvae are of a normal fish shape, and the disk-shaped body of the adult ocean sunfish starts to take form when the baby fish is about ½ in (1.27 cm) long.

OCEAN WATER See SEAWATER.

OCEANIC BIRDS See MARINE BIRDS.

OCEANOGRAPHY is basically the science of the oceans, but more specifically it is the application of many sciences to the study and exploration of the oceans in all their aspects; as such, it is concerned with obtaining a better scientific understanding of ocean-bottom topography, including the structure and composition of the sediments and rocks, the contact zone between the ocean and the atmosphere, the marine plant and animal life, the chemical composition of the water, the behavior of the waters in motion, and the response of ocean waters to external and internal forces. Thus, oceanography utilizes several disciplines and branches of science, embracing, in particular, five primary sciences: geology, biology, chemistry, physics, and meteorology. It may be said that supplementing these are applied mathematics, marine archeology, materials science, as well as environmental science and fluid mechanics. The composite scientific understanding of the oceans, or oceanography, forms the basis for applying engineering principles and practices to the performance of useful work in and/or on the oceans. (See OCEAN ENGINEERING.)

History of Oceanography

Earliest Beginnings Curiosity about the ocean world manifested itself early in human history. Concepts of a great body of water surrounding the habitable land are contained in various ancient myths and legends about the creation of the world. Thales of Miletus (640?–546 B.C.), representing the Ionian school of philosophers, believed that the land world consisted of a huge circular disk. This disk was encircled by a mighty stream of water called Oceanus, the eldest of the Titans. The Babylonians thought of their world as a huge round mountain that stood above and in the midst of a universal sheet of water, while in the ancient Hebrew scriptures, it was stated that the oceans had been brought together in one place by the word of God and then the dry land appeared. (See OCEAN.)

These concepts, and the variations of them held by other ancient peoples, were questioned by some of the early Greek philosophers and geographers. Perhaps the earliest of these were Pythagoras and Anaximander of Greece. Six centuries before the time of Christ, Pythagoras theorized that the world was not disk-shaped but spherical. During this period, Anaximander made the first maps of the Greek known world. Later, Eratosthenes of Alexandria, the astronomer and geographer (ca. 240 B.C.), believed that the oceans consisted of two great streams running at right angles to each other, with one stream following the equator and the other surrounding a spherical earth from pole to pole. Strabo (ca. 20 A.D.) had the same idea, and in his writings about the Earth he estimated its size and described the climatic zones of the globe.

Ptolemy (Claudius Ptolemaeus, ca. 150 A.D.), in his study of natural phenomena, described a factual map of the known world, including an enormous continent in the Southern Hemisphere. Ptolemy, like his distinguished predecessor Aristotle (384–322 B.C.), and accepted authority on scientific matters for over 1000 years, held the view that the inhabited land occupied the major portion of the surface of the earth and that the waters were largely confined to the MEDITERRANEAN SEA.

Even with such meager oceanographic knowledge, the ancients ingeniously designed and courageously used ships to venture out on the ocean waters that surrounded their world. Such ships were employed by the Egyptians, the Athenians, the Sea Kings of Crete, the Phoenicians (reputed to be the first to sail at night guided by the North Star), the Romans, and others to carry on commerce and trade, to discover new lands, to conquer and colonize, and to conduct wars. Practically all these efforts involved travel confined to the Mediterranean Sea. However, a few sailed to unknown destinations, some going as far as Iceland, without knowing they had traveled beyond the confines of the Mediterranean.

The Polynesians were also marine pioneers over 2000 years ago through their explorations of the PACIFIC OCEAN. They used their knowledge of weather, winds, migratory sea birds, and ocean currents in order to travel through their widely separated islands.

Most early examples of ocean-related inquiry, other than those of a geographical nature, can best be traced back to the time of the Greek and Roman cultures more than 1900 years ago. For example, Aristotle, the pupil of Plato and tutor of Alexander the Great, was enamored by all the mysteries of nature and especially those of the ocean. Like many modern-day oceanographers, he was often frustrated by such mysteries, and it is said that Aristotle once threw himself into the whirlpool of the Strait of Euripus, separating Euboea from the Greek mainland, because he was unable to explain it. He also threw himself into his work, and this labor resulted in a treatise on marine biology. Archimedes' (287–212 B.C.) investigations culminated in the physicist's principle that any object wholly or partially immersed in a liquid is buoyed by a force equal to the weight of the liquid it displaces. Euclid (ca. 300 B.C.), the Greek geometer, in his essay *Elementis,* outlined the principles and assumptions with respect to space that have formed the foundations of many aspects of oceanography. Posidonius of Apania, in 2 B.C., correlated the tidal effects at Cadiz with the phases of the moon as well as measuring a depth of 1000 fathoms [6000 ft (1830 m)] in the Sea of Sardinia. Herodotus (484–428? B.C.), writing at Athens, described oceanographic work in the MEDITERRANEAN SEA that assisted navigators by telling them when they were over the toe of the Nile's delta off the coast of Egypt. This was done by comparing their depth measurements with samples of the sea bottom, or, as Herodotus stated in 450 B.C., "The nature of the land of Egypt is such that when a ship is approaching it and is yet one day's sail from shore, if a man try the sounding, he will bring up mud even at a depth of 11 fathoms." Pliny (23–79 A.D.), the author of *The Natural History,* wrote about various investigations of marine life, especially CORAL and its usefulness as a medicine. Incidentally, the early Romans used seaweed to heal wounds, burns, scurvy, and rashes, and many species of seaweed were recognized and characterized, especially in the Orient as early as 3000 B.C., for both their food and medicinal value. (See ALGAE.) MARICULTURE, though relatively new to the United States and other parts of the modern-day world, is one of the oldest of the biological sciences applied to oceanography. Also called fish farming, mariculture basically consists of the rearing of marine plants and animals under controlled conditions. Various forms of mariculture were practiced in China (with carp), in Egypt (with the spiny-rayed tilapia fish), and in Japan (with cultured oysters) as early as 2000 B.C.

In summary, it can be said that many of the thinkers of the Hellenistic Age of Greece, and especially of the Golden Age of the Roman Empire, plus others of the ancient world, passed on to subsequent ages much of value from their thought and experience that has pertained both directly and indirectly to oceanography.

Development of Modern Oceanography Interest in the oceans was essentially lacking between antiquity and the Renaissance. However, in the fifteenth century, it was rekindled again by those who applied their knowledge of astronomy and elementary navigation to explore the oceans of the world. For example, on Columbus' voyages, he became an oceanographer by collecting and studying marine flora and fauna and noting currents and weather, besides discovering the SARGASSO SEA. Ponce de Leon discovered and described the GULF STREAM, Vasco da Gama found the route to India via the Cape of Good Hope, and Ferdinand Magellan's voyage around the world showed human beings how vast was the ocean world they still had to explore and exploit. One such person, Prince Henry the Navigator, established the forerunner of the well-known twentieth century International Hydrographic Office at Monaco when he set up a navigation information and data center in Sagres, Portugal, in the fifteenth century. His idea was the same as today's oceanographic centers—to distribute navigational and other oceanic information to the world. However, many merchants were secretive about information and navigation aids they discovered, and they withheld the valuable data from their competitors.

On the other hand, some early explorers of this era did have some interesting observations to make about marine biology. For instance, according to Sir John Murray, editor of the studies of the famous CHALLENGER EXPEDITION (1872–1876), one seafarer of the times, Sir John Hawkins, is quoted as saying:

> Were it not for the Moving of the Sea, by the Force of Winds, Tides and Currents, it would corrupt all the World. The Experience of which I saw in Anno 1590, laying with a Fleet about the islands of Azores, almost Six Months, the greatest part of the time we were becalmed; with which all the Sea became so replenished with several sorts of Gellies and Forms of Serpents, Adders and Snakes as seem'd Wonderful; some green, some black, some yellow, some white some divers Colours and many of them had Life, and some there were a Yard and a-half, and some two Yards long which had I not seen I could have hardly believed; and hereof are witnesses all the Company of the Ships which were then present; so that hardly a Man could draw Bucket of Water clear of some Corruption.

All such accomplishments, explorations, and the tales that were told about the ocean voyages of the period stimulated many seventeenth and eighteenth century scientists and achievers to further study some aspect of these fabulous oceans.

In the seventeenth century, such Englishmen as Robert Boyle, the originator of the well-known Boyle's law of physics (which states that at constant TEMPERATURE, the volume and pressure of a gas are inversely proportional—or the volume diminishes as the pressure increases), turned his attention without much success, to a study of the temperature and "saltiness" of the sea. Robert Hooke, father of Hooke's law (which states that the stress of a solid is directly proportional to the strain applied to it), improved upon the design of an instrument then used for taking ocean depth measurements. Sir Isaac Newton published the *Principia Mathematica* in 1687, and in this work he applied his universal law of gravitation to the behavior of TIDES. Previous to this in the early 1600s, such scientists as Galileo and Johann Kepler had related tidal phenomena to the orbital and rotational motions of the earth, moon, and sun; however, Newton's "equilibrium theory" gave precision to tidal behavior and has been the basis of all subsequent work on tides.

Edmund Halley, discoverer of the comet in 1682, led the first scientific expedition of the ship *Paramour Pink* to the southern oceans of the world in 1700. The British Admiralty had commissioned Halley to explore the coast of the Terra Incognita, thought to lie between the Strait of Magellan and the Cape of Good Hope, "to measure the magnetic declination and to stop at as many islands enroute as possible." In 1701, Halley recorded the observations of his journey (he was stopped by ice at $52\frac{1}{2}$°S in the south ATLANTIC OCEAN), and in these observations he published a map showing the magnetic variation.

However, in spite of all such accomplishments as these in the early eighteenth century and those which proceeded when the oceans and their inhabitants were the objects of splintered scientific interest, investigation, and exploration, the first really

OCEANOGRAPHY

Underwater lights glow during a dive by *Nemo*, an untethered, self-contained submersible, to the 500-ft level. *(U.S. Navy)*

A scientist in the underwater research craft *Sea-See* records data as divers work in the waters off Santa Catalina Island. *(U.S. Navy)*

The oceanographic research ship *USNS Hayes*, an heir to the duties and challenges of such noted vessels as the *Beagle* and the *Challenger*, sets forth on an expedition. (*U.S. Navy*)

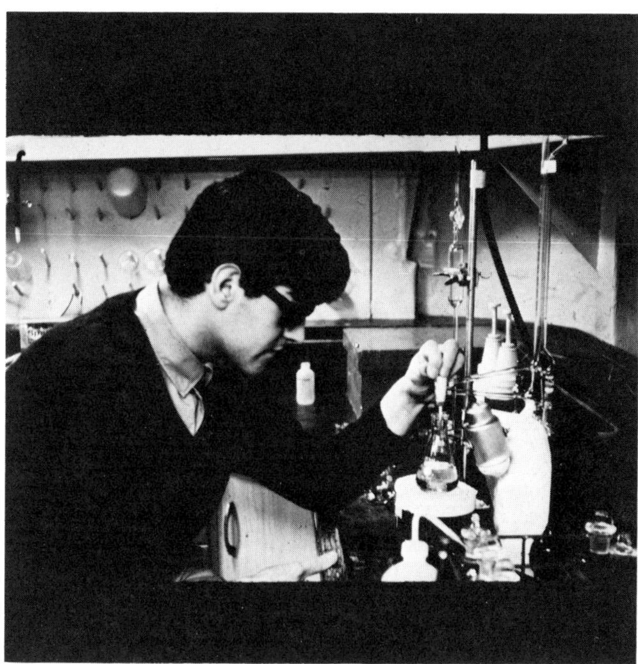

An oceanographer determines the oxygen concentration of a seawater sample by chemical titration. He does this research at sea, shortly after the water sample is obtained. (*Arnold Gordon*)

Pulleys heave the *Sealab III* habitat from the sea after its first test in 1968. (*U.S. Navy*)

organized attempts to study the world's oceans did not materialize until late in the eighteenth century.

The first American to put organized oceanographic knowledge to practical use was Benjamin Franklin. As postmaster of several colonies in the 1750s, he was concerned with fast delivery of mail both to and from Europe. He learned of the whalers' use of the Gulf Stream to speed up trips to Europe and their use of other routes to avoid delays on the return trip. After Franklin assembled data and published his paper on the NORTH ATLANTIC CURRENT (as he called the Gulf Stream), the world took notice. Oceangoing merchant captains, who for years had battled it, began to make the Gulf Stream work in favor of their ships. Franklin's depiction of this route on a contemporary nautical chart also provided the basis for the concept of thermometrical navigation. His work was based upon an observation of currents, sea colors, and surface temperatures. (See NAUTICAL CHART.)

James Cook was one of the earliest of the ocean explorers to bring science into prominence by taking a naturalist along with him on his voyages which began in 1768. (See COOK, JAMES.) In 1768, he circumnavigated New Zealand, and in 1772–1775, the period of his second voyage, he circumnavigated Antarctica and discovered the South Sandwich Islands. On these expeditions Cook, on the British Ship *Endeavour*, took the first recorded subsurface temperatures and deep soundings [4000 ft (1219 m)]. So significant were such contributions, that after that time many ships were sent on voyages of exploration with a scientist or scientists along. Charles Darwin was in such a role when he cruised with the survey ship *Beagle* in 1831. (See DARWIN, CHARLES.) From data collected on his 5-year cruises aboard the *Beagle*, Darwin developed his famed theory on the origin of species. Darwin's collection of oceanic data was an important step in alerting scientists and others interested in the oceans to the vast amount of knowledge that could be gained by a systematic collection and study of data.

The nineteenth century work of physicists and mathematicians, such as Laplace, Vossius, Kepler, and Fourier, as well as naturalists like Ehrenberg, Humboldt, Hooker, and Oersted all contributed to this reservoir of information, either directly or by providing the background for the development of modern theories and oceanic investigations. (See HUMBOLDT, ALEXANDER VON.)

Also, in the nineteenth century, the Survey of the Coast (now the Coast and Geodetic Survey) was established in the United States in 1807 to improve navigation in its coastal waters. Later, in 1824, the U.S. Army Corps of Engineers was established to dredge harbors and navigable channels; the Department of State was established in 1828, and one of its chartered functions is to protect and improve the management of the country's fishery resources; the Depot of Charts and Instruments (later the U.S. Hydrographic Office, now the Naval Oceanographic Office) was established in 1830 to provide charting and routing services to naval and merchant ships; the U.S. Fish Commission (later the Bureau of Commercial Fisheries and the Bureau of Sport Fisheries and Wildlife) was established in 1871.

In the period 1819–1821, the Russian Admiral Fabian von Bellingshausen circumnavigated the southern polar ocean and made valuable observations of the region, as did George Powell, an English sea captain, whose primary mission in 1821 was to hunt for SEALS. Powell observed, among other things on this voyage to the Antarctic, that just as in the Arctic, a warm layer of water existed below the cold surface water. In 1823, another English sealer, James Weddell, sailed to the Antarctic (as far as 74°15′S) and measured water temperatures of the region. James Eights of England, in 1829, after a voyage with the American captains, Pendelton and Nathaniel Palmer, described the natural history of the South Shetland Islands. He discovered a new species of CRUSTACEANS and a marine spider and recognized the fact that rock fragments were transported by icebergs far from their source. Also, about this time, Edward Forbes, as a result of his work at sea, developed a scheme for the geographical and vertical distribution of marine plants and animals.

These men and many others, between the years 1800 and 1870, especially by their explorations and systematic observations at sea, made significant contributions to the mid-beginnings of oceanography. For example, since the collection and analysis of empirical data are the prime requisites of any science, appropriate recognition must be made of the contribution to oceanography by such representative pioneers as Charles Wilkes and Matthew Fontaine Maury. (See MAURY, MATTHEW FONTAINE.)

Charles Wilkes, the man who had been officer-in-charge of the U.S. Depot of Charts and Instruments from 1833 to 1837, conducted the surveys which culminated in the Depot's issuance, in 1837, of some of the first nautical charts published in the United States. In 1839, Wilkes assumed command of a national scientific expedition authorized by the Congress of the United States. This 4-year expedition was charged with "exploring and surveying the Pacific Ocean and the South Seas in order to promote the interests of commerce and navigation, and the extension of scientific knowledge. . . ." The results of the exhaustive investigations of this expedition contributed much to organized knowledge of the oceans in the United States.

Another man who gave oceanography a very significant and lasting impetus was Matthew Fontaine Maury of the U.S. Navy. Maury was injured aboard ship and thus prevented from continuing his active naval career at sea. However, he began collecting information systematically from naval vessels and merchantmen on currents, weather, winds, and other useful data found in ship's logs. His charts and notices to mariners gained immediate popularity and usefulness with both naval and merchant marine people throughout the world. Maury published *The Physical Geography of the Sea* in 1855 and won world acclaim. Later discoveries and studies changed many of the conclusions Maury had made. However, his initiative and systematic methods of compiling and using data formed much of the basis for an organized and structured science of the ocean.

The pioneer who introduced oceanography into the universities was Louis Agassiz (1807–1873), a contemporary of Maury. (See AGASSIZ, LOUIS.) Agassiz arrived at Harvard University from his native Switzerland in 1848 and remained to make important contributions to the development of marine science in the United States. He had broad interests both in marine biology and geology. Professor Agassiz spent his last year at Penikese Island in Buzzards Bay, Massachusetts, developing a summer school of marine biology. Largely because of him, the Marine Biological Laboratory was established at Woods Hole, Massachusetts, in 1888.

In the 1870s, oceanography also received a great inspiration for further growth. This was due to the work of the famous British expedition of the *H.M.S. Challenger* that lasted from 1872 to 1876. Under the scientific leadership of Wyville Thomson, this first true oceanographic cruise covered over 69 000 mi (111 000 km) during which 362 oceanographic stations were occupied. Biological specimens, water, and bottom samples were collected and ocean currents and temperatures measured. The *Challenger* data, reported by Sir John Murray in *Challenger Reports,* consumed 50 volumes and delineated the main outlines of the ocean basins plus providing a huge amount of formerly unknown, fascinating information about marine life. It also led to a wider realization of the importance of the seas. Moreover, coming shortly after Maury's work, it again illustrated to what extent and with what efficiency ocean information could be collected by trained observers and scientists.

After this, several developments of interest to oceanic studies took place. For instance, in the United States, the Smithsonian Institution began to encourage naturalists to perform research for the U.S. Fish Commission. (The first vessel constructed for deep-sea exploration was the 234-ft *Albatross,* built in 1882 by the U.S. Fish and Fisheries Commission.) Deep-sea surveys of the Pacific were conducted by the U.S. Navy, and an oceanographic survey of the velocity and temperature distribution of the Gulf Stream was made between 1885 and 1889 by the Coast and Geodetic Survey of the United States.

Then, as previously mentioned, the first marine laboratory in the Unites States was founded in 1888 at Woods Hole, Massachusetts; this developed into the WOODS HOLE OCEANOGRAPHIC INSTITUTION, formally established in 1930. In 1892, the Pacific Grove Marine Station of Stanford University in California was established, and in 1905 the Marine Biological Association of San Diego (now the SCRIPPS INSTITUTION OF OCEANOGRAPHY and a part of the University of California complex since 1912) was established. Also during this era, the Zoological Laboratory at Naples, Italy, was built. All of these were at first biological laboratories and fisheries stations, and, as noted, the oceanographic institutions established to study the physics, chemistry, and geology of the oceans came into being much later.

During the late nineteenth century, cooperation and strong international ties were established between the small group of marine scientists in the United States with those in Europe. Moreover, innovations and improvements were being made in oceanographic INSTRUMENTATION and techniques. The International Council for the Exploration of the Sea was formed in 1902 by countries of northern Europe to assist in the handling of sea fisheries by solving critical marine biological problems. Finally, important and far-reaching oceanographic theories and instruments were developed by Scandinavian oceanographers Ekman, Sandstrom, Helland-Hansen, Bjerknes, and Nansen. (See NANSEN, FRIDTJOF.)

In the twentieth century, the founding of the International Hydrographic Bureau in Monaco in 1921 formalized the free exchange of various kinds of oceanographic information and resulted in the coordination and standardization of hydrographic surveying and charting by maritime nations.

In the period following World War I, such developments as these, though few in number, emphasized the importance of this type of knowledge of the seas and led to various oceanographic expeditions. A few examples of notable work in this regard include that of the Dutch ship *Wille-*

brord Snellius (1929–1930); the American *Carnegie* (1927–1929); the Danish *Dana* (1920–1922); the Norwegian *Maud* (1918–1925); the German *Meteor* (1925–1938); and the *Princess Anne* of Monaco (1922).

The possibility of another World War in the 1930s led to the development of "active" or echo-ranging sound detection. (By *active* it is meant to send out a signal from a submarine or surface vessel and listen for an echo from the target.) While the study of the oceans was confined to defense purposes, oceanography did gain some fringe benefits from this SONAR research and development.

During World War II, oceanography and how to apply it to the solution of urgent defense problems became vitally important. Then, for the first time, various ocean investigations were pressed and the small number of oceanographers trained in the thirties was augmented by scientists and engineers from other disciplines, many of whom remained connected with marine studies in various parts of the world after the war.

Considering the past, the development that took place in the years between 1960 and 1970 can only be classified as the most significant turning point or the real beginning of the era for marine science. Twenty new oceanographic ships were built and put into operation and eight new laboratories were established in the 1960–1966 period, and more than 50 universities and colleges in the United States offered courses in oceanography.

In education, United States Federal policy to support training and education in technical fields greatly strengthened the marine sciences. Prior to this, professional personnel in these sciences was limited to about 600 persons. With a policy that started in 1961 to expand the student and teacher opportunities at universities, 1967 saw a professional personnel numbering 2600, and approximately 1000 individuals were enrolled in marine science curricula in the growing number of colleges and universities that offered the program. Moreover, the National Sea Grant College and Program Act of 1966 was designed to further strengthen this base of oceanography and ocean engineering training and research. (See SEA GRANT PROGRAM.) In short, the Sea Grant legislation (signed into law on October 15, 1966) made provision for grants and contracts to both public and private institutions of higher education, institutes, and laboratories for the functions of education, applied research, and information transfer designed for marine resource development.

In 1960, the National Oceanographic Data Center was established at the Naval Hydrographic Office to acquire, process, preserve, and disseminate oceanographic data for educational, scientific, and commercial purposes. Shortly afterward, the Hydrographic Office was redesignated the Naval Oceanographic Office, and the Instrumentation Department within the Oceanographic Office became the Navy Oceanographic Instrumentation Center. Later, it became the National Oceanographic Instrumentation Center, expanding its role in the setting of standards for oceanographic equipment and making tests for government agencies and for industry.

On the world scene, several oceanographic ships were built and equipped with modern research equipments such as precision navigational aids, depth recorders, and laboratory instruments.

Over the last few years (1970–present), a great many oceanographic events have taken place. Among others, the submersible has evolved as a significant tool in ocean research and development, and its utilization promises to be increased. There are about 70 manned submersibles in the world that can be considered to be operational or available and ready for operational use on short notice. Also, worldwide, the number of laboratories wholly or primarily devoted to oceanography has expanded and grown in number. A few major ones are listed here:

England

National Institute of Oceanography
Marine Bilogical Laboratory (Plymouth)
Fisheries Laboratory (Lowestoft)

United States

Woods Hole Oceanographic Institution (Massachusetts)
Scripps Institution of Oceanography (California)
Lamont-Doherty Geological Observatory (New York)
Texas A&M College, Department of Oceanography
Oceanography laboratories of the University of Washington (Seattle), University of Hawaii, and Oregon State University
Rosensteil School of Marine and Atmospheric Sciences, University of Miami (Florida)

Germany

Oceanography laboratories (Kiel and Hamburg)

Denmark

Danish Biological Station (Copenhagen)

Others

Other European laboratories include those at Bergen (Norway); Göteborg and Stockholm (Sweden); Helsinki, (Finland); Trieste (Italy); Wormly (England); (Monaco); Paris (France); Madrid (Spain); Edinburgh (Scotland); Moscow (U.S.S.R.). Laboratories are located at Tokyo (Japan); Namaimio and Halifax (Canada); and (Hawaii).

The Work of the Oceanographer Today In the late 1960s. Roger Revelle, the eminent American oceanographer, prophetically said that

> oceanographers do better science when they are not isolated from physics, chemistry, and biology . . . oceanographers tend to feel they are second class citizens. They're sailors, they are not as bright as molecular biologists or physicists who think that everything that does not involve fundamental matter is trivial. . . But oceanography is changing; it is becoming more fundamental, more rigorous and more mathematically oriented. As Clemenceau remarked about war and generals, the ocean is too important to be left to oceanographers. They must forget the notion that it is the monopoly of those who hold trade union cards as working oceanographers.

In the decade following these remarks, oceanography became more fundamental, and the tremendous advances that have been made in the various sciences—chemistry, physics, biology, geology, and meteorology, among others—have been incorporated and utilized by oceanographers in all their studies which relate, one way or another, to the marine environment. Such advances as better analytical techniques and instrumentation, improved materials development, various new concepts, international scientific cooperation, computers, and many other recent insights and developments in science and technology have all impacted on oceanography and been brought to bear directly on this interdisciplinary field.

To obtain a feel for how interrelated the five main faces of oceanography are, a few of the aspects of physical, chemical, geological, biological, and meteorological oceanography are delineated.

Physical Oceanography The physical oceanographer in a broad sense is concerned with research directed to a better understanding of the thermal and dynamic (or kinematic) state of the world's oceans and the control of physical properties within, as well as at, the boundaries of these oceans. Therefore, the theoretical and observational work in this field is strongly oriented toward investigations of ocean circulation processes, the dynamics of WAVES, currents, tides, and tsunamis (see TSUNAMI), the thermodynamics of seawater, various transient ocean phenomena such as UPWELLING, eddies, and other studies designed to provide an understanding of the motion of ocean water, including its physiochemical characteristics, and the behavior of light and sound in the ocean.

Chemical Oceanography CHEMICAL OCEANOGRAPHY, perhaps more than any other one part of oceanography, is a primary contributor to all the other branches of the science, both basic and applied. The chemical oceanographer seeks to determine, at isotopic, elemental, and molecular levels, the complex, organic, and inorganic constituents of seawater—their behavior, their source and mode of formation, and their role in various geological, biological, physical, and meteorological oceanic processes. Advanced techniques developed chiefly in support of science other than oceanography have been applied to a definition of the chemical composition of seawater. Such techniques as NEUTRON ACTIVATION ANALYSIS, isotopic distribution, mass spectrometry, x-ray fluorescence, atomic absorption spectrometry, and radiochemical analysis have been of significant importance. In addition, various analytical methods have been used for *in situ* determinations of chemical compositions.

Geological Oceanography The geological oceanographer is concerned with studies of the physical nature, mineralogy, and fossil remains of marine sediments and sedimentary structures. A knowledge of the history of these formations covered by the world's oceans provides insights into geochemical cycles and processes of organic evolution. Many such investigations are basic for a host of present and future ocean engineering endeavors: the mining of ocean minerals, mariculture, the design and construction of surface and subsurface platforms and structures anchored to the ocean floor, ocean ENERGY exploitation, navigation of surface ships and submarines, and many others.

One traditional method employed by the geological oceanographer for gathering data about the ocean floor is by coring. Cores are obtained by driving a metal tube into the unconsolidated bottom sediment, and then drawing the tube up by cable and removing the core. Certain layers in these cores reveal useful data about geologic history—for instance, the relative dates of volcanic eruptions and periods of glaciation can be compared. Other methods used to obtain samples of dislodged bedrock from submarine outcrops are by dredges dragged along the bottom or by grab samplers. One such sampler collects about 10 ft^3 (283 L) of floor material. When a trip weight hits the ocean bottom, a camera installed on the device takes a photo of the undisturbed bottom sediments and rocks; then the jaws of the sampler "grab" some of the floor material and snap shut. Marine geology, it should be noted, is not concerned only with a study of sediments, but with the entire structure of the ocean basin.

In recent years, the capabilities of performing marine geological investigations have been tremendously extended by the development of instruments and devices of various types. Among these are improved drill coring equipments, more precise echo sounders, magnetometers, gravity meters, navigational aids, the use of submersible research vessels, and the new techniques of chemical, mineralogical, and biological analysis applied to the study of sediments.

Biological Oceanography An understanding of all aspects of living organisms in the marine environment is generally accepted as the concern of the biological oceanographer. Primarily, the work is strongly oriented toward the compositions and ecological relations of marine organisms, how varying oceanic environments affect their distributions, evolution, habits, life processes, structure, and how they impact upon this environment. Studies of the OCEAN FOOD WEB or how ocean organic matter is formed, synthesized, and progressively ingested in the food chain process; determinations of the horizontal and vertical patterns of phytoplankton, zooplankton, fishes, and bottom-dwelling organisms; the dynamics of the production of all such life, as well as the causal factors related to it, are a few representative endeavors. (See FISH; PLANKTON.) Recent advances in biological oceanography include the development and use of microtechniques for analyzing minute amounts of biochemical material present in the seawater environment, the use of radioactive carbon for the *in situ* study of PHOTOSYNTHESIS, biological observations made *in situ* by divers, by underwater photographers, and by scientists in submersibles, some advances in plankton net sampling and collection techniques, and the use of improved underwater sound equipment.

Meteorological Oceanography Truly an interdisciplinary field, meteorological oceanography draws heavily upon the basic concepts and techniques of all the other aspects of oceanography coupled with meteorology. The world's oceans play a significant part in determing the world's weather and CLIMATE. Thus, the meteorological oceanographer is particularly interested in investigating the interacting phenomena where seawater and the atmosphere meet and exchange heat and water vapor. Also of great importance are studies of wind-driven oceanic circulation patterns, the relationship of sea-surface temperature and pressure irregularities to storms and weather changes, energy exchange processes and the several other processes and perturbations that are causal factors in weather and climate change.

Perhaps the most marked improvement that has been made to increase the capabilities of meteorological oceanographic work has been the employment of satellite and aircraft remote sensing devices. These equipments that measure ocean properties from a distance cover much wider areas than is economically possible by surface ships. (See INSTRUMENTATION.) Also aerosols are collected from immediately above the ocean's surface and at a variety of heights up to several thousand feet (1 ft = 0.3 m) are analyzed by sophisticated techniques for trace elements, HALOGENS, and organic compounds to measure fractionation between the air and the sea. This analysis assists in the understanding of processes of exchange between the ocean and the atmosphere. In addition, organic films have been studied to determine the part they may play in the formation and breaking of the bubbles at the sea surface and the injection of material into the atmosphere. It is very probable that surface films have an effect on the processes of evaporation, momentum exchange, and radiative tranfer at air-sea boundaries.

Some Major Events and Contributions to Oceanography

Seventh century B.C.

The Phoenicians as great sea traders navigated the Mediterranean and established thriving colonies in that region.

611–547 B.C. Anaximander, the philosopher, produced a map, probably the first Greek map of the known world.

Fourth Century B.C.

384–322 B.C. Aristotle, the pupil of Plato and tutor of Alexander, published a series of treatises (e.g., *Problema, Meteorologica,* and *Mechanica*) on natural and human philosophy. In these treatises, various species of marine animals were described and grouped by body form into various classes.

Third Century B.C.

ca. 240 B.C. Eratosthenes, the astronomer and geographer, estimated the circumference of Earth to be approximately 25 000 mi (40 000 km).

287–212 B.C. Archimedes, the physicist, evolved the principles of hydrostatics.

First Century B.C.

ca. 100 B.C. Poseidonius measured a 6000-foot depth of the ocean off Sardinia.

Second Century

ca. 150 A.D. Ptolemy (Claudius Ptolemaeus) in his *Geography* produced a factual map of the known world and depicted the earth as a sphere.

Eleventh Century

ca. 1000 A.D. The Vikings sailed to Iceland, Greenland, and North America.

Fifteenth Century

1416 Prince Henry, the Navigator, of Portugal founded a school of navigation.

ca. 1482 Leonardo Da Vinci wrote about the nature of water and fluid flow.

1492 Columbus sailed to America, and Vasco da Gama sailed around the Cape of Good Hope.

Sixteenth Century

1513 Ponce de Leon discovered and described the Gulf Stream.

1521 Ferdinand Magellan sailed to the Pacific Ocean where he attempted to make deep-sea soundings.

1569 Gerhardus Mercator, the Flemish geographer, produced unique nautical charts that facilitated both dead reckoning and the problems of rhumb line sailing.

1576–1587 Martin Frobisher and John Davis, two Englishmen, made several voyages (Frobisher: 1576, 1577 and 1578) (Davis: 1585, 1586, and 1587) in efforts to find a Northwest Passage.

Seventeenth Century

1609–1610 Henry Hudson, in attempting to find the Northwest Passage, sailed up the Hudson River and later to Hudson Bay. Galileo Galilei used the telescopic optical system, thereby opening up a new era in astronomy and navigation.

1616 William Baffin discovered Baffin Bay.

1642 Abel Janzoon Tasman, a Dutchman, circumnavigated Australia.

1663 Blaise Pascal enunciated the law of hydrostatics which states that a confined fluid transmits externally applied pressure uniformly in all directions, without change in magnitude. Isaac S. Vossius (Voss) discussed the circulation of the North Atlantic Ocean and theorized that it consisted of a general clockwise motion.

1686 Edmund Halley wrote "An Historical Account of the Trade Winds and Monsoons Observable in the Seas Between and near the Tropicks, with an Attempt to Assign the Phisical Cause of the said Winds." In this paper Halley attempted to relate primary wind systems to the main ocean currents.

Eighteenth Century

1725 Count Luigi Ferdinando Marsigli wrote *The Physical History of the Sea* in which he described the current system of the Bosporus.

1736 Carolus Linnaeus formulated the binomial system for classifying plants and animals.

1738–1740 Daniel Bernoulli published *Hydrodynamica,* a work dealing with the statics and dynamics of fluids. He also made the first study of the equlibrium tide.

1752 Bathymetric charts were used by Philippe Buache to illustrate the continuity between the topographic features of the land and ocean floor.

1768–1779 James Cook made two expeditions during this period, circumnavigating New Zealand and Antarctica, where he discovered the Sandwich Islands. On each voyage he collected an impressive amount of data about the oceans.

1770 Benjamin Franklin assembled data and published a paper on the Gulf Stream. Franklin's nautical-chart depiction of the best route for mail packet skippers to follow in crossing the Atlantic gave rise to the concept of thermometrical navigation.

1775 Marquis Pierre Simon de Laplace wrote a paper on tides and the hydrodynamic equations of motion.

1780 Horace Bénédict de Saussure measured the thermal gradient in the Mediterranean Sea.

1783 Thomas Earnshaw developed the first modern chronometer.

Nineteenth Century

1802 Nathaniel Bowditch published the book *New American Practical Navigator,* and Franz Joseph von Gerstner proposed the first theory of deep-water surface waves.

1802–1804 Baron Alexander von Humboldt accumulated data about the ocean off the west coast of South America and (in 1814) described the huge northward current that now is often referred to as the Humboldt Current although Peru Current is the presently accepted terminology.

1818 Sir John Ross invented a "deep-sea clamm" for collecting marine life from deep ocean depths; he also utilized a self-registering thermometer for measuring temperatures at deep depths.

1821 William C. Redfield published a work on hurricane movements in the western North Atlantic and made contributions to the study of tides and the prevailing ocean currents.

1828 J.R. Merian published a paper on seiches, stationary wave oscillations having a period varying from a few minutes to an hour or more but somewhat less than the tidal periods.

1835 Gaspard Gustave de Coriolis published a paper on fluid motions on a rotating earth.

1839 Sir James Clark Ross made deep (2677 fathoms) ocean depth measurements off the Cape of Good Hope.

1840 Sir Charles Wheatstone conceived and proposed the idea for the first telegraphic connection between England and France across the Straits of Dover.

1844 Alexander D. Bache, superintendent of the U.S. Coast and and Geodetic Survey, directed the investigation of the Gulf Stream.

1850 Michael and G.O. Sars, two Norwegians, collected deep-ocean marine life.

1855 Matthew Fontaine Maury published *The Physical Geography of the Sea,* a classic book on oceanography.

1856 William Ferrel published "An Essay on the Winds and the Currents of the Ocean" in which he described the distribution of wind-driven ocean currents and the effect of the earth's rotation on such currents.

1857 Lord Kelvin (Sir William Thomson) published "On the Alterations of Temperature Accompanying Changes of Pressure in Fluids," a work that treated the adiabatic temperature gradient in the deep ocean.

1865 Johann G. Forchhammer analyzed seawater samples from several geographical locations and discovered that the ratios of the constitutive ions varied but slightly from place to place.

1866 Nathaniel Bowditch wrote the *American Practical Navigator.*

1868 Sir Charles Wyville Thomson and W.B. Carpenter carried out oceanographic work on the British ships *Lightning* and *Porcupine,* and the temperature observations and deep-sea dredging results led to the *Challenger* expedition of 1872–1876, the expedition that marked the beginning of scientific oceanography.

1872 Anton Dohrn founded the first marine biological station in Naples, Italy.

1873 Louis Agassiz founded a marine biological station at Penikese in near Cape Cod in the United States.

1874 James Croll published papers on ocean currents in which he argued that the heat of the sun is converted to motion in the oceans.

1877 Alexander Agassiz carried out valuable oceanographic surveys with the *Blake,* a U.S. Coast survey ship, in the Caribbean Sea and the Gulf of Mexico. Later (between 1885 and 1889) Agassiz conducted surveys in these areas as well as in the Pacific with the *Albatross,* the first ship designed and built for oceanographic research.

1884 Wilhelm Dittmar demonstrated that the ratio of the principal ions in seawater is essentially constant so that it is possible to estimate the salt content by measuring one ion, usually chloride.

1886–1889 The Russian ship, *Vityaz,* commanded by Stepan Osipovich Makaroff, made a world cruise, taking measurements of the temperatures and specific gravity of seawater, chiefly in the North Pacific.

1889 The German Plankton Expedition carried out valuable investigations of marine biology.

1892 E.W.L. Holt, an Englishman, began systematic studies relating the pattern of commercial fish catches to a fundamental knowledge of fish biology.

1893 Fridtjof Nansen, a Norwegian explorer, drifted across the Arctic Ocean in the *Fram* and made oceanographic soundings and observations of the polar icecap.

Twentieth Century

1902 The International Council for the Exploration of the Sea was founded by countries of Northern Europe to assist in the handling of ocean fisheries by solving critical marine biological problems.

1903 Martin Knudsen published a paper in which the equilibrium temperature for ice and seawater at different salinities was tabulated.

1904 Vagn Walfied Ekman wrote a paper entitled, "On the Influence of the Earth's Rotation on Ocean Currents." This work outlined the concepts that are known as the Ekman spiral and the Ekman transport.

1905 Scripps Institution of Oceanography was founded in California.

1909 Andrija Mohorovičic discovered a major discontinuity in seismic-wave velocity beneath the outer layer of the crust of the earth in central Europe.

1912 Alfred Wegener theorized a continental drift postulate.

1920–1922 First cruises of the Danish ship *Dana* which resulted in important information about deep-sea animals.

1924 The German ship *Meteor* cruised the Atlantic Ocean, and the expedition made several important discoveries about the chemistry of the ocean.

1926 Harald Ulrik Sverdrup wrote *Dynamic of Tides on the North Siberian Shelf* in which he explained tidal behavior as a function of friction effects and the deflecting force of earth rotation.

1930 Woods Hole Oceanographic Institution was founded in Massachusetts.

1932 Albert Defant published a paper on the effects of the earth's rotation on free and forced internal waves.

1934 William Beebe and Otis Barton were lowered to over a 3000-ft (914 m) ocean depth in a bathysphere.

1937 G.E.R. Deacon wrote *The Hydrology of the Southern Ocean.*

1943 Jacques Yves Cousteau and Emile Gagnan

invented scuba (self-contained underwater breathing apparatus).

1946 Harry Hess described guyots, the flat-topped seamounts of the Pacific Ocean.

1946 Börje Kullenberg invented and tested the piston core sampler for taking 70-ft (21-m) long ocean-bottom sediments.

1947 Harold Urey developed a technique for determining the age of ocean sediments by measuring the oxygen isotope ratios.

1947 Harald U. Sverdrup and Walter H. Munk wrote *Sea and Swell: Theory of Relations for Forecasting*.

1947–1948 The Swedish ship *Albatross* made a round-the-world cruise and reported many marine geological findings.

1950–1951 The *Galathea,* a Danish ship, in a global ocean expedition, concentrated on deep-sea dredging and trawling work.

1951 Townsend Cromwell, the American geographer, discovered and studied a huge subsurface current in the Atlantic Ocean (the Cromwell current).

1952 Bruce C. Heezen and Maurice Ewing wrote a paper entitled "Turbidity Currents and Submarine Slumps, and the 1929 Grand Banks Earthquake."

1952 Roger Revelle and Arthur E. Maxwell determined that the heat transfer through the ocean floor is approximately equal to that through the continental crust.

1955 John C. Swallow invented a float with an acoustic signaling system for tracking the movements of deep open-ocean currents.

1957 The Swallow float, utilized by American and British oceanographers, was responsible for discovering a south-drifting subsurface current beneath the Gulf Stream.

1958 The American nuclear submarines *Nautilus* and *Skate* traveled under the North Pole collecting data on the Arctic Ocean and the icecap.

1960 Donald Walsh of the U.S. Navy and Jacques Piccard dove in the bathyscaphe *Trieste* to a depth of approximately 7 mi (11 km) in the Challenger Deep.

1959–1966 • Bruce Heezen, Marie Tharp, and Maurice Ewing, using data accumulated from many voyages, were able to trace out the 40 000-mile-(62 400-km-)long MID-OCEAN RIDGE.

• Harry H. Hess suggested the concept of seafloor spreading.

• F.J. Vine and S.T. Wilson concluded measurements near Vancouver Island, and James R. Heirtzler, W.C. Pitman, C.O. Dickson, and Xavier Le Pichon concluded measurements in the Pacific, Atlantic, and Indian oceans that confirmed seafloor spreading.

• Ronald G. Mason and Arthur Roff discovered that huge areas of the ocean floor were magnetized in long parallel bands so that polarity reversed from one band to the next.

• Allan Cox, G. Brent Dalrymple, and Richard R. Doell, using a radioactive dating technique, were able to demonstrate that the earth's magnetic field has reversed polarity several times in the past 3.6 million years.

• SEALAB 1, a U.S. Navy project, confirmed the viability of saturation diving as well as the ability of divers to perform useful work underwater on the continental shelves.

• International Cooperative Investigation of the tropical Atlantic with coordinated oceanographic surveys made by seven nations.

• The International Indian Ocean Expedition was a forerunner of the International Decade of Ocean Exploration (IDOE) program of the 1970s.

• The U.S. Government passed the Marine Resources and Engineering Development Act and created the National Sea Grant College Program.

1966–1979 • The National Oceanographic and Atmospheric Administration (NOAA) was established.

• The work of the drill ship, *Glomar Challenger,* had tremendous impact on virtually every aspect of oceanography, and huge advances were made in the better understanding of continental drift, submarine geology, and drilling operations at sea.

• International Decade of Ocean Exploration (IDOE), a major ocean research program, was responsible for many discoveries that fundamentally altered several previously held views of ocean phenomena.

• SEASAT-A, the first satellite dedicated to ocean environmental monitoring, was launched by the National Aeronautics and Space Administration (NASA). Although short (90 days), the mission successfully completed many projects, and the data collected provided a proof of concept demonstration.

OCTOPUS is the name applied to any member of the 140 species of cephalopod mollusks of the genus *Octopus* (family Octopodidae); the body of the animal is round with a large head and eight partially

OCTOPUS

A common octopus unfurls its eight tentacles, each of which is equipped with a double row of suckers. The baglike body has an opening on the underside where the gills are located. (*Bureau of Sport Fisheries & Wildlife*)

The fins resemble ears on the cirrate octopus, but are a great aid in swimming. This specimen, one meter long, was photographed in the Cayman Trough. (*Woods Hole Oceanographic Institution*)

webbed arms having one to three rows of disk-shaped suckers. (See MOLLUSK.)

Octopuses are found in all the world's oceans, although their numbers are sparse in the polar regions. They are usually found in natural caves and holes dug beneath rock formations on or near the bottom in coastal areas (from the intertidal zone to depths of approximately 600 ft (183 m). However, octopuses are also found in deep waters; in 1972, a photograph taken by the U.S. Navy at a depth of more than 3 mi (4828 m) in waters some 300 mi northeast of Barbados clearly showed an 18 in (46 cm) octopus (probable species: *Cirrothauma murrayi*). Since the animal has no bones and its tissues are permeated with fluid, it has the ability to withstand tremendous pressures [in excess of 3 tons · in^{-2} (420 kg · cm^{-2}) at this depth]. It also has the ability to squeeze into small openings, and often sunken ships, barrels, jugs, cans, tires, and bottles which lie on the ocean floor furnish environmental niches for its seclusion and protection.

The octopus is a predatory carnivorous animal that feeds especially on CRAB, LOBSTER, CLAMS, and FISH. Its strong-beaked jaws are used to crush this food, some of which (such as clams) it digs by using a powerful water jet that cleans away sand and mud from the bottom and exposes the prey.

Having no external shell like other mollusks, the octopus employs various protective devices against its predators. First, it is highly mobile. While, generally, the octopus moves by crawling lightly along the bottom using its arms, in times of danger, it employs a form of jet propulsion. During the breathing process, water is introduced into its body cavity through a mantle opening. Then, this water is expelled through the siphon or swimming funnel along with an ink cloud, resulting both in considerable thrust and in the action of the ink to confuse and blind the predator. Moreover, the octopus, in some situations, will change color from black to a mottled white. This chromatophoric change makes the animal appear larger and more formidable to its potential attacker.

The body is round with a large head and eight arms. Each arm has from one to three (usually two) rows of disk-shaped suckers. The male possesses a specialized arm, the hectocotylus, which is used to transfer spermatophores to the female.

The life expectancy is about 4–5 years, and both male and female die shortly after breeding. However, the female outlasts the male and broods the eggs for several months until they hatch; she then dies.

Size varies with species, but most weigh less than 70 lb (31.5 kg). The giant Pacific octopus *Octopus dofleini,* and the *O. apollyon* of the Pacific coast, commonly weigh 100 lb (45 kg) or more, while the *O. bairdi* and the *O. vulgaris* are smaller species. The latter is considered a food delicacy in parts of Europe around the Mediterranean Sea.

While the octopus has long been feared as a result of many fanciful stories about its attacks on divers, the animal is by nature rather timid and docile, and records of unprovoked attacks on humans are rare. Of course, this is not to say that the octopus will not attempt to defend itself if molested. One small octopus, the blue ringed octopus, *Hapalochlalna maculosa* common on the Australian reefs, is venomous to humans, and there are cases on record of fatalities caused by the bite of this animal.

Curiously, the eye of the octopus is fully as complex as the human eye. It is a camera-type organ with a light-sensitive retina. Moreover, the chromatophores (saclike structures of pigment) that are used to change the animal's color are controlled by its nervous system rather than by hormonal changes, as is the case with other animals having the ability to change their color. The normal bluish color of its blood is due to the respiratory pigment, hemocyanin, which contains a copper-base material. When the blood circulates, oxygen is consumed, causing the octopus to take on a colorless appearance. When the blood reaches the gills, it takes on oxygen and reverts to its normal bluish color. Finally, the composition of the suckers of the octopuses' arms is such that their feel and discrimination are extremely sensitive. These highly developed sensory qualities are of scientific interest.

OIL SPILLS are caused by the accidental or by-design dumping of oil into the ocean.

It is estimated that in recent times about 1 million (10^6) tons (0.9 million metric tons) of oil a year has been dumped into the world's oceans in standard tanker operations such as tank cleaning and deballasting. In addition, many hundreds of thousands of tons of oil are spilled yearly as a result of tanker casualties. The decade 1967–1977 suffered a rash of spills from wrecked tankers.

- 1967: *Torrey Canyon* ran aground off Lands End, England—820 000 barrels spilled
- 1972: *Red Sea Star* had a collision in the Gulf of Oman—800 000 barrels spilled
- 1976: *Urquiola* ran aground off La Coruna, Spain—about 700 000 barrels spilled
- 1976: *Argo Merchant* broke up off Nantucket Island in the Atlantic Ocean—180 000 barrels spilled
- 1977: *Hawaiian Patriot* burned and exploded in the North Pacific west of Hawaii—670 000 barrels spilled

In March 1978, the tanker *Amoco Cadiz* ran aground

OIL SPILL (*Top left*) Oil spreads from the tanker *Argo Merchant* grounded off Nantucket Island. (*EPA*) (*Bottom left*) An ecology study group captured this view of an oil-soaked shore between San Francisco and Big Sur. (*U.S. Navy*) (*Top right*) A small vessel attempts to approach a larger one for refueling during rough seas. Such a situation can result in the spillage of oil into the waters. (*U.S. Navy*) (*Bottom right*) Oil and oil wastes carelessly or accidentally dumped on the high seas and elsewhere create a problem for birdlife. This oil-soaked gannet on a beach at Nags Head, North Carolina, could not fly and soon died. (*Bureau of Sport Fisheries & Wildlife*)

off the coast of northwestern France, spilling over a million barrels of crude oil. In June 1979, a blowout of an offshore well in the Gulf of Mexico leaked over two million barrels of light crude oil into the sea. (In contrast, the 1969 blowout of an offshore oil well near Santa Barbara, California involved a leak of 6000 barrels.) Then, in July 1979, two VLCC (tankers of more than 200 000 tons), *Aegean Captain* and *Atlantic Empress*, collided off the coasts of Tobago and Venezuela, in a region where the Atlantic Ocean meets the Caribbean Sea. This disaster released about two million barrels of oil into these waters.

Such OCEAN POLLUTION must be dealt with on a global scale as a threat to fisheries, recreation sites, environmental values, and human health. Little is known about the myriad effects caused by such pollution. Meanwhile, the problems are intensifying because populations are expanding, and oceans are being increasingly used for transporting fuel and minerals, for seabed oil and gas drilling, and for seabed mining.

A 25-ship fleet of the Commerce Department's National Oceanic and Atmospheric Administration (NOAA) monitors slicks and other oil pollutants as part of a United Nations program called Integrated Global Ocean Station System (IGOSS), a worldwide pilot program designed to pave the way for the monitoring of marine pollutants on a global scale.

In the case where an oil tanker cleans out the oily ballast water by illegally dumping it into the ocean, the U.S. Coast Guard has developed an "oil fingerprinting" system. This involves the use of infrared spectroscopy techniques for measuring an oil's absorption of this type of radiation. Since each oil cargo varies, the spill of oil can be matched with the oil delivered by a particular tanker.

Several methods have been used, as well as proposed, for coping directly with oil spills. However, the technology is still in its infancy but is growing fast, and it must deal with some 10 000 incidents a year in the United States alone. One method uses polyurethane foam chips to absorb up to 50 000 gal (190 000 L) of oil per hour. The chips are than gathered and the oil is removed. The object is to capture the oil before it reaches the shore since, once it is there, about the only thing that can be done is to haul away the affected sand.

Other methods include the burning off of large patches of the oil slick and the use of oil-water separators, detergents and chemical dispersants, and skimmers.

OKHOTSK, SEA OF (Sea of Okhotsk), is one of the marginal seas on the northern rim of the PACIFIC OCEAN. It is separated from the ocean by the Kuril Island chain, and from the BERING SEA by the Kamchatka Peninsula. The southwestern boundary is the Sea of Japan. (See JAPAN SEA.) On the southeast the boundary is a line running from Cape Noshap (Nosyappu Saki) on the island of Yezo (Hokusyu), through the Kuril Islands, to Cape Lopatka on the Kamchatka Peninsula in such a way that all narrow waters between Yezo and Cape Lopatka are included in the Sea. The Okhotsk Sea covers an area of 613 740 mi² (1 590 000 km²), has an average depth of 2818 ft (859 m), and a volume of 327 477 mi³ (1 365 000 km³). The maximum depth is 11 069 ft (3374 m).

The Sea of Okhotsk is flanked on the west, north, and east by the barren, rugged, and sparsely populated mountains of the East Siberian Highlands. It was here, at the tiny settlement of Okhotsk on the northern coast, that Vitus Bering, in 1728, built the boat which took him eastward across the Sea on his way to the discovery of the Bering Sea and Bering Strait, and the realization that the Siberian and North American landmasses were not connected. The rivers entering the Sea from this borderland, such as the Amur, Inya, Kukhtuy, Okhota, Uda, Ulbeya, Uliya, Togur, etc., discharge around 140 mi³ (586 km³) of fresh water into the Sea annually. Because of its position in the very high latitudes, the climate is harsh and similar to that of the seas of the ARCTIC OCEAN. Mean air temperatures range from (21.2° F) (−6° C) in the winter to 64.4° F (18° C) in summer. In the vicinity of Okhotsk the winter temperature often drops below −13° F (−25° C). Surface water temperatures range from 28.8° F (−1.8° C) in winter to 64.4° F (18° C) in summer. Storms and fog are common, and an ice cover exists for 6–7 months of the year.

Water enters the Sea of Okhotsk through the La Pérouse (Soya) Strait from the Sea of Japan, and through several Kuril Island straits (i.e., Bussol, Friza, Krusenstern, etc.) from the Pacific Ocean. Under the influence of the prevailing wind pattern in the area, the surface current of the sea travels in a cyclonic (counterclockwise) direction at speeds of 0.8–3.9 in/s (2–10 cm/s). The water entering from the Sea of Japan through the La Pérouse Strait produces a strong 20–35 in/s (50–90 cm/s) nearshore current known as the Soya Current.

The CONTINENTAL SHELF of the Sea of Okhotsk covers approximately 40 percent of the total area, with widths varying from 31 mi (50 km) in the vicinity of Sakhalin to 228 mi (367 km) in the north. The depth of the edge of the continental shelf ranges from 328 to 1148 ft (100 to 350 m). From the shelf edge to the north, east, and west, the bottom slopes from 2–20° toward the relatively small ABYSSAL PLAIN, the Kuril Deep, just inside the Kuril Island chain. Here depths in excess of 9842 ft (3000 m) are found, with the maximum yet measured being

11 069 ft (3374 m). The sediment over much of the shelf and slope is composed of coarse sands and gravels trending to silts and clays toward the base of the slope, and to diatomaceous oozes on the floor of the Kuril Deep.

The Sea of Okhotsk supports an abundant plant and animal population. Phytoplankton and Zooplankton are plentiful, and some 300 species of FISH, 30 of which (i.e., SALMON) are of commercial value, are to be found. (See PLANKTON.) The Kamchatka CRAB and the blue crab are also prized.

OSMOSIS is the diffusion through a semipermeable membrane separating two solutions which tends to equalize the concentration of the solutions.

When fresh or pure water and SEAWATER are on opposite sides of a semipermeable membrane, the pure water diffuses through the membrane and dilutes the salt solution. This phenomenon is known as osmosis. The pure water flows through the membrane as though a PRESSURE were being applied to it. The effective driving force causing the flow is called osmotic pressure. The magnitude of the osmotic pressure depends on the characteristics of the membrane, the TEMPERATURE of the water, and the SALINITY of the solution.

The salinity of the waters of the world's oceans is a matter of vital importance to the organisms that live in it. While they are made buoyant by it, they are also presented with a physiological problem, since in many forms of marine life there is a difference between the salt content of the ambient ocean and that of their body fluids.

Biological membranes are selectively permeable (or semipermeable), and water and some ions and molecules flow through them from a zone of low concentration of salt to a zone of high concentration. In ocean fish the salt content of the blood and body fluids is about 30–50 percent of the ocean's salinity. Thus, by osmosis, they receive more salt than they require. This problem is solved by a kidney function whereby small quantities of urine containing a very high concentration of salt are excreted.

In fish such as the SHARK and RAYS, the salt content of the blood is about the same as that of seawater. This is true of many of the invertebrate forms (e.g., SPONGES, JELLYFISH, ECHINODERM, SEA ANEMONE, etc.) in which the gastral cavity is in direct connection with the seawater.

In freshwater fish, the salt concentration in the blood and body fluids is higher than that of the surrounding water. Therefore, the kidney of such a fish works to maintain the salt content in the body while eliminating large amounts of urine having the lowest possible salt content.

In some DESALINATION work the osmosis process is reversed by exerting a pressure on a salt solution or seawater. When this pressure is greater than the osmotic pressure, fresh water diffuses through the membrane of the cell in a direction opposite to normal osmotic flow. In this reverse osmosis, the seawater or saline water is first pumped through a filter where the solid particles that would damage the membranes are removed. The salt water is then raised to operating pressure by a second pump and then introduced into the desalination unit. When desired, some of the brine may be mixed with incoming saline water and recirculated.

OSTRACODS are small [mostly between 0.039 and 0.078 in (1–2 mm) long] bivalved CRUSTACEANS of the subclass Ostracoda. These animals, of which there are some 2000 species, inhabit aquatic environments in nearly all parts of the world. In the oceans they are found from the nearshore shallow waters to the great depths. Some are capable of swimming, and others crawl along the ocean floor. A few of the marine ostracods live commensally (see SYMBIOSIS) on other animals such as ISOPODS, SHARKS, and RAYS. Most ostracods are scavengers, some are herbivorous, and a few are carnivorous. The fossils of these animals are of value in geology (especially petroleum geology) by providing important clues to the age of marine deposits.

OXYGEN CYCLE is the exchange process occurring in the biosphere, whereby oxygen, an essential ingredient for life support on this planet, is used in respiration, combustion, and other oxidation processes, but is also constantly replenished and furnished to the atmosphere.

Oxygen constitutes roughly 21 percent of the air by volume. This concentration has been found to be the same at any level (64 km). Practically all chemical elements, except the inert gases, readily form compounds with oxygen. In SEAWATER, oxygen is chemically bonded with hydrogen. In the Earth's crust, most of the oxygen is combined in the form of silicates, oxides, and water, and this water is composed of about 89 percent oxygen by weight.

Most oxygen in the world's oceans enters from the atmosphere through the air-water interface. The rate of this exchange is rather slow, and the concentration of dissolved gas has traditionally been measured by the Winkler method. The concentrations are governed by solubility coefficients that are essentially functions of TEMPERATURE, PRESSURE, and SALINITY. Thus, the content of oxygen in the oceans can vary on a seasonal as well as geographical basis.

The exchange is assisted by winds, high mixing, and boundary-layer diffusion. Generally speaking, a layer of water, well-mixed by the wind and other ocean processes, exists at the ocean surface where the dissolved oxygen concentration is richest and fairly uniform. However, UPWELLING from the depths below can result in either supersaturation or undersaturation of the surface. Immediately below the surface layer and extending to about 265 ft (80 m) is the euphotic or photosynthetic zone, so-called because of the abundance of sunlight for the photosynthetic processes of PLANKTON. Oxygen supplied to the seawater via this method augments the atmospheric exchange. Oxygen is also lost into the atmosphere at the surface and is also used up in respiration by animals and in the decomposition of organic materials by BACTERIA.

In general, the concentrations of oxygen in seawater at depths of 330 ft (100 m) more or less varies from 4 to 6 parts per thousand (ppt). Below these depths the oxygen content usually tapers off in the THERMOCLINE, chiefly because of the oxidation of ocean organic matter, and reaches a minimum between approximately 1300–3300 ft (400 and 1000 m) in those parts of the ocean from about 20°N to 20°S LATITUDE. Below this, oxygen content often increases linearly with depth. Such an increase is due primarily to circulatory ocean processes that tend to continually renew the oxygen from place to place by the infusion of new oxygen-rich waters. However, most deep ocean areas are poorest in dissolved oxygen, and there are areas in which the oxygen is entirely depleted—the anoxic zones. Oxidation of both organic and inorganic matter may still take place by such oxidizing agents as nitrate, sulfate, and possibly carbonate in areas where elemental oxygen is lacking.

The distribution of marine organisms in the oceans, and a major factor in the survival for many of them, depends on the total level of dissolved oxygen in seawater or the aggregate effect of its depletion and renewal.

OYASHIO CURRENT (KURIL CURRENT) (Japanese for "parent current") originates in the BERING SEA and flows, initially, southwest. As it passes the Kamchatka Peninsula, it entrains WATER from the Sea of Okhotsk and flows on to an intersection with the eastward-flowing KUROSHIO CURRENT in the vicinity of 40° N latitude. (See OKHOTSK, SEA OF.) Part of the Oyashio, being colder and denser, plunges beneath the warmer Kuroshio water, while the remainder turns east to form a cold northern boundary for the Kuroshio. Being heavily laden with nutrients and PLANKTON, the Oyashio water provides a brownish contrast to the beautiful aquamarine of the relatively sterile Kuroshio. The temperature of the Oyashio water is about 39.2–41° F (4–5° C), and its mass transport in the vicinity of the Kuril Islands is about 530 million (530×10^6) ft^3/s (15 million m^3/s.).

OYSTER may be any of the various bivalve mollusks of the family Ostreidae. See BIVALVIA; CLAMS; MOLLUSK.

PACIFIC OCEAN, the largest of the world's oceans, covers 32 percent of the earth's surface, comprises 46 percent of the area of the world's oceans and seas, and occupies an area larger than that of all the land area of the world combined. Clockwise, the Pacific is bordered by Antarctica, Australia, the Indonesian Archipelago, the Malay Peninsula, China, Siberia, and North and South America. The southwestern border is a matter of some controversy, but for most authorities it begins in the STRAIT OF MALACCA along the meridian of Singapore, runs eastward through Sumatra, Java, Roti, and Timor, and joins with Australia so as to include the TIMOR SEA, ARAFUR SEA, and the Gulf of Carpentaria. (Some authorities include some or all of these water bodies in the INDIAN OCEAN). From here the boundary runs along the east coast of Australia to Bass Strait where some authorities place the boundary at the western and some at the eastern extremity of the strait. Below Tasmania the western boundary lies along 147°E. The similar boundary on the east is usually taken as the shortest distance between Cape Horn and the Antarctic Peninsula (68°04′W), but some authorities prefer to follow the outline of the Scotia Arc. In this case the SCOTIA SEA would be included in the Pacific rather than the ATLANTIC OCEAN. The northern border is usually taken as the LATITUDE of the ARCTIC CIRCLE in the Bering Strait, although some prefer to use the east-west line across the narrowest point of that strait. The southern border is Antarctica, unless the SOUTHERN OCEAN is recognized as a separate ocean. In that case, the boundary is 55°S latitude (some authorities prefer 60°S). For convenience, one discusses a North and South Pacific, and the equator is considered the dividing line.

The Pacific waters include a number of marginal seas, such as the Bering, Okhotsk, Japan, East China, South China, Java, Philippine, Flores, Bali, Banda, Sulawesi, Coral, Fiji, Arafura, Timor, Solomon, Halmahera, Sulu, Molucca, and Tasman. Major gulfs include the GULF OF CALIFORNIA, the Gulf of Alaska, the Gulf of Thailand, and the Gulf of Carpentaria. Including these marginal seas and gulfs, the Pacific covers an area of 69 364 200 mi² (179 700 000 km²); it also has a volume of 173 688 000 mi³ (723 700 000 km³), and a mean depth of 13 211 ft (4028 m). The greatest depth is 37 720 ft (11 500 m) in the Marianas Trench. The distance across the Pacific from Panama to the Gulf of Thailand is 10 500 nmi (24 000 km), and the distance from Bering Strait to Cape Adare in Antarctica is 9300 nmi (17 200 km) The Pacific drains a land area of 7 488 400 mi² (19 400 000 km²), or about 25 percent of that which drains into the Atlantic Ocean. Consequently, the Pacific receives water from the Atlantic by way of the Drake Passage and the ARCTIC OCEAN.

Within the open Pacific Ocean lie some 30 000 islands with a total land area of 170 000 mi² (440 130 m²), or about a quarter of 1 percent of the total area of the Pacific. The largest islands are Luzon and Mindanao in the Philippines with an area of about 40 000 mi² (103 560 km²) each. Next in size are New Caledonia (6200 mi² or 16 051 km²), Samar (5000 mi² or 12 945 km²), Viti Lever (4053 mi² or 10 493 km²), and Hawaii (4030 mi² or 10 433 km²). Virtually all of these islands are volcanic in origin, rising as great cones from the deep seafloor. It is also interesting to note that these islands are all being carried along by the Pacific plate as part of the grand pattern of global CONTINENTAL DRIFT.

Early Exploration Balboa was impressed by the placid surface of the broad ocean he first set eyes on after he crossed the Isthmus of Panama from the east in 1513 and therefore gave the name "Pacific"

PACIFIC OCEAN The surf pounds the beach at Onimak Island, Alaska. *(U.S. Coast Guard)*

(peaceful) to the world's largest ocean. Balboa, of course, was not the first to sight the Pacific. There were people living along its western rim, and evidence indicates that those who colonized North and South America moved eastward by means of land bridges (and, perhaps, primitive boats) across the region of the Bering Strait some 30 000 years ago. In doing so they must have gained some familiarity with the Sea of Okhotsk, the BERING SEA, and the Gulf of Alaska (see OKHOTSK, SEA OF.)Evidence also indicates that Japan was colonized from China and Korea in much the same manner and at roughly the same time—during the low stand of SEA LEVEL that accompanied the last Ice Age.

At some indefinite time in the past the Negritos (so named by the Spaniards because of their dark skin and diminutive stature) began moving northward until they eventually populated the Philippine Archipelago. This colonization was complete by the time that Malayan explorers and settlers reached the 3141-island archipelago around 3000 B.C.The long Pacific voyage of the Polynesians, who eventually settled as far east as Hawaii, began some 2000 years ago. And, of course, there is the fascinating speculation by such investigators as Thor Heyerdahl that the exploratory loop of the Pacific may have been closed by natives of South America who drifted westward on balsa rafts to such places as Easter Island and the Tuamotu Archipelago.

Not surprisingly, the twin objectives of trade and adventure greatly accelerated the exploration of the Pacific, beginning with the development of trade with India, then China and the East Indies. These trade routes across the northern waters of the Indian Ocean were initially established by the Phoenicians and the Arabs. The Arab trader, Soleyman, is known to have reached Canton in A.D. 850, and Marco Polo returned home from China via the SOUTH CHINA SEA and the Strait of Malacca in 1295. In 1509, Ferdinand Magellan crossed the Indian Ocean and sailed into the Pacific as far east as Ternate in the East Indies. And then, in the period from 1519 to 1521, he made the first circumnavigation of the globe by crossing into the Pacific through the Strait of Magellan (tip of South America). Magellan was killed in the Philippines and the rest of the voyage through the Pacific and Indian Oceans was completed by Del Cano.

With Magellan's east–west crossing of the Pacific, the pace of exploration quickened. Cortes reached the coast of Mexico in 1521, and returned to explore the Gulf of California in 1535–1536. Pizarro sailed the coast of Peru in 1531, and Alvaro de Saavedra sailed westward from Mexico to New Guinea in 1527–1528. On the other side of the Pacific, Francis Xavier reached Kwantung and Japan on

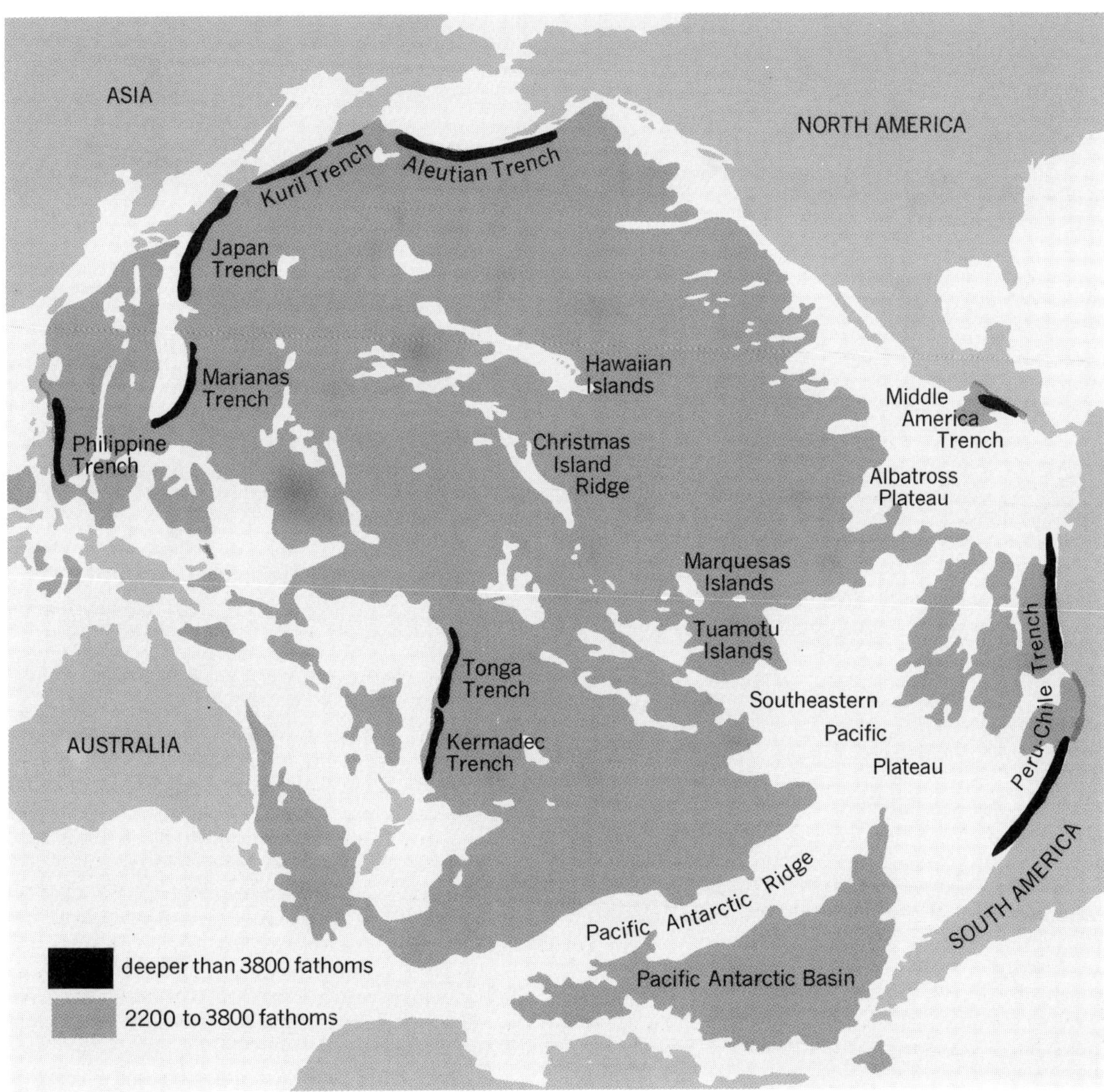

MAP 9 The Pacific Ocean basin contains many trenches

an expedition that lasted from 1540 to 1552. The middle regions of the Pacific were explored by Alvaro de Mendaña in 1567–1568. Mendaña visited the Ellice, Solomon, and Marshall Islands, and returned in 1595 to explore the Marquesas and the Santa Cruz Islands.

In a voyage that lasted from 1577 to 1580, Sir Francis Drake made the second circumnavigation of the globe. Crossing the Atlantic, Drake entered the Pacific through the Strait of Magellan, sailed as far north as Vancouver Island, and then returned home via the Philippines and the Indian Ocean. In 1605, Diego de Prado y Tovar and Luis Vaez de Torres sailed from Callao (Peru) to discover the Tuamotus and the New Hebrides Islands, probably sighting the Great Barrier Reef of Australia, and discovered the Torres Strait between Australia and New Guinea.

The era of Dutch exploration and colonization of the Pacific began in 1616 with the voyage of William Schouten and Jacques Le Maire. Sailing westward through the Atlantic they entered the Pacific via

Drake Passage, thus proving that Tierra del Fuego (tip of South America) was not connected to Antarctica. During their cruise in the Pacific they discovered and named the Schouten Islands off New Guinea. Schouten and Le Maire were followed by Abel Janszoon Tasman who, sailing eastward from Mauritius, discovered Tasmania off the tip of Australia, and sailed on to New Zealand, Three Kings Islands, Tonga, Fiji, and New Guinea before returning to Java. In turn, Tasman was followed by Jacob Roggeveen who rounded Cape Horn to discover Easter Island.

The gross exploration of the Pacific was completed by the great English explorer James Cook during the course of three voyages which took place between 1768 and 1776. (see COOK, JAMES). Cook circumnavigated Antarctica, and though ice prevented him from sighting the continent, he did prove that the legendary "Southern Continent" was much smaller than anyone supposed. He became the first man to cross both the ARCTIC CIRCLE and the ANTARCTIC CIRCLE when, on his last voyage, he passed through the Bering Strait into the CHUKCHI SEA. From his favorite staging area in Tahiti, Cook explored New Zealand and discovered Cook Inlet (between the two major islands), Australia, and Hawaii. Cook was killed by natives on Hawaii on the return voyage from the Chukchi Sea, but his map of the Pacific served for many years to guide those who followed him.

Winds and Currents The low-altitude wind pattern over the Pacific consists of four great belts—two to either side of the equator. The equator itself is an area of slack air and doldrums where the heat causes the air to rise and flow off toward the poles at high altitude. The rising air is replaced by the northeast and southeast trade winds which blow steadily (east to west) throughout most of the year in broad belts that extend to about 25°N and S of the equator. During the age of sail, the cool northeast trades were used to sail from North America to Asia.

To the north and south of the trades—in the LATITUDE of 30–60°N and S—are the westerlies, the northern belt of which was used by sailing ships for the return trip from Asia to North America. In the southern belt the westerlies often blow with such intensity that they have earned the names "roaring forties" and "screaming fifties." The numbers refer to the latitudinal zones in which the severe winds are most frequently found. Storms in these regions often create long-period WAVES that ultimately provide excellent surfing conditions along the California coast.

In the western Pacific during the summer months, moisture-laden winds, called monsoons, blow onto the coast of Asia bringing the rainy seasons for which that area is famous. In winter, these winds reverse and blow out to sea as "dry monsoons."

The Pacific is famous for its typhoons (see HURRICANE) or tropical cyclones. These destructive winds strike in the region of the South China Sea between May and November, along the coast of Mexico and Central America between August and October, and in the North Pacific region between January and March.

The weather in the extreme northern and southern Pacific is largely controlled by the polar regions in which the high-altitude air from the equatorial region is caused to sink and flow seaward. During the winter months (Southern Hemisphere) Antarctica sheds mammoth icebergs (see SEA ICE) which drift northward into the shipping lanes of the South Pacific.

The prevailing wind patterns, the CORIOLIS EFFECT, and the continental barriers act to shape the major currents of the Pacific into two giant gyres that circle the basin in a clockwise direction in the North Pacific and counterclockwise in the South Pacific. The equator divides the two gyres. The northern gyre begins as the westward-flowing North Equatorial Current (see EQUATORIAL CURRENT SYSTEM) which forms up off Central America. In the PHILIPPINE SEA the current turns north to become the swift KUROSHIO CURRENT that flows along the east coast of Japan before gradually turning eastward to flow back across the Pacific as the North Pacific Current. Approaching North America, part of the North Pacific Current turns northward into the Gulf of Alaska as the Alaska Current, while the majority of the flow swings southward to flow along the west coast of the United States and Mexico as the CALIFORNIA CURRENT. Off the coast of Mexico the California Current turns westward to join the North Equatorial Current and complete the gyre.

It has been estimated that within the upper 1500 ft (500 m) of the Pacific the Kuroshio Current flows at the rate of 65 000 000 m³/s, the California Current at the rate of 15 000 000 m³/s, and the North Equatorial Current at about 45 000 000 m³/s. The swiftest of these currents, the Kuroshio, flows at a speed of 2 kn.

The southern, counterclockwise gyre forms up off Ecuador to flow west along the southern edge of the equatorial region as the South Equatorial Current (see EQUATORIAL CURRENT SYSTEM). In the vicinity of the Solomon Islands and the CORAL SEA most of the water turns south to flow along the coast of Australia as the EAST AUSTRALIAN CURRENT before turning east to recross the Pacific as the WEST WIND DRIFT CURRENT. Reaching the coast of South America, a part of the current continues on through the Drake Passage while the rest turns northward as the PERU (Humbolt) CURRENT. Off Ecuador the Peru Current swings west to join the South Equatorial Current and complete the South Pacific gyre.

Between the two westward-flowing equatorial currents—in a belt between 5° and 10°N latitude—is the EQUATORIAL COUNTERCURRENT which flows eastward at a rate of about 25 000 000 m^3/s. Further south, beneath the equator itself, is another eastward-flowing current called the Equatorial Undercurrent or the CROMWELL CURRENT. The core of this current is at a depth of about 300 ft (100 m).

Occasionally, the trade winds slacken and blow fitfully or not at all. Under these conditions the Equatorial Countercurrent may wander southward to spread a layer of warm, low-salinity water over the rich fishing grounds off Peru. At this time, a phenomenon called EL NIÑO occurs; the PLANKTON on which the anchovies feed are killed, and the fish migrate to better feeding grounds. The MARINE BIRDS are left to starve by the thousands.

The surface currents, of course, are not the only currents, only the more obvious ones. At depth, sluggish density currents spread out from areas of DOWNWELLING caused by surface water achieving greater density. The more prominent areas are those around Antarctica where the water is chilled, thereby increasing its density. After sinking, this cold water spreads northward to form the bottom water of the greater part of the Pacific. The antarctic water is thought to spread as far north as Japan and the northeast Pacific.

Where water is piled up by the convergence of currents, sinking also occurs. Thus, water in the vicinity of the Pacific Tropical Convergence, which coincides with the Equatorial Countercurrent, sinks to a depth of about 300 ft (100 m) and spreads out laterally in a slow-moving current.

Water which sinks must be replaced. Much of this replacement is accomplished by UPWELLING which is well developed along the west coast of North and South America.

Tides In contrast to the Atlantic, the Pacific Ocean has both diurnal and mixed TIDES. With diurnal tides a single high water and a single low water is experienced each day (24 hours, 50 minutes). In mixed tides, a mixture of diurnal and semidiurnal tides is experienced. Examples of diurnal tides are to be found in the Gulf of Tonkin, Torres Strait, and Vancouver Island. Mixed tides are found along the Pacific coast of the United States.

Tidal ranges in the Pacific are small: those at Tahiti are about 1 ft (0.3 m); 2 ft (0.6 m) at Hawaii; and 5 ft (1.6 m) at Yokohama. Large tidal ranges of up to 40 ft (12 m) are only found in the Gulf of California and the Gulf of Korea.

Temperature and Salinity Between the equator and 30°N and S the heat received from the sun exceeds that lost by reflection and back-radiation. Therefore, the westward-flowing equatorial currents continually pick up heat during their long voyage across the mid-Pacific. This heat is carried into the colder regions of the Pacific by the Kuroshio Current in the north and by the East Australian Current in the south. Most of this heat, of course, has been lost by the time these currents have recrossed the Pacific as the North Pacific Current and the West Wind Drift Current. Temperatures of the surface water along the equator range up to 55.4° F (28° C), and steadily decline north and south of this zone. Because of the influence of Antarctica the South Pacific waters are somewhat colder. At 60°S the surface temperature has declined to around 35.6° F (2° C), whereas in the same latitude in the North Pacific the temperature is around 46.4° F (8° C).

The highest surface salinities in the Pacific occur toward the center of the current gyres in the North and South Pacific. There, evaporation exceeds precipitation, and the current and wind patterns tend to confine the surface water to these regions. In both regions salinities greater than 35 ppt can be found. Toward the northern and southern rims of the Pacific basin the salinity drops to less than 34 ppt. Low salinities are also found along the North and South American coasts. Low salinities (less than 34 ppt) are also found along the equator because of the heavy rainfall which tends to dilute the surface water and the frequent cloud cover which tends to reduce evaporation.

The Basin The Pacific Basin, in addition to being the largest of the ocean basins, is unique in other ways. For instance, the Pacific Basin is shrinking as the Atlantic Basin grows by the westward migration of North and South America and the Indian Basin grows by the northward migration of Australia. Literally, the Pacific Basin is being consumed at local rates of up to 4 in (9 cm) per year by the encroachment of these continental plates (see CONTINENTAL DRIFT). As a result of these encircling stresses, the rim of the Pacific Basin is known as "the Pacific Ring of Fire" because it contains more volcanoes (both active and dormant) and earthquakes than any other area in the world. The floor of the basin also contains more SEAMOUNTS (approximately 1000) than all the other ocean basins combined. It is little wonder that such early investigators as George H. Darwin (1879) and Osmond Fisher (1889) believed that the moon was ripped from the Pacific Basin by the close passage of a large gravitational body, leaving the basin as a gigantic wound on the earth's surface.

The dominant feature of the floor of the eastern Pacific is the East Pacific Rise, a part of the 47 000-mi- (75 623-km-) long Mid-Ocean Ridge system that runs through all the world's oceans. The East Pacific Rise runs into the Pacific from the west between Australia and Antarctica, and then turns north to intersect North America by way of the Gulf of California. The San Andreas Fault, which runs back out to sea just south of San Francisco, is a land extension of the intersection of the Pacific and North American Plates. If the East Pacific Rise continues northward to the Gulf of Alaska, it is obscured by aprons of sediment that have spread out across the floor. One of the few abyssal plains of the Pacific, the Aleutian Abyssal Plain, exists in the southeastern corner of this region of sediment inflow (see ABYSSAL PLAIN).

The East Pacific Rise, which soars 6000–9000 ft (2000–3000 m) above the surrounding seafloor, is highly fractured by transform faults that radiate at right angles to the axis of the Rise. These faults, which appear to occur at intervals of roughly 200 mi (322 km), are occasionally of such prominence that they have been given names such as the Mendocino Fracture Zone of the northern California coast, and the Eltanin Fracture Zone off the east coast of New Zealand. These faults are quite long; the Clarion Fracture Zone, for instance, runs in a straight line from the coast of Mexico to just south of the Hawaiian Islands.

The topography and structure of the western Pacific Basin differs significantly from that of the eastern. For instance, the eastern basin has many seamounts, particularly off the coast of Mexico and California and in the region south of the Gulf of Alaska. In the western basin, however, the seamounts, including many that pierce the water surface as islands and active volcanoes, are far more numerous.

The most striking features of the western Pacific Basin are the TRENCHES which cleave the ocean floor in arcuate patterns (often connected) from the northern tip of New Zealand to the Kamchatka Peninsula on the eastern rim of the Sea of Okhotsk. Plunging to depths as great as 37 720 ft (11 500 m), these trenches represent the axis along which one giant plate is forced to plunge beneath another. In the western Pacific this subduction process produces a structure known as an *island arc.* The forces produced by the interaction of the two plates typically produce a double line of islands on the continental side, but parallel to the trench. The two belts are 30–90 mi (50–150 km) apart. The belt closest to the trench is nonvolcanic and often contains marine sediments that have been thrust up and folded by the subduction process. The inner belt is made up of a line of volcanoes, both active and dormant, which appear to be triggered by the melting at depth of the downthrust plate. Many scientists believe that this continuing process, combined with the northward migration of Australia, will eventually result in a significant eastward growth of the Asian continent.

Trenches also are evident along the west coasts of Central and South America. It has been suggested that radioactive waste might be safely disposed of in these trenches (and others) since, with time, the subduction process would carry the material deep within the earth's interior.

The sediments which cover the basaltic plates of the Pacific Basin are considered to be less thick than those of the other oceans. Seismic studies indicate that a 0.6-mi- (1-km-) thick layer of consolidated sediments is overlain by about 1000 ft (300 m) of unconsolidated sediment. However, deep-sea drilling off the California coast has found basalt under a layer of only 600 ft (200 m) of unconsolidated sediment. Sediment thickness also diminishes rapidly as one approaches the East Pacific Rise where new plate is being added. (See CONTINENTAL DRIFT).

The detailed distribution of sediment types in the Pacific is not well known. It has been estimated that the density of sediment samples in the Pacific, except around islands and continents, is similar to the result obtained by taking 50 to 60 random samples from across the continental United States. In general, however, the sediments are made up of either red clay, terrigenous materials, volcanic material, or the remains of such sea life as DIATOMS, radiolarians and globigerina (see MARINE SEDIMENTS). Diatom oozes are found in the northern and southern extremes of the Pacific, and a band of RADIOLARIAN OOZE is found just north of the equator. Most of the remainder of the deep basin is made up of red clay and GLOBIGERINA OOZE. Terrigenous sediments, of course, are found in abundance on the floor of the bordering seas and gulfs, and volcanic ash and rubble are found around the volcanic islands of the western Pacific.

All in all, the thickness of the sediment in the Pacific, like that in the other ocean basins, is less than it was supposed to be only a few years ago. (See CONTINENTAL DRIFT.)

See also MANGANESE NODULES.

PACK ICE See SEA ICE.

PALMÉN, ERIK HERBERT (1898–), a Finnish meteorologist and oceanographer, is noted for his work on the interaction between oceans and atmosphere. Oceanographic research prior to World War II led to an abiding interest in extratropical cyclones and the general structure of the atmosphere. Working with C. G. Rossby at the University of Chicago after the war, Palmén was a member of the team that determined the existence of a jet stream in the upper troposphere. Later Palmén showed that two principal jet streams exist in each hemisphere. These great "rivers of the wind," flowing at altitudes of 30–40 000 ft (9144–12 192 m), have a profound effect on global weather and on the planning of commercial air routes.

Through his work on tropical circulation Palmén became interested in the processes by which cyclones, hurricanes, and typhoons are formed and sustained. (See HURRICANE.) He showed why these storms only formed in certain seasons and in certain oceanographic areas. To aid in explaining these processes, he developed a model of the thermal structure of a hurricane and showed how this structure is related to the thermal structure of the tropical atmosphere and ocean temperature.

In recent years Palmén returned to his work on the general circulation of the atmosphere and its energy budget as a function of radiation, flux of sensible and latent heat from the earth's surface, and the mean meridional and vertical energy flux in the atmosphere.

Palmén, the son of a judge, was born in Vasa, Finland, on August 31, 1898. He attended the University of Helsinki, receiving his doctorate in 1927. From 1922 to 1946 he served with the Finnish Institute of Marine Research and was its director from 1939 to 1946. During much of this time he also taught geophysics at the University of Helsinki. In 1947–1948 he was professor of meteorology at the University, and in 1948 received a research professorship from the Academy of Finland. Palmén received the Symons Medal of the Royal Meteorological Society in 1957, the Rossby Medal of the American Meteorological Society in 1960, and the Buys Ballot Medal of the Royal Academy of Sciences of the Netherlands in 1964.

PANCAKE ICE See SEA ICE.

PARROTFISH is the name for some 300 species of fishes belonging to the family Scaridae in the order Perciformes. The parrotfishes are multicolored bony fishes, and they derive their name from the shape of their teeth. These teeth are joined together to form a parrot's beak in the front of the mouth. They inhabit tropical reef areas of the world's oceans. (See CORAL REEF.) The sizes of the adult parrotfishes vary in length on the average from 1 to 6 ft (30 cm to 1.8 m), although some attain a 12-ft (3.6-m) length. Their diet consists of EELGRASS, seaweeds, and CORAL. (See ALGAE.) Certain species of parrotfish (e.g., *Scarus guacamaia*) have the unique ability to secrete a mucous membrane at night which forms a loose shroudlike covering around them. This is open at the area around the mouth of the fish to permit water entry, and an opening in the back allows water to exit. The fish breaks out of this membrane at daylight.

PARTICULATE MATTER consists of the inorganic and biologically produced particles of substances contained in SEAWATER. Such particulate matter ranges in size from colloidal to those which will settle under gravity. (See COLLOID.) The sources of particulates in ocean water can be both external (the atmosphere, landmasses, marine sediments) and internal (those formed biologically as well as those produced by bubbling, precipitation, and coagulation). Fine particulates [particle size = 4×10^{-6} to 2×10^{-3} in (10^{-5} to 5×10^{-3} cm)] consist of mineral substances, DETRITUS, PLANKTON, BACTERIA, and other microorganisms, as well as precipitated and coagulated particles. "MARINE SNOW," for example, is a coagulate composed of phytoplankton and marine bacteria which adhere to nonliving organic matter suspended in the water. (See MARINE SNOW.)

The coarse dispersions [greater than 2×10^{-3} in (5×10^{-3} cm) in size] in seawater consist of coagulates, precipitates, detritus, mineral substances, macroplankton, and organisms of all types. The amount of particulate organic matter, consisting of dead and decomposing plants and animals, found in an area of the ocean reflects the amount of plant and animal activity present there and thus the productivity of that area.

The solid-liquid interfaces provided by particulate matter are the sites for a wide variety of chemical and biological interactions. Also, marine animals, especially in early larval stages, are able to feed on particulate organic matter. (See also LARVAE; BIOLOGICAL OCEANOGRAPHY.)

The combustion products of fossil fuels are also in particulate form. These solids, as well as the carbon dioxide effluent, exist in suspension in the atmosphere. They change the transparency of the atmosphere and, in so doing, change the heat balance. The solids scatter solar radiation, and the carbon dioxide absorbs outgoing infrared rays. Thus, if the concentration of the suspended materials becomes high enough, the earth will cool; and if the carbon dioxide accumulates, it will heat up. The climatic disturbances that could ensue from either phenomenon could seriously affect all life on the planet. While the oceans provide major sinks for the washout of solids and for the solution of carbon dioxide, the residence times of these heat absorbers in the atmosphere are still not well understood. See CARBON CYCLE; ECOLOGY.

PARTS PER THOUSAND (PPT) and PARTS PER MILLION (PPM) represent a method of expressing quantitatively the amounts of various constituents (either major or minor) contained especially in liquid solutions or gaseous mixtures. For example, the salt content, or SALINITY, of the world's oceans is normally expressed in terms of parts per thousand (ppt). This average value is 35 ppt, which can also be written as 35‰; although the salanity of ocean waters ranges from 33 to 37 ppt. (See SEAWATER.)

To place one part per million (ppm) in some perspective, we can say it is equal to

- One inch in 16 mi ($2\frac{1}{2}$ cm in $27\frac{1}{2}$ km)
- One minute in two years
- A 1-g needle in a ton of hay
- One penny in $10,000
- One ounce (28 g) in 62 500 lb (23 312 500 g)

PELAGIC FISH is the name applied to those fish that live primarily away from the ocean shores and in the upper water layers of the oceanic province. (See HABITABLE ZONES.) HERRING, TUNA, MACKEREL, SQUID, SHARK, MENHADEN, and RAYS are some examples of pelagic fish. Most are fast-moving migratory or free-swimming fishes.

Pelagic animals are usually divided into two classes: (1) the smaller forms that drift with the CURRENTS (zooplankton) and (2) the larger and more active NEKTON, or free-swimming animals (e.g., fishes), that are essentially independent of water movements. (See PLANKTON.) Pelagic, deep-sea animals are often termed bathypelagic in contrast to the epipelagic organisms of surface waters. Aside from a few squaloid sharks, the bathypelagic fishes consist of TELEOSTS. The most representative groups are the stomiatoids (Clupeiformes), Myctophidae (lantern fishes), hatchetfish (Myctophiformes), and the ceratioid angler fishes (Lophiiformes). (See DEEPSEA FISH AND ANIMALS; FISH.)

In the pelagic expanses of the open ocean, because of the greater rigor of life, a smaller number and fewer kinds of marine organisms can be found than are found in shallower waters such as on or near the CONTINENTAL SHELF and in the tidal or littoral zone.

PELECYPODS is the name used for the two-shelled mollusks or bivalves. See BIVALVIA; MOLLUSK; SEASHELLS.

PENGUINS is the general name for a separate order, Sphenisciformes, of flightless marine birds of the Southern Hemisphere. The order contains a single family, Spheniscidae, six genera—*Aptenodytes, Pygoscelis, Eudyptes, Megadyptes, Eudyptula,* and *Sphenicus*—and some 18 species.

Penguins are expert swimmers and divers. They propel themselves on the ocean water surface and underwater by utilizing their paddlelike wings as strong flippers and their short tails and webbed feet as rudders. Ashore, their movements are awkward although these animals can still travel at a rather fast pace. Most species feed principally on zooplankton, especially EUPHAUSIIDS, although a few concentrate their diet on FISH or SQUID. Penguins nest together in large rookeries or colonies. They are monogamous and pair for life, and the male takes an active part in rearing the young. Penguins are easy to study at their breeding sites and their breeding biology has become quite well known. The number of eggs can vary from one to three. Highly adapted to the environment of the Antarctic and sub-Antarctic regions, seven species of penguins occur in those areas, and three of these (i.e., the Adélie penguins) are confined primarily within the limits of the pack ice. They make up about 87 percent of the stocks of birds (particularly the penguins, albatrosses, and petrels) in the Antarctic and 53 percent of the stocks of birds in the sub-Antarctic.

The penguin genus *Aptenodytes* is found only in the Antarctic, and this genus includes the two largest species: The emperor penguin (*Aptenodytes forsteri*) and the king penguin (*A. patagonica*). The king penguin reaches a height of 3 ft (1 m) and weighs from 30 to 40 pounds (13.6–18.1 kg) while the emperor may weigh twice as much and stand over 4 feet (1.2 m) tall.

Pygoscelis is another genus found in the Antarctic region. This genus includes the Adélie penguin (*Pygoscelis adéliae*) and the gentoo penguin (*P. papua*). Both of these birds are about 2 ft (60.9 cm) tall.

Conspicuous members of the seven penguin species of the genus *Eudyptes* found in the New Zealand and sub-Antarctic area are the rockhopper penguins (*Eudyptes crestatus*), the macaroni penguins (*E.*

chrysolophus), and the Fiordland crested penguins (*E. pachyrhynchus*). These are birds of medium size—about 1 to 2 ft (30.5–60 cm) tall.

A sole species of the genus *Megadyptes* is the yellow-eyed penguin (*Megadyptes antipodes*). This bird lives in New Zealand.

Eudyptula is a genus characterized by a number of small species of New Zealand penguins such as the fairy penguin (*Eudyptula minor*) and the white-flippered penguin (*E. albosignata*).

The genus *Spheniscus* is made up of medium-sized species, such as Humboldt penguin (*Spheniscus humboldti*) and the magellanic penguin (*S. magellanicus*), which are found in South America. The jackass penguin (*S. demersus*) occurs off the western coast of South Africa, and the Galapagos penguin (*S. mendiculus*) is the only penguin restricted to the tropics.

See also MARINE BIRDS.

PERIWINKLES is a general name for some 80 species of small GASTROPODS in the family Littorinidae. Periwinkles are found widely distributed in temperate and cold ocean waters and are distinguished by their spiral, globular shells. See also LIMPET.

PERSIAN GULF, sometimes called the Gulf of Iran, is an arm of the INDIAN OCEAN that is nearly surrounded by Iraq and Kuwait on the northwest, Saudi Arabia, Qatar, and Oman on the west and south, and Iran on the east. Fresh water enters the gulf in the north by way of the Shatt-al-Arab which incorporates the waters of the Tigris, Euphrates, and Karun rivers. On the seaward end, the gulf connects with the Gulf of Oman, the ARABIAN SEA, and the Indian Ocean through the Strait of Hormuz. The length of the gulf is 615 mi (990 km), and it ranges from 35 to 210 mi (56 to 338 km) wide. The Persian Gulf covers an area of 93 026 mi^2 (241 000 km^2), occupies a volume of 2400 mi^3 (10 000 km^3), and has a mean depth of 131 ft (40 m). Depths as great as 558 ft (170 m) occur in places, and depths of around 360 ft (110 m) are to be found in the vicinity of the Strait of Hormuz.

The Persian Gulf has played an important role in Indian Ocean commerce for at least 3000 years. Today it is one of the busiest waterways in the world as a result of the vast oil deposits in the surrounding area. The Persian Gulf area accounts for two-thirds of the world's proven oil reserves and supplies over half the world's oil requirement. A number of ports and loading facilities exist in the gulf—some capable of handling supertankers of 500 000 dwt (dead weight tons)—and most are under expansion and modernization.

The climate in the area of the Persian Gulf is extremely arid, and evaporation from the gulf exceeds the inflow from the Shatt-al-Arab. The gulf would simply dry up were it not for the inflow of WATER through the Strait of Hormuz. Due to the high evaporation rate, the salinity ranges from 37 ppt in the vicinity of the strait to 41 ppt near the northwestern end of the gulf. Surface-water temperatures range from 60 to 90° F (15.5 to 32° C). Much of the circulation observed in the gulf is due to less saline (lighter) water flowing in through the strait, becoming more saline (heavier) through evaporation, and flowing back through the strait beneath the lighter water.

The floor of the Persian Gulf is covered with silts and clays having a high carbonate content. The latter is derived from the extensive coral reefs that fringe and dot the shallower flanks. The shallows also exhibit many sandbars, mud flats, and small islands. The more important islands are Hormuz, Qeshm, and the Bahrain islands.

PERU CURRENT or **HUMBOLDT CURRENT** is a broad, slow (less than 0.5 kn), relatively cold current that flows northward along the coast of Chile and Peru as part of the counterclockwise gyre that flows around the basin of the South PACIFIC OCEAN. It originates when a part of the current (West Wind Drift) flowing eastward along the southern rim of the basin is deflected northward as it approaches South America. The flow rate of the relatively shallow Peru Current has been estimated at about 15 000 000 m^3/s. The southern part of the current is sometimes referred to as the Chile Current.

In the vicinity of the Peru Current the prevailing wind and the CORIOLIS EFFECT combine to push the surface water to the west. This displacement causes the cold, nutrient-rich water from below to well to the surface. The steady enrichment of the surface layers produces one of the most abundant fishing grounds in the world. Anchovies and TUNA are particularly abundant, as are the MARINE BIRDS which feed on the anchovies. The sea birds produce, in turn, commercial deposits of guano.

During the Southern Hemisphere summer the trade winds that drive the warm, saline equatorial waters westward along the equator sometimes weaken, and occasionally fail. During such periods, upwelling of the cold, rich water ceases and a shallow layer of the equatorial water moves south to replace it. Under these conditions the anchovies find it difficult to feed, and migrate temporarily to other areas. This condition, called EL NIÑO, results in the starvation of many thousands of sea birds, and severely curtails guano production.

PENGUINS (*Top left*) Adélie penguins convene on a rookery in the Antarctic. It has been estimated that the Antarctic and sub-Antarctic regions have a population of about 188 million sea birds with a total biomass of 577 000 tons. The penguins are thought to make up 65 percent of the stocks and more than 90 percent of the biomass for the entire Antarctic area. Penguins in this area consume over 33 million metric tons of food per year. (*U.S. Navy*) (*Bottom left*) Gentoo penguins clamber over mounds of ice near Anvers Island, shown in the background. The Gentoo, which nests among clumps of grass, is found on Antarctica and on the islands of the Antarctic region. (*U.S. Navy*) (*Top right*) An Adélie penguin bears an identification band used in population studies for the Antarctic region. (*U.S. Navy*) (*Bottom right*) An Emperor penguin holds a chick on its feet to keep it warm. Emperor and King penguins breed during the winter season in Antarctica and face very hostile environmental conditions. They incubate their eggs by holding them pressed against the tops of their feet, which are covered by a fold of abdominal skin. (*U.S. Navy*)

Adélie penguins stroll along the ice. An adult pair raises one or two chicks per season, and one parent always stays with the hatched offspring until they are about a month old. Then the young are left to their own devices, and they gather together in large groups for protection against such predatory birds as the skua. *(U.S. Navy)*

Young and adult chinstrap penguins gather on Elephant Island, adjacent to the northern tip of the Antarctic peninsula.

Nesting penguins take care of their chicks at Cape Royds. *(U.S. Navy)*

As the Peru Current approaches the Equator it begins to veer to the west until it joins the South Equatorial Current (see EQUATORIAL CURRENT SYSTEM) and flows across the Pacific under the influence of the trade winds.

PETREL See MARINE BIRDS.

pH is a numerical index used by chemists to describe the hydrogen ion concentration of a liquid system. This concentration, or pH value, is expressed as a logarithmic number (to base 10) and as such indicates the degree of dilution of hydrogen ions in gram-moles per liter of the solution. (See ION.) There is no simple relation between hydrogen ion concentration or activity and pH. Incidentally, the *p* is from the German word *potenz* which means "power" or "exponent."

In pure WATER, which is made up primarily of molecular H_20, there is a slight dissociation to hydrogen (+) and hydroxyl (−) ions in equilibrium amounts at room TEMPERATURE. Therefore, under proper conditions, pure water reacts as a weak base or weak acid, and because of this it is said to be amphoteric. Since the H^+ and OH^- are in balance, the water is electrically neutral; the value of each ion is 10^{-7} or 0.0000001 g'mol, or gram-molecular weights, per liter. Thus, the pH is equal to the logarithm of $1/H^+$, or log 1/0.0000001, or log $1/10^{-7} = 7$.

SEAWATER is measured by a pH electrode meter that is calibrated from 0 to 14; a value of 7 represents a neutral solution, while a pH of 7–0 indicates increasing degrees of acidity, and pH values of 7–14 are measures of alkalinity. There are four main types of pH indicator electrodes: (1) hydrogen (gas), (2) quinhydrome, (3) metal oxide, and (4) glass.

Although coastal waters of the world's oceans may have a pH as high as 9, in most seawater, pH values vary between 7.5 and 8.3, with the general oceanic average value of about pH 8. Near the surface, the alkalinity of seawater is increased by dissolved carbonates from rivers or decreased by the precipitation of lime in the form of CORAL and SHELLS. At greater depths, the alkalinity can vary by the amount of calcareous debris sinking in solution from above.

More important, carbon dioxide occurs in relatively large amounts in seawater as carbonates and bicarbonates, and the carbon dioxide content mainly determines the pH value. For example, at the wind-driven surface of the ocean waters, higher pH values from 8.1 to 8.3 or more will occur, depending upon the partial PRESSURE of carbon dioxide in the atmosphere or the fact that the water at the surface tends toward a state of equilibrium with carbon dioxide and other atmospheric gases. These other gases, such as nitrogen, oxygen, and the rare gases (argon, helium, neon, etc.), are all found in seawater. In fact, seawater contains in solution all the gases found in the atmosphere, but not in the same proportion since they dissolve in amounts determined by the chemical composition of the water and the prevailing physical conditions.

The acidity of localized seawater is important since below a pH of 7.5 many marine animals cannot live, and below a pH of 7 the production of any kind of a skeleton cannot take place due to the presence of carbonate in solution with the seawater. The principal weak acid present is carbonic acid, but significant amounts of boric acid are also present. The salts of these two acids determine the alkalinity of seawater. Boron, whether acid or borate, is in substantially constant proportion to chlorine, at a ratio of about 0.00024. Acid solutions (those with a pH value less than 7) are also, as a general rule, more corrosive to ocean emplaced materials than neutral or alkaline solutions (pH greater than 7). However, normally seawater *per se* contains a number of cations (weak acids), and although there are a lesser number of anions (bases or alkalies), some of these, such as chlorine, are quite strong. The result is that seawater is predominantly alkaline, its pH ranging from 7.5 to 8.3.

Minimum pH values of 7.5 are found in ocean waters of approximately 2500-ft (762-m) depths, and the pH increases gradually at deeper depths except in certain unusual areas where the seawater is diluted or in closed basins characterized by an absence of oxygen. This is true, for example, in the BLACK SEA which is essentially landlocked, having only a narrow, shallow outlet to the MEDITERRANEAN SEA. Here, the water below 600 ft (183 m) contains no oxygen, and the BACTERIAL decomposition of bottom organic matter results in a concentration of hydrogen sulfide such that pH values in the acidic range of less than 7 are present. Similar conditions occur in Norwegian fjords that are separated from ocean waters by shallow passages.

See also CORROSION; SEAWATER.

PHARMACEUTICALS FROM THE OCEAN include drugs, medicines, and vitamins contained in many marine organisms as well as in SEAWATER itself that can be isolated and used in the management of disease in human beings.

Many of the ancient peoples of the world utilized marine plant and animal life to obtain products for use as remedies. For example, the Romans were said

Sea Life	Derived Compounds and Pharmaceutical(s)	Some Actual and Possible Uses Under Investigation
A species of seaweed	Calcium alginate	Stops human bleeding
A species of seaweed	Sulfated lamarin	Prevents blood clotting
Irish moss	Carrageenin	Treatment of gastric ulcers and production of collagen
SPONGES (various species, e.g., *Crytolhethea crypta*)	D-arabinosylcytosine	Inhibits tumor growth
Various species of sponges	Various compounds	Lowering blood pressure, improving muscle tone of the heart, combating virus disease and some forms of cancer
Stonefish (family Synanceiidae), the world's most venomous fish	Various compounds	Lowering blood pressure
Sea cucumber	Various substances	Treatment of tumors
Quahaug clam	Various substances	Treatment of tumors
Sea whip (*Plexaura homomalla*)	Substances similar to the prostaglandins	Source of hormones that are essential to basic biochemical processes
JELLYFISH (*Aequorea aequorea*)	Aequorin	An effective reagent for measuring very small changes in calcium concentrations in humans; such changes often are precursors of cellular and bone diseases
OCTOPUS	Serotonin	Histamine releaser
SEA ANEMONE	Serotonin	Histamine releaser
Octopus (*Eledona moschata*)	Eledosin	A peptide similar to bradykinin—serves as a model for synthesizing new families of drugs in the laboratory
Marine ALGAE, BACTERIA, and fungus, some larger invertebrates (e.g., the sea pork, *Amaroucium stellatum*)	Antimicrobial agents (e.g., Cephalosporin C)	Antibiotics
A blue-green alga	Fatty acid	Combating several microorganisms that cause disease
Pufferfish, sunfish, porcupine fish, and salamanders	Tetrodotoxin	Treatment of cancer
PLANKTON and DINOFLAGELLATES	Paralytic shellfish poisons, saxitoxin	Chemical model for anesthetics and drugs for heart and nervous disorders
Starfish and sea cucumbers	Various compounds	Heart-stimulating and nerve-blocking drugs

to employ the barbs of STINGRAYS in the treatment of toothaches, while the Japanese used a substance from the ovaries of the poisonous PUFFERFISH as a pain killer.

Today, the identification and processing of pharmaceuticals from certain sea life of the world's oceans is a promising field of endeavor. Many antibiotics, steroids, toxins, polysaccharides, and other substances that play an important role in the management of disease in human beings have been produced from marine sources. Others of possible therapeutic value are being investigated. The following chart presents a representative sampling in this field of marine biomedicine.

As noted in the accompanying chart, many of the exotic compounds in marine life, when scientifically characterized, serve as chemical models on which new drugs and medicines can be based. Others can be and are being used directly after extraction and modification or chemical synthesis. It is recognized that benefits to be derived from biomedical research are both immediate and long-range.

Basically, this research includes work in the following areas:

1. TAXONOMY and ECOLOGY of medically important marine organisms
2. Screening of marine organisms for biological activity
3. Investigation of the food web of marine organisms
4. Study of the production of toxicity cycles in marine organisms
5. Use of marine organisms as biomedical research tools
6. Investigation of industrial waste-product contaminants involved in the food web of marine organisms
7. Use of marine organisms as sources of new drugs
8. Investigation of marine biochemical substances as models for the development of new synthetic chemicals
9. Evaluation of health safety standards for new marine-derived foods
10. Clinical aspects, diagnoses, treatment, and prevention of marine biotoxications and other marine-induced diseases
11. Study of disease processes in marine organisms

Studies of this nature indicate the existence of a very wide range of biomarine substances that possess antibiotic, antiviral, and fungicidal properties. Many of these have current commercial uses; others have potential uses. In the latter category are systemic drugs for treatment of human disorders involving the nervous system and the cardiovascular, urinary, gastrointestinal, and other systems. In short, the vast numbers of marine plants and animals which either store or manufacture an array of complex chemical substances that can be utilized as new therapeutic agents represent an untapped source of pharmaceutical wealth. In the search for such wealth, it is entirely possible that the means may be found for detoxifying or coping with and controlling many of the industrial wastes that pose threats to the food economy of the world's oceans.

PHILIPPINE SEA is located off the eastern coast of the Philippine Islands in the western North PACIFIC OCEAN. On the west the sea is bounded by the eastern limit of the East Indian Archipelago, the SOUTH CHINA SEA, and the EAST CHINA SEA. On the north the boundary is the southeast coast of Kyushu, the southern and eastern limits of the Inland Sea (Japan), and the south coast of Honshu. On the east the boundary is a line joining Japan to the Bonin, Volcano, and Mariana islands. And, on the south, the boundary is a line joining Guam, Yap, Palau, and Halmahera islands. The Philippine Sea covers an area of about 400 000 mi² (1 000 000 km²).

The first to explore the 3141 islands that make up the Philippine Archipelago were the Negritos [so named by the Spaniards because of their dark skins and diminutive stature—i.e., 4 ft 9 in (1.45 m) for men and 4 ft 6 in (1.37 m) for women], who arrived at some indefinite period well before the birth of Christ. Evidence indicates that the Negritos occupied much of the archipelago when Malayan explorers and settlers arrived from the south about 3000 B.C. The Malayans (now represented by such tribes as the Igorot, Ifugao, and Tingerian of Luzon, and the Bukidnin, Manobo, and Bagobo of Mindanao) quickly conquered the Negritos and pushed them back into the mountains. The Japanese and Chinese established trade routes to the Philippines after A.D. 700, and the Hindus of Sumatra and Java began to move northward into the area at about the same time. Around 1400, the aggressive Mohammedan Malays (the Moros are the remnant of these people) ravaged the islands and made many long sea voyages to the north and south of the archipelago. These, then, were the peoples encountered by Ferdinand Magellan, the Spanish navigator, when he dropped anchor at Cebu in 1521 in his successful attempt to prove that the Spice Islands (Moluccas) lay within the Spanish half of the world as decreed by the 1493 Demarcation Bull of Pope Alexander VI. Magellan was followed, in 1542, by Ruy López de Villalobos who named the islands "Las Felipinas" in honor of Prince Philip, later King Philip II of Spain.

The floor of the Philippine Sea is occupied by a large basin that is cut down the center by the Parece Vela Ridge. On the west the basin is flanked by the Asiatic continental block on which the East China and South China seas rest, and on the east, by a ridge system that runs south from Honshu until it approaches the southern tip of the Parece Vela Ridge—thus, effectively closing off the basin in the south. Depth within the basin ranges down to 19 000 ft (5791 m), although seamounts pierce the surface in places (i.e., Daito Islands). (See SEAMOUNT.) The sea is also ringed by a system of deep trenches. The Philippine Trench (Mindanao Trench) runs along the eastern edge of the Philippine Archipelago with depths as great as 34 400 ft (10 485 m), and the Ryukyu Trench (Nansei Shoto Trench), running north along the western edge of the Ryukyu Islands, may have depths as great as 21 000 ft (6400 m). On the eastern margin of the sea lies the great trench that begins at the southwestern edge of the BERING SEA in the north and runs south to the vicinity of the southern tip of the Philippine

Trench. This great trench takes on the name of the island chain to its west as it runs south (i.e., Kuril Trench, Japan Trench, Izu Trench, Mariana Trench, and Yap Trench). Near the intersection of the Mariana Trench and the Yap Trench occurs the greatest known depth of the ocean: 36 198 ft (11 033 m). These trenches represent zones of interaction between crustal plates (see CONTINENTAL DRIFT). Red clay (see MARINE SEDIMENTS) covers much of the basin of the Philippine Sea, patches of globigerina ooze are found here and there, especially in the south, and volcanic sediments are found on some sections of the ridges.

As a result of the high-pressure cell (Ogasawara High) that covers the northern part of the Philippine Sea and the intertropical convergence zone (doldrums) in the south, this area is one of the major spawning grounds of the world for tropical cyclones (typhoons). (See HURRICANE.) Once generated, these tropical storms move west and then north, often into the South China and East China seas and the YELLOW SEA.

In the southern part of the Philippine Sea the WATER circulation is dominated by the North Equatorial Current flowing into the sea from the east, and the equatorial countercurrent further south that flows west to east. (See EQUATORIAL CURRENT SYSTEM.) The North Equatorial Current turns northward to feed the KUROSHIO CURRENT which originates in the vicinity of Taiwan and flows northeast along the east coast of Japan. One of the truly great ocean currents, the Kuroshio flows northward at a rate of about 2 kn and transports about 1.4–2.5 billion (2.5 × 10^9) ft^3 [40–70 million (70 × 10^6) m^3] of water each second. In the vicinity of Honshu some of the Kuroshio water spirals southwestward to form the Kuroshio Countercurrent.

PHOSPHATE CYCLE See PHOSPHORUS CYCLE.

PHOSPHORITES are sedimentary rocks composed chiefly of phosphate minerals. Phosphate rock, or phosphorite, occurs on the ocean floor in some areas, and most of the phosphate deposits on land are of marine origin.

Phosphorite nodules have been known since the CHALLENGER EXPEDITION. They occur on the CONTINENTAL SHELF and slope in certain regions where there is a strong ocean UPWELLING, or large and rapid changes of temperature resulting from a meeting of cold and warm CURRENTS. Among the regions are the area under the eastern edge of the CALIFORNIA CURRENT off southern and Baja California and the shelves and banks off western Australia and West Africa. These areas are also characterized by a large production of organic matter related to the transport of phosphate and other nutrients from the deeper ocean to the sunlit upper layer. It is not clear whether the formation of these SEAWATER precipitates is purely an inorganic process or whether BACTERIA and perhaps other organisms are also involved. Phosphate nodule deposits, such as those found in areas like the Thirty- and Forty-Mile Banks off San Diego, California, and the phosphorite sand deposit in Santo Domingo Bay off Baja California, have an important economic potential. However, phosphate rock on the ocean bottom, to justify the operational costs of its recovery, has to compete with present land deposits in purity and grade of the ore body, abundance, cost of transportation, processing, and beneficiation, and cost of exploration at sea versus that on land.

Although most of the phosphate deposits on land are of marine origin, their exposure to weathering processes through geologic time results in an enriched final product with a higher content of phosphate than the marine deposits. Submarine pellets and nodules seldom contain more than 30 percent P_40_{10}, whereas the cutoff grade for fertilizer production is about 31.5 percent, and land deposits now being worked contain around 35 percent.

Nevertheless, the submarine phosphorite deposits appear to have potential economic value since world demand for phosphate products (industry—20 percent; fertilizer—80 percent) has been increasing at an annual rate of 6 percent. In 1975, world consumption totaled 132 million (1.32 × 10^8) tons (120 million metric tons); and in the next decade, the world demand is expected to double. Calculations which show the expected increase in world population indicate that by the year 2000, the total world consumption of phosphate rock will reach 7.6 billion (7.6 × 10^9) long tons, containing about 2.28 billion long tons of phosphate, and total United States consumption is forecast to reach 1.23 billion long tons, containing about 380 million long tons of phosphate.

See also MINING.

PHOSPHORUS CYCLE, or phosphate cycle, is a relatively slow-acting natural ocean system for the replenishment of phosphorus, an element that is structurally important as a skeletal component for marine organisms and is essential for energy conversions in organisms.

Most of the phosphorus found in the world's oceans, estimated at 110 billion (110 × 10^9) tons (100 billion metric tons) is due to land runoff. In its many dissolved and particulate forms, it is

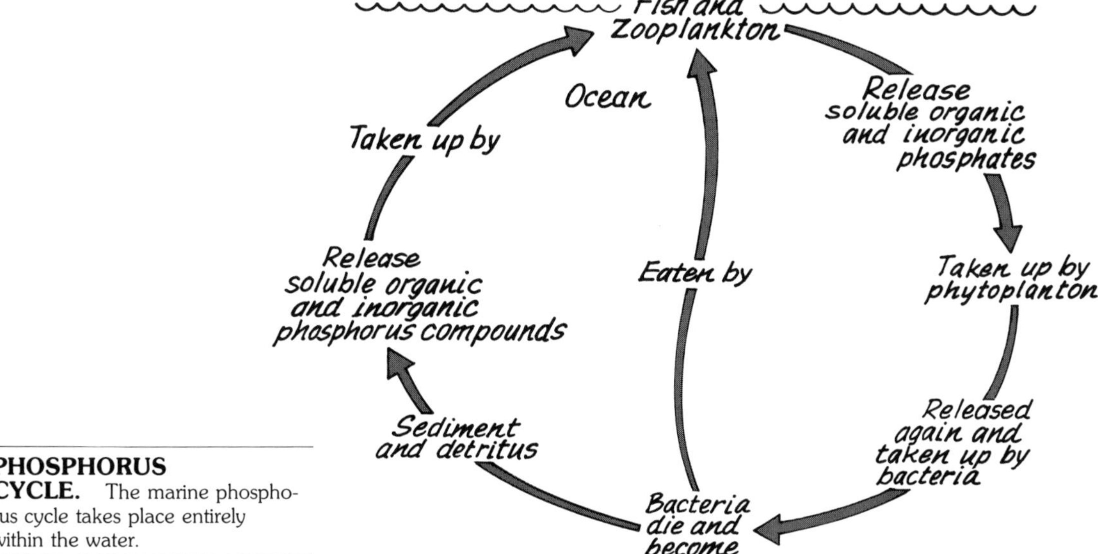

PHOSPHORUS CYCLE. The marine phosphorus cycle takes place entirely within the water.

added, by streams and rivers, to the sediment on the ocean floor at a rate of some 3.5 million (3.5×10^6) tons (3.2 million metric tons) per year. (See PARTICULATE MATTER.) In the oceans it occurs primarily as inorganic orthophosphate (a salt of orthophosphoric acid; the general formula is M_2P_4, where M may be potassium, as in potassium orthophosphate, K_2PO_4). The waters of the North ATLANTIC OCEAN are not as rich in phosphorus as the Southern Ocean, and in tropical seas the amount of phosphorus remains essentially constant, while in the temperate zones it varies.

The reason for this variance is due to many oceanic characteristics that affect the materials' horizontal and vertical dispersion and movement. Among these are the physical effects of ocean currents, winds, WAVES, proximity of landmasses, differences in the DENSITY and TEMPERATURE of water masses, and UPWELLING and downwelling waters. In addition to these factors, the initial phosphorus-rich material that sinks to the ocean floor is acted upon while sinking or when at the bottom and is decomposed by BACTERIA or used as food by higher organisms. All these factors play a role in the phosphorus cycle. While the concentration of phosphorus in ocean sediment is considerably higher than in the waters above, the phosphorus concentrations in this water column are governed by the solubility of the sedimentary calcium fluorophosphate (apatite) at the seafloor–SEAWATER interface. Phosphorus is a major limiting factor in the ocean for the growth of phytoplankton, and where there is intense biological activity in warm, sunlit ocean waters, orthophosphate phosphorus is depleted to very low levels. The phosphorus in phytoplankton consumed by zooplankton herbivores is mostly changed into phosphates and excreted. (See PLANKTON.) Some of the phosphorus in this excrement is then dissolved in seawater as organic phosphorus, while some has already been converted into phosphates by animal metabolic processes. In the deep water, some phosphorus precipitates out as the previously mentioned calcium fluorophosphate material.

PHOTOSYNTHESIS is the process by which electromagnetic energy, present in the visible portion of sunlight, is converted in the presence of the pigment chlorophyll into the chemical energy needed by green plants to synthesize their food materials.

Carbon dioxide gas is a by-product of human beings, animals, combustion processes, etc; WATER exists in air as a vapor and is plentiful as a liquid; sunlight is the energy source for this reaction; sugar produced in plants can then be converted to starch or fat in the plants. By adding nitrogen, the plant can then produce proteins. Oxygen, a by-product of plants, is necessary for humans and animal life, and it is necessary for combustion. The simplified relevant equation for photosynthesis may be expressed as

$$6CO_2 + 6H_2O \xrightarrow[\text{chlorophyll}]{\text{light}} C_6H_{12}O_6 + 6O_2 + \text{Energy}$$

OR

Carbon dioxide + water + sunlight and chlorophyll = Sugar + oxygen + energy

In general, although there are functional differences in plants, the driving force of this process is the absorption of light. The energy is stored in the form of reduced organic compounds which are formed from the carbon dioxide (CO_2) of the ambient environment. There are some 100 or more steps involved in the absorption of light energy, its conversion to chemical energy, and the formation of the stable end product. Some of these steps take place within microseconds of the light absorption. The reactions require a high degree of organization and finally lead to the reduction of CO_2 to carbohydrates and the oxidation of water to oxygen.

In the world's oceans, only a little sunlight penetrates below 262 ft (80 m). Therefore, photosynthesis is carried on by the ALGAE or small marine plants (phytoplankton) chiefly in the upper layers, or euphotic zone, of the ocean where sunlight is available. Some of the red algae, however, can manufacture food in very dim light.

On the primitive earth, the evolution of plants that utilized the photosynthestic process for producing oxygen and organic compounds from carbon dioxide and water doubtlessly provided the way for the development of oxygen-breathing animal life. Now, by assisting in the regulation of the oxygen and carbon dioxide content of the atmosphere, photosynthesis helps to maintain the earth's surface temperature within narrow limits, because carbon dioxide inhibits the loss of heat from the earth to outer space.

Photosynthesis by phytoplankton constitutes the first and most pivotal step in the production of ocean biomaterials. Primary animals of the ocean feed on these plants or on other animals which use the plants as their diet. The larger animals feed upon the smaller animals in this OCEAN FOOD CHAIN.

PHOTOTAXIS is a term applied to the movement of a motile organism in response to light stimulation. The movement away from a light source is called *negative phototaxis,* and movement toward it is termed *positive phototaxis.*

Phototaxis is a strong response in marine organisms. For example, lantern fishes and most other bathypelagic (midwater) fish swim away from a submerged light source. Moreover, in the living strata of the world's oceans known as the DEEP SCATTERING LAYERS, some fishes (particularly the lantern fishes) and myriads of tiny marine organisms (such as EUPHAUSIIDS) migrate varying distances to the surface waters at night to feed. Toward sunrise they return to the deeper, darker waters. The small [less than 6 in (15 cm) long] lantern fishes, such as the species *Myctophum punctatum,* often rise to the surface of some waters in hordes to feed. In the Antarctic, the lantern fish feed on KRILL.

Some pelagic deep-water fishes swim toward the light given off by the luminous light organs (photophores) of various angler fish, dragon fish, and viperfish (see BIOLUMINESCENCE). Such a movement is often a fatal one, since it is by means of this bioluminescence that such fish attract their prey.

PHYSICAL OCEANOGRAPHY is the study of the physical aspects of the ocean, the movements of the ocean, and the variability of these factors in relation to the atmosphere and the ocean floor. See OCEANOGRAPHY.

PHYTOPLANKTON See PLANKTON.

PLAICE See FLATFISH.

PLANKTON include all the floating or drifting small plants, animals, or eggs found in the ocean, in ponds, and in lakes. (The word *plankton* is derived from a Greek word, meaning "wandering.") The plants, principally DIATOMS and DINOFLAGELLATES, are collectively known as *phytoplankton,* and the myriad of animals that live permanently in a floating or drifting state plus the countless numbers of helpless larvae and eggs of animals are called *zooplankton.* (See LARVA.)

Phytoplankton These organisms belong to the most primitive plant groups: various divisions of ALGAE, fungi, and BACTERIA. (See FUNGUS.) Of these, only the algae possess chlorophyll and other pigments, together with the biochemical mechanisms which give them the ability to assimilate carbon in the presence of light. All animal life in the ocean depends upon the growth of this phytoplankton.

Single-celled plants or diatoms constitute more than half of the plankton in the world's oceans. These plants, together with the rest of the phytoplankton, use the nutrient salts and minerals (e.g., phosphorus and nitrates) as food, and, in turn, phytoplankton provide food for the zooplankton and many ocean animals.

Phytoplankton is the name commonly used for the miniscule algal types of plant life that form communities and live most of, if not all, their lives suspended in the water unattached to a substrate. In these plants, energy is stored by carbon fixation;

that is, naturally occurring inorganic carbon is chemically bound through the process of PHOTOSYNTHESIS to form carbohydrates. The forms of plankton that store such carbohydrates are those that form the food base of all the higher levels of the OCEAN FOOD CHAIN. Such organisms are frequently referred to as the "grass of the oceans."

Since sunlight is necessary for photosynthesis, most phytoplankton is found in the uppermost surface layers, 33–330 ft (10–100 m) all over the oceans since sunlight is most available there. However, in warm waters, these autotrophic organisms are also found in appreciable numbers to a depth of 660 ft (201 m), with maximum numbers occurring most frequently between 230 and 400 ft (70 and 121 m). In the cooler temperate, boreal, Antarctic, and Arctic waters the depth range is more restricted, 0–230 ft (0–70 m), with maximum numbers generally found between the first few feet and 165 ft (50 m) in depth. The existence of phytoplankton at great depths in the aphotic zone of the ocean has also been recently demonstrated.

Some important groups of phytoplankton are

- Flagellates and blue-green algae
- Small peridinians (a type of dinoflagellate)
- Siliceous-shelled unicellular plants (diatoms)
- Large shelled dinoflagellates
- Coccolithophores (a group of armored or shelled flagellates, with the shell being composed of minute calcareous plates)

Although the different types mentioned above are considered to be plants, some flagellates are unable to produce carbohydrates in the same manner as plants and have the attributes of both plants and animals. These flagellates (e.g., *Noctiluca*) are carnivores, or predators, and ingest food such as larvae and other forms of tiny floating animal life in the ocean. Moreover, dinoflagellates are capable of self-locomotion. (See PROTOZOA.) The total production of zooplanktonic carnivores in the world's oceans has been estimated to be about 2 billion (2×10^9) tons per year. This is about one-hundredth the estimated annual production for phytoplankton and 10 times the estimated biological harvest for conventional fish and shellfish species.

Very dense concentrations of phytoplankton are called "blooms." During blooming of organisms such as *Trichodesmium, Gonyaulax,* etc., the maximum concentration is often observed in the upper few feet of the ocean. For instance, in the high latitudes (see LATITUDE), phytoplanktonic life flourishes in the ocean from spring (when the amount of daylight increases and when nutrient-rich bottom water is brought to the surface by storms) to fall. During this period of plant growth, the numbers of plants are greatest, and phytoplankton often cover large areas of ocean, coloring the water with shades of yellow, green, or brown. Some species of blue-green algae (e.g., *Trichodesmium erythraeum*), when in sufficient abundance in tropical ocean waters, cause the water to turn red in color. However, the organisms primarily responsible for a RED TIDE in ocean water are the dinoflagellates (Pyrrhophyta). Such dinoflagellate blooms often cause displays of flashing light (or BIOLUMINESCENCE). During this rapid plant growth, many of the planktonic organisms are sensitive to changes in TEMPERATURE and SALINITY of the ocean water. "Red water" is responsible for the destruction of fish because of a marked decrease in the dissolved oxygen in the water, especially in coastal areas. The fish suffocate as a result of this decrease which is the result of the consumption of oxygen by microorganisms. The dinoflagellate *Gymnodinium brevis* and some other species contain a toxin that is lethal to fish. It is the blooming of these particular kinds of dinoflagellates that is generally associated with the massive "red tide" fish kills.

Phytoplankton remains that have fossilized provide scientists with a means to analyze the EVOLUTION of these organisms. Such fossil records also provide valuable clues concerning the past history of the ocean basins. (See BASIN.)

Zooplankton The marine zooplankton is commonly dominated by CRUSTACEANS of the order Copepoda [typified by the species *Calanus finmarchicus*, a 1.4-in (5-mm) long animal with six pairs of swimming legs and two long, jointed feelers or antennae on the head] and by a few types of PROTOZOA. Lesser quantities of other crustaceans, chaetognaths, pteropods, tunicates, ctenophores, and JELLYFISH are also referred to as zooplankton. In addition, among the temporary forms of planktonic animal life are the CLAM larvae, CRAB larvae, various fish eggs and larvae, cladocerans, and the larvae of BARNACLES. Hence, the zooplankton includes animal life ranging in size from microscopic to several meters long. The COPEPODS are the most abundant of the crustaceans and vary in size from about 0.039 in (1 mm) to 0.19685 in (5 mm) long, while the EUPHAUSIIDS, other shrimplike herbivorous crustaceans, are much larger, ranging from 0.78 to 1.97 in (2 to 5 cm) in length. KRILL (*Euphausia superba*), found in the Antarctic waters, is

PLANKTON. A photomicrograph magnifies the passively floating or weakly motile aquatic plants and animals found in the ocean and termed *plankton.* (NOAA)

one species of euphausiids. The copepods and euphausiids occur in fantastically large swarms in the world's oceans and both are considered vitally important keys to the economy and ecological balance of the ocean world. They consume phytoplankton or plant life and convert this vegetable matter into animal protein. These herbivorous zooplankton are in turn eaten by larger planktonic animals or by active swimming animals. The rate at which herbivorous zooplankton, such as copepods, consumes the crop of phytoplankton influences the rate of its production, and that rate influences the rates of production, growth, and survival of the grazing herbivores. These animals themselves are a part of the ocean food chain since the carnivores consume them and carnivores feed upon each other.

As noted above, zooplankton consists mainly of tiny animals, including floating fish eggs and larvae of fish, as well as protozoans which are extremely small. However, the colonial siphonophores (e.g., PORTUGUESE MAN-OF-WAR) are quite large, and the common sea-nettle jellyfish (*Cyanea*) has a variable diameter that may be as large as 6 ft (2 m).

Accordingly, scientists interested in studying population quantities of zooplankton present in various areas as well as in determining the species present in a population sometimes categorize zooplankton by the following size divisions (*Note:* 1 mm = 0.03937 in):

1. Microzooplankton: Zooplankton too small to be retained in the marine biologist's sampling gear of a 0.202-mm mesh net. Instruments used for sampling include a submersible pump, vertical net hauls which have a 103-μm mesh, water bottles, and controlled filtering aboard ships.

2. Small mesozooplankton: Zooplankton retained by nets made of 0.202-mm netting material.

3. Large mesozooplankton: Zooplankton retained by a 1.00-mm mesh netting.

4. Macrozooplankton or micronekton: The larger agile zooplankton and the less-agile nekton captured by such samplers as the Isaacs-Kidd midwater trawl.

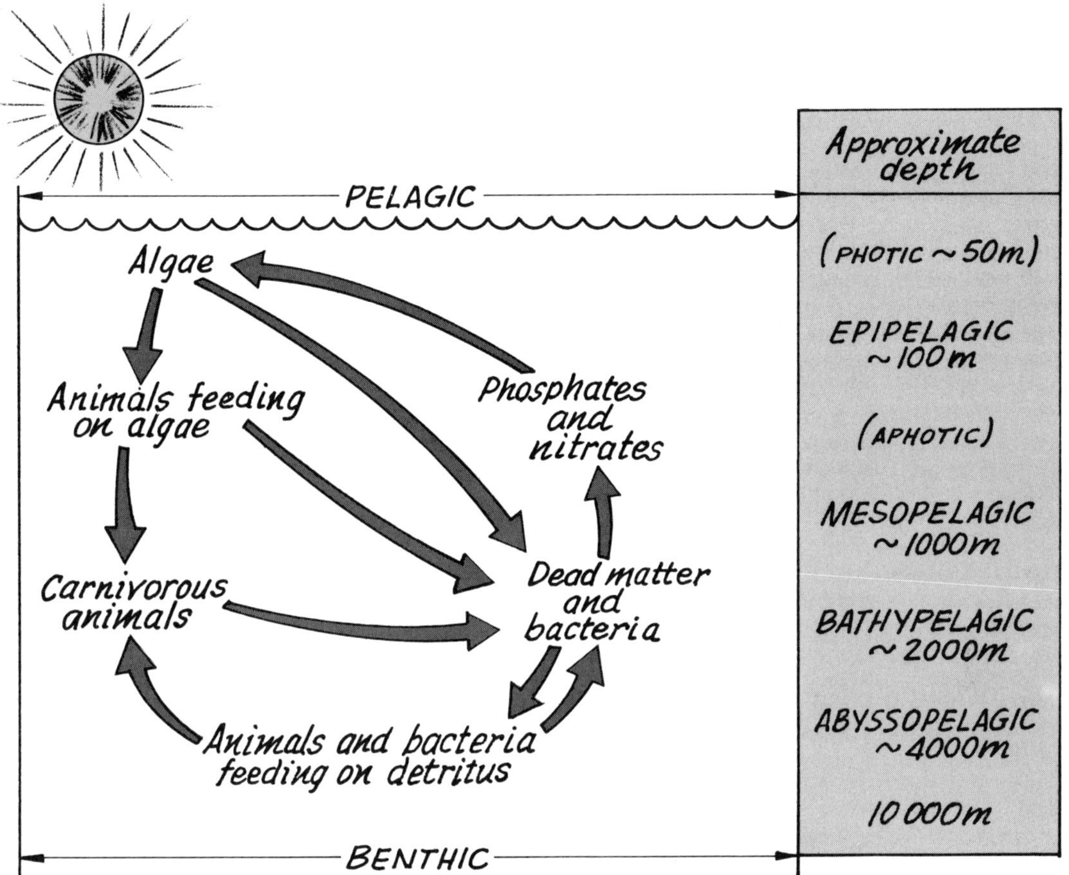

PLANT LIFE IN THE OCEANS. This diagram shows the ocean life cycle and the various ocean zones inhabited.

Zooplanktonic animals are generally able to maintain a preferred depth in the ocean or to perform a vertical migration from a near surface position at night to deeper waters in the daytime, but horizontal movement is largely based on the ocean CURRENTS.

See also BIOLOGICAL OCEANOGRAPHY; HABITABLE ZONES; INSTRUMENTATION.

PLANT LIFE IN THE OCEANS comprises those marine organisms, other than animal life, which are generally characterized by the presence of chlorophyll, a rigid cell wall, and the absence of the ability of locomotion. Ocean plants include the acellular organisms such as blue-green ALGAE of the Kingdom Prokaryota, the DIATOMS and DINOFLAGELLATES of the Kingdom Protista, together with various marine algae and fungi of the Kingdom Metaphyta as well as some of the angiosperms or flowering plants of that kingdom of multicellular plants. (See FUNGUS.)

Compared to the more than 300 000 known species of terrestrial plants and the wide variety of animal life in the world's oceans, the flora in the oceans is quite limited.

Moreover, most ocean plant life is different, in many ways, than that found on land. However, a few ferns and 12 genera (or groups of closely related plant species clearly different from all other groups) of flowering plants (angiosperms) somewhat resemble land plants. Among those that are similar are the anchored flowering seagrasses (e.g., FELGRASS, *Zostera marina*). (See SEAGRASS.) These particular plants are common to many coastal systems throughout the world. They grow and flower when fully submerged in SEAWATER, and their thread-like pollen is distributed by ocean CURRENTS.

However, the vast majority of all ocean plant life comprises the algae, the DIATOMS, and DINOFLAGELLATES. These microscopic plants are called the primary producers of the ocean because they are responsible for most of the carbon fixed by PHOTOSYNTHESIS.

Such plants gain sustenance from the nutrients in seawater: they provide the source of food for all ocean life. Plant forms first become food for small CRUSTACEANS and some fishes, and these in turn are consumed by other animals and MARINE BIRDS. The corpses of plants and animals decompose in the depths of the ocean thus enriching the water, which is recirculated to the surface. An ordinary chain of events would be

Diatom → copepod → FISH (e.g., HERRING) → fish (e.g., COD) or bird→ bacterium → diatom

See also BACTERIA; COPEPODS; HABITABLE ZONES.

PLATE TECTONICS See CONTINENTAL DRIFT.

POGY See MENHADEN.

POLAR BEARS (*Thalarctos maritimus*) are the large aquatic carnivores found in the polar regions of the Northern Hemisphere. Unlike most other bears in the family Ursidae, the polar bear possesses good vision; it also is usually thinner than other species.

Polar bears are powerful predators that roam widely over their barren Arctic habitat in search of food which is principally seal meat. (See SEALS.) So strong is the polar bear that it can drag a 200-lb (90-kg) seal from its breathing hole in the ice. When seals are scarce, it also feeds off fish, bird's eggs, seaweeds, dead WHALES, and any kind of refuse available. (See ALGAE.) It has been ascertained that the polar bear is a resident host for the nematode *Trichinella spiralis*. This same organism causes the trichinosis infection in humans who ingest raw or partially cooked pork. Polar bears are usually solitary animals and travel alone, except when a mother is with her cubs (usually two) or during the mating season. While these animals are often classified as nonaquatic, in reality they are excellent swimmers who use their front legs for propulsion through the water and their hind legs as a rudder.

Unlike other mammals, polar bears do not have a home territory since they do not encounter their food in one place very long. In the summer, when they come on land to moult, they make excursions from place to place in search for food such as LICHENS, berries, and anything they can uproot.

A mature male polar bear ranges in size from 8 to 9 ft (2.4 to 2.7 m) in length and weighs between 800 and 1000 lb (363 and 454 kg). The females are smaller—about 5–6 ft (1.5–1.8 m) in length and around 700 lb (318 kg). However, they are capable of running over rough ice at speeds estimated to be 20–25 mph (32–40 km/h).

Although it has been also estimated that the polar bear population is around 20 000, no accurate census information about these much-hunted animal exists.

POLLUTION See OCEAN POLLUTION.

POMPANO is a spiny-rayed fish (order Perciformes; family Carangidae); these fish are related to scad or horse MACKEREL and jacks. There are over 200 species in the family Carangidae, many of them commercially important, such as the American pompano, TRACHINOTUS CAROLINUS, of the North ATLANTIC OCEAN.

The American pompano is also called the Florida pompano because most are caught in the waters of the South Atlantic and Gulf coasts of the United States. It is a thin deep-bodied fish with forked caudal (tail) and dorsal (back) fins, having a silvery body shading to deep blue above and to a golden yellow on its lower surface. A blue line appears above and in front of the eyes. The upper fins are dark, while the lower fins are yellow shaded with blue. Pompano have the ability to change their coloration in different surroundings from a silvery to a dusk color. Spines and soft rays are found in the dorsal and anal fins. Florida pompano are relatively small and range in weight from $1\frac{1}{2}$ to 3 lb (0.68 to 1.36 kg) on the average. The commercial size is 10 in (25 cm).

The male Florida pompano generally reach maturity in their first year and females in their second year. Spawning is thought to occur offshore in areas where the CURRENTS will carry the young back to shore. Peak spawning is in the spring, although some fish spawn throughout the year. Young pompano, about a month old and less than an inch (2.5 cm) long, first appear off the Florida beaches in April or May. After a 6-week growing period, they migrate northward to New England where they appear in July and remain until October or until cold weather. At that time they migrate south to warmer TEMPERATURE and warmer waters. Lifespan is estimated to be 3–4 years under normal conditions. Pompano feed mainly on bottom organisms such as the CRAB, SHRIMP, and MUSSELS.

Adult pompano are generally netted with commercial trammel or gill nets, but haul seines, runarounds, hand lines, and otter trawls are also used.

Because pompano are scarce, few sports fishermen ever catch them. They are generally considered one of the most delicious of all the marine fishes.

POLAR BEARS. Polar bears normally travel alone except during the mating season or when a mother is with her cubs. These bears are excellent swimmers, and they are a familiar sight to visitors in the Arctic. *(National Archives)*

PORGY See MENHADEN.

PORPOISE (sometimes referred as DOLPHIN) is any of the various species of MARINE MAMMALS of the order Cetacea and family Phocaenidae having a smooth, thick, hairless skin, small flippers, and a highly developed SONAR system. It should be noted that the term, dolphin is also applied to a true fish, the DOLPHIN FISH (*Coryphaena hippurus* and *C. equiselis*). Also, while the terms *porpoise* and *dolphin* are often used interchangeably, strictly speaking the dolphins are cetaceans that not only have basic morphological differences but are classified on a different taxonomical basis from the porpoises. Both are related to the WHALES and are classified in the same superfamily, Delphinoidea, but are divided into two separate families, Delphinidae and Phocaenidae for the dolphins and porpoises, respectively. There are three genera within the family Phocaenidae and four subfamilies and thirteen genera in the Delphinoidea. The common harbor porpoise (*Phocaena phocaena*) of the family Phocaenidae and the Atlantic bottlenosed dolphin (*Tursiops truncatus*) of the Delphinidae are most representative and best known. The largest dolphin is the killer whale (*Orcinus orca*) of the family Delphinidae and superfamily Orcinae. These animals grow to over 30 ft (approx. 10 m) long in the males and 20 ft (approx. 6 m) in the female. They hunt in packs and are the only cetaceans to prey upon other cetaceans, other marine mammals (such as SEALS), and MARINE BIRDS.

The porpoises consist of a few species in the Phocaenidae family in contrast to the some 33 species of cetacean mammals called dolphins that are included in the family Delphinidae. There are differences in the shape of the mouth between the porpoise and the dolphin—a blunt configuration in the former as opposed to a pronounced beak in the latter. Moreover, the teeth of the porpoise are spade-shaped as opposed to the conical teeth of the dolphin.

Although in general it may be said that these animals are found in most all the world's oceans, especially the warm and temperate waters as well as rivers, the specific distribution of the various species of both varies widely.

PORPOISE

A jumping porpoise sprinkles water droplets during his flight through the air. (*U.S. Navy*)

A trained porpoise performs a task by pushing a buzzer on the side of a boat. (*U.S. Navy*)

Representative of the porpoises is the common European porpoise, *Phocaena phocaena,* a black (above) and white (below), approximately 6-ft (app. 2-m) cetacean found in the North ATLANTIC OCEAN, the NORTH SEA, and the BALTIC SEA. Somewhat typical of the dolphins is the bottlenosed dolphin, *Tursiops truncatus,* (usually called a porpoise in the United States). This animal, that attains a 12-ft (3.5-m) length, is found in the MEDITERRANEAN SEA and North Atlantic.

Porpoises and dolphins are known to be excellent swimmers. Most can swim over 30 mph (48$^+$ km/h) for short periods, and speeds of 40 mph (64$^+$ km/h) have been observed when porpoises were swimming before a ship using the bow wave for extra speed. They are also highly intelligent and have the unique capability of imitating the human voice. Interestingly, the earliest descriptions of some of this vocal behavior was recorded by Aristotle (384–322 B.C.) in *Historia Animalium.* Also, Caius Plinius Secundus (23–79 A.D.) wrote that the dolphin makes a moaning or wailing sound similar to the voice of a human being. Latter-day evidence has definitely shown that a dolphin, after a long captivity (e.g., in scientific laboratories and seaquariums), can learn to carry on vocal transactions with human beings in addition to those carried on with other dolphins. As the first step in learning this habit, the dolphin starts clicking, whistling, and buzzing in air with one side of the blowhole open. In its natural habitat, it has also been found that the bottlenosed dolphin in trouble can send out a distress call. This call itself consists of two parts: first, a rising-frequency whistle, and, second, a falling-frequency whistle. A range of the two in the fundamental frequency is from approximately 3000 Hz, (hertz = cycles per second) to at least 16 kHz. In general, the whistle is confined to occasions of physical distress in individual dolphins. The immediate response of surrounding dolphins is to rush to the aid of the dolphin emitting the whistle, pushing it immediately to the surface of the water, and allowing it to breathe. The first breath is usually followed by a whistling and clicking exchange between the distressed animal and the rescuers. The secondary tactics that follow are appropriate to the particular cause of distress.

Most porpoises and dolphins have bones in their two front limbs or flippers which they use for balancing and steering. The tail is unlike that of a fish in that it is moved up and down for propulsion rather than from side to side. On the top of the head is located a blowhole, a respiratory organ about 1 in (2.5 cm) in diameter and provided with inner and outer valves to seal out the water. The lung system is similar to that of humans, and when a porpoise surfaces periodically, as it must for air, it opens the blowhole wide and inhales from 4 to 10 qt (3.8–9.5 L) of air in less than a half-second.

They produce their young alive after a one-year gestation period. Most hunt together at sea in huge schools pursuing SQUID and shoals of HERRING, SARDINES, and MACKEREL.

Unfortunately for them, the dolphins and porpoises have the habit of associating and swimming with schools of the yellowfin TUNA, which are a highly prized commercial fish, especially in the United States. Huge purse-seine nets are used by commercial fishermen to capture the tuna. Hence, the dolphins and porpoises are inadvertently netted, and it has been estimated that over 700 000 of these animals were killed during the period 1972–1977 in this manner. However, the 1978 toll was about 15 000.

See also MYTHS AND LEGENDS.

PORTUGUESE MAN-OF-WAR is the name for several species of the genus *Physalia,* a marine hydrozoan in the phylum Cnidaria (or Coelenterata). It is not a single animal but a colony of separate floating and writhing polyptype organisms numbering from a few to a thousand and containing long filamentous stinging tentacles that can secrete a toxic substance. The common name, *Portuguese man-of-war,* probably originated from the shiplike resemblance of the body of these slow-moving animals to the old Portuguese galleons.

The Portuguese man-of-war normally lives in the warmer high seas of the ATLANTIC OCEAN and is rarely found outside of the current system of the GULF STREAM. However, related forms inhabit the warm parts of the INDIAN OCEAN and PACIFIC OCEAN *Physalia* resembles a hollow, boat-shaped gelatinous mass with the margins of the mass fringed with long trailing tentacles. The colony, which may number up to 1000 individuals, often attains a size of 12 in (30 cm) in length, 6 in (15 cm) in width, and a 6-in (15 cm) height, with colors varying from blue to purple or orchid to shades of scarlet. The filamentous tentacles may trail as far as 40 ft (12 m) behind and are equipped with stinging elements or nematocysts.

The float or bladder is membranous, with the inside filled with more carbon monoxide, nitrogen, argon, and xenon than is found in the surrounding air outside. Muscle fibers and cells on the membrane keep it moist and pliable, and a gland, somewhat akin to the gas gland in fish that have a SWIM BLADDER, secretes the gases into the float.

An expanding mouth and the digestive organs (the gastrozoids) are located on the underside of the float, and the tentacles are used to trap and paralyze prey, such as small CRUSTACEANS and fish. These tentacles convey the entrapped victim to the mouth.

A remarkable feature of these animals, but also one quite characteristic of the JELLYFISH and their relatives, is the presence of stinging cells. These are contained in huge quantities on the surfaces of the threadlike tentacles. Each individual stinging cell or NEMATOCYST is a tiny hollow sphere which, when stimulated, ejects a small harpoonlike tube from within the sphere into the prey. This tube is connected to the sphere by a tiny line and armed with several sharp barbs. At the time these are injected, a toxic fluid, almost as potent as the poison of the cobra, is also forced into the victim from an opening at the end of the tube. The mechanisms which set these discharges off are not as yet well understood in detail by scientists.

As many beachcombers and skin divers can well attest, the stings of the Portuguese man-of-war can be very painful indeed. Moreover, in some allergically sensitive individuals, the reaction can be very serious and at times fatal. Alcohol applied immediately to the infected parts of the skin has some remedial effect since organic solvents reduce the strength of the toxin.

At least two animals endure the stings without any apparent discomfort. Some species of TURTLES (e.g., the loggerhead, *Caretta caretta*) eat these jellyfish. Also, the small fish *Nomeus gronovii* typically is found in association with the Portuguese man-of-war. This fish shares in eating the prey of the jellyfish and is immune to its toxin.

PRESSURE See HYDROSTATIC PRESSURE.

PROJECT FAMOUS See INTERNATIONAL DECADE OF OCEAN EXPLORATION.

PROTOZOA is the name for the diverse phylum of one-celled microorganisms that have been traditionally classified in the animal kingdom. The generally accepted classification of the phylum Protozoa is as follows: subphylum 1—Sarcomastigophora; subphylum 2—Sporozoa; subphylum 3—Cnidospora; and subphylum 4—Ciliophora. The word *protozoa* is taken from the Greek and translates to "first animals." However some protozoa (e.g., phytoflagellates and slime molds) exhibit plantlike characteristics. (See TAXONOMY.)

Protozoa are eukaryotic microorganisms; that is, they contain a nucleus and divide by mitosis. This nuclear division is one in which an exact duplication and separation of the chromosome threads takes place so that each of the two daughter nuclei carries a chromosome complement identical to that of the parent nucleus. Such microorganisms were, like oxygen, a necessary precondition for the EVOLUTION of higher life forms.

Protozoa are very widely distributed in nature, and free-living types occur in the world's oceans. They range in size from less than 1.2×10^{-4} in (3 microns) to a few centimeters (1 cm = 0.4 in.). Although protozoan shapes vary considerably, there is a tendency toward universal symmetry in the floating species, and most swimmers show spiral torsion. Colonies are known in flagellates, ciliates, and sarcodinians.

Among the most common marine protozoa are the sarcodinians (superclass Sarcodina of the subphylum Sarcomastigophora). These organisms have been generally distinguished by the fact that their movement involves protoplasmic flow, often with projections of the protoplast or living portion of the cell, in which cytoplasm flows during extension and withdrawal. Such marine sarcodinians as FORAMINIFERANS and radiolarians are representative. (See RADIOLARIA.) The pelagic foraminiferan *Globigerina bulloides* is found in all the oceans, while the radiolarians *Peridium spinipes* and *Lithocircus magnificus* are among those found in the PACIFIC OCEAN and ATLANTIC OCEAN, respectively.

The flagellates (or members of the protozoan superclass Mastigophora) are also very common in SEAWATER. These types propel themselves by means of one or more tiny whiplike arms or flagella. Such organelles, or flagella, occur also in certain Sarcodina and Sporozoa. A flagellum is comprised of a sheath enclosing a matrix in which a nerve fiber (axoneme) extends from the cytoplasm to the flagellar tip. The DINOFLAGELLATES, *Noctiluca scintillans*, are one of the largest [0.06 in (1.5 mm)] of the species. Dinoflagellates are primarily marine. Silicoflagellidans (or members of the order Silicoflagellida) are marine flagellates which have an internal, siliceous skeleton and a single flagellum. The species *Dictyocha fibula* is found in the Atlantic, Pacific, and GULF OF MEXICO waters. Ocean plants are mostly DIATOMS and flagellates, and the silicoflagellatans (members of the class Silicoflagellata in the plant division Chrysophyta) are a part of the flagellate microorganisms of the marine PLANKTON. Their exoskeletons resemble those of the radiolarians.

The ciliates (class Ciliatea, subphylum Ciliophora) are those protozoa having cilia, relatively short organelles similar in structure and behavior to flagella. Certain species are marine, and *Nycotherus* (order Heterotrichida, subclass Spirotrichida) is found in the digestive tract of amphibians and many INVERTEBRATE species.

In protozoan feeding the methods vary. Some of

the commonest are by ingestion of PARTICULATE MATTER (phagotrophic feeding) and by the passage of dissolved foods through the peripheral layer (saprozoic). In addition, chlorophyll-carrying flagellates are provided nutrition by PHOTOSYNTHESIS.

The study of the role of protozoa in ocean systems is one of the most fundamental problems in marine biology. For example, the role that various heterotrophic protozoans play in marine food webs (see OCEAN FOOD CHAIN) is as yet not well understood.

In seawater there is a substantial production of dissolved organic matter of low molecular weight, such as glucose and amino acids produced both by phytoplankton and by zooplankton. Studies of heterotrophic uptake of these compounds suggest that microorganisms rapidly use them and thereby keep the concentration of these substances low. The relative importance of this aspect of the production-use pathway is not clear.

Additionally, it is thought by some scientists that very large numbers of phytoplankton are consumed by protozoa rather than by the small CRUSTACEANS usually thought to be the primary herbivores in the ocean. This belief, plus the evidence of use of dissolved organic materials by microorganisms, leads many to think that much of the photosynthesis in the ocean is degraded by microorganisms rather than by the crustaceans and FISH. If this is true, microbial aspects of the marine food web may be more important quantitatively than that of most plankton. The determination of the existence of a major microbial energy sink in the world's oceans has important implications in devising a strategy for using them as a source of food.

PROUDMAN, JOSEPH (1888–1975), an English mathematician and oceanographer, devoted most of his long career to the study of tides and storm surges. His work in this area, between 1913 and 1959, earned him the Agassiz Medal of the U.S. National Academy of Sciences in 1946 and the Hughes Medal of the Royal Society of London in 1957.

Proudman, the son of a farmer, was born in Unsworth, Lancashire, England, on December 30, 1888. He received his bachelor of science degree from the University of Liverpool in 1910, and a bachelor's and master's degree in mathematics from Trinity College, Cambridge, in 1913 and 1917, respectively. In 1916 he received his doctorate from the University of Liverpool. In 1919 he became the first professor of applied mathematics at the University of Liverpool and director of the Tidal Institute at that University. In 1933 he became professor of oceanography, and from 1940 to 1946 he was vice-chancellor of the University. He retired in 1954.

While he was at Cambridge in 1912, Proudman was encouraged by H. Lamb to undertake research to determine the theoretical distribution of the tidal oscillations of a flat sea on a rotating earth whose coasts had the form of a sector of a circle. The first general results, published in 1917, transformed the differential equations into equations of Lagrange's type with an infinite number of coordinates. The chief applications of these equations, relating to a hemispherical ocean bounded by a complete meridian, were published by Proudman and A. T. Doodson in 1925 and 1937.

In 1916 Proudman worked with Lamb, at the request of the British Association for the Advancement of Science, to survey the state of research on tides. This work led to a report by Proudman (1920) on harmonic analysis of tidal observations and the establishment of the Tidal Institute (1919) at the University of Liverpool. In collaboration with A. T. Doodson, director of the Institute, Proudman published important papers on the application of the dynamic equations of the tides to determine the distribution of the principle constituted of the tide over the NORTH SEA, and on the "time relations in meteorological effects on the sea" with special reference to storm surges. (See WAVES.)

As professor of oceanography Proudman studied the distribution of temperature and salinity in the IRISH SEA, and effects of turbulence in that body of water. His many studies resulted in a book *Dynamical Oceanography,* published in 1953.

PTEROPOD OOZE See MARINE SEDIMENTS.

PUFFERFISH (puffer, blowfish, globefish, swellfish, among other names) is a slow-swimming fish of the family Tetrodontidae (meaning "four teeth") in the order Tetrodontiformes; it is found widely distributed in the tropical waters of the world's oceans, usually among CORAL REEF, although some species inhabit the open ocean, and some live in such large rivers as the Nile and Congo.

Pufferfishes all have the defensive ability to inflate their bodies more than twice their normal size, which usually is about 1–2 ft (30.5–61 cm) in length. This inflation is performed by taking either air or water into sacs contained in the nonbony body, which then becomes distended by these filled sacs. The skin is normally covered with spines, and these become erect when the fish blows itself up.

Slow swimmers because of their lack of the swimming muscles usually found in the rear half of most

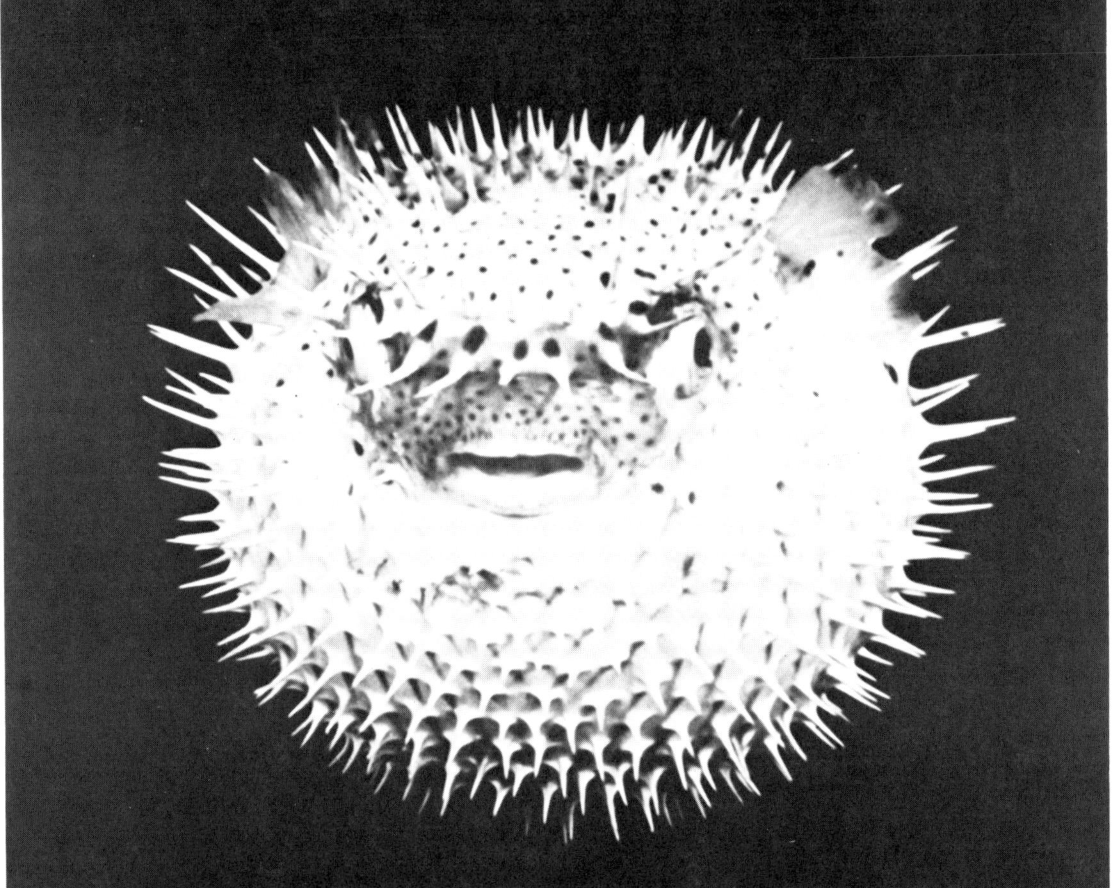

PUFFERFISH. A porcupine fish, closely related to the pufferfish, inflates itself by drawing in water. When this happens, sharp spines all about the body become erect and create a formidable defense weapon. *(Oceanographer of the Navy)*

other fish, the pufferfishes propel themselves by moving their dorsal and anal fins from side to side with some assistance from the pectoral fins; the tail is used only as a rudder.

The teeth of these fishes, like those of the porcupine fish (a member of the same order but belonging to the family Diodontidae), are fused to form a beak. The two upper and lower teeth in each jaw are used to crush mollusks and parts of CORAL. (See MOLLUSK.) The living parts of these are eaten by the pufferfishes along with other marine life such as the CRAB, worms, and BARNACLES.

Pufferfishes are extremely poisonous to humans eating them, although with expert preparation in cooking them, the danger can be lessened appreciably. However, the risk of dying from the powerful nerve poison, tetrodotoxin, contained in various parts of the fish, is still present.

QUAD is a term used to mean 1 quadrillion Btu (10^{15} Btu) (see ENERGY).

QUADRATURE See TIDES.

QUAHAUG is the name for a typical Eulamellibranch CLAM also called the littleneck clam, *Mercenaria mercenaria.*

RADIATION FOG See FOG.

RADIOLARIA is the name for a subclass of the protozoan class Actinopodea. These minute protozoans are noted for the striking forms of siliceous skeletons found in most species. A characteristic feature, the capsule, which separates the outer from the inner cytoplasm, is made up of one or two membranes.

Similar to the FORAMINIFERANS, the radiolarians feed especially in the upper layers, or the photic zone, of the world's oceans and mainly on microALGAE and other protozoa. The shells of the dead animals sink to the ocean bottom where, in conglomerate form, they make up what is known as radiolarian ooze. Such ooze is found particularly on the ocean floor of the temperate and polar oceans.

RADIOLARIAN OOZE See MARINE SEDIMENTS; RADIOLARIA.

RAYS indicates any of the 350 species in the order Batoidea (sometimes called the Hypotremata or Squaliformes), subclass Elasmobranchii, which also include the SKATE and SAWFISH families. These cartilaginous fishes, related to the SHARK, have a flattened body from top to bottom with large pectoral FINS attached to the side of the head, ventral gill slits, and long spikelike tails.

Rays are found in shallow to moderate depths of tropical, subtropical, and warm temperate waters of the world's oceans. They are often difficult to see since they have the ability to change color slightly and blend in with the bottom, especially in sheltered areas such as the sandy bays, lagoons, or muddy estuaries. While some are free swimmers, they are primarily demersal fish, usually lying on the bottom or partly buried for concealment or to feed upon worms, mollusks, and CRUSTACEANS. (See MOLLUSK.)

As swimmers, they are spectacular to watch in their winglike motion through the water. The pectoral fins are joined to the sides of the head, gill openings are ventral, and most have a sharp spine near the base of a long, slender tail. In some rays (stingrays) this spine contains poisonous tissues.

Some authorities divide rays into eight familes:

1. Dasyatidae (stingrays or whiprays)
2. Potamotrygonidae (river rays)
3. Gymnuridae (butterfly rays)
4. Myliobatidae (eagle rays or bat rays)
5. Rhinopteridae (cow-nosed rays)
6. Mobulidae (devil rays or mantas)
7. Urolophidae (round stingrays)
8. Torpedinidae (electric rays)

Representative examples of some of these according to area are

- Diamond stingray (*Dasyatis dipterurus*): British Columbia to Central America. These animals have whiplike tails usually bearing one or more spines which together with a poisonous secretion can inflict a serious wound on the shallow-water swimmer who happens to step on this typically shaped ray. The venom carries a coagulant which prevents the wound from bleeding.
- Bat ray or spotted eagle ray (*Aetobatus narinari*): Tropical or warm-temperate belts of the ATLANTIC OCEAN, RED SEA, INDIAN OCEAN, and PACIFIC OCEAN.
- Round stingray (*Urolophus halleri*): Waters of the Pacific between California and Panama.

The manta or devil rays (e.g., *Manta birostris*) are very large, often as wide as 20 ft (6.1 m) or more

RED SEA. The Red Sea, on the left, connects with the Gulf of Aden between the Arabian peninsula and Africa. (*NASA*)

across, with the anterior part of the pectoral fins projected forward into horns or cephalic fins. When pursued by the shark, or on other occasions, this fish may "fly" out of the water to a height of 5 or more ft (1.5 m). They are viviparous, usually giving birth to only one young at a time.

There are 36 species of electric rays (suborder Narcobatoidea, family Torpedinoidae, genera *Torpedo, Narcine, Hypnarce,* and others). They are all also known as torpedos. The largest, the disk-shaped *Torpedo nobiliana,* attains a 5-ft (1.5-m) length and 200-lb (90-kg) weight and, by large electric organs on either side of the head, is capable of producing an electric current of 200 V in a conductor such as SEAWATER.

RED CLAY See MARINE SEDIMENTS.

RED SEA lies between Egypt, Sudan, and Ethiopia on the African continent, and Saudi Arabia on the continent of Asia. The sea is about 1400 mi (2253 km) long, and nowhere more than 220 mi (354 km) wide. Its northern border is the southern limits of the Gulf of Suez [170 mi (274 km) long and 25 mi (40 km) wide] on the west side of the Sinai Peninsula, and the Gulf of Aqaba [110 mi (177 km) long and 12 mi (19 km) wide] on the east. The southern limit is a line joining Husn Murad and Ras Siyan on either side of the Straits of Babel Mandeb. Through this strait the WATER of the Red Sea mingles with the waters of the Gulf of Aden and the

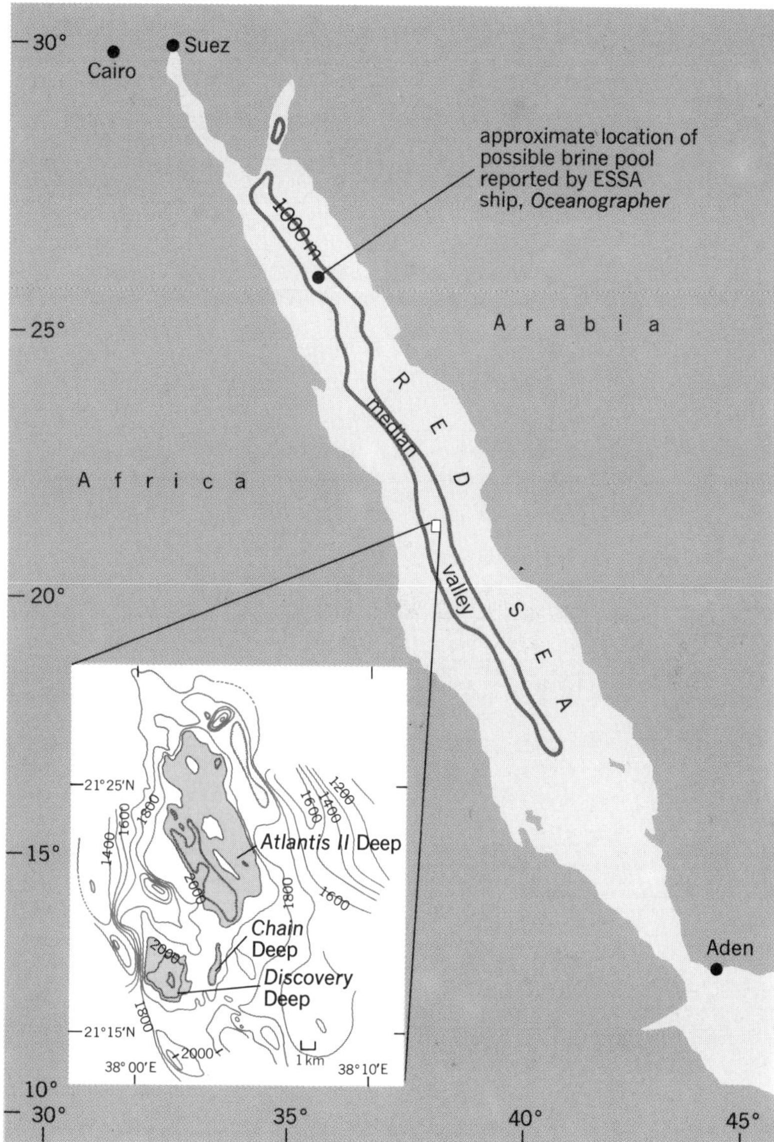

MAP 10 The Red Sea and the Atlantis Deep (inset).

INDIAN OCEAN. The Red Sea covers an area of 173 700 mi² (450 000 km²), occupies a volume of 60 217 mi³ (251 000 km³), and has a mean depth of 1830 ft (558 m).

Since earliest times the Red Sea has provided a valuable trade route between the peoples around the MEDITERRANEAN SEA and those of India, China, and the islands of the western PACIFIC OCEAN. It was used successively by the Egyptians, the Phoenicians, the Arabs, and, during the Middle Ages, the Venetians. The sea declined somewhat as a trade route after Vasco da Gama sailed down the west coast of Africa and rounded the Cape of Good Hope to open an alternative route to India in 1497. The sea regained its commercial importance with the opening of the Suez Canal in 1869. The canal cut the distance from the northern Indian Ocean to northern Europe by 4850 nm (nautical miles) (9000 km).

Through ownership and control of the Suez Canal Company, England and France regulated the traffic through the Red Sea until 1956 when it was expropriated by the Egyptians. Egypt operated the canal successfully until it was blocked during the 1967 Arab-Israeli War. The canal was not reopened until June 1975, and is yet to regain the shipping density it previously carried. In large measure this is due to the severe dislocation of shipping patterns caused by the prolonged closure and the resulting development of a fleet of supertankers that cannot use the canal.

The Red Sea is a relatively new body of water on this planet, and may be an embryonic ocean. Perhaps as recently as 20 million (2×10^7) years ago Arabia split off from Africa and allowed the water of the Indian Ocean to reach within 101 mi (162.5 km) of the Mediterranean Sea. That part of the globe-girdling Mid-Ocean Ridge that runs northward through the Indian Ocean begins angling west in the vicinity of the Chagos Archipelago, and curves into the Gulf of Aden in the form of the Socotra Rift. Another rift valley runs down the center of the Red Sea. It also appears that the fracture takes another near-right angle turn (eastward) to run up the Jordan River valley to the Dead Sea. This may be an extension of the rift valley of Africa that terminates around Lake Tanganyika. The rift valley down the center of the Red Sea lies at depths as great as 7546 ft (2300 m).

Abundant evidence exists to demonstrate that the Red Sea rift is still active and continues to widen at the rate of several inches (1 in = 2.54 cm) each year. In 1947, a Swedish expedition aboard the *Albatross* discovered several "hot spots" in the water above the floor of the rift. Subsequent work by the British *Discovery* and the American *Atlantis II* has shown that in these spots—now named Atlantis II, Chain, and Discovery Depressions—the water temperature reaches 133° F (56° C) and the salinity is as high as 270 ppt (about 35 ppt is normal). Further investigations indicate that molten rock welling up to replace that removed by spreading of the rift heats the water which, in turn, has concentrated the salts and other dissolved minerals both in the water and in the bottom sediment. Chemical analyses indicate that the sediment—300 ft (100 m) deep in places—contains valuable concentrations of iron, manganese, zinc, and lead. The government of Sudan has laid claim to these deposits, but the cost of recovery from depths as great as 7218 ft (2200 m) is, at present, prohibitive.

The CONTINENTAL SHELF of the Red Sea is about 164 ft (50 m) deep, and is, in places, laced with coral reefs, some lying as much as 25 mi (40 m) offshore. (See CORAL REEF.) The shallow shelf supports several islands and archipelagos such as the Farsan Islands near the Arabian shore and the Dahlak Islands near the African. The shallow shelf drops off to shelves that lie at depths of around 1640 ft (500 m). These shelves, in turn, drop off into the rift valley. The bottom sediment is largely derived from the coral reefs and the calcareous remains of other sea life. Since no rivers run into the Red Sea, and its shoreline is flanked by deserts, the only terrigenous sediment reaching the sea is windblown; therefore, it constitutes only a small fraction of the bottom sediments.

Air temperature over the Red Sea often reaches 100° F (38° C), the highest over any body of water except the Persian Gulf. The rate of evaporation is very high—about 83 in/yr (210 cm/yr). Therefore, the dry air moving over the water quickly becomes very humid, producing one of the most uncomfortable climates in the world. The large evaporation rate pulls water in through the Strait of Babel Mandeb from the Gulf of Aden to replace it. Since little or no water flows southward from the Suez Canal, the flow through the strait provides the major surface circulation.

The origin of the name Red Sea has been debated for some time. It has been attributed to seasonal blooms of ALGAE that color the surface water red. The surrounding hills, coral reefs, and seaweeds are also red. Dust deposited by the wind often leaves red streaks on the surface. Perhaps all these together made the choice inevitable.

RED TIDE is the phenomenon resulting in the poisoning of the surface waters when a certain species of DINOFLAGELLATES (single-celled plantlike animals) "blooms" or accumulates in dense numbers; the substances given off into the water by these organisms color it red and are deadly to many kinds of marine animals.

A red tide takes place when physical factors (e.g., sunlight and nutrients) are favorable to the rapid reproduction of dinoflagellates (e.g., *Gymnodinium brevis* and *Gonyaulax sp.*) and when the number of predators is temporarily reduced. The so-called blooming, or what amounts to a population explosion of this type of PLANKTON, is caused by an unusual warming of the surface waters in many parts of the world's oceans. This TEMPERATURE change is brought about by a radical lessening of the prevailing winds and the absence of UPWELLING. In the stagnant water the dinoflagellates thrive vigorously on various salts, chiefly phosphates, which have accumulated at the surface. They then become toxic. In the waters of the GULF OF MEXICO off Florida, the blooms of red tide are due to the species *Gymnodin-*

ium brevis and those in colder waters are caused mainly by *Gonyaulax sp.* which produce the paralytic SHELLFISH poisoning. Normally, the dinoflagellates are yellow-green or golden, but as they become poisonous, the color of the ambient water takes on a reddish brown or orange-brown cast. The substances given off into the water (such as the paralytic shellfish poison, saxitoxin), cause mass deaths of many marine animals. In others (e.g., CLAMS and other shellfish) this neurotoxin is ingested, and although these kinds of animals are usually unaffected by it, people can become quite ill from eating their tainted flesh.

The standard bioassay procedure for measuring the degree of toxicity in shellfish is one in which mice are injected with an extract of the shellfish suspected of having been poisoned. This degree sometimes is expressed in mouse units: one mouse unit equals the amount of poison which, when injected into a white mouse of 0.7 oz (20 g) weight, will cause its death within 15 min. While people vary in their tolerance to shellfish toxins, a mouse unit value of 40 000 is generally considered to be the danger level. More sensitive chemical test methods than this, using a standard laboratory instrument called a fluorescence spectrophotometer, have been recently developed for detecting red tide poisons in shellfish.

Two major compounds given off by the dinoflagellate *Gymnodium brevis* have been separated by chromatography and have been identified as biologically active. One of these compounds, called gymbretoxin, is a potent neurotoxin.

REDFIELD, ALFRED CLARENCE (1890–1974), an American marine biologist, was born in Philadelphia, Pennsylvania, on November 15, 1890. He spent nearly 30 years at Harvard University, first as an undergraduate and graduate student, and finally as professor of physiology and chairman of the department of biology. During this time he also taught for a year at the University of Toronto and conducted research for a year at Christ College, Cambridge University, and another year at the University of Munich.

He joined the WOODS HOLE OCEANOGRAPHIC INSTITUTION in 1930 as a senior biologist, and during the next decade, he devoted his summers to research at Woods Hole, continuing his research and teaching at Harvard University during the academic year. In 1942, he was appointed associate director of the Woods Hole Oceanographic Institution and shortly thereafter moved to Woods Hole to continue his position until retirement in 1957.

Redfield considered the ocean as a vast living organism in which the SEAWATER serves as the "bloodstream" so that changes in its chemistry reflect the multitude of biological processes going on within it. He held the view that life in the sea could not be understood without understanding the sea itself, and thus his research covered a wide range of oceanographic problems. His major work was on the effects that metabolic processes of organisms had on the chemical characteristics of the water. However, he also addressed his attention to such problems as tidal phenomena in narrow embayments and circulation patterns as they related to planktonic populations. In 1942, he was appointed a consultant to the U.S. Navy Bureau of Ships to advise on investigations underway at Woods Hole and in various Navy yards concerning prevention of the attachment of FOULING organisms to the bottoms of ships. This work saved a great deal of money in fuel and docking costs during World War II. Redfield contributed and coedited large portions of the major publication on this work, *Marine Fouling and Its Prevention,* still considered an authoritative review on the problem.

After his retirement in 1957, Redfield concentrated on two major investigations: the use of deuterium as a tracer for both fresh and salt natural waters and the history and development of salt marshes. These salt-marsh studies indicate the breadth of his approach to science, for his work encompassed the biology, geology, chemistry, and physics of the marsh. One of his last publications was an Ecological Monograph entitled *Development of a New England Salt Marsh* issued in 1972.

Redfield served trusteeships for the Woods Hole Oceanographic Institution, the Marine Biological Laboratory at Woods Hole, and the Bermuda Biological Station, where he was also president from 1962 to 1966. He was chairman of the Natural Resources Council from 1946 to 1948 and President of the Ecological Society of America (1946) and of the American Society of Limnology and Oceanography (1956). He was a member of the National Academy of Sciences, the American Academy of Arts and Sciences, and the Boston Society of Natural History, an honorary member of the Marine Biological Association of the United Kingdom, a corresponding member of the Marine Biological Association of India, and a member of a number of other professional societies. He received both the National Academy of Sciences Agassiz Medal and an honorary Ph.D degree from the University of Oslo in 1956. Other honorary Ph.D's were awarded him by Lehigh University (1965), the University of Newfoundland (1967), and the University of Alaska (1971). The

Boston Society of Natural History presented him with the Walker Award in 1973.

REDOX POTENTIAL, eH, or the oxidation-reduction potential, is the voltage difference at an inert electrode that is placed in a reversible oxidation-reduction system; it is a measure of the state of oxidation of the system.

In SEAWATER or any aqueous medium, the redox potential is basically a measure of the ability of the liquid to supply or use up electrons. In such measurements, an inert electrode placed in a large volume of solution or in the natural medium is connected to a hydrogen electrode (or other reference electrode), forming a cell, and the resulting voltage is the eH.

Along with pH, eH is an important variable in seawater since it is a method of determining what molecular or ionic species are stable and what reduction-oxidation reactions are possible. On plots of eH versus pH, the characteristics of the aqueous environment can be shown, the stabilities of minerals and dissolved matter can be determined, and lines representing equilibrium between different species can be located.

REMORA is the name applied to any of 8 species of peculiar ocean fishes of the family Echeneidae (order Perciformes); the unusual feature in these FISH is the presence of a large oval sucker or suction cup on the top of the head where the first dorsal fin usually appears in most fish. (See FINS.)

Remoras live mostly in the tropical oceans. The largest species, *Echeneis naucrates* (or striped remora), is about 3 ft (1 m) long. The remora, using its strong sucker disk, attaches itself especially to the body of a SHARK. However, remoras also will affix themselves on some flat surface area of other fishes such as the marlin, TUNA, and swordfish, as well as to TURTLES, porpoises, and the hulls of ships. (See PORPOISE.) While opinions on this behavior vary, the present view is that it provides a convenient method of travel to its food sources, with the remora releasing itself when it sees a school of small fishes. Most species are somewhat host-specific relative to the kinds of fishes to which they habitually attach themselves. Some of the small remoras, such as *Remorapsis pallidus* and others, also hitchhike in the mouths or gills of large ocean sunfish, manta RAYS, swordfish, and SAILFISH. They probably consume the parasites present in these fishes.

Remoras are usually grayish in color and do not have the countershading which is present in many of the free-swimming pelagic fishes. Spawning occurs in June and July in the mid-ATLANTIC OCEAN and in August and September in the MEDITERRANEAN SEA.

Remoras have been both a subject of curiosity and utility to human beings since ancient times. For instance, it was once believed that they could stop a sailing ship when fastened to its hull. Also, in many parts of the world (e.g., Japan, Africa, Central America) the remora has long been used to catch turtles.

REVELLE, ROGER RANDALL (1909–), an American oceanographer and educator, was one of the first scientists to integrate the fields of geography, geology, geophysics, and meterorology and to focus this powerful tool on the study of the oceans. He led many oceanographic expeditions, particularly to the PACIFIC OCEAN, and his findings were widely published. But, perhaps his greatest contribution to our understanding of the oceans was through his leadership of the SCRIPTS INSTITUTION OF OCEANOGRAPHY from 1951 to 1964.

Revelle was born in Seattle, Washington, on March 7, 1909. He received his bachelor's degree from Pomona College in 1929 and his doctorate in oceanography from the University of California in 1936. He holds honorary degrees from Pomona College, Carlton College, Colby College, Harvard University, and Carnegie-Mellon University. Revelle began his career as an instructor at Scripps in 1936 and became a full professor in 1948. He became acting director of the Institution in 1950 and director in 1951—a post which he held until 1964. In 1964 Revelle became the Richard Saltonstall professor of population policy and director of the Center for Population Studies at Harvard, and professor of population policy in 1970.

During World War II Revelle served as a commander in the U.S. Navy from 1942 to 1947. In 1946 he was called upon to organize the oceanographic investigations before and after the atomic blast (Operation Crossroads) at Bikini Atoll. Seismic surveys of the area showed that the atoll was built of a low-velocity layer down to about 9842. ft (3000 m) and a high-velocity layer below that. Drilling confirmed that the low-velocity layer was composed of reef material which grew progressively older with depth. This and subsequent studies provided strong evidence to support Charles Darwin's theory that coral atolls were built on submerged volcanic peaks, and that the coral was induced to grow upward to remain in sunlight as the volcanic SEAMOUNT subsided under its own weight or by fluctuations in SEA LEVEL. (see DARWIN, CHARLES.)

Measurements made by Revelle's team after the Bikini blast provided the data upon which scientists

concluded that the energy in large blasts, even atomic ones, was insufficient to cause the earthquakes and mudslides needed to trigger tidal waves. The waves created by the Bikini blast were 60–90 ft (18.3–27.4 m) high at the center, and subsided to 6–8 ft (1.8–2.4) within 3 mi (5.55 km).

Over the years that Revelle served as assistant director, acting director, and director of Scripps, he sent out many expeditions and participated in many of them. The expeditions added greatly to our understanding of the oceans. As examples, the Mid-Pacific Expedition of 1950 and the Capricorn Expedition of 1952–1953 demonstrated that heat flow through the ocean floor was roughly the same as that on land and that the flow was greater near the mid-oceanic ridges and less near the TRENCHES—thus adding weight to the convection cell hypothesis (see CONTINENTAL DRIFT) proposed earlier by F.A. Vening-Meinesz and David Griggs; established the thinness of the sediment covering the ocean floor near the mid-oceanic ridges; determined the depth of the Moho, the boundary between the earth's mantle and crust, to be about 4.5 mi (7 km) beneath

REMORA. A remora swims beneath a tiger shark in hopes of attaching itself to the shark's belly. The remora can stick to any flat surface by applying the flat oval disk on top of its head to the surface and erecting a series of 13 to 25 V-shaped transverse plates to create a vacuum. *(Bureau of Sport Fisheries & Wildlife)*

the seafloor; explored the flat tops of guyots discovered by Harry Hess during World War II (see GUYOT); and discovered the surprising depths of Pacific trenches through exploration of the Tonga Trench. These discoveries provided the basis and the impetus for the advancement in many areas of oceanography.

In 1961, Revelle accepted the job of science advisor to the Secretary of the Interior and launched a new career. One of his first tasks was a study of salt accumulation and soil deterioration in West Pakistan—the site of the greatest irrigation project on earth. The panel led by Revelle recommended the building of tube wells to provide more, not less, water. The idea was to flush the salt from the sand.

ROSS SEA. Mount Erebus, the only active volcano on Antarctica, slopes upward on the edge of the Ross Sea. On the right is an unusual cloud formation. *(U.S. Navy)*

The panel also recommended improved farming practices. As a result of the action taken on the panel's recommendations, the productivity of the region increased by about 5 percent per year. For his work Revelle was awarded the order of "Sitara-i-Imtiaz" by the President of Pakistan. In the same year he was appointed by the Indian Parliament to the Education Commission of that government. The report of the Commission (1966) presented a comprehensive plan for educational reform in India. This work, in turn, led to his appointment, in 1964, as Professor of Population Policy at Harvard.

REVERBERATION See UNDERWATER SOUND.

REVETMENTS See BEACH.

RIDGE is a long, narrow elevation on the ocean floor with sides that are steeper than those of a RISE. One of the most impressive examples of a ridge is the Ninety East Ridge, a remarkably straight 3100-mi (5000-km) long ridge that runs north-south up the eastern INDIAN OCEAN from the vicinity of Amsterdam Island to the Bay of Bengal. However, the most spectacular ridge of all is the Mid-Ocean Ridge (see CONTINENTAL DRIFT) that winds for 40 000 mi (64 000 km) through all the world's oceans. The Mid-Ocean Ridge plays a key role in continental drift since the seafloor is spreading outward in both directions from the ridge axis as new material wells up to replace it.

If a ridge is curved, and particularly if it breaks the surface as islands, the ridge is called an *arc*.

RIP CURRENT See BEACH.

RISE is a long, broad elevation on the ocean floor with gently sloping sides. Thus it differs from a RIDGE only in that the ridge has steeper sides. The Lord Howe Rise between Australia and New Zealand is an example.

When a rise has an exceptionally broad top, it is called a *plateau*. The Mozambique Plateau off the eastern coast of Africa is an excellent example.

ROCK COD See GROUPER.

ROSS SEA is located off the continent of Antarctica and is a marginal sea of the South PACIFIC OCEAN. The sea is 600 mi (965 km) wide and extends seaward for 700 mi (1126 km). It is bounded on the west by Victoria Land and on the east by Marie Byrd Land. In this great embayment the Ross Sea lies north of 77°S latitude. South of this latitude lies the Ross Ice Shelf, an extension of the continental glacier which has ground its way out to sea until the outer edge has floated free of the bottom. The ice is tabular in form, has a thickness of about 1000 ft (305 m), only 150 ft (46 m) of which is above water. The ice shelf is 400 mi (644 km) across its seaward edge and covers an area the size of France. Giant pieces of ice break off as the tide lifts the outer portion and float out through the Ross Sea to endanger shipping. The blacksmith aboard the ship that first sighted the Barrier (as the Ross Ice Shelf was originally called) summed up his impression in these words:

Awful and sublime, magnificent and rare,
No other Earthly object with the Barrier can compare.

The Ross Sea takes its name from its discoverer, James Clark Ross of England. In 1840, Ross departed for Antarctica aboard the *Erebus* and *Terror*. In January 1841 these ships became the first to ever breach the Antarctic ice pack. Sailing into what is now the Ross Sea, Ross appears to have witnessed a FATA MORGANA (mirage); a lofty mountain simply disappeared as he sailed through it. Ross landed on Possession Island and named the area Victoria Land. On the same voyage he discovered an active volcano which he named Erebus, and the Ross Ice Shelf which inspired the words of his blacksmith.

The CONTINENTAL SHELF of the Ross Sea is both extensive and unusual. The seaward break in the shelf lies at around 1500 ft (457 m). This unusual depth [600 ft (180 m) is normal] is thought to be due to submergence of the shelf under the weight of ice during the last Ice Age. The shelf has two banks which run roughly parallel to the edge of the Ross Ice Shelf. These are thought to be terminal moraines (sediment deposited at the leading edge of a melting glacier). The sediment on the floor is mostly glacial in origin and is composed of poorly sorted terrigenous boulders, gravel, sands, and clays. The sea is covered with ice for only about 2 months out of the year.

See also SOUTHERN OCEAN.

RUST See CORROSION.

SAILFISH is the name for one species of FISH, *Istiophorus platypterus,* family Istiophoridae, order Perciformes. This fish is distinguished especially by its large first dorsal fin, which forms a "sail" when fully raised, and by a spearlike beak. (See FINS.)

Sailfishes are found in all tropical waters of the world's oceans. They inhabit the rough, well-oxygenated layers of these oceans and are especially plentiful in the CARIBBEAN SEA and the waters off the coast of southeastern Florida in the United States.

Sailfish are very fast swimmers with elongated streamlined bodies. They are large [anywhere from 7 to 15 ft (2 to 4.5 m)] and weigh several hundred pounds; the distinctive blue dorsal fin is folded down into a groove in the back of the fish when it is swimming. Other features include a long, sharp spear that extends from the upper jaw and a pair of pelvic spiny spikes on each side of the body just forward of the tail fin and below the small pectoral fins. The back of the sailfish is blue-black, and its sides and belly are silvery in color.

Their food consists of a varied diet of flying fishes, SQUID, OCTOPUS, and various small fish. These small fish are caught by the sailfish after it swims through schools of them, thrashing violently from side to side with the swordlike beak. The prey is thus killed or stunned by this action. After this, the sailfish eats its victims at leisure.

Sailfish are great game fish, and they put up a tremendous fight for the ocean sports fisherman. Many of the hooked fish are released in the interest of conservation of the species. See BILLFISH.

SALINITY is a measure of all the dissolved salts—not just sodium chloride—in seawater. It is defined as the total amount of dissolved solids in seawater [in parts per thousand (ppt; ‰) by weight] after certain chemical changes have taken place. The determination of salinity by this method is extremely cumbersome.

In the mid-nineteenth century, A.M. Marcet discovered the constancy of proportions in seawater, a fact later confirmed by Johann Forchhammer as well as by Wilhelm Dittmar during his analysis of the water samples collected during the CHALLENGER EXPEDITION. These chemists found that regardless of the absolute concentration of an individual constituent, the relative proportions of the major constituents in seawater are nearly constant. Thus, it is possible to determine the total salts in a sample of seawater simply by measuring only one of the major constituents. The chloride ion has commonly been used for this purpose.

CHLORINITY, expressed in parts per thousand (ppt; ‰), represents the mass of chlorine, bromine, and iodine in 1000 g (35 oz) of seawater if one assumes that the bromine and iodine have been replaced by chlorine. Chlorinity is determined by a titration method in which a solution of silver nitrate—previously standardized against a seawater sample of known salinity (See NORMAL WATER)—is used to produce a precipitate of silver chloride. To make the definition independent of atomic weight, chlorinity is taken as 0.3285233 times the weight of silver equivalent to all the halides. Using this approach,

$$\text{Salinity (ppt)} = 1.80655 \times \text{chlorinity (ppt)}$$

Today, this titration method is rarely used. The current technique capitalizes on the fact that electrical conductivity of seawater is a function of salinity. Therefore, salinity is quickly determined by measuring the electrical conductivity of a seawater sample.

The average salinity of the oceans is usually taken as 35 ppt. However, the horizontal distribution of salinity in the surface layers is quite variable and is influenced by several factors. Evaporation, the formation of SEA ICE , and the UPWELLING of more saline waters tends to increase salinity, whereas precipitation, the melting of sea ice, the upwelling of less saline waters, and the runoff from the continents tends to decrease it. Thus, salinity may drop as low as 5.0 ppt in the Gulf of Bothnia and exceed 40 ppt in the RED SEA. In general, the polar regions have a lower salinity than the lower latitudes with the exception of a narrow equatorial zone where a secondary low is produced by excess precipitation. The higher, open-ocean salinities are found at roughly 30°N and 20–30°S LATITUDE. Also, the ATLANTIC OCEAN is somewhat saltier than the PACIFIC OCEAN.

The vertical distribution of salinity in the ocean is not uniform, and a salinity low or minimum exists at 2100–2400 ft (700–800 m) over much of the southern Pacific, Atlantic and Indian Oceans. (See INDIAN OCEAN.) This is due to the cold, dense, low-salinity waters of the Antarctic Convergence Zone which sinks in low southern latitudes and flows northward as a broad, sluggish current. In the northern Pacific a similar salinity minimum, due to the Subarctic Intermediate Water flowing out of the BERING SEA and the Sea of Okhotsk, can be found (see OKHOTSK, SEA OF). Below this low-salinity zone the salinity increases to a maximum at 4500–12 000 ft (1500–4000 m).

According to legend, a poor farmer traded some strips of bacon to the Devil for a magic mill which, when properly commanded, would grind forth anything its owner wished. Ultimately, the mill was stolen by the captain of a fishing ship after he had tricked the farmer into telling him how the mill was started. When next his ship had caught its limit of fish, the captain ordered the magic mill to grind salt to preserve the fish. Only when sufficient salt had been ground did the captain realize that he had forgotten to ask the farmer how to turn it off. Eventually the ship sank due to the weight of the salt, and to this day the magic mill lies on the bottom of the ocean grinding away—and that is the reason the sea is so salty.

In point of fact, the legend is not too far from the truth. The actual magic mill is a combination of the Mid-Ocean Ridge that runs through all the world's oceans, and the volcanic activity associated with the subduction of oceanic plates (See CONTINENTAL DRIFT). Juvenile water—i.e., water never before in the liquid phase—associated with both these processes is a source of such seawater constituents as chlorine, bromine, iodine, carbon, boron, and nitrogen. The sodium is derived from the weathering of igenous rocks rich in sodium feldspar. This gradual addition of salts to the ocean has compensated for the loss caused by such factors as evaporation of seas cut off from the ocean; the salinity of the ocean therefore appears to have been relatively constant over the past several million years.

SALINITY-TEMPERATURE-DEPTH RECORDER is a device (also known as an STD recorder) that consists of sensing elements and a recorder for simultaneously recording measurements at sea of water TEMPERATURE, SALINITY, and depth. See INSTRUMENTATION.

SALMON is the common name for a number of fish in the family Salmonidae, consisting of 5 genera and about 24 species. Many salmon (the marine species) are noted for being ANADROMOUS FISH (anadromous comes from the Greek word for "running upward") and they ascend rivers to spawn. There is one marine species in the Atlantic Ocean (*Salmo salar*), while some 6 species of Pacific Ocean salmon exist (the Chinook, *Oncorhynchus tshawytscha;* the humpback, *O. gorbuscha;* chum, *O. keta;* silver, *O. kisutch;* masu, *O. masou;* and sockeye, *O. nerka*).

The fish in the family Salmonidae include the freshwater trout, graylings, and whitefish, and perhaps this close relationship accounts for the anadromous spawning behavior of the marine species of salmon. However, the reasons for and the mechanisms involved in this homing ability are not well understood. Some authorities speculate that originally this family of fish were once all of the freshwater type and some of the species migrated to the ocean for food. The evidence is not conclusive on this. However, it is very well established that both the Atlantic and Pacific salmon, which spend most of their lives in the sea (2 or 3 years or more), do return great distances to their birthplace in inland rivers to breed. Among the mechanisms thought to be employed by these fish to accomplish this feat are their sense of smell and the taste of the water. The Atlantic salmon migrates from the ocean mainly in November and December in its fifth year and breeds in the rivers of Europe from Spain to Ireland and in North America from Labrador to the coast of New England. There the females lay their eggs (some 2000 to 5000 in a series of nests) over gravel beds, and the males fertilize the eggs. After some 6 months the eggs are hatched, and the young "parr" grow slowly. When they have reached the end of their third winter, the fish are about 8 in (20 cm) long and become silvery in color. At that time they move downstream to the ocean. Here they spend about 2 years during which they live mostly in the upper layers feeding on the SHRIMP-like crustaceans in the PLANKTON as well as on small HERRING and other fishes. At the end of these 2 years, they are ready to move back upstream and spawn. Like the Atlantic salmon, the Pacific salmon also return to the same river in which they were hatched. Most of these return when they are 2–8 years old and during the period from July through November. Unlike the Atlantic salmon, the Pacific salmon die after breeding.

Both kinds of these oceangoing fish are of great strength and size; they could never be supported by the relatively small food supplies of the streams. To protect these food supplies, nature sends adult salmon back up the rivers supercharged with fat and energy so that they never have to feed in fresh water.

Atlantic salmon often return to the sea after breeding and can breed successfully two or more times. However, the Pacific salmon begin to change chemically as soon as they enter fresh water. This change is irreversible and causes the salmon to waste away and die a few days after breeding has been accomplished.

Although the young Atlantic salmon remain in rivers for three or more years before descending to the ocean, the young of Pacific species migrate to ocean waters a few months after hatching. Thus, there are at least 4 times as many young salmon going to sea from a typical Pacific coast river as from an identical Atlantic salmon river.

The five principal species of Pacific salmon can be briefly described as follows:

- *Chinook salmon, or king salmon.* The largest of the five species, it averages around 30 lb (13.6 kg). A typical Chinook has silvery sides and a bluish-green back marked with small dark spots. The flesh of the Chinook is very rich in oil and ranges in color from deep salmon (orange-pink) to almost white.
- *Sockeye, or red salmon.* These salmon average about 2 ft (61 cm) in length and 3–5 lb (1.36–2.26 kg) in weight. The males, when spawning, have a bright red body and a green head. The flesh is deep salmon in color, firm-textured, and contains considerable oil.
- *Pink salmon.* This salmon is also known as the humpback salmon because of the appearance of the males during spawning. They are common in Alaskan waters but are found in waters as far south as Oregon. Pink salmon, named for its paler flesh, ranges in weight from 3 to 6 lb (1.36 to 2.7 kg).
- *Coho, or silver, salmon.* These fishes weigh from 6 to 12 lb (2.7 to 5.4 kg) and are from 2 to 3 ft (61 to 91.5 cm) in length. The flesh is deep salmon, but lighter than the sockeye.
- *Chum salmon, or keta or calico salmon.* These fish migrate in the autumn and are the last Pacific salmon to run the rivers. They reach an average length of up to 3 ft (91.5 cm) and weigh up to 10 lb (4.5 kg). The flesh is lighter in color than the other species and has less oil.

Salmon have fed the human race since ancient times. The Roman scholar Pliny wrote in 77 A.D. that, ". . . . the river salmon is preferred to all fish that swim in the sea."

Both the Atlantic and Pacific salmon have long been highly prized not only as commercial food fishes, but as game fishes as well. The Pacific salmon are now being exploited more commercially, although fleets from Greenland, Denmark, Sweden, Norway, and West Germany have participated in the offshore fishery for Atlantic salmon off Norway and in the BALTIC SEA. Commercial fishing for salmon is primarily by means of purse seines, gill nets, beach seines, and by trolling. Both types of this fish, however, have been severely threatened as a resource by overfishing and river pollution, as well as by the presence of dams, power stations, etc. There has been a growing awareness of this adverse situation, and steps toward conservation and some husbandry of these valuable fishes are being taken by many individual countries as well as by international agreement.

SALT is a compound formed when one or more of the hydrogen atoms of an acid are replaced by one or more cations of a base. A common example is sodium chloride (NaCl) in which the hydrogen ions of hydrochloric acid (HCl) are replaced by the sodium ions (cations) of sodium hydroxide (NaOH).

See also DESALINATION; ION; SALINITY; SEAWATER.

SALT CYCLE See AGE OF THE OCEAN.

SALT DOME See CONTINENTAL SHELF.

SALT GLAND is a specialized gland located around the eyes and nasal passages of those MARINE BIRDS which spend considerable periods of time at sea (e.g., gulls, petrels, albatrosses, and fulmars), as well as those of marine TURTLES and SEA SNAKES. The function of salt glands is to secrete significant amounts of a watery fluid containing a high percentage of salt, higher than that contained in the urine. As a result of this action, these animals are able to ingest SEAWATER without experiencing the dehydration necessary to eliminate the excess salt via the kidneys.

SALTON SEA See INLAND SEA.

SAND DOLLAR is the common name for a flat, disk-shaped ECHINODERM belonging to the order Clypeasteroida (see SEA URCHIN). The nearly circular body of the sand dollar (*Dendraster excentricus*) may reach 3 in (7.6 cm) in diameter. There is a characteristic pattern of a five-part flower on the aboral (top) surface of the shell.

SAPROPEL is the name for black or brown sediment of marine, estuarine, or lacustrine (belonging to lakes) deposition made up principally of organic debris. (See MARINE SEDIMENTS.) Sapropel is finely divided material that is rich in iron and sulfide and is chemically strongly reducing.

SARDINES are the young of the pilchard, a herringlike fish in the family Clupeidae found in the ATLANTIC OCEAN along the European coasts. The term is also applied to the young of any of various similar and related forms which are processed and eaten as sardines. See HERRING.

SARGASSO SEA is located in the Central North ATLANTIC OCEAN. Though as large as both the CARIBBEAN SEA and the MEDITERRANEAN SEA combined, the Sargasso is a sea without a coast; it is a liquid hill storing enough hydrostatic energy to drive the GULF STREAM, the NORTH ATLANTIC CURRENT, the CANARY CURRENT, and the North Equatorial Drift Current (see EQUATORIAL CURRENT SYSTEM) without the aid of the wind, for approximately 5 years; and yet it is speckled and windrowed with lush vegetation, floating in water nearly 3 mi (5 km) deep. This unique body of WATER covers an area of 2 million (2×10^6) mi² (5.2 million km²) in the vicinity of the horse latitudes (30°N), and is a giant eddy created and bounded by the Gulf Stream in the west, the North Atlantic Current in the north, the Canary Current in the east, and the North Equatorial Drift Current in the south. The combination of sluggish movement within the eddy and weather conditions which cause evaporation to exceed precipitation (classic definition of a desert) produces a lens of water, 3000 ft (914 m) deep near its center, which is warmer (lighter) and more saline (36–37 ppt) than the water within which it floats. The action of currents and winds piles the lighter water at the center so that it reaches a height nearly 3 ft (1 m) above that of sea level along the U.S. Atlantic Coast. The lens is bounded by a sharp temperature gradient (THERMOCLINE), usually taken as the 50° F (10° C) isotherm, which retards the upward migration of rich nutrients from the colder water below. As a result, animal life is sparse and the description "clearest, purest, and biologically poorest ocean water ever studied" is well earned.

The Sargasso derives its name from about eight species of seaweed (Sargassum) that float conspicuously about the sea in clumps and in long windrows. Columbus noted floating seaweed in the vicinity of the Azores as he sailed west to the New World in 1492. Later, becalmed in mid-ocean, as was quite common for sailing ships in those latitudes, Columbus' crew was much disturbed by the abundance of floating seaweeds, thinking the seaweeds indicated shallow water, when the ocean floor lay nearly 15 000 ft (4572 m) below their keel. But it was Portuguese sailors on later voyages who named the sea: They applied the name of one of their grapes—Salgazo—to the seaweed (and later the sea) because the small air bladders by which the seaweed float looked quite similar.

Much of the myth and legend that has grown up around the Sargasso Sea is due to these unique seaweeds once thought to be stripped by storms from reefs and shallows around the North Atlantic Basin and swirled by winds and currents into the center of the ocean. While this may once have been so, sargassum has now evolved into unique and self-sustaining species that reproduce vegetatively. Although they are abundant in places, they represent no hazard to even the smallest boat.

SATELLITES See INSTRUMENTATION; SPACECRAFT OCEANOGRAPHY.

SAVU SEA in the Indonesian Archipelago, is defined by a line running east along the southern limits of the FLORES SEA and the BANDA SEA, south on meridian 125°E between Alor and Timor, turning west through Timor to Roti, Poeloe Dana, and the western point of Sumba Island. The Savu Sea covers an area of 40 530 mi² (105 000 km²), occupies a volume of 42 704 mi³ (178 000 km³), and has a mean depth of 5580 ft (1701 m). The maximum depth is 11 384 ft (3470 m).

The central and eastern part of the Savu Sea is occupied by the Savu Basin which rises rather smoothly to its shallow periphery. The bottom sediments are mostly terrigenous muds, with volcanic muds and globigerina ooze occurring in places. Water flows into and out of the sea through several straits on the north and south (i.e., Ombi and Sumba straits in the north, and Savu and Dao straits in the south). The Savu Sea lies about 10°S of the equator and is alternately influenced by the southeast trade winds and the monsoon belt. Many typhoons are generated in this general area.

SAWFISH is the name applied to any of the six species of marine cartilaginous fishes of the family Pristidae, order Hypotremata, in the class Selachii. These viviparous fishes have a sharklike tail and a flattened, elongated snout drawn out into a long flattened blade with toothlike sidewise projections along each edge.

Sawfishes are strange-looking demersal fishes that live in the coastal zones of all the warm, shallow, and muddy waters of the world's oceans. Being bottom dwellers primarily, their diet consists mainly of mollusks, CRUSTACEANS, and small shoaling fishes. (See MOLLUSK.) They are especially prevalent in the oceans of tropical America and Africa and are often found around estuaries as well as upriver in some rivers that empty into the ocean. Representative species are the common sawfish, *Pristis pectinatus,* found in the GULF OF MEXICO and the Zambesi sawfish, *P. perroteti,* of African waters. *Pristis cuspidatus* goes up the large rivers of India.

With the exception of its unique saw, the fish resembles a cross between a ray and a SHARK, and both are closely related to the sawfish (see RAYS). Sawfishes usually grow to be as long as 20 ft (6 m). They possess a sharklike tail section and a rather flattened forward body with large pectoral fins and gill openings on the underside of the head. The "rostrum," or flat blade of the saw, on the snout is often 6 ft (2 m) long in the large fish. Made of cartilage, this blade contains dermal denticles or skin teeth (rather than true teeth) laterally set in deep sockets. Such teeth are also characteristic in sharks and rays.

Sawfishes are viviparous, with the eggs being incubated and hatched within the body of the female. The saws of the baby sawfish are soft at birth, and until birth these saws are enclosed in membranes. A litter of baby sawfishes often numbers 20 or more.

SCALLOP is the common name for any MOLLUSKS in the class BIVALVIA, family Pectinidae, characterized by radially ribbed valves with undulated margins. See CLAMS.

SCORPION FISH is the name for 300 species of fish of the family Scorpaenidae. Such fishes are found at varying depths of the temperate waters of the oceans although some inhabit tropic waters. The shallow-water species have more poisonous spines than other members of the family, which include the turkey fish, *Pterois volitans,* the barbfish, *scorpaena brasiliensis,* and the California scorpion fish, *S. guttata* among others. The sculpin, often called a scorpion fish, is a member of the family Cottidae. See VENOMOUS MARINE LIFE.

SCOTIA SEA is located in the southwestern ATLANTIC OCEAN. It lies within the 2698-mi- (4344-km long loop made by the Scotia Ridge which connects Tierra del Fuego at the tip of South America with the Antarctic Peninsula on the continent of Antarctica. The Scotia Ridge, which supports the Falkland, South Georgia, South Sandwich, South Orkney, and South Shetland islands, is thought to be an extension of the Andes mountain chain. Drake Passage, which occupies the 621-mi (1000-km) gap between South America and Antarctica, is the western boundary of the Scotia Sea, and it is separated from the WEDDELL SEA on the south by the Scotia Ridge.

Drake Passage is named for Sir Francis Drake, the British buccaneer and explorer. However, Drake did not use the Passage during his voyage (1577–1580) around the world. He first sailed through the Strait of Magellan (between South America and the islands of Tierra del Fuego), but was then blown south by a storm and forced to spend several weeks within the island cluster (Tierra del Fuego). The records indicate that he is not likely to have discovered Cape Horn as has often been claimed. Drake did, however, form the opinion during this enforced stay that the WATER of the Atlantic and the water of the PACIFIC OCEAN rolled together somewhere south of his position.

The Falkland Islands on the northern rim of the Scotia Sea were first sighted in 1592 by John Davis, an English explorer. Another Englishman, John Strong, was the first to set foot on the islands in 1690. Strong named the islands for Viscount Falkland, the treasurer of the British Navy. The islands are now a part of the British Antarctic Territory which includes South Georgia, South Orkney, South Sandwich, and South Shetland. The population of these islands is about 3000 people, primarily of British origin, who derive their living from sheepherding, fishing, and whaling.

The great eastern loop of the Scotia Ridge, one of the more unusual features of the ocean floor, encloses the basin of the Scotia Sea. A minor rise running between South Georgia and South Orkney islands separates the area into an East and West Scotia Basin. The basins are largely covered by oozes made of the tests (skeletons) of DIATOMS and foraminifera. Over parts of the basins, particularly those covered by foraminiferal ooze, are to be found MANGANESE NODULES. Sediment along the edge of the Scotia Ridge and in the Drake Passage tends to be coarser (sands, clays, and silts) due to the winnowing effect of the current flowing eastward between South America and Antarctica. Outside the convex arch of the Ridge is the South Sandwich Trench with depths to 27 132 ft (8270 m).

The major current influencing the Scotia Sea swings up out of the BELLINGSHAUSEN SEA to the west (due to deflection by the Antarctic Peninsula) and flows through the Drake Passage as a mighty river. (See ANTARCTIC CIRCUMPOLAR CURRENT.) Estimates of the water transport through the Passage range from 3.2 billion (3.2×10^9) $ft^{3/5}s$ [90 million (90×10^6) m^3/s] to nearly twice that much. Even the lower estimate is equivalent to 400 Amazon Rivers. The Scotia is also influenced by the Antarctic Convergence Zone which runs through the Drake Passage and circles the Antarctic continent. In this zone the cold (heavy) Antarctic water meets the warmer (lighter) water of the South Atlantic to form a sharp density gradient. The colder, nutrient-laden water sinks beneath the warmer water to flow northward as the Antarctic Intermediate Current. Life flourishes in this zone, and it is here that the blue WHALES come to feed on KRILL.

The surface water temperature in the Scotia Sea ranges from 32–42.8°F (0–6°C). During winter the ice pack extends into the Drake Passage, but the sea is mainly influenced by drifting ice.

SCRIPPS INSTITUTION OF OCEANOGRAPHY is the oceanographic branch of the University of California at San Diego; this institution is world-famous and is one of the most important centers for research and graduate education in the marine sciences.

Scripps Institution, which now occupies 64 buildings on 230 acres, was first an independent biological laboratory dating back to 1903. It became an integral part of the University of California in 1912 when it was designated as the Scripps Institution for Biological Research. The scope and character of the research program ultimately focused on all aspects of the study of the sea. This fact was formally recognized on October 13, 1925, when the name was changed by the Regents of the University of California to the Scripps Institution of Oceanography.

The scientific scope of the Institution's research now embraces physical, chemical, geological, and geophysical studies of the oceans, as well as biological studies.

Scripps' research ships traverse all the world's oceans. Their cruises vary from local, limited-objective trips to far-reaching expeditions designed to gather a variety of oceanographic data.

The graduate department at Scripps offers instruction leading to degrees in OCEANOGRAPHY, marine biology, and earth sciences. The emphasis is on the Ph.D. program, and it is at this level that students are normally enrolled. Usually students will concentrate on one of the following curricular programs: applied ocean sciences, biological oceanography, geological sciences, geophysics, marine biology, marine chemistry, and physical oceanography.

SCUBA See DIVING.

SEA See OCEAN.

SEA ANEMONE is the name for any of the 1000 marine cnidarians (COELENTERATES) that constitute the order Actiniaria in the class Anthozoa; as a class these animals comprise over 6000 marine species such as the CORAL, sea fans, sea pens, sea feathers, and sea anemones.

Sea anemones, so-called because of their flower-like appearance, are widely distributed in the world's oceans. They are most abundant in the tropics; however a considerable number of species are found in the colder waters, including those of the ARCTIC OCEAN. They live in areas ranging from the water levels of TIDES to great depths in the oceans.

Generally solitary, this soft-bodied, cylindrically shaped polyp, in its adult form, possesses at one end a series of tentacles surrounding the distensible mouth. The stomach of the anemone contains powerful juices to digest food. Most species have a pedal disk at the base of the organism, used for attachment to some object (e.g., underwater ledges, coral, etc.).

Other species, lacking this disk, dig into the sediment, and some swim using their tentacles for movement. Water CURRENTS moving into the animal's mouth provide oxygen for respiration. Anemones vary widely in color, and sizes also vary, with some of the smaller species averaging about $\frac{5}{8}$ in (15.9 mm) in length and $\frac{1}{2}$ in (12.7 mm) in diameter. However, some PACIFIC OCEAN species attain an 8-in (20-cm) diameter. The largest, *Stoichactis*, measures 3 ft (1 m) in diameter.

Anemones normally feed on CRUSTACEANS, mollusks, and small FISH which they trap by paralyzing them with virulent nematocysts (harpoon-like cells). (See MOLLUSK; NEMATOCYST.) Some anemones are commensal and live in close association with other forms of marine life such as SPONGES (of the class Hexactinellida) and the hermit crab, among others.

The species, *Edwardsia leidyi, Metridium,* and the MATA-MALU (or *Rhodactis howesii*) are frequently studied by biologists—the first two for their methods of reproduction, which can be both sexual and asexual in anemones, and the MATA-MALU (of Samoa) for the nature of the posion it contains. No pathological, pharmacological, or immunological studies of this lethal anemone have resulted in an antidote, and no one who has eaten matamalu has ever survived. Relative to their reproductive processes, anemones often reproduce asexually by dividing into two parts, each of which becomes a complete animal. Also some species, in moving from place to place in the ocean, leave fragments behind which grow into tiny anemones. They also reproduce sexually, chiefly by discharging eggs and sperm into the water. The fertilized eggs develop into larvae which eventually settle on the bottom and become baby anemones. (See LARVA.) However, fertilization may also be internal, in which case the young anemones develop in the parent coelenteron (or interseptal chambers).

Anemones (such as the species *Actinia equina*) are found in the ATLANTIC OCEAN, the MEDITERRANEAN SEA, and the BLACK SEA, while the rosy anemone (*Sagartia elegans*) inhabits eastern Atlantic Ocean waters from Iceland to the African coast and the Mediterranean Sea.

SEA BASS See GROUPER.

SEA BIRDS See MARINE BIRDS.

SEA COWS is the name generally applied to members of four species of herbivorous, aquatic mammals belonging to two families (Trichechidae or the manatees, and Dugongidae or the dugongs) of the order Sirenia.

Sea cows inhabit the tropical coastal waters and estuaries, and some ascend rivers as well as constructed canals. They subsist primarily on underwater vegetation in these areas. Both the manatees and dugongs have almost hairless, huge, fat bodies, that somewhat resemble those of WHALES although they are unrelated to these animals. Sea cows, or Sirenians, evolved from land herbivores, and their closest living relatives are the elephants.

The forelimbs of the Sirenians are represented by flattened flippers, the hind limbs are absent, and the tail flukes are horizontal rather than vertical like a fish's tail. There are no external ears, and the odd-looking head has a broad snout with small membrane-covered eyes and large rubbery lips bordered with bristles. In reproduction, the females bear one calf at a time after a gestation period of a year.

The dugongs are found in varying numbers in the coastal waters of the RED SEA, the INDIAN OCEAN, the east coast of Africa, the East Indies, Southern Asia, and Australia. All are regarded as being the same species, *Dugong dugon,* an animal that reaches a 10- to 12-ft (3–3.6 m) length and a weight of about 1500 lb (680 kg). They are more marine in habit than the manatees and, unlike them, the male dugongs develop two 10-in (25-cm) tusks from their upper incisor teeth. Dugongs are vegetarians and forage nocturnally on aquatic vegetation. Although they are generally solitary, aggregations of hundreds have been reported.

Slate-gray 1-ton manatees that are found in the CARIBBEAN SEA, in the coastal waters off Florida, especially the GULF OF MEXICO, are the *Trichechus manatus* species, or Florida manatees. Those found from Venezuela to Northern Brazil are called Amazonian manatees (*T. inunguis*). The West African manatee, which has some skeletal differences from the others, is named *T. senegalensis.* All differ from the dugongs in that they have scattered hairs on their bodies, lack incisors, and have hydrodynamic features that make them less completely adapted to an aquatic existence. Both the dugongs and the Trichechids or manatees are mainly nocturnal and usually solitary. They can individually consume 60–100 lb (27–45 kg) of vegetation daily, swallowing the food while submerged.

It is thought that the sea cows gave rise to the mermaid myth originated by the ancient Greek and Roman sailors. Apparently, the sight of a dugong in an upright position in the water holding a calf with one flipper and nursing it in a manner similar to that of a human female was the basis of this belief.

The protein-rich meat of the sea cow has often been praised for its delicacy and flavor. In addition, the skin makes a tough leather, the blubber can be converted to oil, and the rib bones can be used as a substitute for ivory. Because of these attributes, plus the fact that these animals are sluggish and unsuspicious, they were eliminated in great numbers when they were extensively hunted.

The voracious weed-eating habits of these endangered animals serve to act as a control of aquatic plants such as the water hyacynths, which clog canals and irrigation systems in tropical areas. This is important, since infestations of such weeds foster waterborne diseases, inhibit rice cultivation in flood paddies, and render irrigation projects useless.

SEA CUCUMBERS is the name commonly applied to the echinoderms that comprise the class Holothuroidea. They are sometimes called sea slugs, *bêche-de-mer,* or trepangs. An edible type (*Holothuria edulis*) is prized among the Chinese as a delicacy. Physiologically these animals resemble a hollow tube with an opening on each end, a mouth and a vent. See ECHINODERM.

SEA COW. The sea cow, a sluggish animal, can consume 60 to 100 lb (25 to 50 kg) of aquatic plants daily. *(Miami Seaquarium, Florida)*

SEA FARMING See MARICULTURE.

SEA GRANT PROGRAM is a United States national program (started in 1966) that consists of applied research, education, training, and advisory service activities in the fields of OCEANOGRAPHY and OCEAN ENGINEERING. This program of the National Oceanic and Atmospheric Administration (NOAA) of the U.S. Department of Commerce supports a network of sea grant colleges and local sea grant programs through its granting process. Through such programs of research, education, and advisory services, the Sea Grant Program serves as a link between people who use the oceans and those who study it.

Basically the Sea Grant Program in the United States was established for the purpose of

- Initiating and supporting programs at (sea grant) colleges and other suitable institutes, laboratories, and public or private agencies for the education of participants in the various fields relating to the development of marine resources.
- Encouraging and developing programs consisting of instruction, practical demonstrations, publication, and otherwise, by sea grant colleges and other suitable institutes, laboratories, and public or private agencies through marine advisory programs, with the object of imparting useful information to persons currently employed or interested in the various fields related to the development of marine resources, the scientific community, and the general public.
- Initiating and supporting necessary research programs in the various fields relating to the development of marine resources, with preference given to research aimed at practices, techniques, and design of equipment applicable to the development of marine resources.

Programs to fulfill these purposes are accomplished through contracts with, or grants to, suitable public or private institutions of higher education, institutes, laboratories, and public or private agencies which are engaged in, or concerned with, activities in the various fields related to the development of marine resources.

It is generally conceded that the Sea Grant Program has been very successful and that it has made some valuable and long-lasting contributions. This acknowledgement of success has been responsible for a broadening and strengthening of the program especially in education and training activities in order to develop the kind of technical expertise needed for better management, understanding, development, conservation, and utilization of the ocean and coastal resources of the United States. Sea grant research, education, and advisory services have produced a mechanism for bringing about advances in oceanography, ocean industry, and coastal management. Such services have been highly valuable in advising industry of the potential of the world's oceans, for creating new technologies, and for providing counsel on the management of coastal environments, the treatment of OIL SPILLS, and new resources.

See also SPILHAUS, ATHELSTAN FREDERICK.

SEA ICE is defined as that ice which has been formed through the gradual freezing of seawater. As such, it is largely confined to the ARCTIC OCEAN and the SOUTHERN OCEAN, although many northern seas such as the LABRADOR SEA, BERING SEA, Sea of Okhotsk, JAPAN SEA and YELLOW SEA form partial ice cover during the winter. (See OKHOTSK, SEA OF.) The great icebergs produced in both Arctic and Antarctic regions are land ice resulting from the flow of glacial ice and shelf ice (i.e., the Ross Ice Shelf) into the ocean where it calves off to become free floating. Sea ice, though it covers a considerable area during its maximum growth, is insignificant when compared with the ice cover of Greenland and Antarctica. However, because of its ability to reflect solar radiation and to reduce heat exchange between WATER and atmosphere, it has a significant effect on global weather.

Pure water increases in density as the temperature decreases and reaches its maximum density at 39°F (4°C). Since the water density cannot be further increased (the water column is stable), the temperature of the surface layer rapidly drops to its freezing point of 32°F (0°C), and ice forms.

In seawater, conditions are a bit different. As salts are added to pure water, both the temperature of maximum density and the freezing point decrease in a linear fashion until they coincide at a salinity of 24.7 ppt and a temperature of 29.61°F (−1.33°C). (The freezing point of typical seawater at a salinity of 35 ppt is taken as 28.6°F (−1.9°C). The freezing point continues its linear decrease so that seawater freezes before the temperature of maximum density is reached for all salinities greater than 24.7 ppt. In order for ice to begin forming on the sea surface, the depth must be very shallow (along the coast), or the surface layer must be underlain by a zone of warmer, more saline water to prevent the surface layer from sinking before it freezes. The latter condition exists in the Arctic Ocean where there is a sharp increase in salinity at about 61 ft (20 m).

Sea ice begins with the formation of tiny crystals on the surface. In the coastal waters of Antarctica this initial process is aided by the snow blown into the ocean by the katabatic winds that roar down from the interior (see SOUTHERN OCEAN). This now serves as a nucleus for the formation of crystals. As the formation of crystals and spicules progresses, the sea surface takes on a greasy or leaden appearance because of the way light is reflected by the tiny ice surfaces. As freezing progresses, a slush is formed which gradually coalesces into a thin layer of ice. At this point the ice trends easily with the motion of the sea surface. However, as the ice thickens, it becomes more brittle and begins breaking into pieces ranging from 20 to 40 in (50 to 100 cm) across. Through repeated collision with other ice, these pieces become rounded into what is known as pancake ice. With further freezing, the pieces of pancake ice coalesce to form a continuous sheet known as pack ice.

As ice crystals form, salts are excluded because foreign ions or atoms, apparently, cannot be accommodated within the tight water lattice. If sea ice were to freeze very, very slowly under stable conditions, salt-free ice would result. However, because the freezing takes place with relative rapidity, some salt is mechanically trapped between the crystals. New sea ice which freezes rapidly has a salinity of about 20 ppt, but the bulk of older ice has a salinity of only about 4 ppt. This reduction in the salinity of sea ice with time results from a downward migration of the brine pockets under the influence of gravity, and from a flushing action produced by the downward percolation of fresh water from melting snow (or the surface layer of ice) during the summer season. The pools of melt water on the pack ice are sufficiently fresh to be used for drinking water.

In the Arctic Ocean there is a permanent cover of pack ice which occupies the central portion. Only the marginal seas manage to partially or completely

SEA ICE

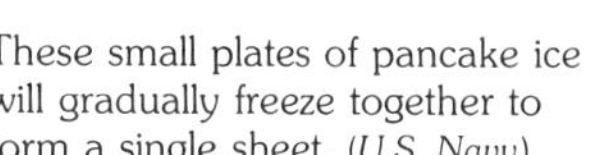

These small plates of pancake ice will gradually freeze together to form a single sheet. *(U.S. Navy)*

Sea ice floats in McMurdo Sound, Antarctica. Biological studies were conducted from the insulated hut (at the bottom of the picture) by way of holes drilled in the ice. *(U.S. Navy)*

A large iceberg in the Southern Ocean looms ahead. The view is from near sea level. *(U.S. Navy)*

Ships wind their way through the ice pack just off Discovery Harbor en route to Devon Island during an Arctic expedition. *(National Archives)*

clear their surface of ice during the summer. However, even the permanent pack ice undergoes some melting during summer. From an average thickness of about 6 ft (2 m) [ice older than 1 year may thicken to 15 ft (4.5 m), and pressure ridges may thicken the ice to 80 ft (24 m)—even thicknesses of 150 ft (46 m) have been noted], 12–16 in (30–40 cm) may be lost by surface melting, and 4–8 in (10–20 cm) may be lost from the bottom by the warming action of the water beneath the ice cover. The only permanent sea ice in the Southern Ocean is around the rim of Antarctica and in the WEDDELL SEA. However, during the winter months (April–November) a pack ice cover with a thickness of around 60 in (150 cm) builds outward in an irregular pattern that covers 30 percent of the Southern Ocean south of LATITUDE 40°S. It is interesting to note that the energy required to melt this ice each year during the Southern Hemisphere summer is equivalent to 16 thousand 100-megaton bombs.

Icebergs, a special form of sea ice or ice at sea, are produced by glaciers and ice shelves on Antarctica, Greenland, Alaska, Spitsbergen, Novaya Zemlya, and other polar landmasses. These great ice sheets grind their way down to the sea and out onto the shallow seafloor until the water depth is sufficient to float them free at the outer edge. Eventually the slow heaving of the tide and its own buoyancy breaks large tabular blocks loose from the parent mass in a process known as calving. The icebergs of the Northern Hemisphere are generally smaller than those of the Southern Hemisphere. The largest iceberg ever sighted in the north was off Baffin Island in 1882—it measured 7 by 3½ mi (11 by 5.6 km). Most northern icebergs are much smaller. The southern icebergs, forming up on the vast ice shelves such as the Ross and Filchner Ice Shelf are the true giants. The Ross Ice Shelf extends out from land as much as 500 mi (805 km) over a 400-mi (644-km) front and is composed of ice 1000 ft (305 m) thick that stands above the water surface as an ice cliff 150 ft (46 m) high. The largest iceberg ever reported in the Southern Hemisphere was located 150 mi (241 km) west of Scott Island and measured 60 by 208 mi (95 by 355 km)—twice the size of the state of Connecticut. This iceberg stood 100 ft (30 m) above the surface of the sea. Since an iceberg floating in water with a salinity of 35 ppt and a temperature of 30°F (– 1°C) will have 87.5 percent of its mass below water, this enormous iceberg had a total thickness of about 800 ft (244 m).

As icebergs move away from their place of origin, they usually encounter warmer water and melt rather rapidly. However, the average age of a Southern Hemisphere iceberg is about 4 years. Under the influence of wind and currents, icebergs often float into shipping lanes and pose a hazard to navigation. Icebergs have drifted as far north as the tip of Africa—2000 mi (3218 km) from their origin. As a direct result of the sinking of the *Titanic* by an iceberg, the International Ice Patrol was formed by a treaty in 1913. Throughout the iceberg season the U.S. Coast Guard maintains surveillance over the affected areas of the North Atlantic and provides iceberg location to ships that request such information. The introduction of radar aboard ships has greatly reduced this hazard.

SEA LEVEL, or more precisely, mean sea level, is defined in the United States as the average height of the sea surface for all stages of the TIDES over a 19-year period. The tide measurements are taken by means of fixed tide gauges along the coast or in the adjacent WATER with free access to the open ocean. The resulting determination of sea level is fixed by means of a benchmark to which all measurements above and below sea level are referred. Sea level serves as the reference plane from which all elevation surveys are made [i.e., Mount Everest is 29 028 ft (8848 m) above sea level]. In meteorology, mean sea level is used as the reference surface for all altitudes in upper-atmospheric work; and in aviation it is the level above which altitude is measured by a pressure altimeter.

In the past, mean sea level determined by a world net of tide gauges has served reasonably well for most applications. However, both the short- and long-term inaccuracies of the technique have long been recognized and have become increasingly bothersome in recent years. The expression "water seeks its own level" leads to the impression that under conditions of dead calm, sea level would be the same at every point in the world's oceans. But, in reality, the surface of the ocean is a pattern of hills, valleys, depressions, and slopes which shift and pulse with changes in a number of environmental factors.

The more important factors affecting local and regional sea level are (1) changes in atmospheric pressure, (2) changes in the total heat content of the water, (3) prevailing wind patterns, and (4) currents combined with the CORIOLIS EFFECT. For instance, a 1-mbar (9.9×10^{-4}atm) increase in atmospheric pressure results in a 0.4-in (1-cm) increase in sea level; the water simply flows laterally to areas of lower atmospheric pressure in order to reestablish equilibrium. The passage of a high- or low-pressure disturbance (see HURRICANE) can raise or lower the local sea level by as much as 3.28 ft (1 m).

Both SALINITY and heat affect the density of seawater which, in turn, affects sea level (the less the density, the higher a column of water stands). It has been estimated that a 1°C rise in temperature throughout the world's oceans would raise sea level by about 23.6 in (60 cm). In part, this effect can be witnessed with the changing seasons. During the Northern Hemisphere summer the water surface north of the equator gradually rises to about 4–6 in (10–15 cm) above normal, while that south of the equator decreases by a similar amount. During the Northern Hemisphere winter the process is reversed. It should be noted that due to the influence of other factors, this effect is not uniform and local anomalies, both negative and positive, occur. This seasonal effect is superimposed on the more basic fact that the Atlantic water is warmer than that of the PACIFIC OCEAN, but the latter is less saline than the former. As a result, the surface of the Pacific stands about 12–20 in (30–50 cm) above that of the Atlantic. A similar gradient exists along the east coast of North America. Sea level off Nova Scotia stands about 15 in (38 cm) above that at Miami, Florida.

Due to the Coriolis effect, currents are deflected to the right (clockwise) in the Northern Hemisphere and to the left (counterclockwise) in the Southern Hemisphere. This imposes a slope to the water surface which raises sea level toward the seaward rim of the current. In the case of the GULF STREAM this amounts to a difference in sea level of about 23.2 in (59 cm).

Over the long history of the earth [4.5 billion (4.5×10^9) years] sea level has remained remarkably stable. On the other hand, transitory swings in sea level have resulted in large areas of the CONTINENTAL SHELF being raised above water, and the sea has repeatedly inundated large regions of the continents. Those factors known to affect sea level over long periods of time are (1) the subtraction and later addition of water to the oceans due to continental glaciation; (2) the depression and later rebounding of land levels due to the accumulation and dissipation of glaciers; (3) the steady accumulation of sediments in the ocean basins due to the wearing down of continents; (4) the deformation of the ocean basin through earth movement and volcanism; and (5) the isostatic changes brought about by evolving densities (gravitational anomalies) in regions of the earth's crust.

The more recent long-term changes in sea level can be traced to the onset and retreat of glaciers during the last Ice Age. About 35 000 years ago sea level was roughly the same as it is today. Then, about 30 000 years ago it began to drop. By 15 000 years ago sea level had dropped by about 426.5 ft (130 m) worldwide. The continental shelves stood largely bare, and streams that had once met the ocean at their landward edge now played across their surfaces depositing vast quantities of sediment released by the glaciers and loosened by the rush of their melt water. Between 11 000 and 8000 years ago the glacial advance reversed and sea level began to rise rapidly. By 5000 years ago it was within 16.4 ft (5 m) of its present level. The rise in sea level over the past 5000 years has been due to the continuing retreat of ice locked away on land as continental glaciers and as mountain glaciers. In some areas this trend has been offset by the rebounding of the land surface resulting from the removal of the ice. For instance, the floor of the Gulf of Bothnia in the northern BALTIC SEA is rising at the rate of about 3.28 ft (1 m) per century, due to rebounding, and the Great Lakes are tilting toward the south. But, worldwide, sea level appears to be increasing at the rate of about 0.05 in (0.12 cm) per year. This is a long-term average with many positive and negative swings in between.

Since much construction by human beings along the coastal areas of the world is sensitive to existing sea level, even a minor long-term increase in that level could be disastrous. For instance, storm waves superimposed on a 1-ft (0.3 m) increase in sea level might well devastate the dykes of Holland as well as many other structures around the world. Today, satellites are becoming increasingly valuable in more accurately measuring the level of the sea's surface and in accumulating the data necessary for long-term predictions.

SEA LILIES See ECHINODERM.

SEA LIONS are those pinnipeds of the family Otariidae that lack a valuable fur such as that possessed by their relatives, the fur seals. They are fairly docile and easily trained to do complicated tasks. See SEALS.

SEA NOISE is the sound in the ocean produced from a number of sources such as marine animals, water currents, motion of WAVES, and shipping traffic. Sound travels about 5 times faster in ocean water than it does in air—4850 ft/s (1480 m/s) as contrasted to 1080 ft/s (327 m/s). The underwater world is a comparatively noisy place, with the average sound level equal to about 10–15 decibels above 1 μbar (which is somewhat comparable to the noise encountered in a busy office). By use of a sound spectrograph, these sea sounds can be measured, a graphic record can be made with the com-

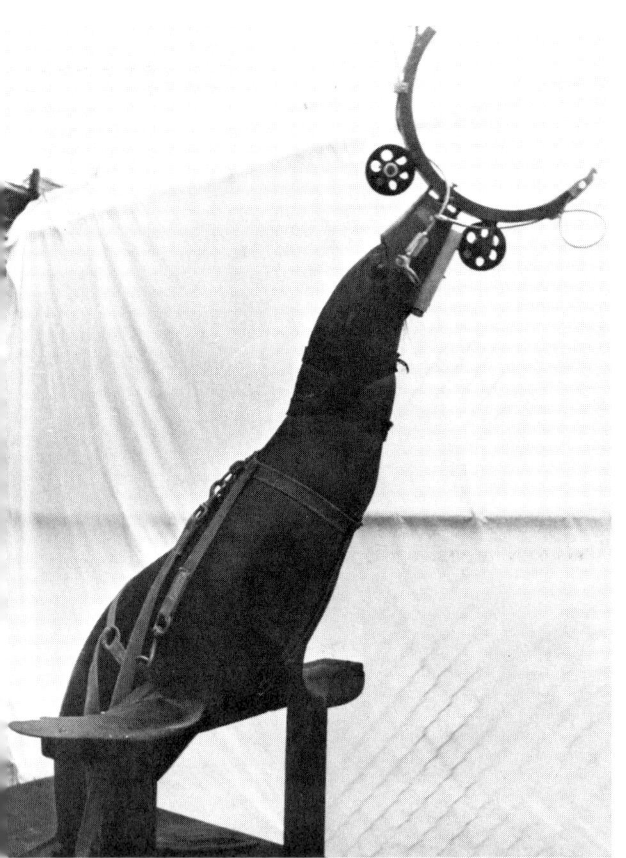

SEA LION. (*Top left*) A sea lion poses with the grabber device he has been taught to press against an underwater target, such as an antisubmarine rocket fired from a Navy ship during tests. The device encircles the target and locks on it when pressed against it. (*U.S. Navy*) (*Top right*) A baby sea lion approaches close enough to nuzzle the camera. (*U.S. Navy*) (*Bottom right*) Sea lions on rookery in Alaska about a week before the height of the breeding season. (*NOAA*)

ponent parts, and various frequencies can be analyzed. (See UNDERWATER SOUND.)

The sources of the sounds are varied. Most obvious are those caused by wave motion on the surface, by shipping traffic, or that caused by ships' engines and screws, by the friction of moving water currents against one another and against the ocean floor, and by marine animals.

Concerning the latter, many CRUSTACEANS (e.g., the CRAB, SHRIMP, and LOBSTER) and mollusks (e.g., CLAMS) produce a variety of sounds with their claws, mandibles, and other parts of their shell-encased bodies. (See MOLLUSKS.) The sounds are emitted both during feeding and moving about.

The sounds emitted by fish are stridulatory (rubbing or rasping), vibratory, and hydrodynamic sounds. Stridulatory sounds are similar to the sounds produced by crickets. SWIM BLADDER noises are drumming sounds resulting from specialized muscles attached to and surrounding the swim bladder. Hydrodynamic sounds are brought about by the sudden movement of fishes, the noise sounding like that of a distant explosion.

Sounds of MARINE MAMMALS are often high-pitched whistles, squeals, and metallic-sounding pulsed signals of the PORPOISE and WHALES. Barklike and pulse sounds from SEALS and sea lions of several species also add to the din. Especially interesting are the underwater noises—whistles, buzzes, beeps, and chirps—that have been identified as being produced by the Weddell seal, an Antarctic species that attains an 11-ft (3.3-m) length and a weight of 1300 lb (585 kg). The Weddells hunt for fish at considerable depths and in total darkness. Thus, it is thought that, as with the whales, the sounds are utilized as a means of SONAR navigation. The hearing ability of seals is different from that of humans, and many of their sounds consist of very high frequencies and rapid pulses very close together. The hearing ability of porpoises is also quite remarkable, and there is evidence that ultrasonic frequencies as high as 100–120 kHz can be detected by these animals. Little is known of the significance of most of the sounds, particularly the INVERTEBRATE and fish sounds, but it is known that porpoises (as well as some WHALES and SEALS) communicate among individuals in a school.

Sound, or acoustic energy, techniques are also employed by oceanographers in studies of the shape of the ocean floor, the nature of the sediments, investigations of the bottom underlying these sediments, etc. To use underwater acoustic techniques reliably, an understanding of the effects of biological organisms on the transmission and reception of sound in the ocean is important. The geographical and vertical distribution of these marine animals, most of which make sounds themselves, and the way that various sound frequencies are scattered by them during their vertical migration in the world's oceans are important areas of investigation. (See also DEEP SCATTERING LAYERS; INSTRUMENTATION.)

Other oceanic phenomena which affect sound transmission are thermoclines, internal waves, currents, and natural sound ducts, which are found at various depths, usually less than 3300 ft (1006 m). (See INTERNAL WAVE; THERMOCLINE.) In these sound ducts, or channels, compressional waves of acoustic frequencies travel great distances. Basically, this is due to the fact that in deep waters the sound waves are reflected upward by pressure (which is more effective in determining the refractive index of these layers than is the decreasing TEMPERATURE). The result is the formation at mid-depth of a wave guide.

SEA OTTER is the common name for one species, *Enhydra lutris,* of a carnivorous marine animal belonging to the suborder Fissipedia and family Mustelidae, a large group that also includes the martens, weasels, badgers, and skunks. Sea otters live in herds, especially along the shoreline of California and northward to areas along the Canadian and Alaskan coasts.

The sea otter attains a length of about 4 ft (1.2 m) and a weight of around 80 lb (36 kg). The body is slender and supple with relatively short legs, large webbed feet, and a shortened cylindrical tail. The animal lacks the fat layer which protects the SEALS and WHALES, but it has a thick fur which it constantly preens to prevent heat loss.

The sea otters' diet consists of the CRAB, MOLLUSK, FISH, and seaweed, much of which they consume, chewing it with their 32 teeth, often while floating on their backs. (see ALGAE.) This back-floating habit carries over even when the female gives birth to a cub in the water, which she then transports on her chest as she floats. The otters also hunt other SHELLFISH, such as ABALONE, which they break with a rock. CLAMS are opened by banging them together until they crack. Sea otters will consume their weight of 45–80 lb (20–36 kg) in shellfish in 3 or 4 days.

Formerly found in large numbers along the Pacific Coast of North America, the sea otter has been hunted extensively for its valuable fur, and the size of the herds has decreased drastically. The animals are now vigorously protected, and in recent years, their numbers have increased significantly.

SEA PERCH See GROUPER.

SEA OTTER. A sea otter floats on its back while breaking shells on its chest. *(Bureau of Sport Fisheries & Wildlife)*

SEA SNAKES are any of the more than 60 species of front-fanged venomous marine snakes of the family Hydrophiidae, one of the 13 living families of snakes, or ophidians. The sea snakes are equipped with long flexible bodies, flattened tails, and a particular lung structure that makes them especially well adapted for the aquatic environment where they are able to swim backward as well as forward.

The average length of a sea snake is about 6 ft (2 m). Two-color patterns found on this length are typical: dark overall on top and light below, or spaced rings or bands of black and light green on the body. They possess a distinctive paddle-shaped tail, a slender forked tongue, and hollow venom fangs. Like the marine TURTLES and MARINE BIRDS, which spend much time at sea, the sea snakes are equipped with a SALT GLAND. This specialized gland enables the snake to consume SEAWATER without experiencing the dehydration necessary to eliminate the excess salt through kidney function.

Found chiefly in the tropical parts of the INDIAN OCEAN and PACIFIC OCEAN, sea snakes are unknown in the ATLANTIC OCEAN. Some species, however, have a wide distribution, and all may at times frequent the open ocean and the coastal areas, as well as estuaries and coastal swamps. Certain islands which are completely free of poisonous land snakes may have sea snakes in their surrounding ocean waters.

Sea snakes tend to float for extended periods of time on the ocean surface. However, they are also capable of remaining submerged for hours. Most of their food is obtained underwater at night and consists of small fish and eels, which they kill quickly with a first bite. (See EEL.) The fish are swallowed head first. Predators of the sea snake are the sharks, some large fish and sea birds. (See SHARK.)

The yellow-bellied sea snake (*Pelamis platurus*) which spends its entire life at sea is especially prevalent off the west coast of Mexico, although it is also found elsewhere and is the most widely distributed of all sea snakes. It exists in East African and Asian waters, northward to Siberia, along the Japanese east coast, the Philippines, Oceania, and eastward to tropical America.

The females are ovoviviparous in that their eggs are incubated and hatched within their bodies. They may bear from 2 to 18 babies, usually giving birth among the rocky inlets of small islands. This

SEA SNAKES. A yellow-bellied sea snake whips through shallow water. *(Smithsonian Institution)*

brilliantly colored yellow and black snake (in the most common pattern, about 80 percent of the body is uniformly black dorsally, yellow midlaterally, and brown ventrally) is sluggish and swims slowly, usually on the surface of the water. In spite of the potential vulnerability arising from such behavior, *Pelamis* appears to have few natural enemies.

The sea snake *Enhydrina schistoso* is a more aggressive species which ranges from the PERSIAN GULF eastward to the north coast of Australia. Other species, nearly 50 in number, inhabit the tropical Pacific and Indian Oceans from the Persian Gulf to Samoa. They also follow the warm CURRENTS northward into the western Pacific to the Ryukyu Islands and Japan.

Unlike these entirely marine snakes, an Asiatic species (*Laticauda colubrina*) native to the Indo-Australian region is partly terrestrial in habit and adaptation. It takes to land areas for reproduction, and this reproduction is of the oviparous type. (In oviparous reproduction the eggs of the female are hatched outside the body.)

Sea snakes are reputed to be generally docile but the fact that they are all venomous and have a tendency at times to gather together in great numbers makes them quite dangerous to humans. While most will not bite a human unless caught or disturbed directly, several people, especially fishermen, are killed each year by them, and no effective serum or vaccine has yet been manufactured to counteract their neurotoxic venom.

Contrary to some public opinion, the coral snakes (e.g., the Harlequin coral, or *Micrurus fulvius*) are land snakes that belong to the family Elapidae which also includes the cobras, kraits, and mambas. There are also land snakes which are rather aggressive and can swim. For example, the world's largest [approximately 16 ft (5 m) long] venomous snake, the king cobra (*Naja hannah*) of India, is a diurnal species which prefers a jungle environment near streams. The common cobra is a nocturnal snake with a decided fondness for water, occasionally entering brackish bays or the ocean coastal waters.

SEA STARS is the name generally used for any of various species of marine animals of the classes Asteroidea (starfish) and Ophiuroidea (brittle stars). Both classes are in the phylum Echinodermata. (See TAXONOMY; ECHINODERM.) Sea stars are distinctive star-shaped seafloor animals which represent the most numerous and one of the most ancient of all types of marine animals; their origins date back 500 million (5×10^8) years.

Sea stars, while never found in fresh water, inhabit practically every known geographical section of the world's oceans. These seafloor animals are found, sometimes in very large numbers, from the shore to the bottoms of the deep ocean TRENCHES. In these areas, starfishes eat mollusks, the SEA ANEMONE, LARVA, and CORAL. (See MOLLUSK.)

Structurally, the starfishes and the brittle stars are somewhat similar. Both have fingerlike and spine-covered arms (five or more) that radiate from a central disk where the animal's mouth is located. In the brittle stars, these arms are smooth, more slender, and more sharply defined from the disk than in the starfishes. Equipped with hollow tube feet at their ends, these arms operate differently in the two types. In the starfishes, the arms are more rigid. Along the lower side of each arm runs a groove (the ambulacral groove) which houses rows of hollow, flexible feelerlike tubes called tube feet. These can be extruded from the groove. Tube feet are often equipped with suckers, and movement from place to place on the bottom is accomplished in a kind of walking motion. The tube feet are also used for trapping animal food. The brittle stars, of which there are more than 1900 kinds, have flexible arms that move from side to side for locomotion. The arms are longer than those of the starfishes and are much more sharply defined. These, too, are used for collecting food on their tube feet. The brittle stars owe their name to the habit of breaking into small pieces if captured or taken out of the water.

In 1969, the tropical crown of thorns, a mauve-colored and large-spined starfish (*Acanthaster planci*), threatened the very existence of some of the PACIFIC OCEAN's most famous coral reefs. (See CORAL REEF.) At that time, a sudden and unexplained population explosion of that species of large [approximately 18 in (45 cm) in diameter] starfishes devoured sizable portions of living coral, especially in areas around Guam and Australia's Great Barrier Reef. This unusual infestation stopped in many places as suddenly as it started. It is unclear whether the destruction was produced by human influence through OCEAN POLLUTION which changed the chemical composition of the animal's environment or because the natural predators of the starfishes had been removed.

SEA STATE, or state of the sea, is a numerical or written description of the degree to which waves have modified the surface of the sea. More precisely defined, sea state is the average height of the highest one-third of the waves observed in a wave train (see WAVES), referred to a numbered code which covers an increasing range of such heights as shown below:

Sea State Code	Wave Height, feet
0	0
1	$0-\frac{1}{3}$
2	$\frac{1}{3}-1\frac{2}{3}$
3	$1\frac{2}{3}-4$
4	4–8
5	8–13
6	13–20
7	20–30
8	30–45
9	Over 45

Winds blowing across the sea surface do not create a single wave train in which each wave is the same height and the wave length and frequency is the same throughout. Rather, a host of such wave trains are created, with each being different from all the rest. In the area of generation all these wave trains are superimposed on each other. Just outside the generating area, only those wave trains moving in the same direction are superimposed. The result, in both cases, is a seemingly random motion of the sea surface in which no two successive waves are of the same height. In such cases the sea state code provides a relatively simple approach to describe a very complex condition.

A practical approach to relating wind speed and sea state is provided by the BEAUFORT WIND SCALE.

SEA TURTLES See TURTLES (MARINE).

SEA URCHIN is the name used for a marine ECHINODERM of the class Echinoidea; the soft internal organs of this animal are encased in a protective shell (test) consisting of a number of close-fitting

SEA STATE DESCRIPTION

SS1	**SMOOTH SEA** Ripples, no foam. Wind*: light air, 1-4 kts. Beaufort 1. Not felt on face.
SS2	**SLIGHT SEA** Small wavelets, no foam. Wind*: light to gentle breeze, 4-10 kts. Beaufort 2-3. Felt on face, light flags wave.
SS3	**MODERATE SEA** Large wavelets, crests begin to break. Wind*: gentle to moderate breeze, 7-15 kts. Beaufort 3-4. Light flags extended.
SS4	**ROUGH SEA** Moderate waves, many white caps, some spray. Wind*: moderate to strong breeze, 14-27 kts. Beaufort 4-6. Wind whistles in rigging.
SS5	**VERY ROUGH** Sea heaps up, with spindrift and foam streaks. Wind*: moderate to fresh gale, 27-40 kts. Beaufort 6-8. Walking resistance high.
SS6	**HIGH SEA** Sea begins to roll, dense streaks of foam and much spray Wind*: strong gale, 40-48 kts. Beaufort 9. Loose gear and light canvas may part.
SS7	**VERY HIGH SEA** Very high waves with overhanging crests. Sea appears white as foam scuds in very dense streaks. Visibility reduced. Wind*: whole gale, 48-55 kts. Beaufort 10.
SS8	**MOUNTAINOUS SEA** Very, very high rolling breaking waves. Sea covered with foam. Very poor visibility. Wind*: storm, 55-65 kts. Beaufort 11.

* Correlation between sea state and wind description is highly variable and dependent on fetch and wind duration. For seas not fully arisen wind speeds may be much higher than indicated.

SEA STATE. (*Top*) The sun rises over a calm sea. (*Woods Hole Oceanographic Institution*) (*Bottom*) Waves disturb the water's surface far out at sea. (*Woods Hole Oceanographic Institution*)

SEA URCHIN. A spiny shell made of close-fitting plates protects the sea urchin's soft internal organs. (*NOAA*)

plates beneath the skin, and the surface is typically covered with bristling spines.

Sea urchins belong to the phylum Echinodermata, which includes the starfishes, sea cucumbers, sea lilies, and feather stars, they are found on the ocean floor in shallow waters to the deep depths of the world's oceans. The sea urchin is an INVERTEBRATE found most commonly in the shallow waters, where it lives on both animal and vegetable matter. There are 850 living species, and sizes range from a quarter of an inch (0.064 cm) in some to more than a foot (30 cm) across in others. However, the average size is about 2–3 in (5–7.6 cm) across the globular skeleton called the test.

The mouth is on the lower side of the test. Five white teeth protrude from the mouth. Each tooth is attached to a set of five jaws which are set in a ring. These jaws and teeth collectively make up what is termed *Aristotle's lantern* because of its resemblance to ancient oil lanterns and the fact that this structure of the chewing mechanism of the sea urchin was first described by Aristotle.

Some representative species are

- Long-spined, or black sea urchin (*Diadema setosum*). They are distributed throughout the INDIAN OCEAN and West PACIFIC OCEAN areas northward to Japan and eastward to Hawaii.
- Sea urchin (*Toxopneustes pileolus*). These are found in the Indo-Pacific waters from East Africa to Melanesia and Japan.
- Sea urchin (*Toxopneustes elegans*). This is an inhabitant of Japanese waters.
- Sea urchin (*Asthenosoma ijimai*). These inhabit the Japanese area southward to the MOLUCCA SEA.

Most species such as these have hollow, brittle spines (radioles) covering the surface of their bodies. These articulating members have a fine [about 6×10^{-4}–8×10^{-4} in (15–20) μm) in diameter] pore structure. Because of this type of porosity, the spines of some sea urchins found in the South Pacific have been investigated for use as molds for making replacement veins and arteries in the human body. In some species, such as the SAND DOLLAR (*Dendraster excentricus*), the spines are very small or lacking. The form of the spines varies with the species, but if only half of the spines are removed, the sea urchin will regenerate new ones to replace them.

Just as in the starfishes, tube feet and small grasping organs, (pedicellariae) are present in these animals. They are guided to immobile organic matter by olfactory receptors around the mouth and tube feet. These tube feet act like suction cups and are used for locomotion as well as for attachment to some substrate. Pedicellariae are employed for securing and holding seaweed over the shell as a camouflage, for cleaning the shell, and for a defensive weapon, since in some species these organs are provided with poison glands. (See ALGAE.) In those species having large-sized spines, the spines are used for movement of the animal.

Reproduction is normally by the ejection of eggs and sperm into the water, with the resulting pelagic LARVA called an *echinopluteus.* The life span of sea urchins is estimated to be about 6 years.

In some areas, such as the waters off the coast of Southern California, where sea urchins (e.g., the white sea urchin, *Lytechinus anamesus,* which travels in herds) are considered pests, many steps have been taken over the years to eradicate them. This has been done especially because the sea urchins have a harmful effect directly on seaweed and young kelp and indirectly on fishes and invertebrates. Sea urchins feed on the seaweed and young kelp and sometimes even destroy mature kelp by feeding on the holdfasts that anchor them to the bottom.

Kelp and other seaweeds provide food and shelter for many species of invertebrates and fish. In addition to this ecological importance, kelp is a useful ingredient in a wide variety of industrial products, such as ice cream, beer, antibiotics, dyes, and welding rods.

In Japan, sea urchins are protected rather than destroyed. A complex system of regulations, imposed by government and by fishery cooperatives, limits the numbers and areas open to harvesting as well as the harvesting methods. In addition, artificial reefs have been created in many areas to provide suitable habitats for sea urchins, which are prized for their roe; this is eaten in Japan and in many European and South American countries.

Sea urchins provide important links in the recycling of nutrients trapped in sediments and furnish food for a wide variety of marine life. They have been the object of much research by paleontologists interested in their evolutionary development and by zoologists, especially because sea urchins have the ability to produce large numbers of reproductive cells. The eggs have been sent into outer space in an effort to determine the effects of that environment on development.

SEAFLOOR SPREADING See CONTINENTAL DRIFT.

SEAFOAM consists of a froth of bubbles found on the surface of SEAWATER in the open ocean or along the shores of the world's oceans; the bubbles are created by breaking WAVES and precipitation.

Seafoam is made up of air bubbles separated from each other by a film of liquid. Bubbles coming together in fresh water coalesce, but bubbles coming together in salt water bounce off each other. Most bubbles in the ocean are caused by wind waves, but they may also be produced by rain and even snow. The bubbles that form along the seashore are very small, mostly less than 0.0197 in ($\frac{1}{2}$ mm) in diameter.

When bubbles rise to the surface, they burst and release salt spray into the air. Each bursting bubble causes a jet of several drops to rise to heights up to 1000 times the bubble diameter. It is believed that most of the airborne salt nuclei for the formation of rain drops comes from bursting bubbles. The jet drops from such bursts carry a significant positive electric charge, and numerous studies have proved that in this way the oceans produce a net positive charge for the atmosphere. It has also been proved that the presence of organic matter in the water results in a reversal of the sign. Therefore, tiny, localized organic-rich areas in the world's oceans probably produce a negative space charge. The exact nature of this charge-separation phenomenon and its significance is being investigated.

SEAGRASS comprises the 12 genera of aquatic angiosperms (flowering plants) that are adapted to the ocean environment. Seagrasses are widely distributed throughout the world and especially in the shallow waters (sublittoral zone) of the coasts.

The value of seagrass-bed environments has become increasingly well understood. They represent a major component of the coastal ecosystem. The beds, which are very productive, provide a needed habitat for a wide variety of organisms, supplement the detrital food chain in estuaries, and stabilize substrates. In this regard, the anchored roots of seagrasses tend to bind the sediments together and preserve the microbial flora of the sediment and sediment-water interface. The leaves retard currents and increase sedimentation of inorganic and organic matter around the plants.

The most important representative of the seagrasses is *Zostera marina,* or EELGRASS, which has a high growth rate and in whose meadows large numbers of marine organisms are supported. Other examples of seagrasses are TURTLE GRASS (*Thalassia testudinum*), shoal grass (*Halodule spp.*), manatee grass (*Syringodium filiforme*), and widgeon grass (*Ruppia maritima*).

SEAHORSE is the name applied to any of the approximately 50 species of peculiarly shaped pipefish of the order Gasterosteiformes, family Syngnathidae, and genus *Hippocampus.* The small [average 8 in (20 cm) long], usually light-brown animal has a horse-shaped head with a tubular snout, a tapering prehensile tail with a caudal fin, a hard outer skeleton, independently moving eyes, and a belly pouch similar to that of the kangaroo. (See FINS.)

Seahorses are found in shallow inshore waters of the coasts of Australia, the Atlantic coasts of Europe, Africa, and North America, as well as the Pacific coast of America. They inhabit the seaweed and EELGRASS areas of the tropical and subtropical waters of these ocean areas, although some species extend into temperate regions. (See ALGAE.)

Hippocampus hudsonius, a typical and common species of seahorse, has a short compressed body, encompassed by bony rings which give it structural rigidity. This bizarrely shaped seahorse swims in an erect position with the head bent at right angles to the body. Propulsion is provided by means of a fanlike motion of the small single dorsal fin. The two pectoral fins oscillate at the same rate as the dorsal fin, and the head is used for steering. The long slender tail can be wrapped around seaweed or similar objects for support when the fish is feeding or at rest. Food, located by turret-mounted eyes that can move independently, consists mainly of small CRUSTACEANS such as COPEPODS.

Reproduction takes place in an unusual way, since the female deposits her eggs (as many as 600 in some species) into the belly pouch of the male. Here they are incubated and ready to be born after 4 to 5 weeks. At birth, the babies, which are perfect miniatures of their parents, are ejected from the pouch of the male one at a time.

SEALS are large aquatic carnivorous mammals of the suborder Pinnipedia which consists of three families: Otariidae, the eared seals, SEA LIONS, and fur seals; Odobenidae, the walrus; and Phocidae, the true or hair seals. These families, a total of 32 species, have as a group a worldwide distribution.

The Pinnipedia pinnipeds are excellent swimmers and divers. Their four limbs are webbed flippers which are used for moving about on land where they mate and bear their young. The bulls (males) are larger than the cows (females) and in most cases establish harems which they aggressively protect from other bulls. Although their teeth are simpler than those of land animals, they are chiefly carnivorous and eat fish, CRUSTACEANS and mollusks. (See MOLLUSK.) The body is usually covered with a heavy coat of short fur that slopes backward expediting easy movement through the water, the eyes are large, and the external ear is small or absent. Their principal enemies are the killer WHALES (grampus) and humans.

The family Otariidae, or the eared seals, includes the sea lions, or haired seals, and the fur seals, both of which are distinguished by having small but well-formed external ears. The neck is rather long, they always have 34 teeth, and the forelimbs are long with the independent, mobile hind limbs turned forward when on land to facilitate locomotion.

The best-known of this group is the pelagic Northern or Alaskan fur seal, *Callorhinus ursinus,* which spends the winter months migrating through the PACIFIC OCEAN. They come to land (or hauling grounds) in great numbers, especially in the remote, fog-bound islands of the Pribilof group (in the BERING SEA). Here, the bulls come in first in May to establish their territories, and as the females move in about June, each bull (with territory) acquires a harem. The males are large, up to 7 ft (2 m) long and over 600 lb (272 kg) in weight, while the females are smaller, about 5 ft (1.5 m) long and up to 150 lb (68 kg) in weight. The bulls remain during the summer, during which time the young pups are born and trained to fish by their mothers for about 4 months. The mothers abandon them at that time to begin their southern migration.

The southern fur seals, genus *Arctocephalus,* make up a few species (such as *Arctocephalus australis*) and, depending upon the species, are found in the Falkland Islands of South America and around Southern Australia, New Zealand, and Southern Africa. The habits of these seals are essentially like those of the northern fur seals, allowing for the fact that the seasons are reversed in the Southern Hemisphere.

The California sea lion, *Zalophus californianus,* is the species usually seen in zoos and trained to perform tricks in circuses. This fast-swimming animal, often 7 ft (2 m) long and 600 lb (272 kg) in weight, ranges along the Pacific coast from Vancouver Island to the Gulf of California. The habits

SEAHORSE. The strange little weed-dwelling seahorse uses its prehensile tail, unique among fish, to cling to a coral branch. *(Miami Seaquarium, Florida)*

of the California sea lion, as well as that of Steller's Northern Pacific sea lion, (*Eumetopias jubata*), approximately 13 ft (4 m) in length and 1300 lb (590 kg) in weight, are similar to those of the Alaskan fur seal of the Pribilof Islands.

The Odobenidae family of Pinnipedia is thought to consist of a single species, the walrus, *Odobenus rosmarus*. Although some walruses are found in the Arctic waters from the KARA SEA to Labrador and Hudson Bay, others inhabit the Bering Sea and surrounding waters. Both the Atlantic and Pacific male walruses are large, averaging up to 3000 lb (1361 kg). Knowledge of the life history of these animals is quite fragmentary and contradictory. Herds exist in remote ice fields, and they do not breed ashore in large rookeries like their relatives, the fur seals and sea lions. The females, or cows, are somewhat similar, about 8 ft (2.5 m) long and weigh up to 2000 lb (907 kg). Neither sex has external ears, but they do possess a definite thick and heavy neck and shoulder region and hind limbs like those of the sea lions. All have only 18 teeth, which, uniquely, include two upper ivories that extend as downward-pointing tusks. These sometimes, in the male, are as long as 2 ft (0.6 m) or more and are used as defensive weapons as well as for digging CLAMS from the ocean floor.

The family Phocidae, which is the largest family of the pinnipeds, represents the true seals in that they have no external ears. All species have 30 teeth. The hair is coarse and short, and the hind limbs are partially fused together with the tail. When swimming, the body is moved by means of side to side strokes of these hind flippers, each acting alternatively. These seals commonly live about 30 years or more.

The true seals consist of 18 species, with the smallest [3 ft (1 m)] being the only freshwater species, *Pusa sibirica,* found in Lake Baikal of Siberia; the largest is represented by the elephant seal (e.g., *Mirounga leonina*) which are characterized by their inflatable pendulous nose and a body size that, in the males, can range up to a 20-ft (6.1-m) length and a 4-ton (3629-kg) weight. The female elephant seals are slightly smaller.

The elephant seals are found in the high northern and southern latitudes, and they breed on the rocky shores of islands, following an uncertain ocean migration route until that time. The cows arrive at these rookeries in the early fall and usually give birth to a single calf. Bulls acquire a harem of cows, and these cows are impregnated soon after the birth of the calves (pups). The gestation period is 11 months. The newborn pups weigh up to 100 lb (45 kg) and possess large luminous eyes and a silvery coat.

SEALS.

A Weddell seal yawns lazily in the Antarctic sun. *(U.S. Navy)*

An Alaskan fur seal family appropriates a large rock on St. Paul Island, Alaska. *(Bureau of Sport Fisheries & Wildlife)*

Crabeater seals lie on an ice floe on the Antarctic Peninsula. *(U.S. Navy)*

A monk seal in Hawaii relaxes with her pups. Twins are a rarity. *(Bureau of Sport Fisheries & Wildlife)*

Young seals roll around contentedly in the rocks and mud.

The population of elephant seals on Guadalupe Island has increased to about 15 000 since the Mexican government set the island aside as a preserve. *(Scripps Institution of Oceanography)*

The cosmopolitan harbor seal, *Phoca vitulina,* is a very slow-breathing species of shallow-water hair seals [approx 5 ft (1.5 m) and 150 lb (68 kg)], well known to many shoreline residents of parts of Europe (France, England, and the Mediterranean) and America (Maine and California). The black and white harp seal, *Phoca groenlandica,* and the dark-gray ringed seal, *Phoca hispida,* are related species of northern waters, while the large leopard seal, *Hydrurga leptonyx,* and the Antarctic Weddell seal, *Leptonychotes weddelli,* are found in the southern hemisphere in the ocean waters around Antarctica, South Africa, and Australia. One of the best-known species is the Atlantic gray seal, *Halichoreus grypus,* which is found along the coastal regions of Europe, Greenland, and Iceland.

Most species of these animals—seals, sea lions, walruses, and elephant seals—have been hunted extensively and without control in the past for hides, blubber oil, and meat. The trade in elephant seals began in the early 1800s and continued into the late 1880s so that in places such as Guadalupe Island off the coast of Baja California, 300 mi (483 km) south of San Diego, California, there were only nine seals left in 1892. In 1911, Mexico prohibited the killing of elephant seals, and the United States put them on the protected list in 1972. In the case of the walrus, which the Vikings, over a thousand years ago, hunted as hvalross or whale-horse, the animal has been relentlessly pursued and slaughtered in great numbers up until very recent times. However, now, for example, at San Miguel Island, a windswept 14 000-acre island with a 24-mi (38.6-km) coastline and located 45 mi (72 km) southwest of Santa Barbara and 100 mi (160.9 km) west of Los Angeles, California, a haven has been established for seals. This environmentally protected area presently is the home for tens of thousands of sea lions and seals which haul themselves onto the beaches to breed. Six different species inhabit the island, the Northern and Guadalupe fur seals, the Northern and California sea lions, the harbor seals, and the northern elephant seals.

Since seals and sea lions have sensory systems and physiologies ideally adapted to the underwater environment, they have been the subjects of much scientific study in an effort to better understand human DIVING physiology. Experiments can be performed by and upon these animals which would be impossible or impractical with human subjects. Such experiments and research have also been concerned with their trainability, visual acuity, and the spectral sensitivity of both sight and hearing. For instance, investigators have found that fur seals, once thought to be vicious and untrainable, are in fact quite tractable. The same species can echolocate and is the first seal species known to do so.

SEAMOUNT is a more or less isolated elevation of the seafloor that is circular or elliptical in cross section and rises at least 0.6 mi (1 km) above the floor on which it rests. The definition fits both those seamounts with pointed tops and those with flat tops that go by the special name GUYOT. The existence of seamounts was first demonstrated by soundings taken by the CHALLENGER EXPEDITION of 1872–1876. Since that time some 1200 seamounts have been detected, and some estimates indicate that the world's oceans may contain as many as 20 000.

Certainly most, if not all, seamounts are submarine volcanic cones. All that have been sampled thus far are made of basalt, and their shape and slope do not resemble any other landform. In addition, detailed studies have indicated that many seamounts have craters at their peak as would be expected of a volcano.

A typical seamount may measure 25 mi (40 km) across at its base, rise some 10 000–15 000 ft (3000–4500 m) above the seafloor, and have sides that slope upward at an angle of about 20°. Seamounts are often grouped in lines or clusters of 10 to 100, and some volcanic cones pierce the surface to become oceanic islands. When the molten material that builds the cone escapes through a linear fracture on the seafloor, the result is an elliptical rather than a circular cross section, and the length of the seamount is about 4 times its width. The Hawaiian Islands are of this type.

SEASAT-A is the name for the United States' first space-research platform designed strictly for oceanographic research.

The specific objectives of SEASAT-A were to:

1. Explore, map, and chart the world's oceans and its living resources.
2. Manage, use, and conserve those resources.
3. Describe, monitor, and predict conditions of the atmosphere, ocean, sun, and space environment.
4. Issue warnings against impending destructive natural events.
5. Develop beneficial methods of environmental modification.
6. Assess the consequences of inadvertent environmental modification over a period of time.

SEASHELLS, the name commonly used for shells or skeletons of marine invertebrates, especially the marine mollusks (or animals such as the oyster and

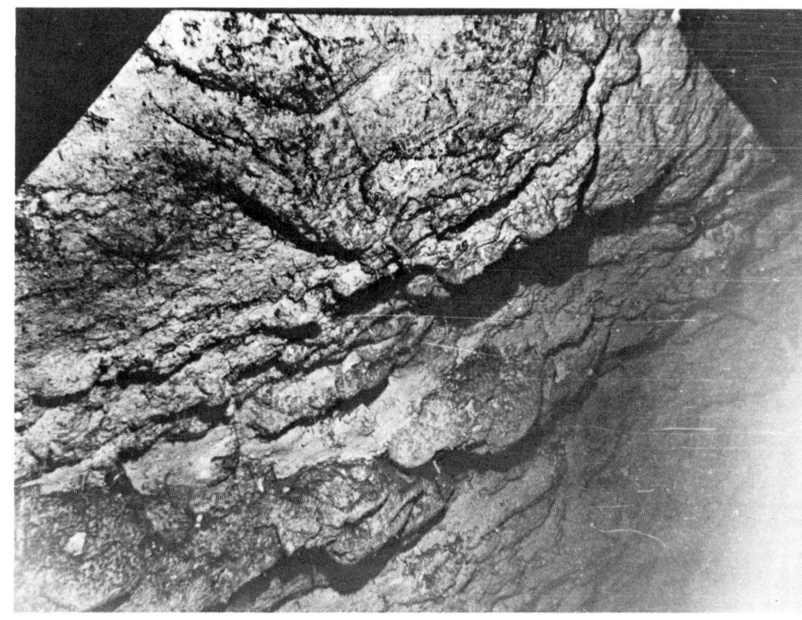

SEAMOUNT (*Top*) Activity in a seamount cluster along the Mid-Atlantic Ridge resulted in the birth of the island Surtsey off the coast of Iceland in 1963. Surtsey was named after Surter, the Icelandic god of volcanoes. *(U.S. Navy)* (*Bottom*) The rocky steep side of a seamount near the 7000-ft depth displays a craggy texture.

CLAMS). (See INVERTEBRATE; MOLLUSK.) About half of the mollusks are found in the world's oceans and the other half on land or fresh water.

Seashells fall chiefly into two main groups: the univalves (GASTROPODS or sea snails) and the bivalves (two-shelled mollusks and pelecypods). Other types of seashells are the cephalopod shells such as the NAUTILUS and the tusk shells or scaphopods (about 200 species).

There are about 80 000 species (100 families) of single-shelled gastropods or univalves, and the marine variety have a great diversity in shape and structure. Among these are the whelks, cones, helmets, conchs, abalones, augers, etc. (See ABALONE.) Many are used for food or ornament.

Most of the 10 000 species of bivalves are marine. These mollusks consist of two plates or valves joined by a hinge, a horny ligament, and one or two muscles. Representatives include the various types of clam shells, oysters, scallops, etc.

The animals that make and live in some seashells can inject a poison fatal to humans. The most poisonous of the mollusks are the cones. The marbled cone (*Conus marmoreus* and *C. marmoreus bandanus*) and the textile cone (*C. textile*) are examples of such toxic species. In these, the animal's radula (slender breathing tube) has been developed so that it contains tiny hollow barbs that are forced into the victim, and a virulent venom is injected through the barb.

The conchs, such as the queen conch (*Strombus gigas*), possess shells of great beauty. The conch lives principally in SEAGRASS beds in less than 70-ft (21-m) depths of water. Its handsome shape and flaring lip of shading colors make it a highly coveted shell for collectors. Cameos and fine porcelain are made from these shells. Also, the meat of the conch is widely used for food in the Bahamas and the Florida Keys.

The abalone (genus *Haliotis*) is a univalve mollusk that is found only in the Pacific coastal waters off the United States and Mexico. The meat of this animal is considered a delicacy, and the multicolored shells (usually red, pink, or green) have long been prized by collectors. The nacreous lining of the shell is used commercially as mother-of-pearl.

A rare and beautiful shell is that of the violet snail, *Janthina janthina*. The shell of this extraordinary animal is paper thin and when empty floats buoyed by the air entrapped inside. When it is occupied by the mollusk, its survival depends upon being held up by a platform or raft of bubbles since the weight of the animal inside is sufficient to sink it. This bubble cushion is formed by the extension of the snail's foot along the surface of the water where it traps air and coats each bubble with a mucus so that several bubbles gather and support the shell.

Another open-sea animal possessing a highly coveted and exquisitely colored shell is the paper nautilus, *Argonauta argo*. This nautilus, or argonaut, as it is often called, is different from any other. The female of the species alone utilizes the shell as a home, a nest for the eggs, and a hydrostatic organ for regulating depth.

SEASHORE PASPALUM See SILT GRASS.

SEAWALL See BEACH.

SEAWATER or **OCEAN WATER** is the water of the seas and oceans and is distinguished as a solution containing many dissolved salts. Seawater is characterized not only by its high SALINITY and complex physicochemical structure but also by its striking biochemical similarity to human body fluids. This water covers over 71 percent of the earth's surface [approximately 139 million (1.39×10^8) mi^2 (361 million km^2)]. The total volume amounts to some 3.62×10^{20} gal (1.37×10^{21} L).

Most scientists believe that the primeval oceans were salty in the beginning, and that salts were dissolved from rocks underlying the ocean basins. Rocks on land broken up by frost and erosion and washed into the sea have also added to this salt content. However, rivers flowing into the oceans contain higher percentages of carbonates than does seawater, in which chlorides predominate.

The Constituents The ability of seawater to dissolve large amounts of solids and gases without chemically reacting with them is one of its most important properties. The dissolved constituents are present in seawater in varying amounts, and the character of many of the substances found can vary with geographical location. The substances can also vary with time and depth at a given location. On the other hand, seawater has an extremely narrow range over which the composition of its major constituents changes. This uniform narrow variation results from ocean circulation and mixing. As a consequence, ratios among more abundant constituents are practically constant regardless of absolute concentrations of total solids. For this reason, and for many purposes, the concentration of the major constituents can be quantified as a single variable, salinity.

Essentially, the constituents of seawater are of three types: (1) the dissolved components such as the salts, organic compounds, and gases; (2) the second phase components such as gas bubbles; and (3) the undissolved inorganic and organic solids. The dissolved materials are either electrolytes or nonelectrolytes. The electrolytes, when dissolved in water, result in an electric charge-carrying species or ION, whereas the nonelectrolytes do not. The major constituents have been determined to be practically uniform in space and believed to have remained more or less constant for the past few hundred million years. Such representative major elements (and their principal species) in seawater are sodium (Na^+), magnesium (Mg^{+2}), calcium (Ca^{+2}), potassium (K^+), strontium (Sr^{+2}), chlorine (Cl^-), sulfur (SO_4^{-2}), and carbon (HCO_3^-, CO_2^{-2}); the organic compounds of bromine (Br^-), fluorine (F^-), and boron [$B(OH)_3$ and $B(OH)_4^-$]. In addition to these elements, it is thought that the ocean contains traces of every naturally occurring element. Since all the dissolved solids present amount to 35 000 ppm (parts per million) parts of water, each single cubic mile (4.7 billion tons) of water holds about 165 million tons of solids. This includes approximately 117 million tons of common salt (sodium chloride), 6 million tons of magnesium, 4 million tons of potash, 300 000 tons of bromine, 200 000 tons of borate, 2200 tons of iodine, 900 tons of iron, and 450 tons of copper plus a multitude of others to make up the difference.

Of these, primarily because of the adverse economics involved in recovery procedures, only common salt, magnesium, and bromine have been extracted from the ocean water in any significant amounts. While the oceans also contain such elements as gold and radium, the concentrations of these present in seawater are negligible.

The biologically important elements such as silicon (Si), nitrogen (N), phosphorus (P); and vanadium (V) are present in seawater in amounts equal to 3 mg/L (3 milligrams per liter), 0.5 mg/L, 0.07 mg/L, and 0.002 mg/L, respectively. Minute ocean plants, phytoplankton, in their growing cycles can reduce these concentrations still lower. (See PLANKTON.) Moreover, many inhabitants of the ocean accumulate large quantities of various elements that are present in extremely minute quantities in seawater (e.g., the sea cucumber accumulates vanadium more than a millionfold above that normally concentrated in seawater). Zinc, iron, and strontium are also found in ALGAE, FISH, invertebrates, and plankton. A species of green algae has the capability of extracting uranium from seawater and concentrating it in its tissues by factors minimal 2000 to 4000 times that of seawater.

Gases Free oxygen, nitrogen, and carbon dioxide are of significant biochemical importance and are relatively abundant in seawater. Other gases constitute about 0.25 percent by weight of the total volume of ocean water.

Oxygen

Oxygen content, which can be determined by the Winkler method or by a gas chromatograph analysis, varies from zero to about 8.5 cm³/L (at standard TEMPERATURE and PRESSURE). The oxygen concentration in seawater is caused chiefly by the infusion of molecular oxygen at the surface and by the PHOTOSYNTHESIS in the ocean. In the gas chromatographic technique for determining this concentration, the dissolved gases in a water sample are stripped from solution by a gas, ordinarily helium, and then measured in a gas chromatograph having thermal conductivity detectors. This method represents a substantial improvement over the Winkler titration way of determining dissolved oxygen content in seawater. In the Winkler titration, an operator-sensitive technique, an oxidation-reduction is made to take place between an added manganous ion and the dissolved oxygen in the sample. Oxidized manganese is then reacted with iodine in an acidic medium, and the amount of iodine consumed is determined by titration with a standardized thiosulfate solution, using starch as an indicator.

Although the oxygen-depth profile is inconsistent from location to location, in general, the oceans are aerobic (contain oxygen) throughout. With increasing ocean depth there is a reduction in oxygen content, with the maximun concentration occurring at the top surface followed by an oxygen-poor layer that can vary between 300 and 4500 ft (91 and 1364 m) in thickness. Ordinarily, because of the lack of circulatory renewal and no wind-mixing action, dissolved oxygen is at a minimum at depths of 2200–3100 ft (660–940 m).

Nitrogen

Nitrogen is essential to ocean living matter since it determines the rate of phytoplankton reproduction. Nitrogen is present in the ocean as a free dissolved gas or in compounds of various kinds (nitrate, nitrite, and ammonia). (See NITROGEN CYCLE.) The bottom sediments contain a small amount of organic nitrogen.

Carbon Dioxide

Carbon dioxide in seawater is found in relatively large amounts as carbonates and bicarbonates. The carbon dioxide content mainly determines its pH (see pH), and near the surface a pH value from 8.1 to 8.3 will occur, depending upon the partial pressure of carbon dioxide in the atmosphere.

Seawater throughout the oceans is predominantly alkaline with pH values ranging from 7.5 to 8.3. Near the surface, the alkalinity is increased by dissolved carbonates from rivers and decreased by the precipitation of lime in the form of CORAL and SHELLS. At greater depths, it can vary by the amount of sinking calcareous debris going into solution from above. The salts of weak carbonic acid (resulting from the hydrolysis of dissolved carbon dioxide) and those from boric acid are the strong bases which constitute the alkalinity of seawater. (See CARBON CYCLE.)

Rare Gases

Rare gases such as argon, helium, and neon are found in seawater. In fact, the ocean contains in solution all the gases found in the atmosphere but not in the same proportions. It is thought that the concentrations of these gases are very close to saturation.

Salinity Since the major constituents are present in the ocean in nearly constant proportions, for many purposes, their concentrations (the total dissolved content) can be quantified as a single variable, salinity. Salinity is a defined quantity that can be derived by salinometer measurements of the electric conductivity of seawater. The conductivity value can be related to the CHLORINITY (Cl or the concentration or chloride ions) of the water, which, in turn, approximately relates to salinity (S = 1.80655, where S and Cl are expressed in parts per thousand by weight). Chlorinity also has been traditionally determined, rather accurately, by titration with silver nitrate. The chlorinity accounts for about 55 percent of salinity, the average being about 19 ppt. (The average salinity equals about 35 ppt.) The approximate ratios of various elements such as potassium, sulfur, magnesium, and iodine to chlorinity are 0.02000, 0.1395, approximately 0.06801, and approximately 3.32×10^{-6}, respectively.

Various Properties

Density Seawater, like pure water, exhibits many anomalous properties—properties that are quite unlike those of other materials. (See WATER.) One such property was probably crucial to the development and sustenance of life on earth. In this regard, seawater and fresh water have in their solid state a mass per unit volume, or DENSITY, less than that of their liquid state. If the reverse had been the case, the ice would sink to the bottom and the world's oceans would be converted largely to ice. Of course, freezing ordinarily takes place at and in the near surface of water bodies, and the life-sustaining liquid remains at depth.

The density of seawater depends upon salinity, temperature, and pressure. At constant temperature and pressure, density varies with salinity. [A temperature of 32° F (0° C) and atmospheric pressure are considered standard for density determinations.] The effects of thermal expansion and compressibility are used to determine the density at other temperatures and pressures.

Temperature Temperature in the world's oceans varies widely both horizontally and with depth. Maximum values of about 90° F (32° C) are found in the PERSIAN GULF in summer, and the lowest possible values of about 28° F (−2° C) (the usual minimum freezing point of seawater) occur in polar regions.

The vertical distribution of temperature in the oceans nearly everywhere shows a decrease of temperature with depth. Generally, the ocean operates as a system where 80–90° F (26° C–32° C) water temperatures at the surface drop to around 40° F (4° C) at around 1300 ft (400 m). Ordinarily a mixed layer of isothermal water exists below the surface where the temperature is the same as that of the surface. Below this layer is a region of rapid temperature decrease, called the THERMOCLINE, where the temperature falls to that of the deep oceans. In the deeper and denser layers, which are fed by cooled waters that have sunk from the surface in the polar regions, temperatures as low as 33°F (1° C) exist. However, in the deepest ocean BASIN, the temperature increases slightly with depth, the increase being about 1° F at 18 150 ft (5532 m). This warming is believed to be caused more by a slight compression of the seawater rather than by heat from the earth's crust.

Pressure One of the most dramatic features of the ocean environment is the spectacular hydrostatic pressure. Pressure is usually given in atmospheres, cgs units, or SI units. The fundamental cgs pressure unit is the dyne per square centimeter, and 10^6 dyn/cm² = 1 bar (0.1 bar = 1 decibar). Pressure is measured in newtons per square meter in SI units; cgs units can be converted to SI units, since 1 dyn/cm² = 0.1 N/m², or 0.00005 N/cm². The decibar is the hydrostatic pressure exerted on one square centimeter of surface by one meter of seawater. Hydrostatic pressure increases by one decibar with every meter of depth, or about 1 psi of pressure is added for each $2\frac{1}{4}$ ft of depth. Thus, at 9900 ft (3018 m) (80 percent of the PACIFIC OCEAN depth), the hydrostatic pressure is equivalent to about 4300 psi (approximately 30 million N/m^2), while at 5000 m (or 27 percent of the Pacific), the depths exert an approximate pressure of 7000 psi (49 million N/m^2).

Electric Conductivity Fresh water is a very poor conductor of electricity. However, when salt is in solution in water, the salt molecules are ionized and become carriers of electricity. Therefore, the electric conductivity of seawater is directly proportional to the number of salt molecules in the water. For any given salinity, the conductivity increases with an increase in temperature. Seawater, whose characteristic impedance is 3600 times that of air, is a fair conductor of electric current [4 ohms per meter (Ω/m)].

Specific Heat Specific heat is the amount of heat, in calories or joules, required to raise the temperature of a unit mass (one gram) of seawater one degree Celsius. Specific heat of seawater decreases slightly as salinity increases. The ratio of specific heat at constant pressure to that at constant volume has a direct relationship to the speed of sound in seawater.

Compressibility Seawater is nearly incompressible its coefficient of compressibility being only 0.000046 per bar of pressure under standard conditions. This value changes slightly with changes of temperature or salinity. The effect of compression is to force the molecules of the substance closer together, causing it to become more dense. Even though the compressibility of seawater is low, its total effect is considerable because of the vast amount of water involved. If the compressibility of seawater were zero, sea level would be about 90 ft (27 m) higher than it is now.

Viscosity Seawater's resistance to flow is slightly greater than that of fresh water. Its viscosity increases with greater salinity, but the effect is not nearly as pronounced as that occurring with decreasing temperature. The rate of change is not uniform, becoming greater as the temperature decreases.

Thermal Expansion The rate of expansion with increased temperature is greater in seawater than in fresh water. At a temperature of 59° F (15° C) and atmospheric pressure, the coefficient of thermal expansion is 0.000151/° C for fresh water and 0.00214/° C for seawater having 35 ppt salinity. The coefficient of thermal expansion increases with salinity and also with increased temperature and pressure. At 35 ppt, the coefficient of surface water increases from 0.000051/° C at 32° F (0° C) to 0.000334/° C at 86° F (30° C). At a constant temperature of 32° F (0° C) and a salinity of 34.85 ppt, the coefficient increases to 0.000276/° C at a pressure of 14 666 psi (10 000 decibars) [at a depth of approximately 33 000 feet (10 000 m)].

Radioactivity Radioactive isotopes (or radioisotopes) are found in seawater. Several radioactive isotopes occur naturally. They come from various terrestrial and extraterrestrial sources. They include those that have persisted since the earth's formation (and their radioactive daughters) and those that are being produced constantly in the earth's atmosphere. For example, of at least 14 isotopes produced by cosmic rays, three, ^3H (tritium or hydrogen 3), ^{14}C (carbon 14), and ^{39}Ar (argon 39), are radioactive gases that enter the oceans from the atmosphere. They are returned to the atmosphere as the gaseous molecules $^3H^1HO$, $^{14}CO_2$, and ^{39}Ar. The radioactivity of seawater from these and all the other natural radioactive isotopes is approximately 750 dpm/L (disintegrations per minute/liter), 97 percent of which is derived from ^{40}K (potassium 40). Many of the natural radioisotopes have long half-lives, and because of this all living ocean organisms have been and will continue to be exposed to an essentially constant level of background radiation.

In addition to the natural radioisotopes, there are those atoms of radioactive elements that also enter the oceans from various artificial sources. Such sources include fallout from nuclear detonations, nuclear power plants, nuclear-powered ships, radioisotope energy devices, radioactive wastes from medical, scientific, and industrial uses, and any other nuclear energy operations. The unstable forms of the ordinary atoms of strontium, cesium, cerium, ruthenium, cobalt, iodine, zinc, phosphorus, manganese, iron, chromium, and others occur in very small amounts in seawater because of such operations.

Many ocean plants and animals are efficient concentrators of radioisotopes, but the detailed biological effects of such a concentration phenomenon are as yet not well understood. However, present evidence indicates that the dosage of radiation from radioisotopes in seawater that enter food chains leading to human beings are not harmful overall. (See OCEAN FOOD CHAIN).

Carbon 14 and other radioisotopes are purposefully introduced to the ocean environment as tracers of physical, chemical, or biological processes. For example, tracers have been successfully used for geochronology dating measurements of organic and calcareous sedimentary components. Also experiments conducted by injecting the surf zone with batches of sand tagged with radioactivity have proven of value in understanding the dynamics of shoreline coastal waters and harbor erosion and buildup. Although the use of radioisotopes to study diffusion and mixing processes in the ocean has been considered many times, the technology of fluorometers has advanced to the point where it is possible to measure dye concentrations of the same low order as radioisotopes—0.5 ppb. Hence, dyes have largely superseded isotopes in such measurements of water mixing processes.

While much work has been carried out to measure various radioactive isotopes in ocean waters, especially ^{90}Sr and ^{137}Cs, studies also have shown that many artificial radioisotopes cannot be determined accurately due to the lack of a present understanding of the complex relationship that exists between these nuclides (i.e., the lanthanides, zirconium, and niobium) and their stable isotopes in the marine environment.

Thermal Conductivity Fresh water is a poor conductor of heat, having a coefficient of thermal conductivity of 0.00139 cal s^{-1} cm^{-1} $°C^{-1}$. For seawater, it is slightly less but increases with greater temperature or pressure, and its conductivity varies as the reciprocal of the viscosity. However, if turbulence is present, which it nearly always is to some extent in the ocean, the process of heat transfer are altered. The effect of turbulence acts to increase the rate of heat transfer. The *eddy coefficient* used in place of the stillwater coefficient is so many times larger and so dependent upon the degree of turbulence that the effects of temperature and pressure are not important.

Surface Tension Surface tension of water in dynes per square centimeter is approximately equal to $75.64 - 0.144T + 0.0399\text{Cl}$, where T is temperature in degrees Celsius and Cl is the chlorinity of the water in parts per thousand. Thus, because surface tension increases with chlorinity, seawater has a higher surface tension than does fresh water. However, the presence of impurities causes the surface tension to be somewhat less than indicated by the formula.

Transparency Transparency varies with the number, size, and nature of particles suspended in the water and also with the nature and intensity of illumination. The rate of decrease of light energy with depth is called the *extinction coefficient.* The earliest method of measuring transparency was by means of a secchi disk, a white disk a little less than 1 ft (30 cm) in diameter. This was lowered into the sea, and the depth at which it disappeared was recorded. In coastal waters this depth varies from about 16 to 83 ft (5 to 15 m). Offshore, the depth is usually about 148–198 ft (45–60 m). One of the greatest depths at which the disk has disappeared is 218 ft (66 m) in the SARGASSO SEA.

Color Seawater varies greatly in color from place to place. Water of the GULF STREAM is a deep indigo blue, while a similar current off Japan was named KUROSHIO CURRENT (black stream) because of the dark color of its water, especially where the nutrient-rich brownish-colored waters of the Kuril Current intersect the Kuroshio which is largely indigo blue in color at other locations. Along many coasts the water is green, and in certain localities a brown or brownish-red water is typical.

Offshore, shades of blue are common, particularly in tropical or subtropical regions. This is due to scattering of sunlight by tiny particles suspended in the water, or by molecules of the water itself. Because of its short wavelength, blue light is more effectively scattered than light of longer waves, and the ocean appears blue for the same reason that the sky does. The green color often seen near the coast is a mixture of the blue due to scattering of light and a stable soluble yellow pigment associated with phytoplankton. Brown or brownish-red water receives its color from large quantities of certain types of phytoplankton, microscopic plants in the sea. (See MARINE OPTICS.)

Summary The basic physical properties of seawater, such as density, specific heat, specific volume, electric conductivity, compressibility, sound velocity, viscosity, and surface tension, depend on its salinity, temperature, and pressure. The colligative properties (osmotic pressure, freezing and boiling points, and vapor pressure) are determined largely by the salinity of the seawater.

An understanding of the physicochemical properties of seawater must take into consideration the fact that seawater is a dilute solution of almost everything. This complex mixture consists of dissolved constituents, strong interactions between the ions of the dissolved salts and the water molecules, ion-to-ion interactions, many forms of naturally produced and artificially introduced organic compounds (such a compound can form a chelate with a metal ion), and bubbles and other active surfaces such as inorganic particles that can concentrate organic matter. Moreover, biochemically, seawater is quite analogous to body fluids. Seawater is also an active transport system intricately charged with living cells, fats, proteins, carbohydrates, hormones, enzymes, vitamins, minerals, and even antibiotics.

While the gross chemical composition of seawater

is well known, new analytical techniques and devices have been responsible for more precise determinations of the amounts of constituents present. Such continued usage of these new and improved devices will also assist in clarifying the many diverse and as yet ill-understood chemical and physical actions that take place in the seawater system. Such actions all impact on the biological productivity of the sea; the physical ocean processes such as stirring and mixing; the hydration of ions and ionic complexes; and the various colligative and biochemical properties of the seawater. A better understanding of these will form the foundations for a more extensive thermodynamic treatment of seawater systems (e.g., DESALINATION processes, heat-exchange processes, and pollution-abatement practices), the recovery of materials from the sea, and the solution to a myriad of other problems that face all peoples of the world.

SEAWEEDS See ALGAE; KELP.

SEAWORMS or annelid marine worms are members of the various elongated soft, segmented bodied animals of the phylum Annelida, primarily of the large class Polychaeta (meaning many setae or tufts) which consists of about 68 families, 1600 genera, and over 10 000 species.

Seaworms are distributed worldwide in all marine habitats, and almost all these families tend to be represented in any major geographical area. These animals are mostly free-living, although some are commensal and live in close association with other marine forms of life. Reproduction is highly involved and can be sexual or asexual.

The seaworms, or Polychaetes, range in length from less than a millimeter in the syllids to about 12 ft (3.65 m) in the Eunicea (rockworms) and Amphinomidae (group Errantia). Most possess tufted, silky, chitinous bristles in a row along each side of the body. Upon contact or stimulation of any kind, in some of the animals, the bristles rise on edge and the worm contracts and presents a defensive armor of tiny spears to the intruder. The fine bristles penetrate and are difficult to remove from the skin of humans who encounter them. A burning sensation is first produced, and later the area becomes inflamed and may swell or become numb. Other worm species have strong jaws which inflict a painful bite. These worms may be encountered under rocks or coral.

Representative species of these types are

- Bristleworm, *Eurythoe complanata,* is found in the GULF OF MEXICO and throughout the tropical Pacific area.
- Bloodworm, *Glycera dibranchiata,* is found on the Atlantic U.S. coast and northward into Canadian waters.

The bite of the bloodworm may or may not penetrate the skin. The bite appears round with a red dot in the center and is usually surrounded by a pale area, later becoming hot and swollen and then turning numb or itchy. A bristleworm contact, on the other hand, produces inflammation, swelling, or numbness which may persist for several days.

Some of the serpulids, or members of the family Serpulidae (e.g., *Hydroides norvegica* and *Mercierella enigmatica*), are calcareous-tube-building animals that attach themselves to the hulls of ships. (See also FOULING.)

SEICHE See WAVES.

SEISMIC SEA-WAVE WARNING SYSTEM (SSWWS) is a network of seismographs and tide stations, tied together by a communications net, which stretches across the PACIFIC OCEAN from the west coast of the Unites States to Japan and serves to warn the people of this vast area against the arrival of destructive seismic sea waves, or tsunamis.

Established in 1946 [after 55-ft (16.8 m) waves created by the April 1, 1946, earthquake at Unimak, Alaska, left 159 people dead in Hawaii], the SSWWS is headquartered in Honolulu, Hawaii, and is operated by the Coast and Geodetic Survey of the U.S. Department of Commerce. The communication net is made up of facilities provided by the Federal Aviation Agency, the National Aeronautics and Space Administration, the Defense Communications Agency, and the U.S. Weather Bureau. In addition to the seismographs and tide stations operated by the Coast and Geodetic Survey, those operated by other government agencies, private institutions, and participating Pacific nations are also utilized. The SSWWS is credited with saving many lives and many millions of dollars in property damage since its establishment.

Seismic sea waves are also known by the Japanese word TSUNAMI, which means "harbor wave." More commonly, these waves are called *tidal waves,* although they have nothing to do with tides or tide-producing forces. Seismic sea waves are definitely associated with earthquakes—particularly those exceeding 6.5 on the Richter scale—and may be caused by the displacement of large volumes of WATER through such earthquake-related mechanisms as vertical faulting or vast mudslides on the seafloor.

The waves generated by a seismic disturbance affecting the seafloor are unique in many respects. They radiate from the source much as the ripples from a stone dropped into a pool. In very deep water the height of a seismic sea wave is generally no

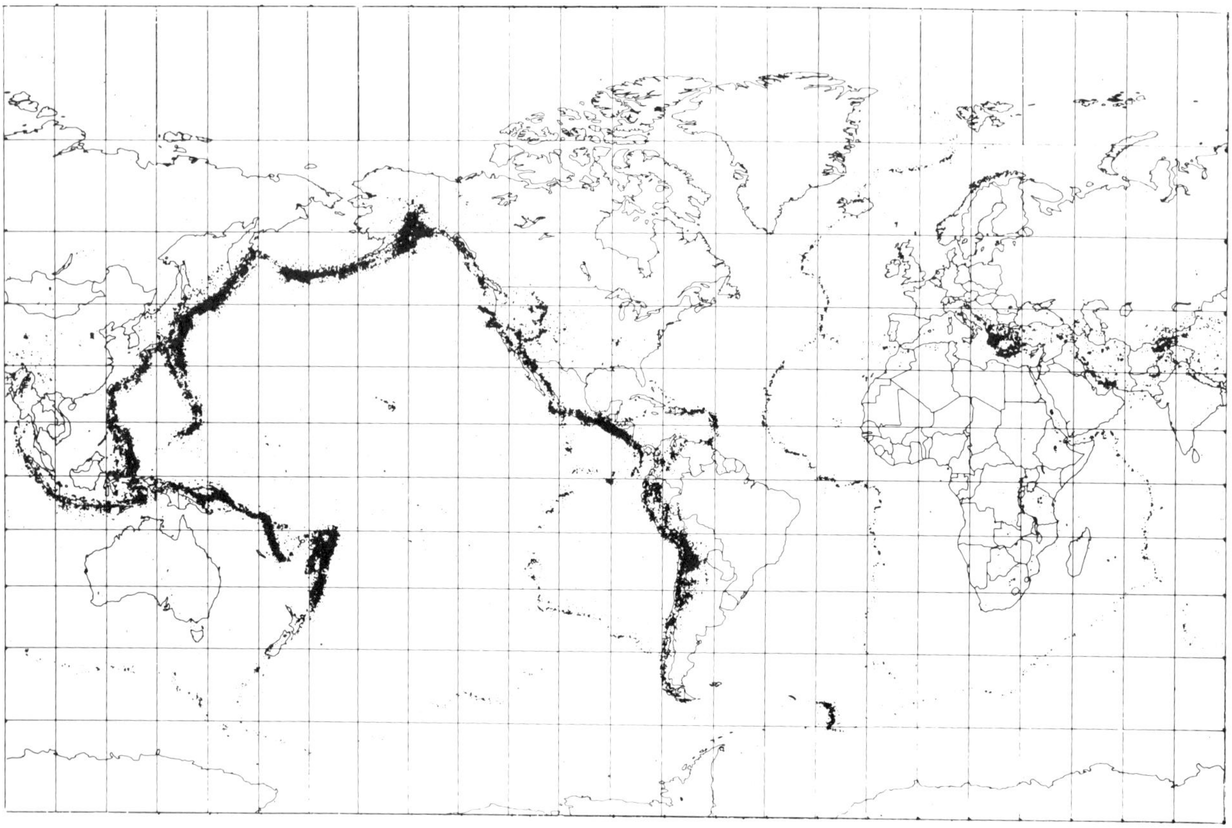

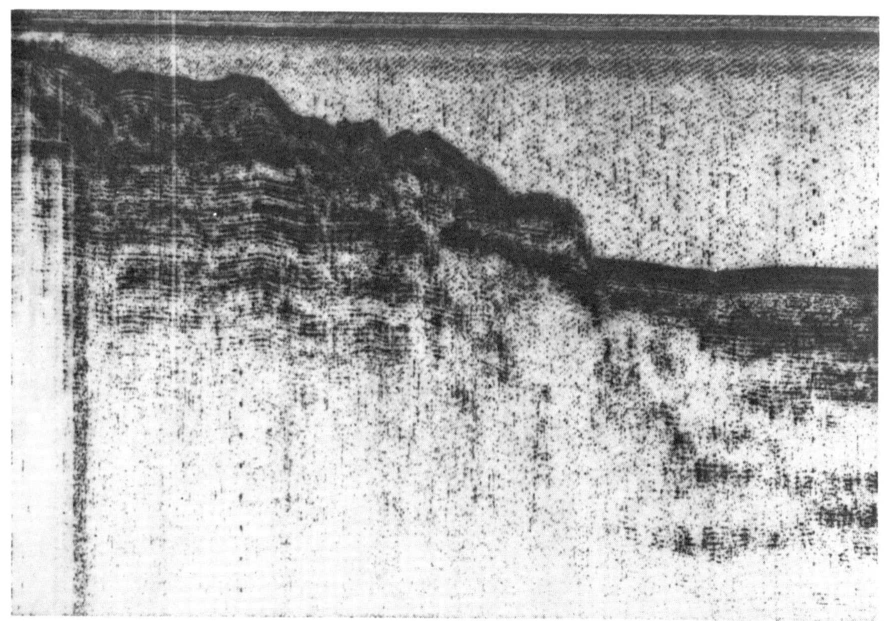

SEISMIC SEA WAVE WARNING SYSTEM. (*Top*) Clusters of major earthquake epicenters for a ten-year period mark the edges of the earth's tectonic plates as posited in the continental-drift theory. The principal seismic zones are the circum-Pacific seismic belt, the earth's most active seismic feature; the mid-Atlantic Ridge; and the Alpide belt, which links up with the circum-Pacific belt at New Guinea. (*NOAA*) (*Bottom*) A continuous seismic profile illustrates the structure beneath the sea floor. (*NOAA*)

more than a foot or two (less than 1 m), but the wavelength measured from crest to crest may be in excess of 100 mi (160 km). Because of the great wavelength, these waves "feel" the bottom even in the greatest ocean depths. (By rule of thumb, a wave disturbance begins to be interfered with by the bottom when the water depth becomes equal to one-half the wavelength.) Because interference with the bottom robs energy from the wave system, the speed of these waves is dependent on the water depth over which it travels. For instance, in 30 000 ft (9144 m) of water the speed of a seismic sea wave is around 670 mph (1678 km), around 212 mph (341 km) for a depth of 3000 ft (914 m), and approximately 30 mph (48 km) as it comes ashore in shallow water. As the wave approaches land, its energy is concentrated in an ever-narrowing wedge of water such that its height begins to grow. As a consequence, seismic sea waves of 125 ft (38 m) in height were reported to be associated with the explosion of the volcanic island of Krakatoa in 1883. These characteristics all have implications for the SSWWS.

In practice, seismic activity throughout the Pacific Basin is continuously monitored by the seismograph net. When a disturbance of sufficient magnitude is recorded, its distance from the seismograph station is determined by recording the arrival times of the P (primary) and the slower S (secondary) waves associated with the earthquake. In other words, since the characteristic speed of both waves is known, the amount of time by which the S wave is late gives an indication of the distance—though not the exact direction—of the earthquake. To determine the exact location (epicenter) of the disturbance requires that at least two seismograph stations that have recorded the P and S waves draw an arc on a chart about their own location, the radius of which is equal to the calculated distance to the epicenter. The epicenter of the earthquake is found to be where the two arcs intersect on the chart. At this point an advisory warning of the possibility of an earthquake that could cause seismic sea waves is flashed throughout the net.

Having located the epicenter of the earthquake, tide stations in the vicinity which have tide gauges designed to filter out all waves not having the proper characteristics are alerted to be on the lookout for seismic sea waves. If a wave is detected, its travel time to various populated areas throughout the Pacific Basin is calculated through a knowledge of the bathymetry over which the wave must travel, and its speed as a function of that bathymetry. At this point a full warning, containing arrival time information, is broadcast throughout the net, and the various stations, in turn, notify the populated areas likely to be affected.

At present the SSWWS consists of 15 seismograph stations and 30 tide stations and serves the United States, American Samoa, Canada, Chile, the Fiji Islands, French Polynesia, Hong Kong, Japan, New Zealand, the Republic of the Philippines, Taiwan, and Western Samoa.

SELF NOISE See UNDERWATER SOUND.

SEMIDIURNAL TIDE See TIDES.

SERGEANT MAJOR See DAMSELFISH.

SEVEN SEAS See OCEAN.

SHAD See HERRING.

SHADOW ZONE See UNDERWATER SOUND.

SHALLOW-WATER WAVE See WAVES.

SHARK is the name for any of about 250 species of Chondrichthyes or cartilaginous fishes belonging to the subclass Elasmobranchii (the sharks and RAYS). Sharks have well-developed, hinged jaws; five to seven gill openings or slits on each side of the head; a cartilaginous skeleton covered by a thin nonbony calcified layer; the presence, in the head region, of sensory organs (called the Ampullae of Lorenzini); and rigid fins. Sharks possess eyes that can be shielded from physical or chemical irritants by an opaque (nictitating) membrane. The scales or dermal denticles of the animal are essentially toothlike in structure.

Sharks, as a group, occur in all the world's oceans, although the majority are found in tropical and subtropical regions. Some species frequent coastal waters, and a few enter fresh waters, while others inhabit the open sea. All are carnivorous, with ravenous appetites. Most are scavengers but prefer fresh, oily fish. However, sharks have been observed to eat SQUID, SEALS, MARINE BIRDS, other sharks, TURTLES, LOBSTER, garbage, and humans.

Characteristics Sharks vary in length—from 1 ft (30.5 cm) to more than 40 ft (12 m)—and in their habits, depending upon species. Some of the more interesting characteristics of typical sharks are as follows:

- The fins of a shark, other than the caudal (tail) fin, are not as flexible as those of bony fishes and cannot be utilized for forward propulsion. Such a fin structure dictates that the shark cannot back down once it has overshot its target; it must swing out and make a new figure-8 approach while searching or feeding.

- Sharks have a small brain but possess a well-developed sense of smell which enables them to detect small amounts of blood in the water. Despite the inferior anatomy of its retina, the shark probably has very acute vision. It also has a type of hearing sense that permits it to pick up sounds (or pressure-wave vibrations) such as those given off by a struggling fish in the water. This detection is carried out by sense organs and by the Ampullae of Lorenzini, which are specialized sense organs, each with small sensitive pores, on the snout of the animal. These pores lead to a network of fluid-filled nerve canals that act as receptors for the brain. By means of these organs and ampullae, the animal is apparently capable of detecting not only changes in the temperature of the water but also the slightest fluctuations in water pressure. (See AMPULLAE OF LORENZINI.)

- Sharks have very well-developed hinged jaws, with the upper jaw having a long overhang. The gaping, open mouth of a feeding shark often has been compared with a steel trap. The shark's bristling, sharp-honed teeth are arranged in four to six rows around its jaws. The front functional row (or rows) of teeth can easily sever many types of fishing cable leaders. Behind are several reserve rows, all pointing backward. These reserve rows of teeth are used, via a strong membranous tissue on the inside of the jaws, to replace the outer teeth as they are periodically shed. One shark tooth alone can apply a force of 132 lb (60 kg) over an area of 0.3 in^2 (2 cm^2). The teeth in the lower jaw are used to hold the prey, while the teeth in the upper jaw rip it apart.

- Sharks, with their incredible musculature, are capable of a display of power and vitality that is hard to estimate and analyze. They are unpredictable creatures and, when hooked, can be docile one minute and aggressive the next.

- Of the four major groups of fishes, sharks and their relatives, the skates and rays, are the most primitive, having descended with only superficial changes from the Devonian Period which began about 310 million (310×10^6) years ago. (See SKATE.) In contrast, bony fish dating from the same time have developed many specialized forms. Shark skeletons are cartilaginous, only the jaws and a few of the fin spines being of bone; hence, these are the parts which alone may be preserved as fossils. (Cartilage is the flexible structural substance in the ears and nose of humans.)

- Owing to its extreme adaptability, the shark has survived where other marine species have become extinct. It developed much earlier than the bony fish, and in contrast to the latter's habit of laying eggs in the water at random, the young are born alive—a half-dozen to sixty at one time. Most sharks have a gestation period of 8–9 months; however the spiny dogfish sharks require 22 months. Sharks are equipped with a complete set of teeth at birth. They leave their parents right after they are born—a fortunate move, for the parents often eat them soon after their birth. Some sharks are oviparous (e.g., the European dogfish, *Scylliorhinus caniculus* and *S. stellaris*) and lay their eggs in what is often described by the term *mermaid's purse,* a leathery case in which the embryo is contained.

- The sexes of sharks are distinguishable because the males have paired copulatory organs called *claspers.* These tube-shaped appendages develop along the inner edge of each pelvic fin. Supported by internal cartilages and having a groove covered with a skin fold on the inner side, the claspers act as dilators to direct the sperm into the female opening (the cloaca or oviduct).

- Unlike most bony fishes, sharks possess no SWIM BLADDER, an internal balloonlike bag which can be filled with gas to make the fish lighter and help it to float. Thus, the shark ordinarily needs to be constantly on the move in order to stay afloat. Some species overcome this by gulping air into their stomachs, while a few others (e.g., some species of requiem sharks) can remain comatose for several hours by actively pumping high-oxygen–content water over their gills. Certain lesser-known species that inhabit the mid-depths of the ocean have developed large livers that contain hydrocarbon oils. These lighter-than-water liver oils assist these sharks in staying at the desired depths.

- Although most sharks possess a mouth full of exceptionally strong teeth for seizing, shearing, and crunching their prey, the largest sharks, the whale sharks (*Rhincodon typus*) and the basking sharks, feed on PLANKTON—floating microscopic plants and animals. Also, the thresher (*Alopius vulpinus*) has a tail as long as the rest of its body, and it uses his tail to thresh through schools of fish. The stunned fish are then consumed by this voracious animal.

- Sharks are often accompanied by what are termed pilot fish (*Naucrates ductor*). Legend has it that they guide their host to prey, but probably they just clean parasites from the shark's body. Another companion is the REMORA (*Echeneis naucrates*) which attaches itself with a sucking plate to the body of the shark.

- Sharks have an unusually short intestine: a 10-ft (3-m) shark has only 9 ft (2.7 m) of intestines, while a 6-ft (1.8-m) human being has 25 ft (7.5 m) of intestine. Food is kept in it long enough to be digested by a special valve, the spiral valve. This is a fold of the intestine which is so shaped that it makes the food in it travel in a spiral or coiled route instead of passing straight through. This route increases the absorptive surface very appreciably. Also, the digestive enzyme of the stomach, called *shark pepsin,* is effective at temperatures much lower than those in warm-blooded animals. Indigestible material is easily regurgitated by the shark to make room for another meal.

- Many species of sharks are highly territorial and each animal has a personal space into which it is unsafe to venture, since the intruder will invariably be attacked.

Sharks Dangerous to Humans Sharks have always fascinated and terrified human beings. In 492 A.D., Herodotus wrote that survivors of shipwrecks were killed and eaten by monsters off the Thessa-

SHARK.

One of the oldest species of fish, sharks have remained almost unchanged for more than three hundred million years. They are found in all the oceans of the world, feeding along the bottom near the coast, and close to the surface when in deep waters. All are carnivorous.

A fisherman grasps a hammerhead shark caught in a shrimp trawl net. This animal is a maneater, and is found in large numbers in the warmer waters of all oceans, including the Mediterranean Sea. *(NOAA)*

lian Coast. Pliny the Elder called the shark the dogfish because of its tearing teeth and method of attack. However, it was not until the middle of the sixteenth century that the word *shark* was added to the vocabulary of English-speaking people. While its origin is uncertain, it is generally believed that English sailors, hearing the German word *schurke* which means "villain," applied it to the shark.

While only about 27 "villains" of the more than 250 known species of sharks will attack humans, all sharks over 3 ft (0.9 m) long are potentially dangerous. Members of the four families that are believed to be of greatest concern to divers or swimmers are

1. The mackerel, or man-eater shark family (Isuridae)
2. The requiem shark family (Carcharhinidae)
3. The sand shark family (Carchariidae)
4. The hammerhead shark family (Sphyrnidae)

Representative of these families are the following sharks:

- The white shark, *Carcharodon carcharias* (Isuridae), is savage, aggressive, and fast-swimming. Several human attacks are accorded to this species. Distinctive features are the beefy body, the moon-shaped tail fin, and coarse serrated teeth. Color is slate brown, slate blue, gray, or almost black above, shading to dirty white below. The tips of the pectoral fin are also black, usually with some adjacent black spots. These sharks may reach a length of 36 ft (10.9 m) or more, and are widespread in tropical, subtropical, and warm temperate belts of all the world's oceans.

- The tiger shark, *Galeocerdo cuvier* (Carcharhinidae), when stimulated, can be a vigorous and powerful swimmer. Attacks on humans are credited to this shark, although it normally tends to be a scavenger. Distinctive characteristics are the very short, blunt snout; wide, bladelike irregularly triangular saw-edged teeth; and sharply pointed tail. It is a mottled gray or grayish brown, and darker above than on the sides. It can attain a length of 30 ft (9.1 m), but more commonly is about half that length. The tiger shark is widespread in the tropical and subtropical belts of all oceans, both inshore and offshore. It is the commonest large shark of the tropics.

- Sand shark, *Carcharias taurus* (Carchariidae), lives on or close to the bottom near shore and has a voracious appetite. It has gill openings in front of the pectoral fin. The second dorsal fin is about as large as the first, with the first dorsal entirely in front of the pelvic fin. The body is colored a bright gray-brown above, white on the belly. The body rearward from the pectorals is marked with round-to-oval yellow-brown spots. The sand shark may reach a length of 10 ft (3 m). It is an inhabitant of the western ATLANTIC OCEAN from the Gulf of Maine to Florida and Southern Brazil, as well as the MEDITERRANEAN SEA, tropical West Africa, the Canaries, and Cape Verde Islands in the Eastern Atlantic.

- Hammerhead shark, *Sphyrna diplana* (Sphyrnidae), is a powerful swimmer that may frequently be seen at the surface, inshore, or far at sea; it is known to attack humans. It is ashen-gray above, fading to white below, and distinguished by the widely expanded head with eyes at the outer edge. It may attain 15 ft (4.5 m) in length or more. Closely related species are distributed throughout tropical warm temperate zones of all oceans, including the Mediterranean Sea.

Also prominent among those sharks that attack humans are the great blue, or, occasionally, blue fin (*Prionace glauca*); the white-tips (*Carcharhinus longimanus*); the mako, blue pointer, or sharp-nosed mackerel shark (*Isurus oxyrhynchus*); the gray nurse of the Indo-Pacific (*Carcharias arenarius*); and the black whaler (*Galeolamna macrurus*). The record of unprovoked attacks on humans by these species is notorious and extensive (about 900 in the past 50 years). Moreover one might easily add to the list such dangerous species as the dusky shark (*Carcharhinus obscurus*), the bay shark (*Carcharhinus lamiella*), the Ganges (*Carcharhinus gangeticus*), the Lake Nicaragua (*Carcharhinus nicaraguensis*), the large black-tip or spinner (*Carcharhinus maculipinnis*), the lemon (*Negaprion brevirostris*), the porbeagle or mackerel shark (*Lamna nasus*), the reef or gray reef shark (*Carcharhinus menisorrah*), the salmon shark (*Lamna ditropis*), the white-tip reef shark (*Carcharhinus albimarginatus*), the Zambesi (*Carcharhinus zambezensis*), the wobbegong (*Orectolobus maculatus*), the scalloped hammerhead (*Sphyrna lewini*), the smooth or common hammerhead (*Sphyrna zygaena*), and the bonito or Indo-Pacific mako (*Isurus glaucus*).

Classification All these anatomically unique creatures of the sea can be divided into two major groupings: the galeoid and the squaloid. Sharks belonging to these different groups vary somewhat in such features as the size of their jaws, tooth shape, and fin structure. For example, the squaloids do not possess an anal fin.

The galeoids consist of the isurids, which include the mackerel sharks and the carcharhinids, including the tiger sharks.

The isurids comprise the families Orectolobidae (e.g., the carpet and nurse sharks), Carchariidae (e.g., the sand sharks), Scapanorhynchidae (e.g., deep-sea goblin sharks), Rhinoodontidae (e.g., whale sharks), Cetorhinidae (e.g., basking sharks), Isuridae (e.g., mackerel sharks), and the Alopiidae (e.g., thresher sharks).

The carcharhinids include the families Scylliorhinidae (e.g., cat sharks), Pseudotriakidae (e.g., dogfish and leopard sharks), Carcharhinidae (composed of some 60 species, such as the blue sharks, tiger sharks, etc.), and the Sphyrnidae (e.g., the hammerhead sharks).

The other main shark group, the squaloids, comprises the families Squalidae (spiny dogfishs), Dalatiidae (spineless dogfishes), Pristiophorideae (saw sharks), and Squatinidae (angel sharks).

The common spiny dogfish (*Squalus acanthias*) is a small shark, up to $3\frac{1}{2}$ ft (1 m) in length, found along both coasts of the Atlantic and Pacific oceans. Closely related spiny dogfishes are found throughout the temperate and tropical seas of the world.

Dogfish sharks are distinguished from the larger species by the absence of the anal fin. Another feature peculiar to all dogfish is the presence of two short, stout poisonous spines, one situated immediately in front of each dorsal fin. Although dogfish are somewhat sluggish, this does not prevent them from thrashing around with much vigor when caught in commercial fishing nets, and the spines are capable of causing painful wounds to humans. The venom is a glistening white substance in a shallow groove of each spine. The poison secretion enters the skin with the spine.

In late 1976, a new species, genus, and family of shark was captured at a depth of 500 ft (152 m) of water northeast of the coast of Hawaii. Somewhat resembling a sand shark, this particular 15-ft (4.6-m) shark possessed a very peculiarly shaped enormous head and mouth, a short snout, and peculiar gill structures. The fish was nicknamed "megamouth."

Practically every part of the shark has a commercial use. For example, the liver is often rich in vitamin A; the meat of most species is used in many places of the world as human food, being sold as fillet of sole, sturgeon, cuban cod, etc.; the hide, with its shagreen (calcareous coating) produces a good leather; and the remainder of the fish can be converted into animal feed and fertilizer.

SHEARWATERS See MARINE BIRDS.

SHELF CHANNEL See SUBMARINE CANYON.

SHELLFISH is a name for an aquatic INVERTEBRATE, such as a MOLLUSK or any of the CRUSTACEANS, that has a shell or exoskeleton. See CLAMS; LOBSTER; MARICULTURE.

SHELLS See SEASHELLS.

SHEPARD, FRANCIS PARKER (1897–), an American educator and geologist, is most closely identified with his studies of submarine canyons and his theories regarding their origin. (See SUBMARINE CANYON.) He argued that these great canyons were cut by rivers that ran out over the CONTINENTAL SHELF during the low stand of SEA LEVEL brought about by the last Ice Age.

Shepard was born May 10, 1897, in Brookline, Massachusetts. He received his bachelor's degree from Harvard in 1920 and his doctorate from the University of Chicago in 1922. He remained at Chicago, first as instructor and then as professor of geology until World War II brought him to the SCRIPPS INSTITUTION OF OCEANOGRAPHY in 1942. At Scripps he served as research associate in the division of war research until 1945 when he became principal geologist. Shepard became professor of submarine geology in 1948, and professor of oceanography in 1964.

Shepard's study of submarine geology began while he was at the University of Chicago. Using his father's small yacht off the Massachusetts coast, he found the sediment of the continental shelf to be much more varied than was then depicted in most textbooks. During a sabbatical leave in 1933–1934 he studied the submarine canyons along the California coast. This work led, in 1948, to the publication of *Submarine Geology*, the first textbook on the subject. His continuing interest in submarine canyons led him to publish with Robert F. Dill, in 1966, *Submarine Canyons and Other Sea Valleys*. The book contained an analysis of seafloor canyons and valleys in many parts of the world's oceans. Shepard has also added much to our knowledge of fluctuation in sea level around the world.

SHOAL is a detached elevation over which the WATER is so shallow that it poses a hazard to surface navigation, but which is not composed of rock or coral. The danger of navigation in shoal waters, particularly before the advent of fathometers, was dramatized by the repeat line ("Shoal! 'ware shoal!") in Rudyard Kipling's 1896 poem *The Bell Buoy*. The term *shoal* (from "school") is also used to mean a mass of PLANKTON or FISH.

SHRIMP is the common name applied to several thousands of species of CRUSTACEANS, primarily the edible and economically important decapods (order Decapoda) consisting of several families, genera, and species.

Shrimp, in general, may be characterized as small crustaceans, principally of the malacostracan group (which also includes the LOBSTER and CRAB, among others), all having a common basic body plan. The body of the shrimp is slender, with the thorax (or the section between the head and abdomen) composed of eight segments and the sharply bent abdomen usually of seven. The relatively long thoracic legs, a pair for each of the eight upper body segments, are utilized for various functions, such as feeding and capturing prey, while the well-developed appendages (or pleopods) on the abdomen are used in swimming as well as for egg carrying by some species.

In the suborder Natantia of decapod shrimp, there are three main types or sections: Penaeidea, Caridea, and Stenopodidea; and each of these consist of several families and species. In one such family, the Hippolytidae, a species (*Merguia rhizophorae*) has been found to be semiterrestrial. This particular shrimp is found in the MANGROVES of Brazil and seems to have evolved behavioral adaptations to this environment.

Many of the penaeid shrimp (family Penaeidae) are of substantial commercial importance. In the United States, large shrimp fisheries are located along parts of the coasts of the GULF OF MEXICO. Here, great quantities of such species as brown shrimp (*Penaeus aztecus*) and others (the white and pink species) are caught yearly by trawling. Some species (e.g., the Pacific, *P. monodon*) reach 1 ft (30 cm) in length.

A pattern of life history common to most of the penaeids is this: They enter estuaries and inshore brackish water areas at an early age and live in these nursery grounds until they are ready to spawn. Just before maturity, they migrate to the ocean to breed. The eggs are deposited there by the females, and these eggs hatch into larvae. Each LARVA undergoes a series of changes until a young shrimp resembling the adult emerges. To illustrate this cycle, the pink shrimp (*P. duorarum*), indigenous to south Florida and Mexico, are spawned in the ocean waters around Dry Tortugas at depths of about 120 ft (36.3 m). The newly hatched shrimp undergo a distinct series of growth changes called the nauplius, protozoea, mysid, and postlarval stages. Such a cycle requires approximately a month to complete. At a young age, the shrimp migrate to the mainland of Florida, some 100 mi (161 km) away, where they inhabit canals, inlets, and bays of Florida tidal estuaries around the Everglades National Park. Here, for about 7 months, they feed on vegetation and small crustaceans until they are 3–4 in (7.5–10 cm) in length. About halfway to maturity, which requires approximately 2 years, they return to the Dry Tortugas ocean area. It is thought that here they die after spawning; in any event, they do not return to the nursery grounds.

The Caridea types of shrimp consist of some 22 families and about 1600 species. Representative species are the 2–3 in (5–7.6 cm) long common brown European species, *Crangon vulgaris;* the snapping shrimp, e.g., *Alpheus pachychirus;* and the parasite-eating shrimp, e.g., *Periclimenes pedersoni.*

The Stenopodidea types consist of only one family, the Stenopodidae, which are all marine. One species, *Spongicola venusta,* enters the canal system of certain SPONGES as a young shrimp and spends its entire life in that environment. Another 3-in (7.6-cm) long tropical species, *Stenopus hispidus,* performs the same function of removing parasites from the skin of FISH that the aforementioned *Periclimenes pedersoni* species carries out.

The so-called seed shrimps belong to four orders of marine and freshwater ostracods. These vary in size from microscopic forms to some of about $\frac{1}{2}$ in ($1\frac{1}{4}$ cm). The Stomatopoda, or mantis shrimp, are warm-water marine Malacostracans [largest species is approximately 1 ft (30.5 cm) long], while the opossum shrimp are most minute cold-water animals, forming much of the PLANKTON in ocean polar waters.

Shrimps of the family Palaemonidae, better known as the freshwater shrimps, are composed of several genera and many species, most of which are marine. These marine types usually live in close permanent association with another INVERTEBRATE of some form. This intimate symbiotic association of unrelated animals living with one another, wherein one is benefited and the other neither benefited or harmed (a relatively uncommon phenomenon for marine life in cold waters) is termed commensalism. Several commensal shrimp species (e.g., *Anchistus custos, Paranchistus brunguiculatus*) of the subfamily Pontoniinae are found in the warm CORAL REEF waters of the Indo-Pacific region where they live with mollusks (e.g., CLAMS, oysters, etc.). (See MOLLUSK; OYSTER.)

Various species of shrimp form a major part of several countries' protein harvest from the world's oceans. In addition to the United States, other countries such as Germany, Canada, Spain, Brazil, France, Japan, and Thailand also produce a high yield of these crustaceans, while Denmark and Norway have deep-water shrimp fishing activities off Spitsbergen and in the waters off west Greenland.

SIBBALD'S RORQUAL is the name sometimes applied to the largest living animals known to evolutionary history, the blue WHALES (*Balaenoptera musculus*). This whale is also called the sulfur-bottom because of the heavy yellowish growth of DIATOMS that cover the underside of this huge [100 ft (30 m) long, 160 tons in ight] animal.

SILICA or **SILICON DIOXIDE** is a naturally occurring compound of the elements silicon and oxygen (SiO_2). It is found in crystalline forms (e.g., quartz, tridymite, cristobalite, coesite, and stishovite), in cryptocrystallines (those which have an extremely fine grain structure e.g., chalcedony), in amorphous and hydrated forms (e.g., opal), and combined in silicates (compounds containing silicon, oxygen, and one or more metals as well as hydrogen). (See ELEMENT.)

The element silicon exceeds every other element in natural abundance except oxygen. Twenty-eight percent of the solid crust of the earth is composed of silicon (oxygen constitutes 48 percent). In fact, all of the solid crust of the earth, except for carbonate and phosphate rocks, contains some siliceous material.

The concentration of silica in waters entering the oceans is around 13 parts per million (ppm). This is over twice the average concentration found in the oceans, although SEAWATER concentrations of silica vary widely. Silica enters the oceans by infusion from rivers, by volcanism, and by the submarine alteration of comminuted glacial silicate materials. Also, the material is transferred from ocean to ocean by circulation. While the deep waters of the PACIFIC OCEAN and INDIAN OCEAN are richer in silicon than the Atlantic, precise inflow and outflow figures are presently not known, nor have the rates of seawater transfer and silicon concentrations in the world's oceans been determined accurately.

However, silicon is present in seawater as PARTICULATE MATTER and as dissolved silicates [principal species are $Si(OH)_4$ and $Si(OH)_30^-$]. Investigations have demonstrated that silicon presents in the oceans can have concentrations ranging from undetectable amounts to several parts per thousand. While the silica profile in the ocean tends to be characteristic of a given ocean, the distribution of content varies with depth, the biological community, and the physical circulatory processes. An example of this is that significantly higher silicate values for the deep waters of the CARIBBEAN SEA have been observed compared with those of the open ocean. This has been attributed to the fact that these depths are not being continually renewed.

Silicon is indispensable for marine life. In their life cycle process, ocean DIATOMS and radiolarians can deplete the silica content of surface water to almost zero. (See RADIOLARIA.) The tiny ALGAE [35 ft^3 (1 m^3) of seawater may contain 8 billion (8×10^9) diatoms] have their cell wall made up of pectin impregnated with a silicon compound which is often equal to two-thirds the total weight of the organism. These organisms are eaten by ocean animals which, in turn, are eaten by others in the OCEAN FOOD CHAIN. In this regard, it is interesting to note that diatoms contain a fat that has an odor quite like that of fats derived from fish, so that the characteristic "taste" of fish may be due to fats stored up by the diatoms. Diatoms that escape consumption by animals eventually die, and their siliceous shells fall to the ocean floor where they accumulate, forming diatomaceous earth (a substance used as an abrasive in industry, a filter in sugar manufacture, a heat insulation material and a filler in the production of rubber and dynamite). During the downward flight to the ocean bottom and after reaching it, the siliceous material of the dead shells is dissolved in the seawater. The re-solution process destroys about 90 percent of the shells and replenishes the supply of silicates in the ocean water. This cycle of replenishment is essential not only because silicon is an important nutrient in the oceans but also because siliceous materials, like carbonates, play a role in buffering seawater.

SILICON DIOXIDE See SILICA.

SILL refers to a submerged elevation between two basins. Sill depth is the greatest depth at which there is a horizontal water flow between two basins. The depth of the sill that separates the Atlantic Basin from the Mediterranean Basin in the Strait of Gibraltar is about 1000 ft (304.8 m).

SILT GRASS, *Paspalum vaginatum* (often called seashore paspalum, sheathed paspalum, saltwater couch, etc.), is a halophytic (or salt-tolerant) plant that grows wild on the seacoasts of both hemispheres from Australia to southern Spain and from Argentina and Chile to Baja California and North Carolina.

A good sand-binding plant, silt grass grows prolifically in dry, sandy beach soils of the coastal regions around the world. However, it has been domesticated only in Australia, Florida, and the Netherlands Antilles where it is used primarily as forage for animals and as a lawn grass for residences.

SINKING See UPWELLING.

SKATE is the name for any of the various batoid elasmobranchs in the family Rajidae. (See ELASMOBRANCH.) There are more than 100 species of skates, and these fish are characterized by their broad, flat bodies with winglike pectoral fins and a slender tail having two small dorsal fins. Both the skates and RAYS are related to the sharks and often the names skate and ray are used interchangeably, especially since they both have a similar batlike shape. (See SHARK.)

Like the rays, the skates are found mostly in shallow temperate and tropical waters of all the world's oceans except for the South PACIFIC OCEAN and a portion of the northeast coast of South America.

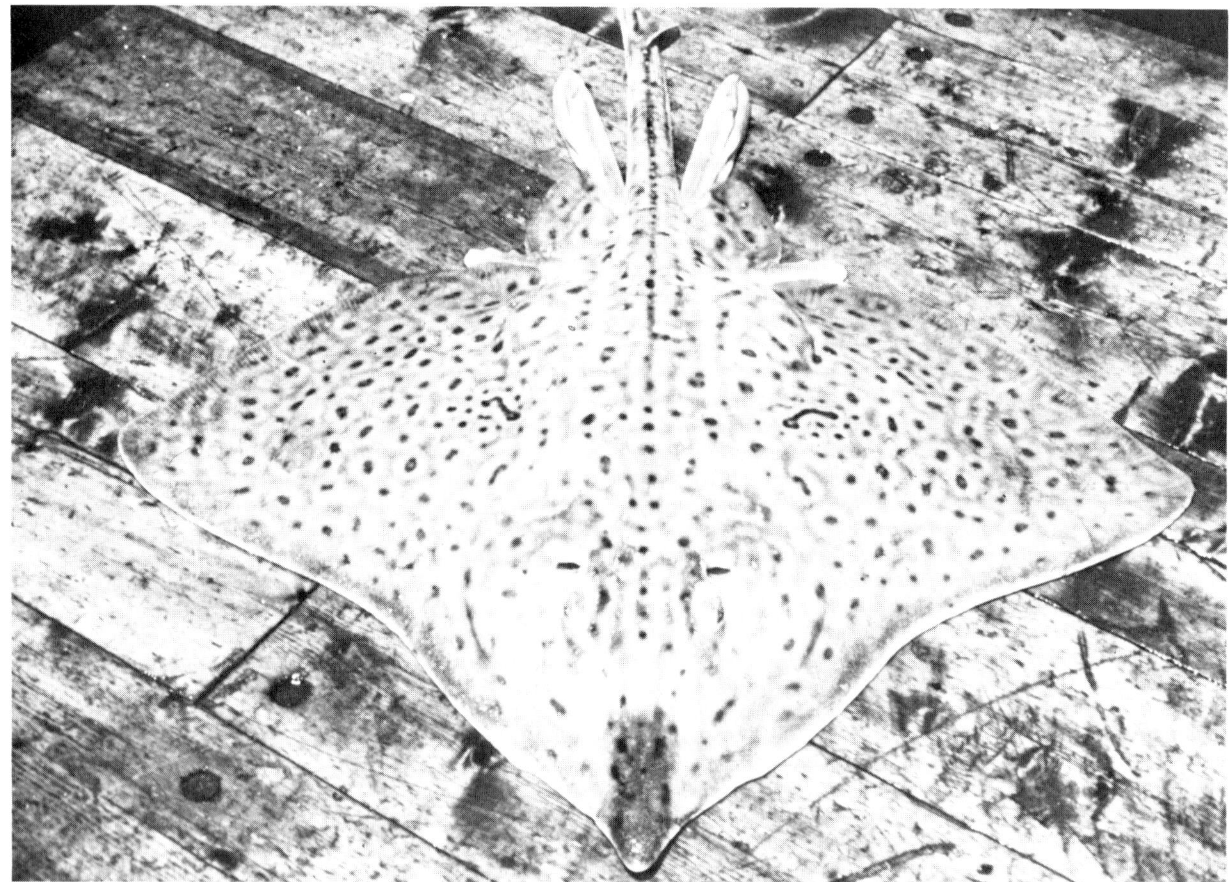

SKATE. This flattened fish lives mainly near the bottom of shallow temperate and tropical waters. When it does occasionally migrate into deeper water, it swims well, using its large pectoral fins. *(NOAA)*

The common skate, *Raja batis,* attains a length of over 6 ft (2 m) and, when young, exists on a diet of various bottom-living animals such as the CRAB and LOBSTER. Later, they also eat SQUID and other fish, especially HERRING, which they hunt mainly by scent rather than by sight. Most species, like the common skate, are demersal and live on or near the bottom in shallow waters, although some are found at the 600-ft (182-m) depths, and one species, the deep-sea skate (*Raja abyssicola*), lives at a depth of over a mile (1.6 km) in the water of the Pacific Ocean off the coast of North America. All species are good swimmers when the occasion demands, and all, when lying on the bottom, utilize a specialized method of closed-mouth breathing in order to minimize the swallowing of mud or sand. In this method, water is taken in through passages in the top of the head and eventually passes out through gill slits on the underside of the fish. The eyes are on top of the snouted head of a diamond-shaped body. The male has small pelvic fins equipped with a pair of claspers or organs for injecting sperm inside the female. The eggs of the female are contained in unique fiber-reinforced cases called MERMAIDS' PURSES.

The batlike wings of the skate are the part generally used for cooking when this fish is eaten. The flesh is gelatinous and has a distinctive flavor.

SKIPJACK See TUNA.

SNAPPER is the common name for any of over 250 species of fishes of the family Lutjanidae in the order Perciformes. These fish are chiefly found in tropical coastal waters around coral reefs. (See CORAL REEF.) They attain a maximum size of about 2 ft (61 cm) in length and are distinguished by their deep body, large flattened head, and a spiny front dorsal fin that is soft-rayed in the back portion. (See FINS.) The name snapper is derived from the way the captured fish quickly and forcibly opens and closes its mouth.

The greatest variety of species and abundance of snappers occurs in the Indo-Australian ocean waters, although many are found in the tropical waters of the ATLANTIC OCEAN off the coasts of North and South America.

Although comparatively little is known about the life history of the snapper, commercial fishermen have learned that concentrations of them are usually found over certain types of bottom. Irregular bottom formations of rock and limestone covered with live coral and grass, called *lumps* or *gullies,* are especially preferred by the snapper, as is the case with the GROUPER. Fish schools are usually located several feet off the bottom of these areas where food material brought in by eddying CURRENTS settles out. Snappers are euryphagous, which means they will eat anything and everything edible. As a consequence they have a varied diet consisting of FISH, crabs, lobsters, OCTOPUS, BARNACLES, mollusks, etc., as well as marine plants such as ALGAE. (See CRAB, LOBSTER, MOLLUSK.) Due largely, perhaps, to this form of eating behavior, snappers are among the several species of ocean fish that at times can cause ciguatera poisoning in humans when the snapper is eaten by them. Such poisoning is unpredictable, and it is thought to be caused by a fish eating a particular blue-green alga or by a fish who has eaten another fish that has ingested the alga. (See also BARRACUDA.)

The red snapper, *Lutjanus aya,* is caught in the GULF OF MEXICO, and a red snapper industry is centered in Florida. As the name implies, this fish is brilliant red in color. The red snapper stays in shallow water during the summer months and moves offshore again as fall arrives. It is thought that these snappers spawn in deep water during the fall.

SNELL'S LAW See MARINE OPTICS.

SOAPIES See GROUPER.

SOFAR CHANNEL See UNDERWATER SOUND.

SOLE See FLATFISH.

SOLOMON SEA is located on the northwestern corner of the South PACIFIC OCEAN. It is flanked by the islands of New Britain to the north, New Guinea to the west, the Solomon island chain to the east, and the Louisiade Archipelago and the northern limit of the CORAL SEA to the south. The Solomon Sea covers an area of 277 920 mi² (720 000 km²) and occupies a volume of about 335 874 mi³ (1 400 000 km³). The maximum depth is 29 987 ft (9140 m).

The Solomon Sea gets its name from the Solomon island chain running down its eastern flank. In 1567 the Spanish explorer Álvaro Mendaña de Neyra discovered the islands and gave them their present name because he thought he had discovered the source of the gold in King Solomon's temple. New Britain to the north was discovered in 1700 by the English navigator William Dampier, and New Guinea on the west was first sighted by Antonio de Abrea, a Portuguese navigator, in 1511.

The bathymetry of the Solomon Sea is dominated by two basins: the New Britain Basin in the north and the Solomon Basin in the south. Depths of these basins range from 13 123 to 22 966 ft (4000 to 7000 m). Between the northern basin and the island of New Britain is the New Britain Trench which exceeds 16 400 ft (5000 m) in depth. The Trench turns south to parallel the northwestern flank of Bougainville (Bougainville Trench). This extension contains the greatest depth in the area—29 987 ft (9140 m). Between the outer and inner row of Solomon Islands is the Santa Isabel Trough with depths as great as 14 173 ft (4340 m). To the south of the New Britain Basin is the Kiriwina Trough, running east-west, with depths exceeding 16 000 ft (5000 m) in some places. The deep basins and trenches are covered with globigerina ooze grading to terrigenous sediments on the flanks. The shallow rim of the sea consists of calcareous muds and sands derived from the extensive reef structures around the periphery. The entire area is seismically active, and many earthquakes have been experienced.

The circulation in the Solomon Sea is strongly influenced by the South Equatorial Current. (See EQUATORIAL CURRENT SYSTEM.) In the winter the current enters the sea from the north and flows generally southward. It swings around to the southeast in the summer and flows out through northern passages.

SONAR or **ECHO-RANGING SONAR** is a term applied both to the application of underwater sound for a variety of oceanic endeavors and to the electromechanical system used. In its most common form, active sonar, a sound pulse is emitted into the water and its return is recorded when it is reflected or scattered by a submerged object. The time required for the pulse to go out and the echo to return indicates the distance, or range, of the object. In the passive system noises emitted by the target are detected and characterized. The term *sonar* is the acronym used in the United States and is derived from *so*und, *na*vigation *a*nd *r*anging. The British have called the device ASDIC (from Anti-Submarine Detection Investigation Committee). The method is widely used in naval warfare for the detection of submerged sea mines and submarines, and, in principle, for a variety of nonmilitary oceanographic efforts such as mapping the sea bottoms, the detection of schooling FISH, location of submerged wrecks, in ocean oil-drilling work, and underwater navigation and communication.

Sonar was originally designed for the purpose of detecting icebergs after the *Titanic* disaster. However, the system was quickly adapted for many other purposes, especially by the military for submarine detection.

Sonar can be used in two different modes—active (echo-ranging) and passive (listening). Identification and tracking of a target in either mode depends on the strength of the signal received from the target.

The intensity of acoustic energy is lessened as sound travels through SEAWATER, or as it is reflected from either the ocean floor or surface. This loss through the water occurs as (1) the acoustic waves spread with increasing distance from the source and/or because some sound energy is converted to heat (absorption), (2) sound waves are refracted because of differences in sound speed gradients, or (3) the sound waves are scattered by marine life, particulate matter, air bubbles, and by both the surface and the bottom interface. Some energy also may be lost at the ocean floor (bottom loss) because of absorption and reflection. A sound wave will be reflected whenever it strikes a boundary between masses of two different densities and sound speeds, for example, the interfaces of ocean and bottom or air and ocean. If the surface and bottom were completely smooth and the water clear, acoustic signals would reflect in the form of perfect echoes. However, the ocean has irregular upper and lower boundaries and contains many particles and organisms. Whenever a sound pulse hits a rough surface, a rocky hilly bottom, suspended particles, or marine organisms, much of the sound is returned as many small, unwanted echoes called reverberations. Reverberation from scatterers within the water mass is known as volume reverberation. Reverberated sound energy is sometimes strong enough to mask other desired signals and often limits the performance of active sonars.

In sonar work, sonobuoys, or modified electronic devices that activate on contact with seawater in order to deploy an omnidirectional hydrophone to a predetermined depth, are often used. Such sonobuoys and hydrophones are launched from aircraft or surface ships, and a radio link between the sonobuoy and the launching platform is utilized to monitor acoustic information received by the hydrophones. See INSTRUMENTATION: UNDERWATER SOUND.

SONOBUOY is the name for a device consisting of an acoustic receiver and radio transmitter mounted in a buoy which can be placed in the sea to receive underwater sounds and transmit them to an aircraft or other remote location. In addition to their widespread use to detect and track submarines in antisubmarine warfare (ASW), sonobuoys have been used as geophysical miniature broadcasting stations, providing a convenient means of analyzing underwater acoustic returns in ocean-bottom profiling. See INSTRUMENTATION; OCEANOGRAPHY; SONAR.

SOUND See UNDERWATER SOUND.

SOUNDING is the measurement of the distance between a given point on the surface of the water and the bottom directly beneath. This measurement may be taken by a weight attached to a line or by sonic means.

Soundings or depth measurements in the ocean date back to ancient times (see OCEANOGRAPHY). However, the first successful deep-sea soundings were made in 1840 in the South ATLANTIC OCEAN by Sir James Ross. Later, Matthew Fontaine Maury collected depth measurements in the Atlantic Ocean on a systematic basis, and these data furnished the information for a bathymetric chart produced by him in 1854. (See MAURY, MATTHEW FONTAINE.) Prior to the introduction of SONAR, soundings were made by lowering a weighted wire or rope line over the side of a ship and measuring the amount of line played out when the ocean bottom was reached. With the echo sounder, or sonar, continuous deep-sea soundings are now made and automatically recorded. Such measurements to be precisely accurate are corrected for HYDROSTATIC PRESSURE and TEMPERATURE effects in the ocean area being measured. These effects vary the speed of sound transmission in SEAWATER.

See also INSTRUMENTATION.

SOUTH CHINA SEA (NAN HAI) is a marginal sea of the western PACIFIC OCEAN which lies along the east coast of China, Vietnam, and the southeastern extremity of the Gulf of Thailand. The southern boundary is provided by the Malay Peninsula and Sumatra. Borneo, the Philippines, and Taiwan

SOUTH CHINA SEA. A diver swims about the top of an undersea mountain in the South China Sea, about 90 miles west of Borneo. *(NOAA)*

form the eastern boundary. On the north the South China Sea merges with the EAST CHINA SEA along a line crossing the Formosa Strait from Fukien on the mainland of China to the north point of Taiwan. The sea covers an area of 1 422 410 mi² (3 685 000 km²), occupies a volume of 937 328 mi³ (3 907 000 km³), and has a mean depth of 3478 ft (1060 m). The maximum depth is 16 456 ft (5016 m).

The floor of the northern half of the South China Sea is dominated by a deep basin from which rise numerous seamounts, rises, and banks. The central portion of this basin is known as the China Sea Basin which has a maximum depth of 16 456 ft (5016 m). The remainder of the seafloor is occupied by shallow continental shelves. (See CONTINENTAL SHELF.) In the north and northwest the shelf includes the Gulf of Tonkin and is about 150 mi (241 km) wide. The dominant feature of this shelf is the Chinese island of Hainan. To the south is the great Sunda Shelf, one of the largest in the world. Beginning in the Gulf of Thailand the shelf extends to a line running between the Lesser Sunda Islands and eastern Borneo. The shelf depth is about 131 ft (40 m), deepening to 328 ft (100 m) at its center. The central part of the Sunda Shelf contains a network of drowned river valleys ranging up to 3 mi (4.8 km) in width and known as the Molengraaff River System. These valleys were cut during the low stand of sea level associated with the last Ice Age.

In the north, access to the East China Sea is through the 100-mi (160-km) wide Formosa Strait. In the southwest the South China Sea mingles its waters with that of the INDIAN OCEAN through the strategic STRAIT OF MALACCA which is only 17 mi (27 km) wide and 98 ft (30 m) deep. Access to the Pacific is mainly through the Gaspar and Karimata straits between Borneo and Sumatra, the Balahac Strait north of Borneo, the Mindoro Channel west of Mindoro Island, and the Bashi Channel between the Philippines and Taiwan. Under the influence of the winter and summer monsoons, water flows into the South China Sea through the Formosa Strait,

Bashi Channel, and Balabac Strait in the north, producing a weak southerly current. In the summer the winds are reversed, and water flows in through the Gaspar and Karimata straits. The inflow of fresh water from rivers is mainly from the Red and Mekong rivers of Vietnam and the Si Kiang River of southern China.

SOUTHERN OCEAN, or Antarctic Ocean, is that unique body of WATER that encircles the Antarctic continent. Officially the Southern Ocean is simply the southern extremity of the ATLANTIC OCEAN, PACIFIC OCEAN, and INDIAN OCEAN, and their tributary seas (AMUNDSEN SEA, BELLINGSHAUSEN SEA, ROSS SEA, SCOTIA SEA, and WEDDELL SEA). However, scientists have found it increasingly useful in their studies to set these waters apart, although the proper positioning of the northern boundary is still under debate. Some believe that the boundary should be placed at 45°S LATITUDE, which is about the southern extremity of Australia and possibly the northern limit of icebergs. Others believe the boundary should be drawn on the basis of water masses; namely, the zone of Antarctic Convergence at about 52°S latitude, where the Antarctic and subtropical waters meet. Still others believe that 55°S latitude, roughly the tip of South America (Cape Horn), is a more realistic boundary. If the latter boundary is taken, the Southern Ocean covers an area of 12 352 000 mi² (32 000 000 km²) and occupies a volume of 28 788 000 mi³ (120 000 000 km³). If 40°S is taken as the boundary, the area of the Southern Ocean expands to about 28 950 000 mi² (75 000 000 km²).

Early Exploration Speculation about the existence of a southern continent (terra australis) was evident 2000 years before the actual discovery of Antarctica. In part, the existence of such a continent was held necessary in order to balance the landmasses to the north and prevent the earth from toppling over. Many voyages were made to the regions where the southern continent was thought to exist, but only empty, rolling seas were found. Thus the size of the continent was gradually whittled away by negative findings.

In 1520, Ferdinand Magellan, the famous Portuguese navigator, discovered the Strait of Magellan between the mainland of South America and Tierra del Fuego, and sailed into the Pacific on his voyage around the world. Magellan correctly believed that Tierra del Fuego (Fireland) was part of South America rather than the of Southern Continent. In 1577, Sir Francis Drake also sailed through the Strait of Magellan, but was blown south before he could resume his voyage up the Pacific Coast. It has often been incorrectly claimed that Drake discovered Cape Horn during this enforced detour, but that honor belongs to the Netherlanders Willem Schouten and Jakob Le Maire in their voyage of 1616. Drake did, however, form the opinion that the waters of the Pacific and Atlantic rolled together somewhere south of his most southern position. He was, of course, correct, and the area between Cape Horn and Antarctica is now called the Drake Passage.

During the 1700s and early 1800s the Southern Continent was finally proved to be fact rather than mere legend, although its size was considerably diminished. In 1700, Sir Edmund Halley, of Halley's Comet fame, was sent on a voyage to measure the magnetic declination as far south as possible. He reached $52\frac{1}{2}$°S, or just inside the boundary of the Southern Ocean if we accept the Antarctic convergence as the northern limit. Then, during a voyage that lasted from 1772 to 1775, James Cook, the most capable explorer of his time, circumnavigated Antarctica, becoming the first man to cross the ANTARCTIC CIRCLE. Though Cook and his crew respectfully played "cat and mouse" with icebergs during the voyage, they never actually sighted Antarctica. It was Cook's report of plentiful whales and seals in Antarctic waters that shortly turned the Southern Ocean into a favorite hunting ground for these animals. (See COOK, JAMES.)

The Russians claim that Baron Fabian von Bellingshausen, an admiral in the Russian Navy, was the first to sight the Antarctic continent during his voyage of 1819–1821. However, Bellingshausen did not lay claim to that honor. The record indicates that credit for the first sighting must be shared by Captain William Smith of England and a Connecticut sealer by the name of Nathaniel Palmer. In 1819 both men sighted land that now bears the dual name Palmer Land and Graham Land, which is also now known as the Antarctic Peninsula. Due to ice conditions, neither Smith or Palmer were able to reach the land they had sighted. Thus, James Davis, another Yankee sealer, became the first man to go ashore on Antarctica during a voyage in 1821.

Unfortunately, with rare exception, the discoveries made by adventurous sailors who roamed the Southern Ocean in search of profitable whaling and sealing grounds were never reported. The fact that Palmer had sighted land in 1819 was not revealed until a century later because the ship's log had been kept secret to protect any profitable finds. An admirable exception to this practice was the British firm of Enderby which encouraged its captains to seek new discoveries, and, even more praise-

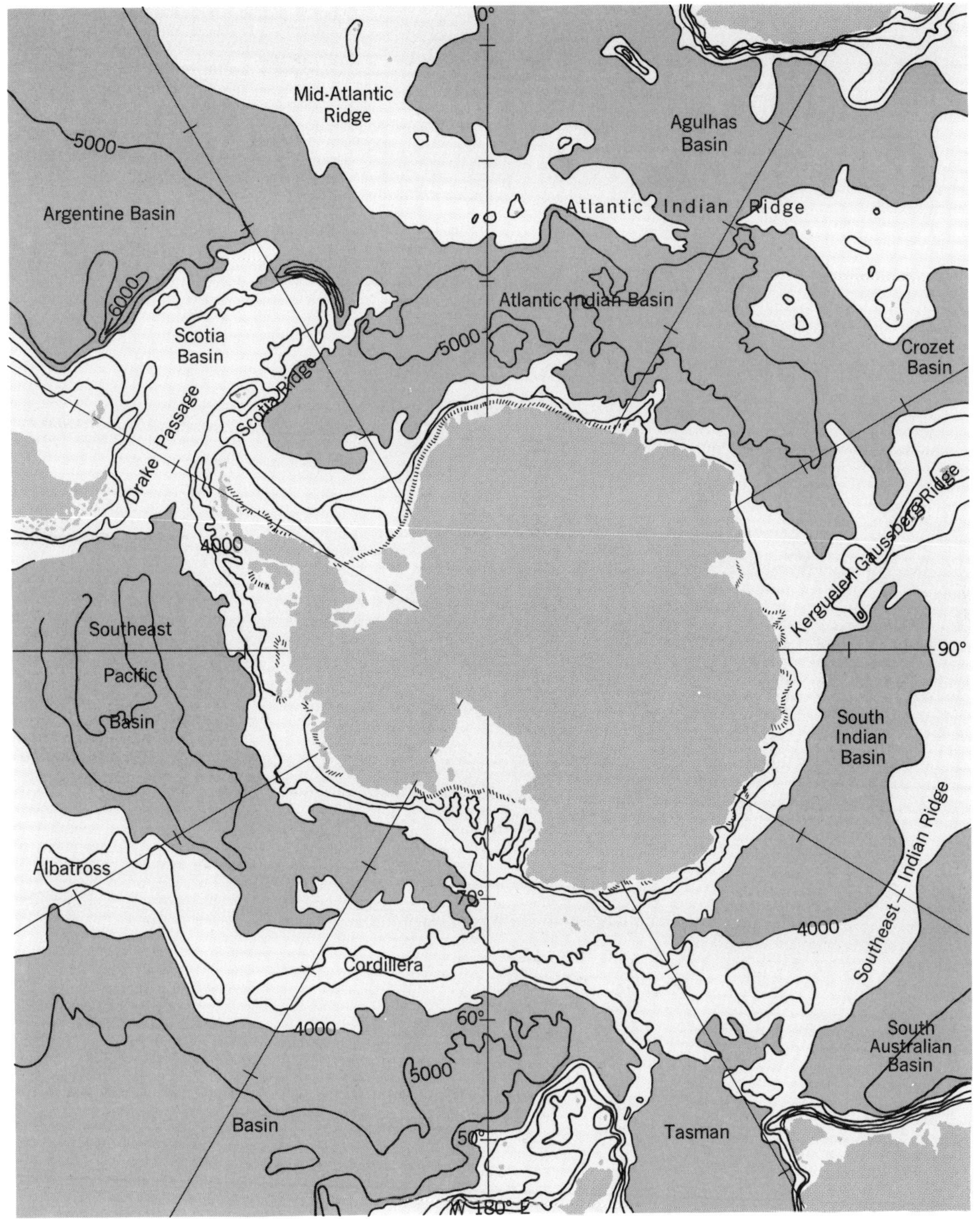

MAP 11. The Southern Ocean surrounds the least-known continent of Earth.

SOUTHERN OCEAN.

The ship nears a small iceberg in Kainan Bay, Antarctica. Such icebergs break away from the shelf ice and may be up to 1000 ft in overall thickness. *(U.S. Navy)*

The *USS Glacier* plows its way through the ice in McMurdo Sound, in the Ross Sea. *(U.S. Navy)*

A fish peeks out from its hiding place in a sea anemone at the bottom of McMurdo Sound. Most of the 100 species of fish found in the Southern Ocean are slow-moving bottom feeders. *(U.S. Navy)*

Elephant seals share the beach with chinstrap penguins on Elephant Island near the northern tip of the Antarctic Peninsula. Antarctica and its surrounding waters abound in animal life, principally because the nutrient-rich Circumpolar Deep Water near the edge of the continental shelf provides the beginning of a food chain that moves up through several kinds of plankton to fish, seals, birds, and whales.

Weddell seals poke through an ice hole in the McMurdo Sound. One of the six species of seal found in Antarctica, the Weddell population ranges between 200,000 and 500,000. *(U.S. Navy)*

Chinstrap penguins cavort on ice floes in Arthur Harbor, Antarctica. The Southern Ocean supports roughly 50 species of birds, which feed mainly on krill, squid, and fish. *(U.S. Navy)*

worthy, to publish their findings. In 1822, an Enderby captain by the name of James Weddell established a record for "fartherest south" of 74°15′. The Weddell Sea is named in his honor. Another Enderby captain, John Brisco, circumnavigated Antarctica in a latitude more southerly than either Bellingshausen or Cook. It was Brisco who named Enderby Land and gave Graham Land (after the First Sea Lord) on the Antarctic Peninsula its name.

Currents, Climate, and Ice Unraveling the complex movement of the various layers of water which make up the Southern Ocean has taxed the best efforts of oceanographers for many decades, and the problem is still short of a complete solution. A general description might begin with the cold winds that roar down to the sea from the ice-covered Southern Continent. These katabatic winds are caused by the chilling of that layer of air in contact with the continental ice sheet, thus increasing its density. Under the influence of gravity the air flows downslope to the sea. These winds, often reaching storm velocities, transport vast quantities of loose snow into the coastal water where it serves as the freezing nucleus for the formation of ice which effectively doubles or triples the size of the continent each year. In the formation of the vast ice sheet that encircles the continent some of the salt is excluded from the water as it freezes. This process not only chills the underlying water but increases its salinity. This dense water with a temperature of around 29.5° F (−1.4° C) follows the CONTINENTAL SHELF and plunges into the deep sea like a gigantic, circular waterfall in slow motion. The volume of this flow has been estimated at around 28 billion (28×10^9) ft³/s [800 million (800×10^6) m³/s] during its maximum production. This layer of Antarctic Bottom Water flows northward into the Atlantic, Pacific, and Indian Oceans and has been traced as far north as the ARABIAN SEA and the BAY OF BENGAL. The Antarctic Bottom Water has a significant effect on the total heat content of the three major oceans into which it flows.

The outflow of Antarctic Bottom Water must be compensated by the inflow of an equal amount of water. This appears to be accomplished by an overlying layer of warmer water that flows southward from more northerly latitudes and is known as Circumpolar Deep Water. This warmer layer rises to relatively shallow depths around the coast of Antarctica. Being high in nutrients, it represents an area of flourishing sea life.

The Antarctic Surface Water, chilled by melting ice and low air temperatures, meets the 46–50° F (8–10° C) warmer Subantarctic Water flowing south in a belt that lies between 50 and 60°S latitude. The zone around Antarctica where these two surface layers meet is known as the Antarctic Convergence Zone, or the Antarctic Frontal Zone. In some areas the cold Antarctic Water sinks beneath the warmer Subarctic Water to form a layer known as the Antarctic Intermediate Water. The surface waters on either side of this frontal zone stand in marked contrast to each other because of their temperatures and the effect on air temperature. Moving south across this zone has been likened to moving from spring to winter in one step. Not only are the salinity and temperature different on either side of the frontal zone, but so is the sea life.

The major current pattern around the Southern Ocean is the mighty Antarctic Circumpolar Current which is under the influence of the strong westerly winds which circle the ocean from west to east. The Antarctic Circumpolar Current—often called the West Wind Drift—is centered at the Antarctic Frontal Zone and is one of the largest ocean currents. The amount of water transported by this current has been studied and debated for years, and disagreement still exists. One estimate suggests that the flow between Antarctica and Cape Horn is 5.3 billion ft³/s (150 million m³/s); the flow between Antarctica and Tasmania is 6.4 billion ft³/s (180 million m³/s); and, the flow between Antarctica and Africa is 6.7 billion ft³/s (190 million m³/s). If these figures are representative of the actual case, then the Antarctic Circumpolar Current is roughly equivalent to 800 Amazon Rivers flowing continuously around the continent of Antarctica.

The Antarctic Circumpolar Current does not follow a smooth circular path around the continent but is diverted north or south wherever it comes under the influence of submarine topography. In addition, substantial quantities of the water are diverted northward into the TASMAN SEA and into the Indian Ocean.

Close to the Antarctic continent a weak and intermittent current—called the Western Coastal Current—is produced by a wind pattern that blows from east to west.

Aided by the influx of snow driven by the katabatic winds mentioned earlier, pack ice begins to form in the coastal regions of the Southern Ocean around March or April and reaches its maximum extent in September or October. At its maximum the pack ice covers an area equal to about 6 percent of the area occupied by the world's oceans and around 30 percent of the Southern Ocean, if the larger boundary of 40°S latitude is accepted. The extent of the pack ice is not uniform around Antarctica, but in some sectors it may extend as much as 100–1000 mi (160–1600 km) out from the coast. The pack ice is augmented by glacial ice which originates primarily from the Filchner Ice Shelf (Weddell Sea), Ross Ice Shelf, and the Shackleton Ice Shelf. These great sheets of ice that move out from the land glaciers calve off periodically at their seaward end to form tabular icebergs of impressive dimensions. They may extend above the surface a distance of 100–150 ft (30–46 km) and below the surface as much as 1000 ft (305 m). Lengths of about 5 mi (8 km) are fairly common for icebergs, although giants of 90 mi (145 km) with an overall area "the size of the state of Connecticut" have been recorded. Most icebergs stay locked in the ice pack south of the Antarctic Frontal Zone but may escape to pose a hazard to navigation as far north as the tip of Africa.

The combination of ice, currents, and water masses of the Southern Ocean has a profound influence on global weather that is just now beginning to be fully appreciated. In the extreme, surges in the production of ice around Antarctica could provide the key to the onset of a new Ice Age.

Sediments and Bottom Topography Geologic and geophysical studies indicate that the Antarctic continent has been in southern polar latitudes for, perhaps, as long as 200 million years. The existence of fossilized plants indicates that, prior to 250 million years ago it was positioned in more temperate latitudes. With the breakup of the giant continent of Gondwanaland about 200 million years ago, the major continents and subcontinents began drifting away to leave Antarctica isolated. South America, Africa, and India apparently drifted away as the break up began, and an initial Southern Ocean was formed at that point. New Zealand separated 60–80 million years ago, and Australia broke away about 45 million years ago. Fragments left behind by these massive movements are represented today by the Scotia Arc between Tierra del Fuego and the Antarctic Peninsula which supports the South Georgia, South Sandwich, and South Orkney Islands, and the Kerguelen Plateau which runs northward into the Indian Ocean and supports the Kerguelen Islands at its northern tip. The most impressive evidence of the process which split the continent away, and the dominant feature of the Southern Ocean floor, is the Mid-Ocean Ridge. (See CONTINENTAL DRIFT.) The ridge makes its way for 10 000 mi (16 000 km) along the seafloor around Antarctica as it connects with its northern extension into the Atlantic, Pacific, and Indian oceans. The ridge, which rises 0.6–1.8 mi (1–3 km) above the seafloor and ranges up to more than 932 mi (1500 km) in width, approaches the Antarctic continent closest in the general vicinity of the Ross Sea, and is farthest away to the east and west in the Pacific and Indian Oceans.

The combined effect of these several ridges is to divide the floor of the Southern Ocean into three major basins. The first of these is the Valdivia Basin which is bounded by the Scotia Arc on the west, the Kerguelen Plateau on the east, and the Mid-Ocean Ridge on the north. The basin has a maximum depth of 19 265 ft (5872 m). The Knox Basin lies off the Knox coast and has a maximum depth of 17 907 ft (5458 m). The Bellingshausen Basin which lies off the Bellingshausen Sea has a maximum depth of 21 043 ft (6414 m).

The sediment covering the floor of the Southern Ocean is rather straightforward in its distribution. From the coast to about the limit of winter pack ice, the sediment is dominated by glacial debris deposited both by a TURBIDITY CURRENT and by melting icebergs. This material is derived from land and ranges from gravel to very fine muds. The Antarctic Frontal Zone acts as a boundary line between two distinctive sediment types. To the south of the zone, the sediments are principly diatom ooze (siliceous), and to the north, foraminiferal (calcareous) ooze. This separation of sediment types results from a preference in water masses between the two species.

Biology In spite of its frigid waters and howling storms the Southern Ocean abounds in many forms of animal life. Perhaps the single most important factor contributing to this unlikely abundance is the Circumpolar Deep Water in that belt between the Antarctic Frontal Zone (between 50 and 60°S) and the edge of the continental shelf. This water is rich in those nutrients required for large populations of phytoplankton—particularly DIATOMS, dinoflagellates, and silicoflagellates. The coastal waters are also areas rich in phytoplankton. It is on this "grass of the sea" that the swarms of zooplankton feed. In turn, it is the zooplankton—primarily copepods and euphausiids—that form the food base for the FISH, MARINE BIRDS, WHALES, and SEALS which inhabit or migrate to the Southern Ocean. KRILL, a shrimplike euphausiid, is the most important of the zooplankton for this purpose.

In 1775, James Cook reported on the large population of whales that he found in the Southern Ocean and triggered an industry that has persisted to this day. These waters still contain the greatest concentration of whales of any ocean area, with the blue, humpback, fin, sei, southern right, minke, and sperm whales being represented. The baleen whales (all the above except the toothed sperm whale) feed primarily on krill, although other euphausiids and copepods are consumed. Conveniently (for the whale) the krill tend to collect in dense swarms that may reach more than 1900 ft (600 m) in diameter. The baleen whale has only to swim through a swarm with its mouth open.

In the late 1800s the whale population in the Southern Ocean was estimated at about 1 million. By 1930 the figure had been reduced to an estimated 340 000 and has been further reduced over the years. This has prompted the speculation that the krill population must have increased dramatically as the whale population declined. This has led to discussions by several nations about harvesting krill for human consumption. It has been estimated that the present surplus of krill is around 165 million tons (150 million metric tons), or about twice the annual world fish catch. On an experimental basis the U.S.S.R. took 200 000 tons (180 000 metric tons) of krill in 1974, and Japan took 650 tons (590 metric tons). The total impact of krill harvesting on the ecology of the Southern Ocean is not fully known at present.

Six species of seals are found in the Southern Ocean: Weddell, Ross, leopard, crabeater, southern elephant, and southern fur seals. The crabeater is thought to be the most abundant seal in the world with a population of around 30 million. The Weddell population is between 200 000 and 500 000, the leopard population is estimated at 200 000–300 000, and the Ross at about 100 000. Seals feed on krill, fish, squid, and crustaceans. The leopard seal includes penguins in its diet, making it the only seal to feed on warm-blooded animals.

The fish population of the Southern Ocean consists of about 100 species, most of which belong to the Notothenïiforms group—a rather sluggish bottom feeder. Little commercial use is made of these fish, although the U.S.S.R. has recently begun marketing *Notothenia rossi* in several Soviet cities.

The Southern Ocean is home, or temporarily so, to some 50 species of birds. The flightless penguin is best identified with the harsh Antarctic climate, but others, such as the skua, which feed on the penguin young, appear equally well adapted. The favored diet of most Southern Ocean birds is krill, squid, and fish.

SPACECRAFT OCEANOGRAPHY uses satellites to gather and interpret various oceanic data. The first satellites employed to obtain information pertaining to the world's oceans were meteorological satellites used in the late 1950s. Since that time, the concept of monitoring marine environmental conditions from above has become a part of marine science—known as space or spacecraft oceanography.

Spacecraft oceanography, led in the United States by the National Oceanic and Atmospheric Administration (NOAA), involves the use of satellites equipped with appropriate sensing equipment to improve upon the study of oceanic processes and life forms from such characteristics as WATER color, TEMPERATURE, texture, topography, nutrient distributions, and other identifiables that can be observed from satellite altitudes.

In the use of color photography at these altitudes, the colors can be related to ocean depth, weather, and biological activity in the sea. A deep blue indicates a sterile, clear open ocean, while increasing amounts of chlorophyll-bearing PLANKTON add to the green and red portions of the spectrum. MARINE SEDIMENTS tend to be brownish. In shallow water, a white seafloor lightens the general color of a region and vegetation darkens it. High-resolution color photographs also show physical and marine biological information such as areas of UPWELLING, water mass boundaries, and dispersion patterns of coastal effluents. (See MARINE OPTICS.)

Temperature measurements from space are based on the fact that objects on the ocean and earth emit ENERGY in the infrared and microwave portions of the spectrum—the wavelengths extend from the longer, invisible side of red into the range of radar. These are the regions in which heat can be seen by spacecraft sensors. Thus, the global distributions of heat energy that drive the atmosphere are monitored and mapped. For oceanographers, high-resolution radiometers aboard satellites have brought an improved view of oceanic processes. Infrared data from the satellite can be used to measure CURRENTS, sea-surface temperatures, life-rich areas of upwelling, and ice thickness.

In addition, by means of sensors operating in the microwave radio-frequency range, it is possible to observe both ocean temperatures and the texture of the ocean surface—SEAFOAM, small-scale roughness, differences between new and old ice, etc.

SPEARFISH is a name commonly applied to those ocean fishes that have a spearlike bill, such as the marlin. See BILLFISH.

SPECIFIC HEAT is basically the capacity of a substance to store heat. It can also be defined, with some technical oversimplification, as the ratio of the amount of heat required to raise the TEMPERATURE of a unit mass of a substance one degree (under specified reference conditions) to the amount of heat required to raise an equal mass of a reference substance, usually WATER, one degree in temperature. In most oceanographic work, specific heat at constant pressure (C_p) is usually the quantity given for the heat capacity of SEAWATER. This value, which represents the amount of heat [expressed in units of calories or joules (1 cal = 4.1868 J)] required to raise a gram of seawater of 35 parts per thousand (35 ppt) SALINITY one degree Celsius (C) in temperature at one atmosphere and at 17.5° C, equals 3.898 J/g°C. (See PARTS PER THOUSAND.) In comparison, pure water under the same reference conditions (1 atm and 17.5° C) has a specific heat (C_p) of 4.182 J/g°C. The capacity for heat in seawater, as this comparison implies, diminishes slightly as the salinity increases.

The specific heat of seawater at constant volume (C_v) is defined as the ratio of its thermal capacity to that of pure or distilled water. However, these values are normally estimated and used, while reliably proved values for such a quantity are lacking because of the current paucity of fundamental thermodynamic data on pure water and seawater compositional effects, especially under high HYDROSTATIC PRESSURE.

Such data as these would assist in the understanding of the molecular structure of seawater and the effects of temperature, pressure, and electrolytes on that structure as well as supply information needed for the calculation of the adiabatic temperature gradients, heat budgets, the heat of mixing waters, and the transmission of sound in seawater.

Perhaps from what is known about the heat capacity of seawater, which is higher than all solids and practically all other liquids, many aspects of the variegated physical-biological ocean behavior can be explained. For example, because of this property, extreme ranges in water temperatures are prevented, and heat transfer by water movements is large. Such phenomena, among other factors, are of significant importance to the sustenance of all forms of ocean life. However, of most importance is the fact that the oceans can be thought of as a vast heat engine and as such absorb the major portion of the heat ENERGY delivered to the earth as solar radiation. Indeed, the top 30-ft (9-m) layer of the oceans has as much heat absorption capacity as our entire atmosphere. The oceans can also release a great deal more heat to the atmosphere than can the land, and they can do so without greatly changing the ocean temperature. Thus, the ocean is a giant flywheel that significantly influences CLIMATE.

SPECIFIC VOLUME is the volume of a substance, such as SEAWATER, per unit mass; it is the reciprocal of the DENSITY.

Currently used values for the specific volume (sp vol) of seawater at elevated pressures are based upon the results of measurements made by V.W. Ekman (1908). Ekman determined the isothermal compression at various seawater temperatures and hydrostatic pressures up to 600 bars of pressure. [600 bars = 10^5 pascals (Pa) or 10^5 newtons per square meter (N/m^2) or 10^6 dynes per square centimeter (dyn/cm^2), or 8700 pounds per square inch (psi).]

The specific volume, expressed in cubic centimeters per gram of seawater of 35 parts per thousand (35 ppt) at sea level is 0.9727 at 0° C (32° F), 0.9738 at 10° C (50° F), 0.9757 at 20° C (68° F), and 0.9784 at 30° C (86° F). (See PARTS PER THOUSAND.) At 25° C and in water of 35 ppt SALINITY, the specific volume varies linearly with pressure from 0.9771 cm^3/g at 0 pressure to 0.9412 at approximately 1000 bars or 14 500 psi. In water of approximately 5 ppt salinity, for the same temperatures and pressures the values are 0.9988 and 0.9599, respectively.

Specific volume is of great importance to the physical oceanographer because of its relation to salinity, CHLORINITY, and conductivity of seawater. Moreover, the specific volume of the ocean water is a measurement highly desirable in the study of water masses and current flow (geostrophic) computations.

SPIESS, FRED NOEL (1919–), an American oceanographer, is most commonly identified with the development of the research platform knows as FLIP (Floating Instrument Platform), which is used primarily for research on WAVES and UNDERWATER SOUND. FLIP is a 355-ft (108-m) platform that can be towed in the horizontal position like a ship, and then it can "flip" to the vertical by adjusting the ballast in the aft sections. In the vertical position, 300 ft (91.4 m) of the "vessel" is submerged, resulting in a very stable platform for sensors. In its first 10 years of operation, FLIP completed 35 expeditions and contributed significantly to our knowledge of wind waves and internal waves, acoustic transmission in the sea, temperature structure of the ocean, and the structure of the earth's crust.

Spiess was born in Oakland, California, on December 25, 1919. He received his bachelor's degree (1941) and his doctorate (1952) from the University of California at Berkeley, and his master's from Harvard University (1946). After serving briefly on the staff of the Knolls Atomic Power Laboratory, he joined the staff of the Marine Physical Laboratory at the University of California, San Diego, becoming the director of that laboratory in 1958. From 1961 to 1963 he served as acting director of SCRIPPS INSTITUTION OF OCEANOGRAPHY and director during much of the 1964–1965 academic year. He also served that institution as professor of oceanography (1961–).

As an independent researcher, and as head of a large research organization, Spiess is responsible for many advancements in underwater sound and in instrumentation; FLIP and Deep Tow, a towed instrument pack which permits rapid topographic, substructure, and magnetic surveys of the deep ocean floor, are but two examples of contribution in the latter category. Spiess' work has been recognized by the Franklin Institute's Wetherill Medal (1965), the Marine Technology Society's Distinguished Achievement Award (1971), and the U.S. Navy's Captain Robert Dexter Conrad Award (1974).

SPILHAUS, ATHELSTAN FREDERICK (1911–), an American scientist, inventor, and author, is best known as the "Father of the Sea Grant Program" (see SEA GRANT PROGRAM) and the inventor of the bathythermograph. He was born November 25, 1911, in Cape Town, South Africa. He received his bachelor's degree (1931) and doctorate (1948) from the University of Cape Town, and his master's (1933) from the Massachusetts Institute of Technology. He also holds 10 honorary degrees from universities in the United States and England.

Spilhaus began his career as a research assistant at the Massachusetts Institute of Technology in 1933. In 1936 he became research assistant to the WOODS HOLE OCEANOGRAPHIC INSTITUTION, investigator in physical oceanography in 1938, and associate in 1960. He became an assistant professor at New York University in 1937, associate professor in 1939, and professor of meteorology in 1942. From 1949 to 1966 he served as dean of the Institute of Technology, University of Minnesota, and professor of physics in 1966. From 1967 to 1969 Spilhaus was president of the Franklin Institute, and of Aqua International, Incorporated, from 1969 to 1971. From 1971 to 1974 he was a fellow of Woodrow Wilson International Center for Scholars at the Smithsonian Institution. And, in 1974 he became the Special Assistant to the Administrator of the National Oceanic and Atmospheric Administration.

Spilhaus' research interests and his writings have ranged over the fields of oceanography, meteorology, astronomy, and the space program. In 1938 he invented the bathythermograph, a rugged instrument that allowed oceanographers to obtain a record of water temperature as a function of depth from a moving ship. During World War II he was awarded the Legion of Merit for the development of meteorological instruments for upper atmospheric sensing. During the keynote address at a symposium held in Washington in 1964, Spilhaus first proposed that support for oceanographic research through the universities be patterned after the Land Grant College program that had done so much for American agriculture. The successful National Sea Grant Program grew out of that recommendation.

SPONGES is the name commonly applied to all members of the phylum Porifera (which means pore-bearers). A sponge is an aquatic, mostly marine, sessile (permanently attached to a substrate) INVERTEBRATE that occurs in all the world's oceans, in a great range of depths, and in a wide variety of forms, sizes, and colors.

There are thousands of species of sponges within the three classes of the phylum: the Calcarea, those inhabiting shallow waters and having little spines (spicules) based on a three-ray pattern and composed of calcium salts; the Hexactinellida, the typically deep-sea glass sponges having spines based on a four- to six-ray plan, and composed of siliceous compounds; and the Demospongiae, the tropical and subtropical sponges characterized by skeletons composed of an organic substance called spongin. Only a dozen species of the latter class have commerical value, and these are collected in the MEDITTERANEAN SEA, the GULF OF MEXICO, and near the West Indies. However, in recent years, the development of synthetic sponges has been responsible for the downfall of the once prosperous sponge-fishing industry.

Sponges are nevertheless valuable to the support of various forms of ocean life and they are an object of considerable scientific interest. They possess no organs but have a multicellular body permeated with small pores with a canal system through which water constantly passes. In the majority of species, the body is supported by a network of small spikes or by spongin fibers or by a combination of both. Reproduction by purely vegetative or asexual methods is a common manner of multiplication. Essentially a growth and differentiation of the cells forms on the side or bottom of the parent. These buds or growths may either remain attached or fragment off, but in either case each bud is capable of developing into a new sponge.

Biochemical studies on the composition of sponges have disclosed some interesting and potentially valuable information from a medical and industrial point of view. For instance, in the red sponge, *Microciona prolifera*, large amounts of iodine incorporated in an aromatic amino acid are accu-

SPONGE.

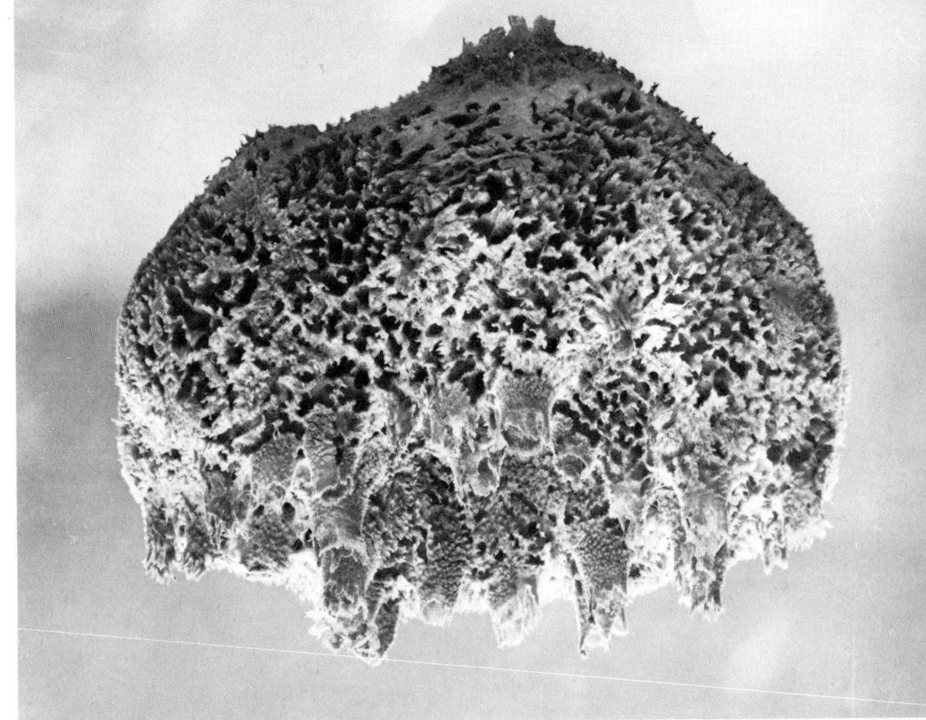

Sheepswool sponge from the Florida Keys. Sponges are found at all depths throughout the world's oceans; there are thousands of species of these invertebrates in all sizes, shapes, and colors. (*NOAA*)

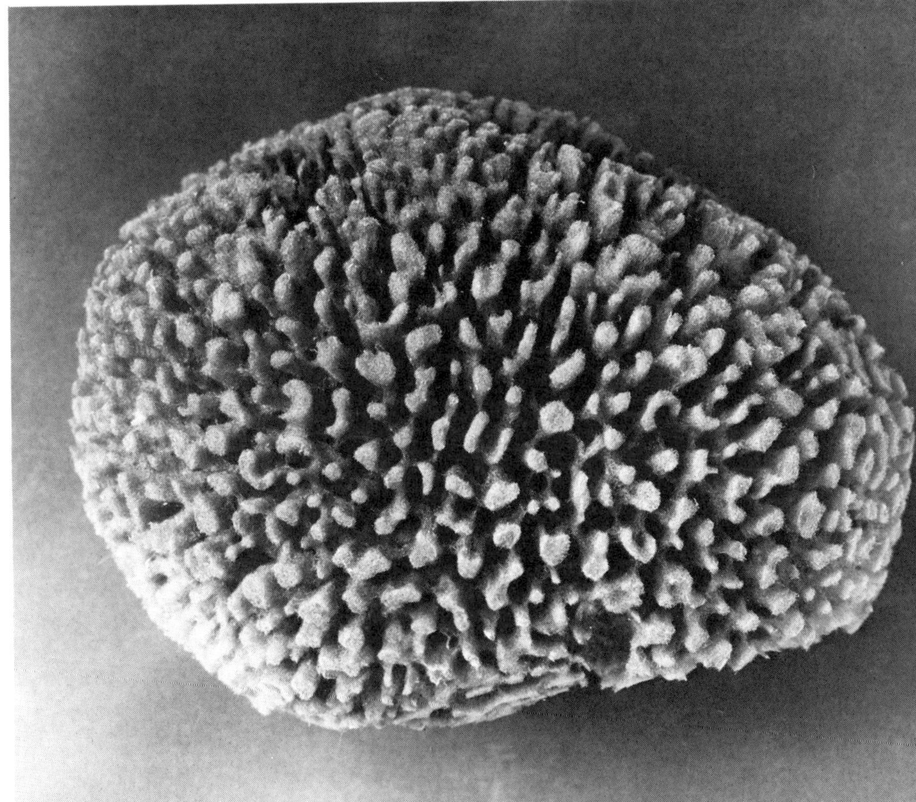

Velvet sponge from the Bahama Islands. Sponges have no organs; their multicellular bodies are perforated with small pores through which life-giving water constantly passes. (*NOAA*)

mulated in the skeleton. The method of accumulation of this element is not well understood. In addition, several organic substances, such as the nucleosides, are contained in the body of the West Indian sponge *Crypotothia crypta*. The *Spheciospongia vesparia* species contains methanethole, while chalinasterol is found in the common spongs *Chalina arbuscula* of the coastal waters of New England. Sponges are also a good source of fatty acids of high molecular weight.

See also PHARMACEUTICALS FROM THE OCEAN.

SPRING TIDE See TIDES.

SQUALOID is the name of a major group of sharks; the squaloids comprise the families Squalidae, (spiny dogfishes), Dalatiidae (spineless dogfishes), Pristiophorideae (saw sharks), and Squatinidae (angel sharks). (See SHARK.) The other major shark group is the GALEOID. Sharks belonging to these different groups (squaloid and galeoid) vary in such features as the size of their jaws, tooth shape, and structure of FINS. For example, the squaloids lack an anal fin.

SQUID is the name applied to various species of marine cephalopod mollusks. (See MOLLUSK.) Squids are distinguished by the possession of an internal shell, 10 tentacles, an ink sac, and chromatophores. They are pelagically distributed in ocean waters throughout the world, usually over the continental shelves from the coastline to the shelf edge. (See CONTINENTAL SHELF.)

The squid possess an internal shell and 10 appendages, of which one pair (called *tentacles*) is long and has suckers at the ends, while the other four pairs (called *arms*) are half the size of the tentacles and have suckers their entire length.

CUTTLEFISH, also cephalopods, are similar to squid, but differ in having a heavy calcified shell or "bone" rather than the thin "pen" of the squid. Squid, like the OCTOPUS, are equipped with an ink sac plus chromatophores which provide protective coloration. Also, like the octopus, reproduction is affected by transfer of male spermatophores to the female by means of the hectocotylus tentacle. The fertilized eggs are discharged to the ocean floor for development.

Most of the smaller species range between 3.9 and 7.9 in (9.9 and 20.06 cm) in length. These may be typified by the long-finned squid, *Loligo pealii*, the common squid encountered in the ATLANTIC OCEAN off southern New England. The greatest concentrations are on inshore grounds where spawning occurs.

In addition to *Loligo pealii*, there are *Loligo opalescens*, a Pacific Coast species, and *Loligo vulgaris*, a Mediterranean species. Some other squid species, e.g., *L. abralia*) are phosphorescent, with light-producing organs on the tentacles and on the body near the eyes. Others, such as the tropical and temperate-water *Ommastrephes bartrami*, swim so fast that they are able to skim across the surface of the water.

The giant squid, *Architeuthis principes*, the largest INVERTEBRATE, whose 10-ft (3 m) body length plus its tentacles can measure up to approximately 51 ft (15.5 m), are sometimes seen at the surface. Such sightings have given rise to some of the sea serpent MYTHS AND LEGENDS. While it is known that such giant squid, the Cranchidae, definitely exist, one can only speculate that titanic battles must have taken place in the sea between them and their enemies, the sperm WHALES (*Physeter catadon*).

Squids feed on small FISH, CRUSTACEANS (e.g., the LOBSTER, CRAB, and SHRIMP), and other marine organisms. In turn, they furnish food for the pinnipeds (SEALS, SEA LIONS, and walruses), large fish, whales, and MARINE BIRDS (see WALRUS).

The smaller squids are highly sought after as food in many parts of the world, particularly among the peoples of the Orient and the Mediterranean. In addition, the squids, with their highly developed sensory organs and nervous systems, are of great interest to scientists. For example, the eye of the squid is being actively used to explain certain aspects of photoreception. The giant nerve fiber of the squid is also unique in the diverse ways in which it can contribute to knowledge of nerve conduction.

STANDARD WATER See NORMAL WATER.

STANDING WAVE See WAVES.

STEAM FOG See FOG.

STINGRAYS See RAYS.

STOMIATOIDS is the identifying name for fishes of the suborder Stomiatoidei of the order Clupeiformes; they include the lightfishes and their relatives, which are of small size and often bizarre shape and are equipped with photophores (luminous glands and organs).

STOMMEL, HENRY MELSON (1920–), an American physical oceanographer, made many significant contributions to the understanding of how the ocean behaves and why. His greatest contribution has been in the field of ocean currents, with

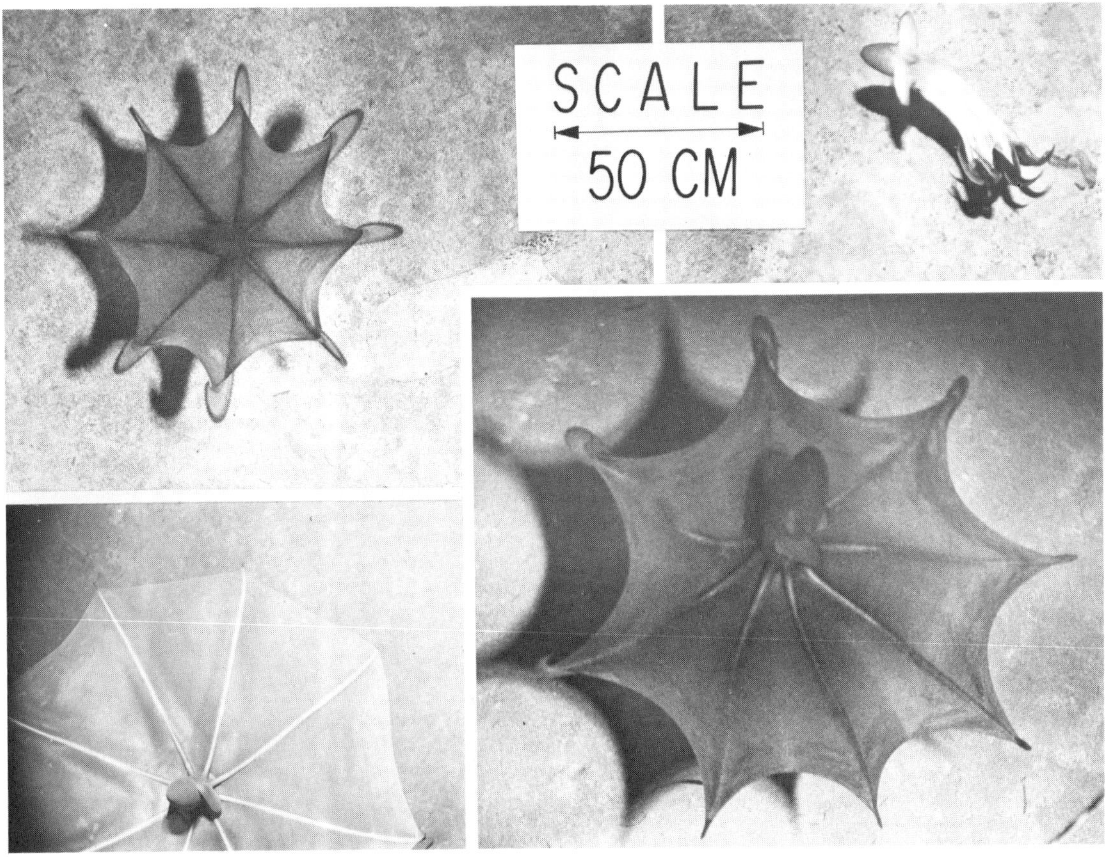

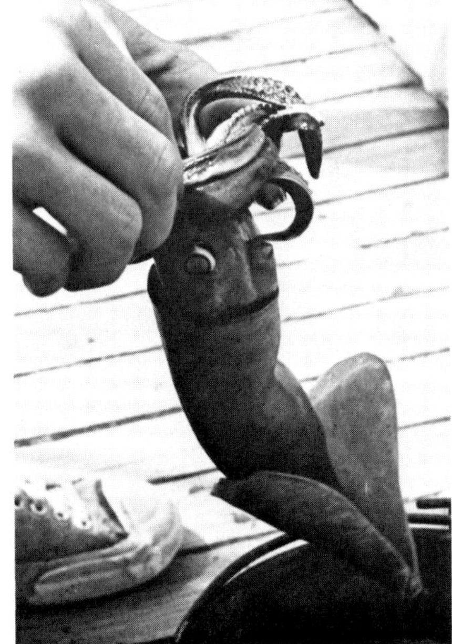

SQUID. (*Top*) The flying squid, also known as the vampire squid, has eight arms united by a web. These animals live at depths of 3300 to 6600 ft (1000 to 2000 m) in the tropical and subtropical oceans. (*U.S. Navy*) (*Bottom*) A Midwater squid undergoes examination; little is known at present about the biology of the cephalopods. The squid's large eye, visible here, has been used by scientists to clarify certain aspects of photoreceptivity, and its large nerve fiber has contributed greatly to our knowledge of sense conduction.

special attention to the GULF STREAM. However, his broad research resulted in many publications in such areas as the general theory of ocean circulation, mixing processes in salt water, the flushing of estuaries, and the interaction between the atmosphere and the ocean surface. He was extremely influential as a teacher and in gaining international cooperation on ocean research projects which exceed the capacity of individual institutions. His research during the 1950s is generally credited with being the blueprint in physical oceanography for the advancements of the 1960s and 1970s.

Stommel was born in Wilmington, Delaware, on September 27, 1920. He received a bachelor of science degree from Yale University in 1942 and holds honorary degrees from Harvard University (M.A., 1961), Göteborg Universitet (Ph.D., 1964), Yale University (Ph.D., 1970), and the University of Chicago (Ph.D., 1970). After graduation he became an instructor in mathematics and astronomy at Yale from 1942 to 1944. From 1944 to 1960 he was a research associate at WOODS HOLE OCEANOGRAPHIC INSTITUTION and served there as physical oceanographer (nonresident) from 1960 to the present. He also served as professor of oceanography at Harvard from 1960 to 1963 and held the same position at the Massachusetts Institute of Technology from 1963 to the present.

STORM SURGE See WAVES.

STRAIT See OCEAN.

STRAIT OF MALACCA, one of the most important waterways in the world, is located between Malaysia on the northeast and Sumatra on the southwest. The southeastern end of the strait cuts through a narrow channel between Singapore and the offshore islands of Palau Batam and Palau Bintam, and is known as the Singapore Strait. The Malacca-Singapore Strait is over 200 mi (322 km) wide at its northwestern boundary with the ANDAMAN SEA, but tapers down to as little as 6 mi (10 km) in the Singapore Strait. The shipping channel itself narrows to 2½ mi (4 km) in the Malacca Strait and 1 mi (1.6 km) in the Singapore Strait. The depth at some locations in both channels is as shallow as 75 ft (23 m), which restricts passage to vessels no larger than about 225 000 dead wight tons (dwt). The total length of the Malacca-Singapore Strait is 550 mi (885 km).

Despite its limitations the Malacca-Singapore Strait is one of the most heavily traveled waterways in the world, with up to 1000 ships working their way past each other every week. Cargo liners, bulk carriers, and tankers make up 90 percent of the traffic. In addition, the strategic importance of the strait has increased significantly since the British Navy pulled out of the INDIAN OCEAN in 1971, and the navies of the United States and the Soviet Union have gradually filled the vacuum. The straits are in international waters, but this is now being challenged by Indonesia and Malaysia. Both countries argue that ships using the straits should do so under the right of innocent passage (subject to some control by the bordering state) rather than free transit, and have asked that no ships larger than 200 000 dwt use the straits. This has gone unheeded in spite of the large oil spill in 1975 caused by a 238 000-ton Japanese tanker. These countries also argue that pollution could endanger the fragile fishing industry and that the unrestrained passage of warships might jeopardize their national security.

The Malacca-Singapore Strait serves as a link between Pacific and Indian Ocean WATER by way of a current flowing out of the SOUTH CHINA SEA, through the straits, and into the Andaman Sea. The floor of the straits also serves as a link between the great Sunda Shelf on the east and the Mergui Plateau on the west.

SUBMARINE CANYON is a term often applied rather loosely to a variety of valleys and elongate depressions found on the ocean floor. However, after a great deal of research spanning more than a century, geologists have been able to place these valleys into several specific categories which relate to physical characteristics and possible origins.

Submarine Canyons Such canyons are characterized by a V-shaped valley with high, steep, rocky walls and a floor that slopes continually outward so that, in overall characteristics, the valley resembles those land canyons (i.e., the Grand Canyon) that have been cut by streams and widened by gravitational creep on their flanks. All submarine canyons head up on the CONTINENTAL SHELF, and most debouch at the base of the CONTINENTAL SLOPE. Many have alluvial fans (broad sedimentary aprons) which spread out from their mouths to cover the continental rise and terminate on the ABYSSAL PLAIN. Most canyons have numerous tributaries, and the main axis of the canyon has several meanders along its course.

Submarine canyons vary considerably in length, wall height, and slope of the valley floor. The average length of those canyons studied thus far is 34.5 mi (55.5 km), and they range in length from 4.1 mi (6.6 km) for some of the Hawaiian Canyon group to 265 mi (426 km) for the Bering Canyon. Wall heights average just over 3000 ft (915 m), with

SUBMARINE CANYON.

A "sandfall" spills sand about 30 ft into the Cape San Lucas submarine canyon, Baja California. Recent research has shown that longshore currents feed sand from the nearby beaches into the canyons. *(University of California, San Diego)*

Some 450 ft below the surface, 6-in shrimp and an 18 in anemone thrive on the floor of a submarine canyon. *(Scripps Institution of Oceanography)*

the greatest yet discovered, the Great Bahama Canyon, having a wall height of 14 060 ft (4280 m). The slope of the floor of submarine canyons is, in general, greater than that for land canyons, with an average of 190 ft/0.6 mi (58 m/km), and ranging from 26 ft/0.6 mi (7.8 m/km) for the Bering Canyon to 472 ft/0.6 mi (144 m/km) for the Hawaiian Canyon group.

Submarine canyons are found throughout the world, and many of them equal or exceed the Grand Canyon in size. [The Grand Canyon is 5500 ft (1700 m) in depth or wall height.] Notable of these are the Great Bahama Canyon with a wall height of 14 060 ft (4285 m) and a length of 125 nautical mi (232 km); the Bering Canyon with a wall height of 6000 ft (1829 m) and a length of 265 mi (426 km); the nearby Pribilof Canyon with a wall height of 7000 ft (2134 m) and a length of 86 nautical mi (159 km); and the Manila Canyon with a wall height of 6000 ft (1829 m) and a length of 31 nautical mi (57.4 km). Maximum wall heights are difficult to measure, and the figures noted above have been adjusted to the nearest 1000 ft (300 m).

The first discussion of record regarding submarine valleys was in a paper by J.D. Dana in 1863 in which he described the valley off the Hudson River estuary. The first contour charts of submarine canyons were published in 1887 by G. Davidson. It was these charts which created a worldwide controversy over the origin of submarine canyons that has raged ever since. Over the years many of these theories have fallen into disfavor. Those that remain can be summarized as follows:

- The drowning by subsidence of valleys cut while the continental shelves stood above sea level. Much evidence suggests that the last Ice Age resulted in a substantial lowering of sea level, and that the cutting action of streams was extended out onto the present shelves. And, equally abundant evidence shows that the continental shelves have undergone considerable subsidence as the sediment load on them has increased. The fact that many submarine canyons today lie off the mouth of rivers, or where rivers once entered the sea, and that submarine canyons have features similar to stream-cut canyons on land is offered as further evidence. Therefore, the theory holds that the canyons were cut by streams flowing across an exposed continental shelf, that rising sea level plus subsidence has submerged these valleys to their present level, and that currents, sediment creep, and flow have kept the valleys clear.
- Erosion by a TURBIDITY CURRENT. It has been demonstrated both in the laboratory and in nature that when great quantities of sediment are injected into the water in the presence of a slope, the mixture will run down the slope as a narrow and relatively swift current. It has not, however, been demonstrated to the satisfaction of all geologists that significant erosion results from such currents. In any event, the theory holds that submarine canyons were cut by turbidity currents resulting from the massive and frequent slumping of sediment at the head of the canyon. Presumably such erosion began at the shelf edge and worked back to the site of the present canyon head.
- Erosion by the slow mass movement of sediment down the canyon by creep, progressive slumps, and sandfalls. Research over the past 20 years, particularly that off the California and Baja California coasts, has demonstrated that a great deal of the sand which moves along the coast under the influence of longshore currents (see BEACH) is trapped at the head of submarine canyons and eventually creeps and flows down the canyon floor.
- Erosion by bottom currents other than turbidity currents. Temperature and salinity gradients, tides, and internal waves (see WAVES) are all capable of producing currents along the bottom. This theory proposes that these currents, perhaps aided by the sediment they carry, are responsible for the erosion which resulted in the cutting of submarine canyons. Such currents have been observed operating in canyons (both up and down canyons) and are strong enough in some cases to move sediment along with them.
- Persistence of an initial break in the shelf basement or offshore dam (see CONTINENTAL SHELF). Most geologists now agree not only that most submarine canyons keep their floors clear of significant sediment accumulation but that many of them are gradually eroding and extending their heads landward. They also agree that this process is due to the combined effect of slumping and undercutting of wall material, turbidity, and other currents, and sediment creep. This theory holds that the edge of the initial surface upon which the present continental shelves are built, or the offshore dams which serve to pond the sediment, are likely to have had steep-faced irregularities or breaks in their surfaces from place to place. Such breaks would, if steep enough, prove unstable to the accumulating sediment in their vicinity. Therefore, the initial break would persist and, under the influence of subsidence and the processes mentioned earlier, would extend itself headward. In other words, the theory suggests that submarine canyons were, in effect, built from the bottom up as sediment accumulated, rather than from the top down after most of the continental shelf sediments were in place.

Fan Valleys The sediment fan that spreads out from the mouth of many submarine canyons represents an enormous quantity of material; far greater, in many cases, than would result from the erosion of the canyon itself. These fans often have a V-shaped, or in some cases a trough-shaped, valley that represents an erosional continuation of the floor of the adjoining submarine canyon. Fan valleys have steep sides that frequently reach heights of 600 ft (183 m) and are all cut in unconsolidated sediment. Some valleys show natural levies along their sides, indicating that great sediment-carrying currents have periodically overflowed their high walls and dropped their sediment load as the current velocity dropped in the absence of the confining valley walls.

Shelf Channels A shelf channel is a relatively shallow type of valley that extends across the continental shelf, rarely exceeds 600 ft (183 m) in wall height, often has basins along its course, and has never been observed to intersect a submarine canyon. Actually, shelf channels are not very common. The Hudson Channel off the coast of New York, the Hurd Deep in the English Channel, St. George's Channel in the IRISH SEA, and the channels of the Sunda Shelf are among the better known. The Hudson Channel, as an example, begins near the entrance to New York Harbor and extends across the entire continental shelf. The channel exhibits a number of curves in its course, averages 100 ft (30 m) in depth below the surrounding shelf, and has a width of 3–4 mi (4.8–6.4 km). Several elongate basins with depths of around 60 ft (18.3 m) are found along the course.

Glacial Troughs These trough-shaped valleys are found off glaciated coasts, have depths that are often greater than 600 ft (183 m), exhibit basins on their floors, and have both tributaries and distributaries. Some glacial troughs are deepest near the shore and become shallower with distance offshore until they disappear. The Laurentian Trough which runs for 650 mi (1046 km) from the Gulf of St. Lawrence to the edge of the shelf 150 mi (241 km) beyond the Saguenay River is certainly the most impressive of the glacial troughs. It often has a width of 50 mi (80 km) and a depth as great as 1700 ft (518 m). Evidence for the cutting of these valleys by glaciers moving across an exposed shelf is overwhelming, but the glacial cause is not universally accepted.

Deep-Sea Channels These channels on the deep ocean floor are trough-shaped in cross section, have few tributaries, and may or may not serve as a continuation of fan valleys. Some deep-sea channels run parallel to the continental margin, while others run at large angles to it. One of the best known of the deep-sea channels is the Mid-Ocean Canyon which begins off the west coast of Greenland and terminates on the deep ocean floor off the tip of the Grand Banks. The depth of the channel below the bordering seafloor is around 300 ft (91 m), and it has a width ranging from 1.5 to 4 mi (2.4 to 6.4 km). Turbidity currents are thought by many geologists to have cut the deep-sea channels. It is, perhaps, equally valid to think of them as zones of no sediment deposition because they serve as the axis of the current flow. Thus, the current velocity is highest in the channel and less likely to drop its sediment load.

It is, perhaps, surprising that after well over 100 years of thought, study, and investigation, substantial disagreement should still exist over the cause of these great features on the seafloor—some of which dwarf their land counterparts in both length and depth. But one must remember the difficulty of studying the one as compared with the other. Most of what we know of submarine valleys comes from fathometer soundings (acoustic pulses that are bounced off the bottom; the two-way travel time indicates the depth) made by ships hundreds, and even thousands, of feet above the seafloor, from sediment cores obtained by dropping a weighted tube from a ship to the bottom of the sea, and from pictures taken by cameras towed on the end of a long umbilical cord. Very few valleys are so situated that they have been studied by scuba divers and research submersibles. Under similar conditions it is not unlikely that the origin of land canyons and valleys would still be under debate.

SUBMARINES See OCEAN ENGINEERING.

SULAWESI SEA (Celebes Sea), located in the western PACIFIC OCEAN, is ringed by Mindanao on the north, a row of islands connecting Mindanao with Celebes (Sulawesi) on the east, by Celebes on the south, and by Borneo and the Sulu Archipelago on the west. The Sulawesi Sea covers an area of 182 192 mi² (472 000 km²), occupies a volume of 372 580 mi³ (1 553 000 km³), and has a mean depth of 10 797 ft (3291 m). The maximum depth is 20 407 ft (6220 m).

The Sulawesi Sea occupies a deep basin surrounded by very narrow continental shelves. (See CONTINENTAL SHELF.) The bottom sediments are mostly terrigenous muds, with volcanic muds and globigerina ooze in localized areas. An active volcano is located at 124°E longitude and 4°N latitude.

Surface currents in the Sulawesi Sea flow southward from Mindanao and out through the Makassar Strait between Borneo and Celebes. Deep water from the Pacific flows into and out of the Sulawesi Sea basin along the same general path. Surface temperatures range from 82° F (28° C) in summer to 81° F (27° C) in winter.

SULFUR CYCLE is the natural cycle in which the ELEMENT sulfur is recycled from the oceans to the land by way of the atmosphere.

Sulfur, a nonmetallic element, is found in the world's oceans as sulfates, sulfites, and sulfides.

PLANT LIFE IN THE OCEANS utilizes the sulfates which are produced by chemical or biological oxidation. When plants and animals decompose in the absence of air (anaerobically), hydrogen sulfide is released. It may be produced chemically, or the reduction of sulfur, like that of nitrogen, may be carried out by microbes. In the oceans, the primary locus for such a process is in oxygen-deficient muds where sulfate-reducing anaerobes reside. These heterotrophic BACTERIA break down various sulfates in the mud to such compounds as hydrogen sulfide, methane, and formaldehyde. Of these, hydrogen sulfide tends most to escape from the SEAWATER environment to the atmosphere. There it is oxidized to sulfur dioxide, and after dissolution in rainwater, sulfates (and sulfuric acid) are formed. These are then delivered to the land and the world's rivers and then back to the oceans.

It is fortunate that in most parts of the oceans hydrogen sulfide is released to the atmosphere (the BLACK SEA is one exception where hydrogen sulfide is present), for it is a hostile material to all marine life except bacteria. In this regard, not all bacteria in the sulfur cycle produce hydrogen sulfide; the true sulfur bacteria produce pure sulfur which is put back in the cycle by oxidation.

While some sulfur compounds (i.e., sulfur dioxide) in the atmosphere originate from industrial sources, the large percentage, or about two-thirds, take the route from the world's oceans.

The sulfur cycle is important since sulfur is a vital biosphere material. Its basic function in living matter is the role it plays in the formation of the protein molecule.

SULU SEA is located in the western PACIFIC OCEAN and lies between Borneo and Palawan Island on the southeast, Mindoro and Panay on the north, and Negros, Mindanao, and the Sulu Archipelago on the east. The Sulu Sea covers an area of 162 120 mi² (420 000 km²), occupies a volume of 114 677 mi³ (478 000 km³), and has a mean depth of 3737 ft (1139 m).

In the southwest and north the Sulu Sea is underlain by wide continental shelves, but the central portion has a deep basin. (See CONTINENTAL SHELF.) The basin is divided by several ridges which, in some areas, break the surface as reefs (e.g., Tubbataha Reefs) and islands (e.g., Cagayan Islands). The floor of the basin is largely covered by globigerina ooze.

During the summer the surface currents in the Sulu Sea are southerly and assume a counterclockwise pattern in winter. Deep-water exchange with the SULAWESI SEA and SOUTH CHINA SEA appears to be restricted by the shallow sill depth of connecting passages. The surface WATER temperature ranges from 82° F (28° C) in summer to 81° F (27° C) in winter.

SUSPENSION CURRENT See TURBIDITY CURRENT.

SVERDRUP, HARALD ULRIK (1888–1957), a Norwegian oceanographer, is best known for the book *The Oceans: Their Physics, Chemistry, and General Biology,* written with Martin W. Johnson and Richard H. Fleming. The book, published in 1942, represented a balanced survey of existing knowledge about ocean waves and currents, the chemistry of seawater, plant and animal life in the sea, and the geology of the seafloor. (See JOHNSON, MARTIN WIGGO.) The book had a profound effect on oceanographic research throughout the world, partly because of its timeliness. The naval operations of World War II and the submarine threat which dominated the Cold War years demanded a better understanding of oceanographic processes; Sverdrup's book served as the foundation for the resulting research effort.

Sverdrup was born November 15, 1888, in Sogndal, Norway. He received his doctorate from the University of Oslo, and served as research associate at Oslo and Leipzig from 1911 to 1917. From 1918 to 1925 he was director of scientific work on the Norwegian north polar expedition aboard the *Maud.* After the *Maud* expedition Sverdrup served as professor of meteorology at the Geophysical Institute in Bergen for 2 years. In 1931 he participated in the Wilkins-Ellsworth submarine arctic expedition on board the *Nautilus.* He was a research professor at the Christian Michelsen Institute in Bergen for 4 years. In 1936 he was made the director of the SCRIPPS INSTITUTION OF OCEANOGRAPHY in La Jolla, California. He completed his career as professor of geophysics and director of the Norwegian Polar Institute in Oslo.

Among Sverdrup's many contributions to oceanography was his theory, now generally accepted, that the ANTARCTIC CIRCUMPOLAR CURRENT through the Indian Ocean was the source of the salinity in the PACIFIC OCEAN. He also contributed to the theory that accounts for the equatorial countercurrent. (See EQUATORIAL CURRENT SYSTEM.) In recognition of his work, he received the Agassiz Medal from the U.S. National Academy of Sciences in 1938 and the Bowie Medal of the American Geophysical Union in 1951.

SWELL See WAVES.

SWELLFISH See PUFFERFISH.

SWIM BLADDER is a gas-filled cavity that normally occupies about 5 percent of the volume of most bony fishes (see TELEOSTS). The swim bladder serves primarily as a flotation organ to give ocean animals a neutral buoyancy or an average DENSITY equal to that of SEAWATER, thus saving them the labor of continuous swimming.

A FISH having a swim bladder is capable of precisely regulating the amount of gas (largely oxygen) to this bladder, an airtight sac equipped with glands for taking the gases from the bloodstream to the bladder. This remarkable regulatory mechanism in the fish's circulatory system, which obeys the gas laws and changes volume when PRESSURE is changed, gives the fish just the amount of buoyancy it requires to remain at whatever level it may dwell. As fishes evolved with this bladder development, extensions to it later developed, and an "inner ear" labyrinth was affected deep within the skulls of these fishes. This permits them to locate any source of sound waves.

CUTTLEFISH and the cranchian SQUID have developed quite different kinds of flotation organs from the swim bladder. In the case of the squid, a fluid-filled cavity lowers their average density and enables them to float.

The very deep-dwelling fishes as well as most bottom inhabitants have no use for a swim bladder and therefore have not developed one. However, certain fishes such as the MACKEREL have either a very small swim bladder or none at all. Usually a fish with a heavy skeleton has a swim bladder. The buoyancy imparted by the bladder makes the heavier skeleton biologically feasible.

See also DEEP SCATTERING LAYERS; DEEP-SEA FISH AND ANIMALS; HABITABLE ZONES.

SYMBIOSIS is the living together of two organisms (plants or animals) for the benefit of one or both members. If one member cannot live without the other under natural conditions, the relation is called *mutualism*. The most common example of mutualism is the actual union of ALGAE and fungi to form LICHENS, the crusty gray-green plants found on rocks and trees (see FUNGUS). Another form of symbiosis is *commensalism*, the arrangement in which one member benefits without harming the other. Commensalism is very common in the world's oceans where SPONGES, SHELLFISH, and burrowing worms support other forms of life almost without exception.

SYZYGY See TIDES.

TARAUANA See AMA.

TASMAN SEA is located off the southeastern coasts of Australia and Tasmania in the South PACIFIC OCEAN. The western boundary is provided by the Australian coast from latitude 30°S (near Grafton), southward through Gabo Island near Cape Howe, East Sister Island, Flinders Island, Barren Island, and along the coast of Tasmania to South East Cape, its southern point. The remaining boundary is defined by a line running eastward along the parallel of 30°S, through Elizabeth Reef and South East Rock to Three Kings Island, and on to North Cape, the northern tip of New Zealand. The line then runs southward down the east coast of New Zealand, crossing Cook Strait between North Island and South Island, and across Foveaux Strait to Steward Island. Finally, the line turns westward through the Snares and Auckland Island to South East Cape on Tasmania. The Tasman Sea covers an area of about 900 000 mi² (2 330 100 km²).

The Morioris were probably the first people to cross the Tasman Sea since they were well settled on New Zealand when the Maoris arrived in the 1300s. The first European of record in the area was the Dutch explorer Abel Janszoon Tasman. In 1642, Tasman discovered the island that bears his name (although he named it Van Diemen's Land in honor of the governor of the Dutch East Indies) and continued eastward to discover New Zealand. Incredibly, Tasman continued on to circumnavigate Australia without ever sighting the continent. The same applies to the Spanish navigator Luis Vaez de Torres who, in 1606, sailed through the strait bearing his name which separates Australia from New Guinea. Although it is likely that the Portuguese discovered Australia (an account of its existence was published in Portugal in 1542), the Dutch navigator Willem Jansz became the first European of record to sight the continent when he explored the eastern shoreline of the Gulf of Carpentaria (Australian north coast) in 1606. Although the area was visited often, Australia stirred little interest until James Cook explored the east coast and claimed it for England in 1770.

The CONTINENTAL SHELF around the Tasman Sea is of limited development, being largely restricted to the east coast of Australia and the west coast of New Zealand. The continental shelf is cut in several places by a SUBMARINE CANYON that transports sediment to the deep seafloor. The deep floor is occupied by the Tasman Abyssal Plain (Tasman Basin) along the western flank, the Lord Howe Rise down the center of the sea (north-south), the New Caledonia Trough east of the rise, and the New Zealand Plateau which supports the islands of New Zealand on the east. The first three of these features extend northward into the CORAL SEA. The basins are 15 000–16 000 ft (4572–4877 m) deep, and the rise lies at a depth of 3000–4000 ft (914–1219 m) but rises to the surface in several locations to form small islands and reefs. Numerous seamounts dot the deep basin floors. (See SEAMOUNT.) Sediments range from terrigenous materials mixed with calcareous sands and muds on the shallow flanks of the basin to clays on the deep floor. Some volcanic rubble and ash are to be found on the Lord Howe Rise.

Surface circulation in the Tasman Sea is largely controlled by the EAST AUSTRALIAN CURRENT which flows southward from about latitude 20°S. This current, one of the strongest in the South Pacific, is fed by the South Equatorial Current (see EQUATORIAL CURRENT SYSTEM) which flows in from the north, and to a lesser extent by the West Wind Drift Current.

TAXONOMY or **SYSTEMATICS** is the study directed at producing a hierarchical system of classification of organisms which most accurately reflects the totality of similarities and differences. Some 350 000 different kinds of plants and more than a million kinds of animals have been described.

As begun by Aristotle and Theophrastus, and developed in the eighteenth century by Linnaeus and others, systematics (or taxonomy as it is known interchangeably) involved little more than a simple grouping of plants and animals on the basis of structure. But after the publication by Darwin, in 1859, of the *Origin of Species*, natural system classification came to be based upon evolutionary relationships. (See DARWIN, CHARLES.) Any natural system can be expected to change from time to time as further knowledge becomes available.

In both animal and plant classification, the species is taken as the basic unit of classification. Here, the species is expressed as a binomial or double name consisting of the genus and species. Determination of the species is often difficult, but, in general, a species may be characterized as a self-perpetuating natural population distinct and distinguishable from other such populations. A group of similar species is termed a genus; a group of genera, a family; a group of families, an order; a group of orders, a class; and a group of classes, a phylum. The phyla are the main subdivisions of a kingdom. Intermediate categories, such as subdivisions, subgenera, subclasses, and subfamilies are also recognized, but these are optional groupings not required in the formal hierarchy. One way that the early student of taxonomy remembers the ordering of the units of classification is by the phrase "*K*ings *p*lay *c*hess *o*n *f*ine *g*rains of *s*and" K being for kingdom, P for phylum, etc.

Some Problems Ocean fishes are the most numerous of the recent vertebrates, and estimates of the number of species range from 15 000 to 40 000, although 25 000 appears to be the figure most often quoted. The discrepancy exists since fish species are sometimes named more than once because of inadequate descriptions and variations due to environment or geographical distribution. Also, for some species, the male has been described as belonging to one species and the female to another because of a difference in body form or color pattern. This phenomenon is called sexual dimorphism. Other fishes have been named more than once because the young look different than the adults. In addition, most scientists agree that not all fishes have yet been named; the estimate of 25 000 allows for this unknown. The species of fishes with bony skeletons are more numerous than those with skeletons of cartilage (see SHARK and RAYS). Bony fishes number around 20 000 while the cartilaginous fish number only about 600.

The question of the validity of classifying only two major groups of organisms—the photosynthetic, rooted, higher plants and the food-ingesting motile higher animals—has been raised over the years by many investigators. Arguments about the limitations of such a two-kingdom (plant and animal) categorization especially revolve around the treatment in such classifications of various unicellular organisms (see PROTOZOA). Some of these organisms have been claimed both for the plant kingdom by botanists and for the animal kingdom by zoologists. Consequently, new concepts of kingdoms and organisms have been suggested in order to better represent evolutionary relations in the living world and more effectively classify the division involved. (See also DIATOMS; DINOFLAGELLATES; PLANT LIFE IN THE OCEANS.)

There has been considerable effort in taxonomy to develop a more definite concept of species, analyzing the organism not only on the basis of isolated characteristics, but more from the standpoint of its being the product of many interacting characteristics such as fundamental structural features, i.e., the type and anatomy of the respiratory systems, the openings in the wall of the heart, the relations of the legs to the body, etc.

The aim of the taxonomist is to trace the pathways by which organisms have evolved and to discover the genetic interrelations of biological groups. Thus, systematics is not only the most elementary branch of biology—since organisms cannot well be discussed or treated in a scientific way until some taxonomy has been achieved—it is also the most inclusive—since in its various aspects it gathers together, summarizes, and utilizes everything that is known about living things, whether morphological, physiological, behavioral, or ecological.

See also EVOLUTION.

An Abbreviated Example of a System of Classification of Living Organisms

Viruses Infectious agents characterized by a total dependence on living cells for reproduction and by a lack of independent metabolism.

Kingdom Protista Unicellular organisms, with or without nuclei

Phylum Schizophyta: Bacteria

Phylum Cyanophyta: Blue-green algae

Phylum Euglenophyta: Euglenoids

Phylum Protozoa: Unicellular animals

Subphylum Plasmodroma

- Three classes: Flagellata, Sarcodina, and Sporozoa

Subphylum Ciliophora

- Two classes: Ciliata and Suctoria

Phylum Pyrrhophyta: Dinoflagellates and cryptomonads

- Two classes: Dinophyceae and Cryptophyceae

Phylum Chrysophyta: Diatoms

Kingdom Metaphyta Multicellular plants.

Subkingdom Thallophyta No true roots

Phylum Chlorophyta: Green algae

Phylum Phaeophyta: Brown algae

Phylum Rhodophyta: Red algae

Phylum Myxomycophyta: Slime molds

- Two classes: Myxomycetes and Acrasiomycetes

Phylum Eumycophyta: Fungi

- Four classes: Phycomycetes, Ascomycetes, Basidiomycetes, and Deuteromycetes

Subkingdom Embryophyta

Phylum Bryophyta: Mosses and liverworts

- Two classes: Musci and Hepaticae

Phylum Tracheophyta: Vascular plants

Subphylum Psilopsida
Subphylum Lycopsida
Subphylum Sphenopsida
Subphylum Pteropsida

- Three classes: Filicae, Gymnospermae, and Angiospermae
Two subclasses: Dicotyledoneae and Monocotyledoneae

Kingdom Metazoa Multicellular animals.

Subkingdom Parazoa No nervous tissue

Phylum Porifera: Sponges

- Three classes: Calcarea, Hexactinellida, and Demospongiae

Subkingdom Eumetazoa

Phylum Cnidaria (Coelenterata)

- Three classes: Hydrozoa, Scyphozoa, and Anthozoa

Phylum Ctenophora

Phylum Platyhelminthes

- Three classes: Turbellaria, Trematoda, and Cestoda

Phylum Nemertea

Phylum Aschelminthes

- Two classes: Nematoda and Rotifera

Phylum Ectoprocta

Phylum Brachiopoda

Phylum Annelida

- Three classes: Polychaeta, Oligochaeta, and Hirudinea

Phylum Mollusca

- Five classes: Amphineura, Gastropoda, Scaphopoda, Pelecypoda, and Cephalopoda

Phylum Arthropoda

Subphylum Chelicerata

- Three classes: Xiphosurida, Pycnogida, and Arachnida

Subphylum Mandibulata

- Class Crustacea
Five subclasses: Branchiopoda, Ostracoda, Copepoda, Cirripedia, and Malacostraca
- Class Chilopoda
- Class Diplopoda
- Class Insecta
Two subclasses: Apterygota and Pterygota
Two suborders: Exopterygota and Endopterygota

Phylum Echinodermata

Subphylum Pelmatozoa

- Class Crinoidea

Subphylum Eleutherozoa

- Four classes: Asteroidea, Ophiuroidea, Echinoidea, and Holothuroidea

Phylum Chordata

Subphylum Tunicata
Subphylum Cephalochordata
Subphylum Vertebrata
Superclass Agnatha

- Two classes: Ostracodermi and Cyclostomata

Superclass Gnathostomata

- Class Placodermi
- Class Chondrichthyes
  Subclass Holocephali
- Class Osteichthyes
  Subclass Actinopterygii
    Superorder Chondrostei
    Superorder Holostei
    Superorder Teleostei
  Subclass Choanichthyes
    Superorder Crossopterygii
    Superorder Dipnoi
- Class Amphibia

- Class Reptilia
- Class Aves
- Class Mammalia
  Subclass Theria
  Infraclass Eutheria

TELEOSTS is the common name for fishes belonging to the Teleostei, a superorder of the subclass Actinopterygii, or ray-fin fishes. Teleost fish represent an extremely large and diverse assemblage constituting about 90 percent of all living fish. They are distinguished by paired bracing bones in the support structure or skeleton of their caudal (tail) fins, thin cycloid scales, and a SWIM BLADDER used as a hydrostatic organ.

Teleosts are believed to have appeared in the Jurassic Period and to have undergone a rapid adaptive radiation. They are fishes which represent end products of fish specialization and have not evolved types of vertebrates beyond their own.

TEMPERATURE is a property of an object which determines the amount of heat produced by the average kinetic energy or speed of its molecules and the direction of heat flow when the object is placed in thermal contact with another object. Heat flows from a region of higher temperature to one of lower temperature. It is measured either by an empirical temperature scale (expressed in degrees Celsius or Fahrenheit), which is based on some convenient property of a material or instrument, or by a scale of absolute temperature, for example, the Kelvin (K) scale ($K = °C + 273$). T mperatures measured in Fahrenheit may be converted to Celsius by using the formula $°C = (°F - 32)/1.8$. Temperatures measured in Celsius may be converted to Fahrenheit by using the formula $°F = (1.8 \times °C) + 32$.

In the oceans, heating of the waters is due primarily to solar radiation, and the amount of heat transferred from the earth's interior to the world's oceans is relatively negligible. Ocean surface and near-surface temperatures vary widely with LATITUDE, and the top few hundred meters of the ocean are highly variable in heat content from one geographical area to another. The reason for this is that heat that is added to, or taken from, this layer is dependent, to a large extent, on the nature of the atmosphere above it. If the atmosphere is overcast or clear, the radiation input varies; if it is cold or warm, there is a difference in the exchange of sensible heat; if it is moist or dry the evaporation is variable; if it is windy or calm, the exchange of momentum is variable.

Ocean-surface temperature extremes may range from 28 to 90° F (−2 to 32° C). The high temperatures are encountered in the PERSIAN GULF, and the low temperatures occur in the water masses of the polar regions where the sun's rays strike the surface at a lesser angle than that in tropic waters. As previously mentioned, surface temperatures vary as a function both of latitude and of the characteristics of the overlying atmospheric layer over the surface of the water. However, the temperature fluctuations that occur daily [about 32° F (0.2° C) in oceanic surface water] and on a seasonal basis are much less than those of equivalent latitude land areas. This fact is due to one of the unique characteristics of WATER, namely, its very high SPECIFIC HEAT. This means that water requires more heat to raise its temperature than any substances found on land, and it also requires a longer period to cool down than any other substance.

Just as ocean temperatures vary greatly with latitude, they also vary greatly vertically, with the temperatures decreasing with increasing depth [at 600 ft (182 m) the temperatures are about 68° F (20° C); at 4000 ft (1200 m), about 46° F (5° C); and below this depth, temperatures fall to about 28° F (−2° C) in the abyssal regions]. In general, cold is one of the diver's worst hazards.

In general, in the ocean there is usually a mixed layer of isothermal water below the surface, where the temperature is the same as that of the surface. This layer is best developed in the trade-wind belts, where it may extend to a depth of 600 ft (182 m); in temperate latitudes in the spring, it may disappear entirely. Below this layer is a zone of rapid temperature decrease, called the THERMOCLINE, to the temperature of the deep oceans. At a depth greater than 1200 ft (364 m), the temperature everywhere is below 60° F (16° C), and in the deeper layers, fed by cooled waters that have sunk from the surface in the Arctic and Antarctic, temperatures as low as 33° F (0.5° C) to 28° F (−2° C) exist.

In the deepest ocean basins, the temperature increases slightly with depth, the increase being about 1° F at 18 000 ft (5486 m). The warming is believed to be caused more by the slight compression of SEAWATER than by heat from the earth's crust.

The constancy of temperature with depth enables sea life to live under stable conditions. Of importance to the ocean engineer is that over the temperature range of 54 to 68° F (12 to 20° C) the conductivity of seawater, an important consideration in the CORROSION of materials, almost doubles.

TEREDO or **SHIPWORM** refers to any of several bivalve MOLLUSK species of the family Teredinidae; these ocean animals (especially *Teredo navalis,* a species familiarly known as the shipworm), are particularly destructive to the wood in boats, barges, bulkheads, piles, docks, and bridges.

Teredos superficially resembles earthworms, although they are not worms but elongated sorts of CLAMS. The shell covers only the front end; the plates of this shell are hemispherical and equipped with ridges of microscopic filelike teeth. An opening formed by two deep adjacent notches in the plates permits teredo to extend its long and muscular foot to explore the outside environment. The animal attaches itself with this foot to any wood in the tidal cycle of the marine environment and proceeds to burrow into it by digesting and metabolizing the cellulose fibers that have been already partially broken down by the action of ocean fungi. (See FUNGUS.) In the adult stage, teredo grows to a length of about 6 in (15 cm), and in subtropical waters the average life span is 10 weeks. During this time as an adult, it reproduces by self-fertilization, or by first taking on one sexual form and then another. As a male, it discharges sperm into the water. Later, the same animal functions as a female, and the spermatozoa in the water liberated by closeby teredos in the male form are taken up to fertilize the ova. The LARVA of a teredo (each teredo gives birth to tens of thousands of young) swims by means of a ciliated rotorlike organ, and if by 3 days it does not find a piece of wood to attach itself to, it dies.

Some other bivalves of the subclass Lamellibranchia, such as various species of *Rocellaria* and *Tellina,* can bore not only into wood but into rocks and cement as well.

See also FOULING.

THERMOCLINE, or discontinuity layer, is the name for a layer of ocean water where TEMPERATURE changes take place drastically with depth; the term is derived from the Greek words *therme,* or "having to do with heat," and *klinein,* meaning "gradient." One water column may have more than one layer in which temperature changes are quite pronounced.

In the world's oceans there are both permanent and temporary thermoclines. The latter are usually near the surface, although they may vary in depth or disappear completely due to seasonal and daily fluctuations in the atmosphere above the water. Also, mixing of the top water layers caused by wind-driven and tidal currents can cause this thermocline to change drastically by the production of an isothermal or equal-temperature water condition. The permanent thermocline normally occurs at deep depths [e.g., around 500 ft (152 m) in the ATLANTIC OCEAN at the middle latitudes above and below the equator]. (See LATITUDE.) However, a change in this temperature zone takes place as one approaches the equator, and here the permanent thermocline is found nearer to the surface. This is due, among other factors, to the fact that at the equator ocean currents diverge and tend to cause the thermoclines to go upward, while at the middle latitudes the currents converge and push them down.

Thermoclines affect the propagation of underwater sound, a tool widely used in oceanographic work for a variety of applications. (See also INSTRUMENTATION; OCEANOGRAPHY; SONAR; UNDERWATER SOUND.)

Usually the thermocline is found at the interface of the warm and lighter-weight waters above it and the cold and heavier waters below. However, when temperature inversions occur, the reverse is true. These inversions can take place especially in areas where the surface waters are colder than 39° F (4° C) and where the currents keep these cold surface layers separated from the warmer deep water which is actually heavier than the overlying cold water [less than 39° F (4° C)].

TIDAL BORE See TIDES.

TIDAL SPECIES See TIDES.

TIDAL WAVE See TSUNAMI.

TIDE WAVE See TIDES.

TIDES are the rhythmic, worldwide rising and lowering of SEA LEVEL which, along most coastal areas, occur twice daily in response to the gravitational attraction of the moon and sun. All bodies of water experience a tidal effect to a degree which depends upon their size and the configuration of their basins. Lake Superior, for instance, has a tidal range (vertical distance between the WATER level at low and high tide) of about 2 in (5 cm), whereas the Bay of Fundy in Canada has the greatest tidal range in the world at 44.6 ft (13.5 m). The world average is about $2\frac{1}{2}$ ft (0.76 m). Currents associated with the ebb and flow of tides play a significant role in flushing and freshening our harbors and bays and in helping to keep their channels free of silt. And, it is the daily fluctuation of sea level in conjunction with wave action that alternately depletes and replenishes the sand on our beaches. (See BEACH.)

Pytheas, the Greek explorer, was the first to associate the rise and fall of the tides with the moon. In the fourth century B.C. he departed Massilia (Marseilles) and sailed out through the Strait of Gibraltar to circumnavigate the British Isles. Since the MEDITERRANEAN SEA has a tidal range of only 4–6 in (10–15 cm), it is likely that Phytheas made the connection between the moon and tides along the English coast where tidal ranges as great as 43 ft (13 m) (Severn River estuary) are observed. Both Pliny the Elder and Curtius Rufus (the biographer of Alexander the Great) made similar observations during the first century A.D. But, it was not until Isaac Newton laid down the conceptual foundation of universal gravitation in his *Principia Mathematica* of 1687 that any real progress could be made in understanding the relation between the tides and the gravitational attraction of other bodies in the solar system.

The moon and sun are the two gravitational sources that control earth tides. The major planets in the solar system also exert an influence, but the effect is so small that it can be neglected in most calculations. The sun, even though its gravitational attraction is 27 million (2.7×10^7) times that of the moon, is so far away from earth [93 million mi (149.6 million km)] that its effect on the tides is only 46 percent that of the more neighborly moon [240 000 mi (386 160 km)]. This results from the fact that although gravitational attraction is proportional to the mass of the body exerting the attraction, it is also inversely proportional to the cube of the distance of that body.

If we visualize, for the moment, a stationary earth-moon system, it is obvious that the water directly below the moon is 4000 mi (6436 km) closer to the moon than is the center of the earth, and that on the other side of the earth the water is 4000 mi further away from the moon than is the center of the earth. Therefore, keeping in mind the relation of mass and distance expressed earlier, the moon exerts a greater pull on the water directly beneath it than it does on the earth. This results in a slight bulge in the water on the near side of the earth. On the opposite side of the earth the reverse is true but the result is the same. The earth, being closer to the moon than the water, moves slightly away from the water, producing a bulge on the far side of the earth as well. As a result, a high tide exists directly beneath and directly opposite the moon. The resulting egg-shaped form causes a slight thinning of the water surrounding a plane perpendicular to the earth-moon axis, and an earth-girdling low tide is produced.

The combination of the earth's rotation about its own axis every 24 hr, and the moon's revolution about the earth every 27.3216 days, produces an apparent revolution of the moon about the earth every 24 hr and 50 min. Thus, the world's oceans experience two high tides and two low tides each day. This is called a *semidiurnal period.* The range of these tides is influenced by the position of the sun with respect to the earth and moon. When the earth, moon, and sun lie at or near a straight line (full moon or new moon—syzygy), the gravitational attractions are amplified and exceptionally high tides —called *spring tides*—exist. On the other hand, when the earth, moon, and sun are at right angles to each other (quarter moon or quadrature), the gravitational attractions are conflicting and exceptionally low tides—called *neap tides*—prevail.

Unfortunately, for those who have the job of predicting the tides, the actual case is not so simple as is being portrayed here. For one thing the sun's declination (relationship between the plane of the earth's equator and that of its revolution about the sun) varies by 23.5° north and south of the equator each year, and the moon's declination varies by as much as 28.5° north and south of the equator each month. Therefore, at the extremes of declination one high-tide bulge is in the Northern Hemisphere and the high-tide bulge on the opposite side of the earth is in the Southern Hemisphere. This offset with respect to the equator causes a point at any latitude to experience one *higher high tide* and one *lower high tide* as well as one *higher low tide* and one *lower low tide* each day. In other words, the two tides experienced each day can only approach equality when the moon is in the plane of the equator (ascending or descending node) and syzygy occurs.

Tidal predictions are complicated by additional factors such as the CORIOLIS EFFECT, the placement of continents, and the detailed shape and dimensions of the basins in which the water rests. Each basin, because of its geometry, is tuned to a particular frequency of oscillation. Therefore if the basin is subjected to several astronomical periods, it will favor the one that falls closest to its resonant frequency. The Mediterranean Sea, for example, is very poorly tuned to tidal periods and has a tidal range of only 4–6 in (10–15 cm). The ATLANTIC OCEAN is best tuned to semidiurnal tides so that two high tides and two low tides are experienced every day. Some parts of the Pacific Basin are best tuned to *mixed tides* in which two high tides, with a slight ebb in between, are experienced, followed by a single low tide. Other places such as Saint Michael, Alaska, and certain places around the GULF OF MEXICO experience only a *diurnal tide,* or a single high and a single low tide. And, certain locations such as Puerto Rico

TIDES. A tidal bore (the crest of a tidal wave rushing up an estuary) sweeps into the Bay of Fundy. In some rivers, tidal bores may reach a height of 25 ft. (*NOAA*)

and the Solomon Islands appear to be near a null or *amphidromic point* where little tidal fluctuation is experienced.

Because of the complexity of the tidal problem, predictions cannot be made by mathematical formulation alone. It is necessary to build up actual observations and measurements (using tide gauges) over a long period of time at many locations around the world in order to be able to extrapolate into the future. This comes only after all the *tidal species* (various unique tidal periods involved) have been identified and subjected to harmonic analysis to determine the amplitude of each species. Once this has been done, predictions can be made that are good to about 0.1 ft (0.03 m) over a predicting period of 1 yr, provided, of course, that fluctuations in water level due to other causes are neglected.

The wavelength of the tidal wave is as much as 12 500 mi (20 113 km) at the equator and travels at the rate of 1000 mi (1666 km) per hour. Both gradually fall off with increasing latitude until they reach zero at the poles. Thus, tidal waves are shallow-water waves and interact with the bottom (see WAVES) even in the greatest ocean depths. As a result of this interaction, the wave lags a bit behind the moon as it races across the ocean. As the wave approaches a coast, the bottom interaction increases with decreasing depth and the wave begins to steepen its crest. In certain river estuaries the rate of river flow may be sufficient to stop the advance of the tidal wave for a time. Eventually, however, the tidal wave increases the water depth until it overrides the river water. When this happens the crest sweeps up the ESTUARY as a conspicuous wave called a *tidal bore.* The most pronounced tidal bore is that in the Tsientang Kiang estuary in China where the wave may reach 25 ft (7.6 m) in height. Other well-known tidal bores occur in the Amazon River, the Severn River of England, the Petitcodiac River of Canada, and the Seine, Orne, and Gironde rivers of France.

TIMOR SEA is located northeast of Australia in the northeastern corner of the INDIAN OCEAN. Its boundaries are defined by a line running clockwise from the southeastern limit of the SAVU SEA, the island of Timor, and the southern limit of the BANDA SEA, southward along the western edge of the ARAFURA SEA to the north coast of Australia, and then across to Roti Island. The Timor Sea covers an area of 237 390 mi² (615 000 km²), occupies a volume of 59 978 mi³ (250 000 km³), and has a mean depth of 1332 ft (406 m).

Much of the Timor Sea rests on the Sahul Shelf (see CONTINENTAL SHELF) off the north coast of Australia. The shelf also contains the Gulf of Bonaparte between Darwin and Cape Londonderry and, at the northern extreme of the Gulf, the Bonaparte Basin which has a maximum depth of 459 ft (140 m). Between the edge of the shelf, which breaks at about 394 ft (120 m) in depth, and the island of Timor is the Timor Trough running northeast-southwest. The trough has a maximum depth of 10 498 ft (3200 m). The sediments covering the shelf are fine-grained carbonates mixed with some glauconite, whereas the trough area is covered with clays mingled with calcareous sands.

The Timor Sea lies 10°S of the equator and is alternatively influenced by the southeast trade winds and the monsoon belt. It lies in the generating area for many typhoons. Surface-water temperatures range from 84° F (29° C) in summer (Southern Hemisphere) to 73° F (23° C) in winter. A surface current running southwestward prevails for most of the year.

TORNADO See HURRICANE.

TRANSDUCER See SONAR; UNDERWATER SOUND.

TRANSPONDER See UNDERWATER SOUND.

TRANS-TIDAL WAVES See WAVES.

TRENCHES, among the most spectacular features on the deep ocean floor, are long and narrow depressions running parallel to the margin of some continents and island archipelagoes. Trenches are thought to be the zone of interaction between an oceanic and a continental plate. (See CONTINENTAL DRIFT.) The oceanic plate, having the greater density, plunges beneath the continental plate at an angle of about 30°. This process, and the frictional forces between the two plates, results in a V-shaped depression that runs the length of the zone of interaction. As an example, the Mariana Trench, running in a north-south arc along the eastern margin of the Mariana Islands, has a maximum depth of 36 198 ft (11 033 m), a length of 1584 mi (2550 km), and a width of 44 mi (70 km). Such trenches circle the PACIFIC OCEAN in what is known as the "ring of fire" (due to associated earthquakes and volcanoes) and are also found in both the ATLANTIC OCEAN and INDIAN OCEAN. None have been identified in the ARCTIC OCEAN.

As depicted on topographic charts, trenches give the impression of having near-vertical walls. This is due to the scale of the map. Actually most trenches begin with a slope of about 6° which deepens to about 14° further downslope. Occasionally slopes as great as 45° are encountered. The bottom of a trench is often flat as a result of sediment filling by turbidity currents. (See TURBIDITY CURRENT.)

It is understandable that the gradual submergence of one section of the Earth's crust beneath another does not take place smoothly. It has been demonstrated that the plane down which the oceanic plate (30°) plunges is the site of both shallow and deep focus earthquakes. In the case of island arcs or archipelagoes, plate interaction is further manifest by a characteristic surface structure. This consists of an inner belt of volcanic islands and an outer belt of strongly deformed material. The two belts are separated by a trough ranging from 31 to 62 mi (50 to 100 km) wide. The trench lies seaward of the outer ring by about the same distance as separates the rings. The inner volcanic ring lies roughly over the area at which the plunging plate will have melted due to the increasing heat of the earth. Whether this melting plate escapes to the surface through volcanic action is a matter of speculation. The outer ring appears to be simply an upward bulging of the earth's surface to compensate for the subduction of the oceanic plate. This structure is known as an island arc or island archipelago. Such structures are common in the Pacific Ocean (i.e., Java Arc). On continents the volcanic belt is represented by the mountain range which parallels the coast. In South America, for instance, the Andes parallel the Peru-Chile Trench of the west coast.

TROPICAL CYCLONE See HURRICANE.

TROUGH is a long, broad depression on the ocean floor. An example is the Bounty Trough, running eastward between the Chatham Rise and the Bounty Islands off the east coast of New Zealand.

TSUNAMI is a wave produced by the cataclysmic displacement of a large volume of water. (See WAVES.) Commonly referred to as tidal waves, although unrelated to TIDES or tide-producing forces, tsunamis (Japanese, meaning "harbor wave") can be extremely destructive of life and property in coastal areas. Since the precipitating event usually occurs on the seafloor, the detailed process by which tsunamis are generated is imperfectly understood. Volcanic explosions, vertical faulting associated with earthquakes, and vast underwater mudslides are thought to be among the causative processes.

Once generated, a tsunami behaves similarly to the waves produced by dropping a pebble into the middle of a shallow pond; i.e., the waves radiate in an ever-widening circle from the source. Beyond this, however, the similarity rapidly vanishes. The radiating waves of a tsunami, while typically about 2 ft (0.6 m) high in deep WATER, may measure over 100 mi (160 km) from crest to crest and travel at the maximum speed for open ocean depths, i.e., around 500 nautical mi (924 km) per hour. The disturbance may consist of a single crest, or the initial crest may be preceded by a broad trough and followed by a succession of smaller waves. The latter type, combined with human curiosity, has been responsible for the loss of many lives. The wave trough, impinging on a shore, causes the water level to fall at a steady but fairly rapid rate. Many people, curious about this strange phenomenon, have wandered onto the exposed "tidal flat" only to be inundated by a towering wall of water as the crest rushes ashore with startling speed.

While still at sea, the tsunami can be detected only by the most careful observation. Also, since the energy in the system is finite and nearly constant, the wave height slowly diminishes as the energy is increasingly diluted by an ever-widening circle. But, as the wave races onto a CONTINENTAL SHELF, the remaining energy is rapidly concentrated by an ever-narrowing wedge of water; the result is an increase in wave height that may reach several tens of feet. As an example, the Unimak (Alaska) tsunami of April 1, 1946, consisted of a succession of crests 2 ft (0.6 m) in height, 122 mi (196 km) between crests, and moving at better than 400 kn in deep water. Reaching the Pololu Valley on the coast of Hawaii, these waves crested, at about 15-min intervals, to a height of 55 ft (16 m). By the time the tsunami had reached Bikini Atoll in the Marshall Islands, the deep-water wave height had dropped from 2 to 1.5 ft (0.6 to 0.4 m).

Fortunately, the more disastrous tsunamis occur at infrequent intervals. Lisbon, Portugal, was devastated by 50-ft (15-m) waves in 1755. Over 100 years later (August 27, 1883) the volcanic island of Krakatoa exploded in the Sunda Strait, causing a tsunami whose waves, some reaching 125 ft (38 m) in height, destroyed 300 towns and villages and killed 36 380 people on the surrounding islands. Nine hours after the explosion, the waves smashed 300 river boats in the harbor at Calcutta, India, and Australia was reported to have lost 6000 boats to the waves that battered her coastline. But, so great a loss of life because of any future tsunami was made less likely by the installation of the SEISMIC SEA-WAVE WARNING SYSTEM following the destruction caused by the Unimak tsunami of April 1, 1946. Headquartered at Honolulu and covering the Pacific, the system consists of 15 seismic stations to detect earthquakes [tsunamis are more commonly associated with seismic disturbances that exceed 6.5 on the Richter scale and have focal depths shallower than 31 mi (50 km)], and 30 tide stations to determine whether a tsunami is associated with the disturbance, all connected by a rapid communication link. Many lives were saved by this system on March 28, 1964, when the extremely destructive Alaska Earthquake sent waves surging into towns along the coast from Alaska to Chile.

TUNA is the name for any of the large pelagic ocean fishes of the family Thunnidae (or Scombridae), genus *Thunnus,* which includes species that rank among the most valuable of commercial and game fish. The principal species are the great albacore, also called bluefin tuna (*Thunnus thynnus*), the albacore (*T. alalunga*), the blackfin (*T. atlanticus*), bigeye (*T. obesus*), and the longtail (*T. tonggol*).

Tunas are voracious carnivorous fish that have a worldwide distribution. These fish typically have a streamlined and rather deep, smooth body with a crescent-shaped forked tail in front of which are finlets on both the upper and lower surface of the aft body. There are no scales on the posterior, and those in the front are fused to form an armored covering. The fins are set in grooves on the body which ranges from 8 to 14 ft (2.4 to 4.2 m) in length. The color is usually dark blue on the back, the belly is white, or near white, and the flanks white with silvery markings.

Swift and strong migratory swimmers, tunas are usually found in schools of varying sizes, depending upon the size of the fish. The larger the tuna, the smaller the school, and vice versa. They tend to swim near the surface of the northern waters in summer and return south where they live at depths of between 100 and 600 ft (30.5 and 183 m) in winter. Tunas are unusual fish that seem to be warm-blooded and have a high digestive rate. Under the skin of the flanks, a network of blood vessels feeds the red muscles, and there is a similar elaborate circulation in the liver. The skin of these fish consists of fibrous layers separated by oily spaces. Accordingly, they can conserve their body heat with marked efficiency and are capable of moving into water which is oxygenated but too cold for long-term comfort, as long as there is water above in which they can warm up afterward. Because of their tremendous muscular activity in swimming, both their oxygen and food requirements are very high. Their diet consists mainly of shoaling fishes such as HERRING and MACKEREL. They also eat SQUID, CUTTLEFISH, flying fish, and sand EEL. In the fight for survival in the world's oceans, the chief predator of the large tunas is the KILLER WHALE.

The following is a list of some representative species.

- Albacore (*Thunnus alalunga*). These fish are also known as "longfins" and are distinguished by long, sabre-sharp pectoral fins. They are further recognized by the metallic steel-blue color on the tops and sides of the body and a silvery color on the bottom side, as well as by the absence of stripes.
- Yellowfin (*T. albacares*). The yellowfin is considered one of the most commercially valuable of the tunas, and is also popular as a game fish. The name yellowfin describes these fish well since they are distinguished by elongated, yellowish dorsal and anal fins, and yellowish coloring on the sides.
- Bluefin (*T. thynnus*). These fish have a history as a game fish that goes back to Greek and Roman times. They are distinguished by the deep-blue or green color on the tops and sides of the body. Unlike most fishes, because of its high metabolic rate, the bluefin tuna maintains a body TEMPERATURE warmer than the water.
- Skipjack (*Katsuwonus pelamis*), or striped tuna. Skipjacks are distinguished by parallel black to dusky stripes on the lower sides of the body. They are dark metallic blue on the tops and sides, shading to a silvery color on the bottom surfaces. Skipjacks are the smallest of the tunas mentioned here and are found in tropical waters, similar to the yellowfin.

The bluefin tuna (*T. thynnus*), found on both sides of the North Atlantic, often attains a weight of over 1000 lb (454 kg), while the yellowfin (*T. albacares*) is considerably smaller [approximately 300 lb (136 kg)] and is prevalent off the coasts of South and Central America. The albacore (*T. alalunga*) migrates from waters off the Pacific Coast of North America to Japan and Hawaii. Its counterpart is the Atlantic albacore [up to 4 ft (1.2 m) long and 65 lb (30 kg)]. The albacore is one of the world's most sought after tunas. It is also one of the most valuable. More than half of all albacore landed in the world are consumed in the United States. However, U.S. production does not meet the demand. Albacore rarely form compact surface schools, and so purse seines, which have become the primary gear used by U.S. fishermen for yellowfin and skipjack tunas, are not suitable for catching albacore. They are caught on the surface primarily with live bait and by trolling; larger fish living well beneath the surface are taken by longlines.

The tunas are all robust game fish and have been a valuable commercial food fish for centuries. However, some species have been threatened by overfishing. Moreover, in California, the center of the commercial tuna fishing industry where the dollar value of the catch amounts to approximately 30 percent of all the fin fish caught by U.S. boats, the industry itself has been threatened. The fact that huge numbers of porpoises are killed by tuna fishermen has been the cause of the problem. (See PORPOISE.) Something like 50–70 percent of the annual U.S. catch of yellowfin tuna in the eastern tropical Pacific comes from schools that are associated with porpoises. Tuna primarily associate with two species: the so-called spotters and spinners. At times, they may be found in the company of a third species, which the fishermen call "whitebellies." Once the seine has been set and pursed around a school of porpoise and the associated tuna, the problem arises of releasing the porpoises unharmed without losing the fish. In 1976, these fisheries killed over 78 000 porpoises when the mammals were drowned after being trapped in the tuna nets which are used for catching some species. The 1972 Marine Mammal Protection Act sets limits on the annual size of these porpoise kills.

The tuna, because of its migratory habits and unique metabolic system, is a subject of considerable scientific interest and study. As a food fish, it is a rich source of proteins, vitamins, and minerals.

TURBIDITY CURRENT, also called suspension current and density current, is a gravity current set in motion by the sudden entrainment of sediment by the overlying water in the presence of a slope. The sediment may be injected into the water by an earthquake, by mudslides down the CONTINENTAL SLOPE or SUBMARINE CANYON, or by a sediment-laden river running into a larger body of WATER. If a slope is present, the resulting mixture, now more dense than the surrounding water, will flow downhill as long as the slope persists or until the gradual loss of sediment brings the density of the moving water back to that of the water through which it is flowing. Mathematical calculations, laboratory experiments, and indirect observations in nature indicate that turbidity currents may reach a speed of 50 mph (80 km/h) and persist for several hundred miles.

The possibility of turbidity currents was first suggested in 1885 by F. A. Forel, a Swiss physician and naturalist. Forel speculated that a canyon in the bed of Lake Leman might have been cut by the sediment-laden water from the Rhone River flowing in along the bottom of the lake as an abrasive current. Forel's idea was revived in 1936 by R. A. Daly, a geologist at Harvard University, who suggested that such currents might have cut the canyons observed in the oceans. Stimulated by Daly's interest, Philip H. Kuenen, a marine geologist from the Netherlands, demonstrated in the laboratory that density currents could be produced by the sudden introduction of a sediment mixture into a tank of water with a sloping bottom. Kuenen coined the term *turbidity currents* to describe currents produced in this manner. The first turbidity currents observed in nature were in Lake Mead after the completion of Boulder Dam. To date only small-scale turbudity currents have been observed directly in the ocean. (See KUENEN, PHILIP HENRY.)

Evidence for the existence of turbidity currents in the deep ocean comes, primarily, from sediment samples taken from the seafloor. Sediment which has been transported to the ABYSSAL PLAIN by turbidity currents has unique characteristics. These sediments are composed of the sands, silts, and clays commonly found on the shallow CONTINENTAL SHELF rather than the fine-grained muds and oozes (see MARINE SEDIMENTS) which accumulate in the deep ocean by other means. Fragments of plant and animal life which only exist in shallow water are common. Also characteristic is the existence of graded bedding; i.e., the grain size of the sediment increases with depth in any bed laid down by a turbidity current (heavy particles settle first). The grading is in both the vertical and the horizontal directions since heavier particles are carried less far than lighter ones.

Another indication of the existence of turbidity currents in the ocean comes from the breaking of submarine cables. Cables laid across submarine canyons off the mouth of the Magdalena River of Colombia and the Congo River of western Africa have broken repeatedly over the years. Each of these breaks can be correlated with flood stages of the rivers or with an avalanche at the head of the canyon. The most spectacular example of cable breaks was associated with the earthquake on November 18, 1929, which shook the Grand Banks area south of Newfoundland, dislodging an estimated 24 mi^3 (100 km^3) of sediment which flowed downslope as a giant turbidity current. The area contains one of the denser concentrations of submarine cables in the world. Those cables near the epicenter of the earthquake broke instantaneously, whereas those further downslope broke one after the other as the turbidity current reached them. The turbidity current was later estimated to be moving at about 45 kn (83 km/h).

Turbidity currents are thought to be the key agent by which sediments derived from the continents are carried to the deep seafloor. The smooth abyssal plains found off the continents and in the bottom of deep-sea TRENCHES are thought to be formed by turbidity currents which periodically flow across them seeing the lowest point in which to drop their sediment load. The amount of material that can be transported by this mechanism is truly astounding. The greatest construction project in all human history was the Great Wall of China, begun in 215 B.C. by Shih-huang-ti and running for 1400 mi (2253 km) around northern and western China. It has been estimated that if the material devoted to the construction of the Wall were transported to the equator, it would result in a wall 8 ft (2.4 m) high and 3 ft (0.9 m) wide circling the globe. The material transported to the deep ocean floor in a few hours by the turbidity current which followed the Grand Banks earthquake would build 100 such walls.

The role of turbidity currents in the cutting of submarine canyons as suggested by Forel and Daly is still under debate. Most oceanographers do not believe that such currents have sufficient erosional power to account for these great canyons cut in the continental shelf and slope. However, it is generally conceded that such currents are instrumental in keeping the canyon floor clear by transporting to the deep sea that sediment which slumps from the canyon head and walls.

TURBOT See FLATFISH.

TURTLE GRASS, *Thalassia testudinum,* family Hydrocharitaceae, is a submerged marine plant that received its common name because it is a favorite food of sea TURTLES. (See PLANT LIFE IN THE OCEANS.)

Turtle grass is a bay-bottom vegetation that plays a vital part in bay ecosystems. It provides food and protection for sport and commercial fishes during various stages of their life cycles.

TURTLES (MARINE) or **SEA TURTLES** is the name commonly applied to various species of marine reptiles which belong to the order Chelonia; they are mostly similar in overall shape to other chelonians (e.g., tortoises and land turtles), especially in that they have a characteristic solid, roof-type shell, although, unlike the others, the marine turtles possess flattened and widened flipperlike limbs adapted for swimming.

TURTLES. Although they can be transported all over the world by currents (specimens have been sighted near the Scandinavian coast), sea turtles prefer the surface layers of tropical oceans. The loggerhead turtle shown here is carnivorous, dining principally on crustaceans, mollusks, fish, and jellyfish. (*U.S. Fish and Wildlife Service and Rex Gary Schmidt*)

Sea turtles normally live in the surface or near-surface layers of the tropical and subtropical waters of the world's open oceans, although they are sometimes transported by CURRENTS, especially to the European and American coastal waters of the North ATLANTIC OCEAN. Sea turtles reach an average age of 150 years and some also reach monstrous sizes—up to a ton [2000 lb; 907 kg] in some species. They can travel at speeds of 20 mph in the ocean and can remain underwater for as long as 4 hours.

Among the most important species are the green turtle, *Chelonia mydas,* the common loggerhead, *Caretta caretta,* and the hawksbill, *Eretmochelys imbricata.*

The green turtle is an edible species and is prized for its cartilaginous undershell (or calipee), an important ingredient of turtle soup. Hunters also seek this turtle for its meat, skin, and shell. This large migratory animal may attain a 4-ft ($1\frac{1}{4}$-m) length and weigh up to 400 lb (181 kg). It feeds chiefly on ALGAE and other marine plants, such as SEAGRASS. Green turtles are in danger of extinction not only because of their use as a food, but also because of the loss of some of their natural habitats for laying eggs. The eggs (about 100 per female) are usually laid at night and buried beyond the reach of high water in the warm incubating sands of isolated Caribbean, Central American, and South American beaches by large hordes. As is the case with most sea turtles, the male never goes on land but waits offshore for the female to place the eggs into the sand. After this is completed, several hours later, the female returns to the ocean where she is immediately mounted by the male.

The newly hatched turtles head straight for the surf but en route are preyed upon by birds and crabs. The young turtles that escape to the sea do not reappear until they are a year old. Where they go and how they live during their first 12 months is not known. What is known is that sea turtles are born obsessed to get into the ocean water as soon as they are hatched, and there they keep swimming out to sea, thus evading coastal predators.

Various scientific organizations and local governments have taken steps designed to protect and conserve the green turtles from these predators as well as from hunters. Their migratory habits and the "compass sense" mechanisms they employ to navigate long distances from their home pastures to the natal beaches are also being studied by scientists.

The common loggerhead turtle is about 3 ft (1 m) long, brown in color, and unlike the green turtle it is carnivorous, living on crustaceans, mollusks, fishes, and jellyfishes.

The hawksbill turtle is slightly smaller than the loggerhead and has a hooked beak. It is brown in color with yellow markings and has essentially the same diet as the loggerhead.

The largest of the sea turtles is the leatherback (*Dermochelys coriacea*) of the family Dermochelidae. This predatory, heart-shaped, shell-backed turtle measures (as an adult) up to 6 ft (2 m) in length and up to 800 lb (363 kg) in weight. It is sometimes carried by the GULF STREAM as far as the Scandinavian coasts.

TYPHOON See HURRICANE.

ULTRA-GRAVITY WAVES See WAVES.

UMBELLULA is a primitive deep-ocean, multicelled animal of the phylum Cnidaria, class Anthozoa, subclass Alcyonaria, and order Pennatulacea.

These strange, red-tentacled animals with their slender stalks anchored to the deep-ocean floor are somewhat akin to the HYDROIDS, SEA ANEMONE, and living CORAL. They have been known to exist since the 1870s when scientists aboard the English oceanographic research vessel, the *Challenger,* first found them. (See CHALLENGER EXPEDITION.)

UNDERWATER ACOUSTICS See UNDERWATER SOUND.

UNDERWATER PHOTOGRAPHY is the technique of using photographic equipment under water.

Underwater photography can be divided into two categories: photography with a diver-held camera and photography with a camera operated remotely. In the first category, the photographer, wearing some type of DIVING equipment such as SCUBA gear, is submerged with the photographic equipment and takes pictures directly. In the second, the photographer is able to record events at underwater depths that the diver cannot usually reach or can only reach with complex surface support. The remote camera also has the advantage of disturbing many subjects less. On the other hand, in shallow underwater photography the skilled diver can exercise a degree of mobility and precise positioning in relation to the subject that cannot be achieved remotely.

In both types of underwater camera work, many different factors control the effectiveness of the effort. Some of these are the available light, water clarity, differential absorption of light at various wavelengths, film capability, and the construction of the underwater camera. It is because of these that there is less room for error than in photography on land. For instance, underwater photographs taken by the diver, in most cases, depend on natural light for film exposure. Flashbulbs and electronic flash units are often used as supplemental light sources, but, because of the differential absorption of light by water, considerable care and experience are required if true color balance is desired. Light and color go hand in hand under water. Color films are relatively blind to the color subtleties which the eye can distinguish in the blue or green spectra of water. Filters are somewhat useful, especially in shallower depths; a color-correction filter over a lens will break up the blue, enough to restore some of the color that the unfiltered film cannot record. However, because of the usually low level of light and actual lack of light in certain wavelengths of the light-energy spectrum underwater (such as those of red), filters tend to do more harm than good.

The selection of lenses for the housed camera (a camera designed for air use but housed in a watertight casing) is dictated by the clarity of the SEAWATER and the required field of view. Since the distance from the camera to the subject must be short, wide-angle lenses are required. Such lenses are also necessary because of the magnification of objects underwater, owing to the difference between the indexes of refraction of the water and the air in the camera casing.

In deep-sea underwater photography the requirements for both camera and lighting equipment are stringent. Auxiliary lighting is essential because sunlight is rapidly absorbed with depth, usually becoming nonexistent at depths of a few hundred meters. Watertight cameras and lamp

UMBELLULA. This primitive, multicelled animal, first observed by scientists aboard the *Challenger* in the 1870s, makes its home on the deep ocean floor. *(U.S. Navy)*

housings must be designed and built to withstand the HYDROSTATIC PRESSURE of the ocean. The camera must also be positioned accurately and actuated either from a cable (with SONAR sensing equipment) or from deep-diving submersibles. This fact poses its own varied set of problems. For example, when a steel cable is used to lower the camera, the weight of the cable becomes significant and the cable may kink or break. Also, since the cable is so heavy, it is difficult for the winch operator to determine when the camera is at the desired depth or on the bottom.

In spite of the difficulties involved in underwater photography, its usefulness as a recording system has grown rapidly. All forms of photography are now in use: single pictures, motion pictures, stereo, shadow, elapsed time motion pictures, and so forth. These various photographic techniques are used in the study of MARINE SEDIMENTS, underwater biological activity, oil-drilling, underwater salvage work, and others.

See also DIVING; INSTRUMENTATION; MARINE OPTICS; OCEANOGRAPHY.

UNDERWATER SOUND is one of the most valuable tools available to scientists for the study of the ocean—the characteristics of its waters, its bottom and subfloor structure, and the life within it. Sound is used in the detection of submarines and sea mines, in avoiding ice pendants when transiting under the Arctic ice pack, in underwater communication, and in navigation. It has provided a means of continuously recording the depth of the ocean from a moving ship, rather than the 2 h or more it once took for a single measurement with a weighted steel wire. Also, it is increasingly useful in detecting schools of FISH, and in studying the "language" used by many sea creatures. But, underwater sound is subject to many mechanical and environmental influences which make it a difficult tool with which to work.

Characteristics Underwater sound, like sound in any elastic medium, is a longitudinal wave (particle motion is back and forth along the line of wave travel) which propagates in all directions from the source (unless focused) as a series of pressure crests (compression) and pressure troughs (rarefaction). A train of such waves may be depicted on a sine wave similar to the surface waves on the ocean (see WAVES). The distance between any two wave crests (pressure crests) is the wavelength, and the number of crests that pass a given point in one second is the frequency. Frequency is measured in cycles per second and reported as so many Hertz (i.e., 100 cycles per second is reported as 100 Hz). The amplitude of a sound wave is one-half the distance between the trough and the crest. The relative intensity of sound is measured in decibels (dB). Since the sound levels or intensities of interest cover an enormous range, the decibel, being plotted on a logarithmic scale, serves a very practical purpose. For instance, a 10-dB change is a factor of 10, and a 20-dB change is a factor of 100 in sound level. For purposes of analyzing sound waves and comparing one sound wave with another, a complete wave cycle (crest-to-crest) is divided into 360 equal increments of 1° each. This phasing of the wave allows one to identify any of 360 positions on a single wave. If two intersecting wave trains are in phase, they reinforce each other and the sound level is doubled; if the trains are 180° out of phase, they cancel each other and no sound is produced.

The speed of sound in WATER increases with increasing temperature, pressure, and salinity, with temperature being the most important. An increase in temperature of 1° C will produce an increase in sound speed of 10–13 ft/s (3–4 m/s), whereas a salinity increase of 1 ppt or a depth increase of 180 ft (55 m) only increases the sound speed by about 3.2 ft (1 m) per second. The average speed of sound in the ocean is usually taken to be around 4850 ft/s (1480 m/s), as compared with about 1082 ft/s (330 m/s) for the speed of sound in air.

The influence of temperature, pressure, and salinity on the speed of sound in the ocean has profound implications. If these three factors were constant throughout the depth and breadth of the ocean, sound waves would travel in a straight line. Since this is not the case, the waves must bend as the speed increases or decreases in the several layers of the water column. The first 394 ft (120 m) or so of the ocean comprises what is known as the mixed or isothermal layer. Here the effects of wind, wave action, and sun create conditions in which the temperature is nearly constant, though there are times and places where the mixed layer does not exist. Below the mixed layer lies the THERMOCLINE which extends downward to about 3937 ft (1200 m) in the Atlantic and 1968 ft (600 m) in the Northeast Pacific. In the thermocline the temperature drops so rapidly that it more than offsets the influence of increasing pressure. As a result, the speed of sound decreases until it reaches a minimum value. Below the thermocline the temperature becomes fairly constant, but the pressure continues to increase and, therefore, so does the speed of sound.

The bending effect caused by the variability of sound speed in the various ocean layers is important in sound communication, underwater exploration, and detection of submarines. When a ship projects sound ahead of itself into the mixed layers, the increasing pressure causes some of the energy to curve back toward the surface before it reaches the thermocline. Reflected back by the surface, this curve is repeated as long as enough energy remains. That part of the sound which does not curve to the surface enters the thermocline and is bent downward (velocity decreases) because of dropping temperature. The area in front of this split in sound energy is called the shadow zone, and it is here that submarines attempt to hide from searching ships. A large part of the sound energy that is projected downward at an angle greater than 20° to the horizontal will be carried back to the surface by increas-

UNDERWATER PHOTOGRAPHY. A Navy photographer uses a movie camera to film underwater life near the edge of the 3000-ft underwater cliff which surrounds much of Andros Island, Bahamas. *(U.S. Navy)*

ing pressure, or reflected back by the bottom. The point at which the sound reaches the ocean surface is called the *convergence zone* and occurs at intervals of about 36 mi (58 km). If the incident angle is great enough, the sound energy will penetrate the bottom and be reflected back to the surface by subsurface sediment layers of differing hardness. It is this property of sound transmission that is used by oil companies in exploring for oil and gas, and by geophysists for studying the substructure of the ocean floor.

The region near the base of the thermocline represents a unique zone in sound transmission. The increasing temperature above and the increasing pressure below cause sound waves to be bent back toward this plane. Thus, if sound is projected horizontally, or nearly so, along this plane, it will curve and recurve about the plane with little loss of energy. This is called the deep-sound channel or the Sofar Channel and acts like a voice tube in channeling the sound energy over tremendous distances. For instance, depth charges exploded in the deep-sound channel off Australia in 1960 were detected near Bermuda—a distance of 11 804 mi (19 000 km).

In addition to the bending effect, the efficiency with which sound can be used in the ocean is influenced by frequency and reverberation. The higher the frequency of sound, the more quickly the ocean will absorb its energy. For instance, a frequency of 100 Hz can be transmitted for thousands of miles, whereas a frequency of 1 000 000 Hz can only be transmitted for a few tens of feet. In a way, this is unfortunate because the higher frequencies are necessary to achieve the resolution needed in certain types of bottom exploration.

Reverberation is the scattering of sound energy by the roughened surface of the ocean (surface reverberation), the particulate matter, bubbles, and thermal discontinuities in the water column (volume reverberation), and the irregularities of the bottom (bottom reverberation). The sound energy that is bounced off a distant target for detection purposes is depleted by the amount lost to reverberation during the two-way transmission.

The equipment most commonly used to utilize underwater sound is a projector which produces sound by mechanical, chemical, or electric energy. The echo that returns from a target is received by a hydrophone, or underwater microphone. The combination of a projector and a hydrophone is called a transducer or SONAR. These devices can be designed to produce an omnidirectional or directional sound field, and to produce continuous or pulsed signals.

Noise A sound signal in the ocean must be detected in the presence of a chorus of other sounds that are lumped under the heading of background noise. Background noise, in turn, may be divided into self-noise and ambient noise. Self-noise is produced by the circuits in the sonar receiver, noises generated within the ship (engines, etc.) and radiated through the hull to the water, and the noise made by the interaction of the ship and the water.

Ambient noise includes that received from all other sources. Ambient noise begins with the sound produced by the thermal agitation of the water molecules and the salts that are dissolved within it. It includes the noise produced by the wind strumming on the sea surface, rain, surf, whitecaps, and bursting bubbles. Traffic noise produced by other ships and industrial noise from harbor and nearshore activities are substantial contributors. Also important is biological noise. Early in World War II it was found that as one approached shallow water in tropical and subtropical regions, the ordinary ambient noise was replaced by a sound like the sizzle of frying fat; on coming closer to shore, the sound resembled that of the crackle of burning twigs. It was found that the noise was produced by snapping SHRIMP in the process of closing their pincers with an audible click. With hundreds of thousands of shrimp clicking their pincers, the din was enough to make listening very difficult. Among the FISH, certain species of croakers and drumfish make a noise similar to four to seven rapid blows on a hollow log, and then the cycle is repeated. During the period from May to July croakers assemble in large schools in certain localities along the U.S. east coast. During this season the evening chorus of croakers lasts for several hours, ending just after dark. In addition to the shrimp, croakers, and drumfish, a multitude of clicks, squeaks, honks, groans, barks, gobbles, whistles, beats, and moans are heard in various localities. Some of the animals producing these noises have been identified as the GROUPER, CRAB, LOBSTER, pompano, PORPOISE, WHALES, SEA LIONS, SEALS, and sea robins.

Efforts to reduce self-noise have long been a continuous activity, particularly within the Navy. Ambient noise is best approached by signal-processing techniques in which the wanted signal is extracted from the background.

Early History The properties of underwater sound do not seem to have greatly attracted the interest of early investigators. Aristotle noted that sound could be heard in water as well as in air. And then, 2000 years later, Leonardo da Vinci (1452–1519) observed that; "If you cause your ship to stop and place the head of a long tube in the water and place the other extremity to your ear, you will hear ships at great distance." This was a remarkable observation when you consider that ships of the time had no engines. The first well-publicized measurement of the speed of sound in water was made in 1826 by Daniel Colladon, a Swiss physicist, and Charles Sturm, a French mathematician. The measurement was made in Lake Geneva, using a submerged bell as the sound source and a visual signal to indicate when the bell was struck. The underwater receiver was the ear of the observer. The speed thus determined was 4708 ft/s (1435 m/s)—very close to the 4850 ft/s (1480 m/s) often used as an average figure today.

The determination that underwater sound had a measurable speed led to numerous attempts to build instruments that would measure the depth of the ocean, detect such objects as ships and icebergs, and communicate by sound signals in the sea. Thomas Edison invented a means of sound communication between ships but found no interest among the maritime powers. In 1912, R. A. Fessenden built a successful transducer for the purpose of underwater communication and the detection of underwater objects. Improved versions of the transducers were able to detect icebergs at a distance of several miles and were used on American submarines during World War I. In 1919, M. Marti invented a recorder which provided a continuous record of the sound signal returning from a reflecting object or surface. Coupled with the progress already made in transducer technology, this invention led to the fathometer that is in use today. The fathometer measures the time interval between the outgoing sound pulse and the receipt of the return signal, and records the arrival as a continuous contour on a paper tape. No other signal instrument has been more valuable in mapping and understanding the world's oceans.

The rise of the submarine as a significant instrument of naval warfare beginning in World War I greatly accelerated the research and development in underwater sound. Today a vast array of sound transmitting and receiving devices exist for a multitude of purposes, and our understanding of sound transmission in the sea has been greatly advanced. However, it is a fitting testimonial to the complexity of the environment that much is yet to be learned.

Major Uses Certainly, the greatest use of underwater sound today is by the world's navies in their attempt to achieve and maintain a capability to meet the challenge posed by the submarine. Sonars for this purpose are both active (create a sound pulse in the water that is reflected back to the receiver by a target) and passive (listen only), and are used by aircraft, helicopters, ships, submarines, and as fixed, unmanned sensors. In the case of the submarine and ship, the sonar is mounted on the hull, usually near the bow, in order to better project the sound in front of them and to escape the sound produced by water flowing along the hull (flow noise) and machinery noise (radiated sound). Helicopters use a dunking sonar which is lowered into the water from a hovering position. Aircraft, on the other hand, must rely on sonobuoys: a small buoy with a deployable hydrophone and a radio transmitter. The sonobuoys, usually in a pattern, are dropped into the water from several thousand feet, the hydrophone is deployed, and any sound it receives is radioed to the circling aircraft.

The location and neutralization of sea mines also depends, in part, on underwater sound. Mines either rest on the bottom or are moored to the bottom, with the mine itself floating some distance below the surface. Bottom mines are about the size of a home hot-water heater, and moored mines range up to about 3 ft (1 m) in diameter. Therefore, relatively high-frequency sound energy is required to detect these small objects, particularly in the presence of sound scattered back by the bottom. The sonars for mine detection are fixed to the bow of ships, lowered beneath them in order to reduce the sound produced by the disturbed surface layer (variable depth sonar—VDS), and towed behind them in a "bird" that glides at a fixed distance above the bottom.

Homing torpedoes also use sound to seek out a target. They are either passive and home on the noise radiated by the target, or they are active and home by sound pulses bounced back by the target. The torpedo uses a sound receiver on either side of the nose to achieve the binaural effect necessary to determine the direction from which the signal is being received and to avoid chasing its own engine noise.

Increasingly, underwater sound is being used as a navigational aid by oceanographers as well as by commercial groups to provide greater accuracy in determining position for certain kinds of activities. The use of sound for this purpose falls into two general classes. The first requires a very accurate and detailed map of the ocean floor of interest, and a bottom which has numerous distinctive features. The ship, submarine, or submersible then projects sound vertically downward to "paint" a picture of the bottom and to determine its position by the bottom feature directly beneath it.

The second method requires the accurate positioning of either sound beacons or transponders (usually three) on the bottom. The beacon emits a sound pulse at precisely timed intervals which is synchronized by a clock both on the beacon and on the ship. The transponder, on the other hand, only emits a pulse when it is "queried" by a sound pulse from the ship. For identification purposes each transponder has its own frequency but responds in a frequency common to all transponders being used. Knowing the speed of sound in that particular body of water, the ship can position itself by triangulation. The determination of the distance from one beacon or transponder tells a submersible, for example, that it is somewhere on a hemisphere having a radius of the determined distance. The signal from two beacons places the position somewhere on a circle formed by the intersection of two hemispheres. With three beacons the intersection of the three hemispheres gives a single precise position.

Since the speed of sound is influenced by temperature, salinity, and pressure, the accuracy of sound (or acoustic) navigation falls off with distance over which navigation is attempted.

The use of sound, in the form of the fathometer, to map the ocean floor has already been alluded to. The use of similar techniques to study the rock and sediment layers below the bottom is of interest to marine geologists and seismologists. The idea is to project sound downward in such a way that a large fraction of the energy will penetrate the bottom to the desired depth, with some of it being reflected back to a receiver by any layer which has a density different from the layer above it. Since the higher frequencies are more quickly absorbed in rock and sediment (as in water), very low frequencies must be used for this purpose. Thus, by using a frequency of 40 Hz, a penetration of approximately 1 mi (1.6 km) can be obtained, whereas at a frequency of 3500 Hz (3.5 kHz), half the energy is lost in the top 100 ft (30 m) or so of sediment.

The sound energy for geologic and seismic studies is produced by a number of different devices which are capable of emitting a sharp and repetitive burst of energy. These devices include the ignition of high explosives, propane-oxygen mixtures, or the activation of electric sparks, steam bubbles, or hydraulic plungers. The preferred device appears to be an air gun which releases a large volume of compressed air into the water in a near-explosive fashion. The compressed air is supplied by compressors aboard the ship. Since the return signals do not return vertically to the ship, a series of receivers are towed behind the ship or in sonobuoys which radio the return signals to the ship.

One of the most exciting uses of underwater sound is just beginning to emerge—the study of density structures within the water column itself. Acoustic probes and tracking techniques promise a saving in worker hours and ship time similar to that realized by replacing the wire-sounding technique by a fathometer. For instance, a float has been developed at the University of Rhode Island and the Woods Hole Oceanographic Institution. The buoyancy of the float can be set so that it will remain at any desired depth and move with any currents that operate at that depth. The float has a sound emitter by which it can be tracked by distant stations, using triangulation to accurately fix its position at any time. Since much is yet to be learned about the details of ocean currents, this development has already proved a valuable tool. The use of sound also appears promising for the measurement of the speed of major currents, the study of internal waves (see WAVES), and the detection, tracking, and temperature sensing of the large eddies that occasionally spin off from currents.

UNDERWATER SOUND. *(Opposite page, top)* Underwater sound analysis has proven invaluable in mapping the ocean floor and its subsurface features. Pictured here is a composite of echo-sounding profiles of the Cayman Trough in the Caribbean Sea. *(Woods Hole Oceanographic Institution)* *(Opposite page, bottom)* A schematic of critically refracted ray paths through the ocean floor shows the path and velocity of sound in the various layers of the bottom. Techniques like this are used to study the substructure of the ocean floor and to detect oil and gas deposits. *(Woods Hole Oceanographic Institution)*

UPSLOPE FOG See FOG.

UPWELLING is the result of any of several processes by which WATER is induced to rise from a lower to a higher level. Thus it is the opposite of downwelling or sinking. Upwelling occurs from place to place throughout the world's oceans and is important in its effect on local climate and in renewing the nutrients in the surface layer where it occurs.

Upwelling most commonly occurs at the point where ocean currents either diverge or converge and where the prevailing wind blows parallel to the coastline. Where the surface diverts the flows away from each other, the subsurface water must rise to replace it. In the case of converging currents, particularly where they impinge on landmasses, the water tends to sink to accommodate the added volume and the surrounding deeper water must rise in compensation. In the case of the wind blowing parallel to the coast, the surface water is driven ahead of the wind. However, the CORIOLIS EFFECT—in the North-

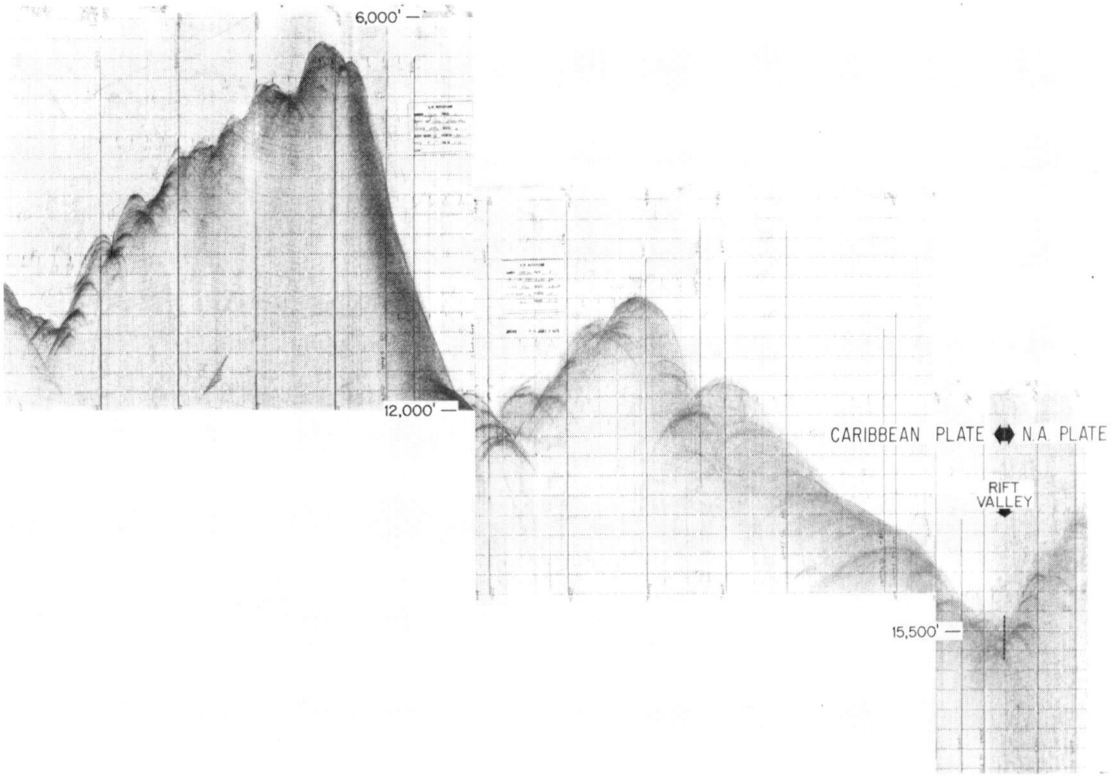

6,000' —
12,000' —
CARIBBEAN PLATE ↔ N.A. PLATE
RIFT VALLEY
15,500' —

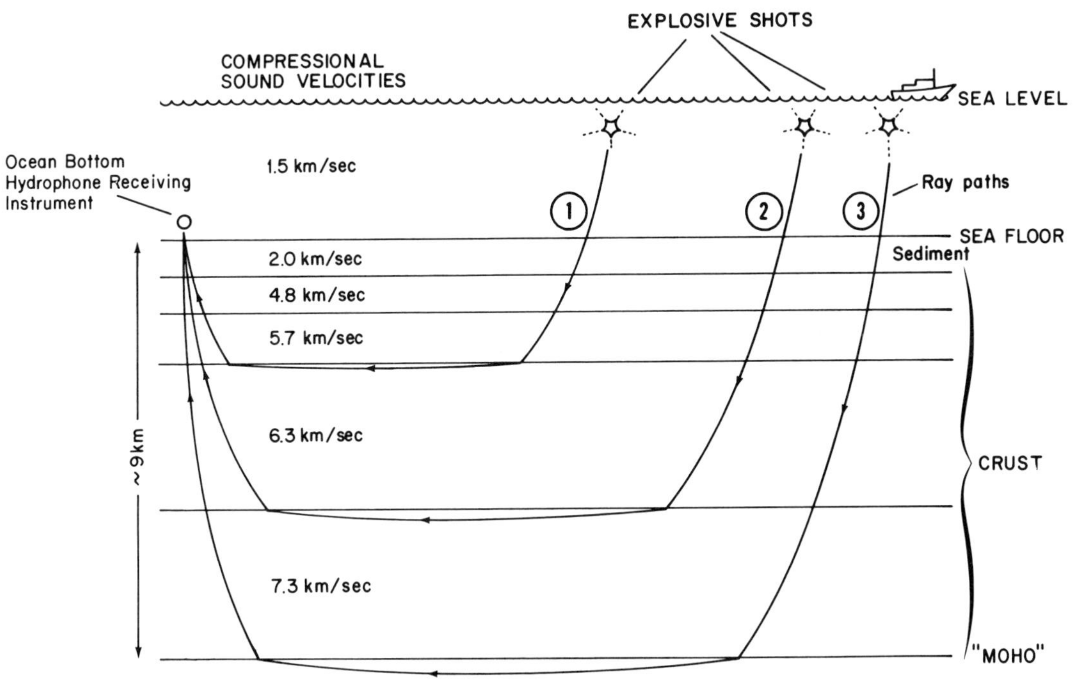

ern Hemisphere—causes the resulting surface current to turn to the right. The result is that the surface water is pushed out to sea and that the deeper, colder water is induced to rise to replace it. The rate at which this type of upwelling takes place is very slow, being on the order of 2 ft (0.6 m) per day.

One of the better examples of upwelling as a result of wind action occurs along the California coast. The prevailing wind is from the northwest. The combined effect of the wind and the Coriolis effect causes the warmer surface water along the coast to drift off to the southwest. This, in turn, causes the colder water at depth to well up to replace it. In the summer the warm, moisture-laden wind is chilled by the cold surface water—the result being the famous California fog. (See FOG.)

In areas where upwelling is relatively constant, the sea life not only becomes dependent upon the ready supply of high nutrient water but flourishes in profusion in its presence. The coast of Peru is an excellent example. There the anchovy catch has consistently placed Peru among the top fishing nations, and the guano "crop" from birds that feed on FISH is an important source of fertilizer. This prolific belt results from upwelling that is caused by the offshore deflection of the PERU CURRENT. At about 7-year intervals there is a major shift in the pattern of the trade winds which allows a tongue of the Equatorial Countercurrent (see EQUATORIAL CURRENT SYSTEM) to flow southward along the coast. This warm surface water replaces the cold water resulting from upwelling. The result is a phenomenon known as EL NIÑO (the Child) because it occurs around Christmas time. El Niño causes catastrophic destruction of sea life. The anchovies appear to migrate to better grounds, leaving the MARINE BIRDS to starve by the thousands.

Downwelling or sinking occurs whenever the density of the surface water is adjusted so that it is more dense than the water layer on which it rests. This can occur through a reduction of temperature, or can increase in salinity, or both. In winter this occurs in the central portion of the LABRADOR SEA. The central water, resulting from a mixture of Arctic water and Atlantic water, has a salinity of about 34.9 ppt. In winter the temperature of this water drops to around 35° F (2° C) and sinks to a level where its new density is equalized. This cold, saline water flows southward and can be traced well south of the equator. Areas of upwelling also exist around the sinking area, particularly along the "front" between the Labrador Current and the Atlantic water in the Davis Strait, and off the west coast of Greenland. The result is a highly productive fishing ground.

U.S. FISHERY CONSERVATION ZONE is the zone adjoining the territorial sea (the 3-mi limit); its outer boundary is 200 nautical mi (370.4 km) from the coast.

Under the Fishery Conservation and Management Act of 1976 (Public Law 84-265), the United States exercises exclusive fishery management authority over

1. All FISH found within the fishery conservation zone.
2. All ANADROMOUS FISH that spawn in U.S. waters, through their migratory range beyond the zone, except during the time they are in another nation's territorial sea or fishery conservation zone that the United States recognizes.
3. All U.S. CONTINENTAL SHELF fishery resources that extend beyond the zone, such as CORAL, CRAB, LOBSTER, CLAMS, and SPONGES.

The enforcement of the U.S. Fishery Conservation and Management Act is carried out by the officials of the U.S. Coast Guard and the National Marine Fisheries Service.

VAPOR PRESSURE is the pressure exerted by the molecules of a given vapor. For a pure, confined vapor, it is that vapor's pressure on the walls of its containing vessel; and for a vapor mixed with other vapors or gases, it is that vapor's contribution to the total pressure (i.e., its partial pressure).

Vapor pressure effects are important in the interchange that takes place between the ocean and the atmosphere. These fluids of different energy states—the liquid (SEAWATER) and the gaseous (atmosphere)—have a free surface boundary layer between them which inhibits but does not completely stop the transfer of mass and energy between them. In this transfer, both heat and water vapor migrate from the boundary—from the ocean to the atmosphere. Chiefly, the main exchange takes place by evaporation of the water, and this evaporation depends upon the difference between the partial pressure of water vapor in the atmosphere and the vapor pressure of seawater. Vapor pressure increases with TEMPERATURE, and partial pressure increases both with temperature and humidity. Accordingly, the difference is greatest when the ocean surface is warm and the air above it is cold and dry. In winter off the east coasts of continents such a condition for ideal transfer exists, and very large quantities of water are absorbed by the air.

See also WATER CYCLE.

VEMA is the name of the oceanographic research ship of Columbia University (U.S.A.) and is operated by the Lamont-Doherty Geological Observatory.

This 202-ft- (61-m-) long, 734-ton ship was originally a privately owned, three-masted schooner called the *Hussar* when she was christened in 1923. After service as a transatlantic yacht in the 1930s and a floating training barracks during World War II, the ship was acquired by Columbia University in 1953. It was completely overhauled from engine to hull and superstructure and refitted with electronic devices and oceanographic equipment. Since that date, the *Vema,* called the queen of the oceanographic fleet, has made many outstanding scientific voyages covering over a million (10^6) miles (1.6 million km) of the world's oceans. Major discoveries of ocean geological and physical behavior have ensued as a result of such *Vema* explorations.

VENOMOUS MARINE LIFE comprises ocean organisms that are toxic or poisonous to humans.

Some 1000 or more species of marine organisms widely distributed in almost all the world's oceans are poisonous, in one way or another, to human beings. They range from the simplest acellular protistan, *Gonyaulax* of the DINOFLAGELLATES to certain FISH.

Several of the dinoflagellates, which are tiny [39×10^{-5} to 39×10^{-3} in. (10 micron to 0.1 cm)] animallike organisms widely distributed throughout the world's oceans, contain a lethal toxic substance at certain times and under certain conditions. When large numbers of these organisms collect together, the poison is transferred to other marine organisms and there may be a mass mortality of this life in the area. Mollusks and echinoderms naturally feed upon the dinoflagellates, and when the infected animals are in turn ingested by humans, paralytic SHELLFISH poisoning often occurs. (See ECHINODERM; MOLLUSK.) The minimal lethal oral dose of the toxin for humans is believed to be as low as 3.5×10^{-5} oz (1.0 mg). This type of poisoning almost immediately causes sensory changes about the mouth and lips, and these changes spread to the rest of the face, neck, and fingers and toes. If death does not ensue in 10 hr, the victim usually survives.

A few of the tropical SPONGES are recognized as dangerous to humans. The as yet uncharacterized poison is transferred to humans via fine sharp spikes of the organism. While the stinging and subsequent effects of this particular toxin can be exceedingly unpleasant (viz., burning and swelling of the infected area, followed, in the more severe cases, by nausea and loss of consciousness), they are rarely fatal.

In the echinoderms, a form of marine life that includes the SEA URCHIN, SEA STARS, and sea cucumbers, among others, some 80 species are known to be venomous. Certain sea urchins are the most dangerous to humans. All these animals possess calcareous spikes, many with a poison apparatus consisting of small grasping organs called *pedicellariae*. The globe-shaped pedicellariae, armed with pincer jaws for holding, serve as the venom organ. On contact, a sense bristle causes a small muscle to contract, releasing the venom. A function of the pedicellariae is defense of the sea urchin. If an object touches the extended organ, it is immediately seized and poisoned. If the object is large and strong, it will tear away from the urchin, but the pedicellariae remain fast on the object and continue to poison it for several hours after being parted from the sea urchin.

Penetration of human skin by the spines usually produces an immediate and intense burning sensation. This is followed in a short time by redness, swelling, and a generalized aching sensation. Muscular paralysis has been reported, and secondary infections may ensue.

The sting from sea urchin pedicellariae may produce immediate distress and, in some severe cases, death. The sting produces intense radiating pain, faintness, muscular paralysis, loss of speech, and respiratory distress.

Not much is known about the chemistry of this sea urchin toxin, although it is known that it has a direct action upon the heart of mammals by causing a pronounced reduction in systemic arterial pressure.

The OCTOPUS can inflict a poisonous bite with its powerful parrotlike beak. A well-developed venom apparatus exists. This venom tends to also retard the clotting of the usually profuse bleeding from the victim's wound. The bite consists of two small puncture wounds, according to the size of the particular specimen. A burning sensation with localized discomfort may later spread from the bite. Swelling and redness commonly develop in the immediate area. While recovery is fairly certain, fatality has taken place from the bite of a small variety known as the blue-ringed octopus (*Hapalochlaina maculosa*).

There are approximately 100 000 species of mollusks, of which approximately 85 have been identified as being involved in human poisoning. In addition to the shellfish poisoning, the sting of the GASTROPODS of the genus *Conus* can be fatal. Stings of the cone SEASHELL are brought about by a venom which is forced into the victim by the animal within the shell by means of tiny hollow barbs. Numbness or burning of the wound takes place immediately, followed by a tingling sensation, particularly around the mouth and lips. Paralysis and coma may follow, and death may ensue as a result of heart failure.

Some 500 species of ocean fish are known to be poisonous to humans if they are eaten or if contact is made with the venomous spines that some fish possess for their protection.

The kinds of poisoning incurred by eating the flesh or liver of some of these species are usually termed *ciguatera, tetradon,* and *scombroid*. The ciguatera type is most prevalent, and it usually follows the eating of certain, mostly carnivorous, nonmigratory, and bottom-dwelling fish such as the BARRACUDA, GROUPER, surgeon fish, parrot fish, jacks, WRASSE, and others. However, some of these 300 species that have been implicated in the often fatal ciguatera poisoning of humans have tissues that are toxic at certain times only. Some, on the other hand, are poisonous at all times. Tetradon poisoning is usually fatal and has been caused at times by the ingestion of such fishes as PUFFERFISH and OCEAN SUNFISH as well as about 100 others. Scombroid poisoning, rarely serious, is sometimes brought on by eating inadequately preserved MACKEREL and TUNA.

Some fish with venomous spines are the scorpion fish, ocean catfish, and stingrays. The most virulent of these are scorpion fish, a family of fish that also includes the stonefishes (*Synancea*). The sting from the spines of any of these fishes can produce death or very serious results, and victims who recover from paralysis and other effects after months of treatment usually have impaired general health.

The SEA SNAKES are all deadly to humans. They are closely related to the cobras. Fatalities are common among fisherpeople in the Philippines and other areas of the world's oceans, especially in the tropics.

A few cnidarians (COELENTERATES), such as JELLYFISH, SEA ANEMONE, and CORAL, are capable of inflicting very serious injury on humans by means of stinging units called *nematocysts*. The lethal and paralyzing effect produced is caused by the presence in the toxin of a low-molecular-weight protein substance which effects the nervous system. The most poisonous is the Australian sea wasp, *Chironex fleckeri,* a box jellyfish; the sting from any one of its 15 tentacles has been known to kill a human being in a few minutes.

The beaches around Queensland, Australia, in particular, have been subjected at times to large-scale invasions of this poisonous jellyfish, and more deaths are caused by these animals than by the notorious man-eating sharks of the area. (See SHARK.)

VITYAZ is the name of the ship used by the Russians in their 1886–1889 oceanographic expedition. Seventy-six years after the remarkable SOUTHERN OCEAN voyage of F.G. von Bellingshausen of the Imperial Russian Navy, the *Vityaz II,* under Stepan O. Makarov, started a 3-year cruise around the world's oceans. During this voyage, many oceanographic observations were made. Of particular importance were TEMPERATURE and DENSITY measurements of the waters in the North PACIFIC OCEAN.

Vityaz II is an oceanographic research vessel placed in service in 1948 by the Russians. This 360-ft- (110.7-m-) long, 5710-ton ship, operated by the Institute of Oceanology of the U.S.S.R. Academy of Sciences, has a scientific staff of 73 and a crew of 64. It is capable of extended cruising to waters of the world's oceans. Since 1953, this Soviet ship has continued the work of the Danish GALATHEA and has carried out comprehensive biological sampling in the deep ocean TRENCHES. For example, in its exploration of the Kuril-Kamchatka Trench, the *Vityaz II* brought up many samples of ocean life that exist at a 29 700-ft (9053-m) depth. Many of these belong to the Pogonophora group (wormlike animals which live in tubes; they have no mouth openings or intestines, and it is thought that food is absorbed through the skin of their tentacles). Others encountered in such hadal depths are the ISOPODS.

WALRUS *Odobenus rosmarus,* is the single species of the pinniped family Odobenidae, Walruses are distinguished by their 18 teeth, with the upper canines prolonged into downward pointing tusks.
See SEALS.

WATER is a colorless, odorless, and tasteless chemical compound (hydrogen oxide, or H_2O) composed of two parts of hydrogen and one part of oxygen which are joined together to form a water molecule; pure water, or water with no salts in solution, freezes at 32° F (0° C) and boils at 212° F (100° C). Almost three-quarters of the earth is covered with water,—about 326 000 000 mi³ [1.36 billion (1.36×10^9) km³]. However, less than 1 percent of this is available as fresh water. This percentage is very unequally distributed, and the large lakes in North America, Africa, and Asia contain over 75 percent of the world's fresh water.

Water has been studied by scientists from almost every discipline. Although it would appear that the properties of one of the apparently simplest molecules should be well understood, this is not the case. The basic science underlying the structure of water (how its molecules are actually arranged and hydrogen-bonded), and, specifically, why water exhibits the most atypical behavior of any liquid are still not completely understood. Scientists have grappled for years with the problem, and many theories have been advanced in an attempt to account for anomalies in the properties of water. The most familiar of these is that the DENSITY of water's solid form (ice) is less than that of the liquid. This unusual property (and one shared only by the elements of bismuth, antimony, and gallium) was probably of primary importance to the development and existence of life on earth. For, if ice had been denser than water, then wherever freshwater bodies (those containing less than 1000 parts of dissolved salts per million parts of water) froze, freezing would have progressed from bottom to top, as newly formed ice at the cooled surface sank. Such a phenomenon would have eliminated life-sustaining liquid water at depth.

In the oceans, where many authorities believe that life on earth originated, SEAWATER (that by definition contains approximately 35 000 parts of dissolved salts per million parts of water) is 1.026 times as dense as fresh water due to the added weight of the dissolved salts.

Just where the water on earth and in the world's oceans first came from when the planet was formed is still a moot and provocative question. There are several theories. One suggests that the water was distilled from the inside of the earth by volcanoes. The revolving newly formed planet Earth, about $2\frac{1}{2}$ billion ($2\frac{1}{2} \times 10^9$) years ago, had its heavy metals attracted to the center of the molten core, while the lighter substances, including water, were centrifuged to the surface through volcanic valves. The water then participated in the hydrologic cycle: it was evaporated from the surface, carried by air masses, precipitated as rain and snow, carried back to the ocean by surface flow and percolation, and laden with mineral salts from the dissolved and eroded rocks. (See WATER CYCLE.)

FISH that imbibe seawater are so constituted by nature that they can secrete the salt from the water and retain the water itself for survival. Marine mammals, such as the WALRUS, WHALES, SEALS, and PORPOISE are also able to drink seawater because their kidneys produce urine that is more concentrated than seawater, and so they excrete all the salt they have swallowed in a smaller volume of water.

The human kidney does not have such an ability as that of these marine mammals, and when a person drinks seawater, he or she must excrete more total volume than that taken in, so that actually the

WALRUS. A bull walrus rears its head in Togiac Bay, Alaska. The walrus family has only one species, distinguished by elongated tusks, pointing downward from the upper jaws. *(U.S. Fish and Wildlife Service)*

person suffers a net water loss. Thus, humans must have fresh water or water containing no significant amounts of salts. This fresh water is called *hard* when dissolved minerals are present, and *soft* when it is relatively free of minerals.

Water in an unpolluted form plays an essential role in the mechanisms of biological processes. Moreover, the adequate supply of good quality water to the population and all parts of a nation's economy is as vital a problem as the supply of ENERGY and various kinds of raw materials is to agriculture, industry, and commerce.

WATER CYCLE or **HYDROLOGIC CYCLE** or **WATER EXCHANGE** is the natural cycle by which WATER, the medium of life processes and the source of their hydrogen, is moved from the world's oceans to the atmosphere, then to the land, and back again to the oceans.

In the water cycle, the supply of water is fixed and stable. This supply consists of the water in the world's oceans, ice caps, glaciers, lakes, rivers, water soils, water in living organisms, and water in the atmosphere. The total amounts to some 360 million (3.6×10^8) mi³ [1.5 billion (1.5×10^9) km³], and the amounts of water in each of these does not change in any appreciable amount.

Over half the solar radiation reaching the earth's surface is absorbed by the oceans. Some of this heat is used for surface evaporation, while the rest is stored in surface layers of the sea or moved downward into deeper water by various dynamic and thermodynamic processes. The evaporated water from the oceans plus that from the sun-irradiated land surfaces, including water vapor released from plants in transpiration, all enter the atmosphere. Dissolved salts in SEAWATER are also introduced into the atmosphere by evaporation of the fractionated sea spray during turbulent activity in the ocean; it is estimated that 1 billion tons of sea salt particles enter the atmosphere from the ocean each year. While about 90 percent of these salts are probably carried back to the oceans, nearly all the chlorine and about half the sodium in river waters is due to this ocean-atmosphere-land transfer.

Winds are also caused by solar energy, and wind currents are important in determining the local distribution of atmospheric moisture. Warm air masses carrying moisture from the oceans are moved by the winds and, when cooled, lose their capacity to retain this moisture. The moisture leaves the atmosphere as rain or snow. Based upon the PRESSURE and TEMPERATURE gradients aloft, the precipitation back to the oceans is less than the amount evaporated from that source. The reverse is true of the land, and it is estimated that 9000 mi³ (38 000 km³) of water falls on the land surfaces each year. However, excess land precipitation may result in ice caps and glaciers, may reinforce the land-water table, or may augment rivers and lakes subsequently returning to the oceans as runoff. This runoff depends upon depth, porosity, and compactness of the soil, and the underlying material, steepness, and geometry of the surface, and the type and density of the vegetation.

WATER EXCHANGE See WATER CYCLE.

WATER SPOUT See HURRICANE.

WATER TAGGING is the process of introducing foreign substances (tracers) into the ocean to detect

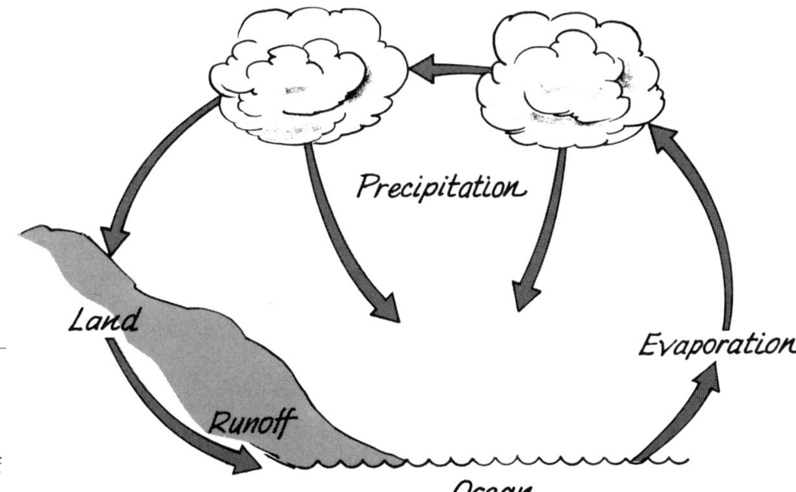

WATER CYCLE. Between the marine water cycle and the terrestrial water cycle, the total amount of water on Earth has changed very little over millions of years.

the movement of its waters by subsequent measurement of the location and distribution of the introduced substance. See INSTRUMENTATION; OCEANOGRAPHY.

WAVE PERIOD See WAVES.

WAVELENGTH See WAVES.

WAVES may be defined as time-varying quantities that are also a function of position. In other words, waves represent a disturbance propagating in a medium in such a manner that at any point in the medium the displacement is a function of time, and the displacement of a point in the medium, at any given instant, is a function of the position of that point in relation to some undisturbed reference plane. This definition applies equally to waves in a solid, such as the seismic waves that are continually threading their way through the earth's crust, mantle, and core; waves in a liquid, such as the wind waves that pound our beaches to the delight of bathers and surfers alike; and waves in a gas, such as the sound waves in air that permit voice communication.

In the ocean there are many different types of waves, ranging from the surface waves generated by the wind to the tide waves (see TIDES) caused by the gravitational attraction of the moon and sun. Less obvious are internal waves because they propagate along sharp density gradients below the surface, and such waves as *tsunamis,* storm surges, and seiches because they are uncommon in the experience of most of us. (See TSUNAMI.) All these waves can be placed into useful classes and categories which are derived from the forces that generate them, the forces that restore their vertical excursions, their wavelength (distance from the crest or trough of one wave to the crest or trough of the next), their period (time it takes for two crests to pass a given fixed point), and their height (vertical distance measured from the trough to the crest of a wave).

The distinction between wind waves and tide waves has already been made. The next useful distinction is that between capillary waves and gravity waves. The vertical oscillation of waves shorter than about 0.68 in (1.73 cm) in wavelength is controlled by the surface tension of the water. Such waves are known as capillary waves and are characterized by rounded crests and V-shaped troughs. The speed at which capillary waves progress increases with *decreasing* wavelength. Therefore, the minimum speed of capillary waves is that found at maximum wavelength [0.68 in (1.73 cm)], or 9 in/s (23 cm/s).

The vertical oscillation of gravity waves is under the restoring force of gravitational attraction, and their speed increases with *increasing* wavelength. Thus, the speed of wind waves ranges from 9 in/s (23 cm/s) to as much as 78 mph (125 km/h) for a 22.5-s (period) wave—the longest yet measured—and over 600 mph (965 km/h) for tsunami waves. Gravity waves may be further divided on the basis of their period as follows: ultra-gravity waves are those with a period of less than 1 s; ordinary gravity waves have a period of 1–30 s (this category contains most of the waves we observe on the ocean surface); infragravity waves have a period of 30 s to 5 min; long-period waves (excluding tide waves) have periods greater than 5 min; and trans-tidal waves have periods greater than 12 or 24 hr.

Another useful distinction is that between deep-water waves and shallow-water waves. The words deep and shallow do not refer specifically to the depth of the WATER in which the waves are traveling but to the relation between water depth and wavelength. In practice, a deep-water wave exists when the water depth divided by the wavelength is greater than one-half, and a shallow-water wave exists when the depth divided by the wavelength is less than one-twentieth. Those waves which fall into the deep-to-shallow transition zone (one-half to one-twentieth) are known as intermediate waves. It should be noted that under this definition the waves associated with a tsunami, even when passing over the greatest ocean depth, are shallow-water waves because their wavelength may exceed 100 mi (160 km). For deep-water waves the speed is a function of wavelength, and the greater the wavelength, the greater the speed. The speed of shallow-water waves, on the other hand, is controlled by water depth and is independent of wavelength. It is also important to recognize that for deep-water waves the speed of an individual wave (phase speed) is twice that of the wave train (group speed) of which it is a part. This is due to the fact that waves are continually dying out at the leading edge of the train while new ones are being added at the rear. For shallow-water waves, however, the phase and group speeds are the same.

Wind Waves For capillary waves and ordinary gravity waves the principal generating force is the wind. One might say that the wind "plucks" at a smooth sea surface until it becomes roughened by capillary waves, ripples, and wavelets. Having achieved this roughened state, the wind now has the backside of these irregularities to push against, and the more energy it imparts to the sea, the higher these irregularities become. But the wind does not produce waves with a single period and wavelength. In the generating area (area in which the wind blows with considerable force and duration), waves with periods ranging from a fraction of a second to the limit for ordinary gravity waves (22.5 s is the longest period yet measured for such waves) may be found. Thus, in the generating area, there is no regularity to the waves as one is accustomed to seeing along the beach. Rather, the sea surface is in wild and seemingly random motion, and no single wave can be identified and followed with the eye. If the wave crests steepen to an angle of less than 120°, they break to produce whitecaps, and if the wind is too strong, it simply shears the tops from the waves, thus increasing the time required to produce a fully developed sea.

The height of the waves produced in a generating area is determined by three parameters: the speed of the wind, the length of time that it blows (duration), and the distance over which it continually acts on the water surface (fetch). For each wind speed, a point exists at which the waves will grow no further because energy is being dissipated through breaking waves and the like at a rate equal to that at which it is being added. At this point a fully developed sea or a fully arisen sea is said to exist. The duration and fetch vary widely with wind speed in the generation of a fully arisen sea. For instance, a 10-kn wind must act on a sea surface 10 nautical mi (18.5 km) long for 2.4 hr to develop a fully arisen sea (for that wind speed), while a 50-kn wind requires a fetch of 1420 nautical mi (2630 km) and a duration of 69 hr. Fortunately, there are few places in the world where a fully arisen sea can be achieved by wind speeds in excess of 50 kn.

The term SEA STATE is used to describe the condition of the sea surface both inside and outside a generating area. The term is used with a numerical code ranging from 0 to 9 to denote the average height of the highest one-third of the waves in a train. A rule of thumb long used by mariners says that the wave height is equal to one-half the speed of the generating wind.

The apparent confusion within the generating area is due to the fact that many trains of waves with different wave heights, periods, wavelengths, and, to some degree, directions are effectively superimposed on each other. In some places wave crests momentarily coincide, each adding its height to the other, while in others, crests coincide with troughs to level out the surface. However, as this assortment of wave trains moves outside the generating area (waves move in the direction of the generating wind), it begins to sort itself out; those with longer wavelengths outrace those with shorter wavelengths (dispersion), and the latter tend to die out before the former. Waves outside the generating area are known as *swell*, whereas those within the generating area are referred to as *sea*. The longer swell may propagate for many hundreds of miles. For instance, storms off the Antarctic coast are frequently represented on the beaches of California by long period swell. A great deal about these storms can be deduced by monitoring the period and arrival times of these waves as the shorter and shorter wavelengths reach the beach.

The local effect of a wave propagating through the water column is to cause individual water particles to move in a circular, vertical orbit whose diame-

WAVES.

In the vertical position this manned spar buoy, called FLIP (for Floating Instrument Platform), is resistant to vertical oscillation as a result of wave action. For this reason it has been a valuable tool in the study and measurement of waves in the open ocean. FLIP is towed in the horizontal position and floods the lower end of the spar to assume the vertical position shown here. *(U.S. Navy)*

Coast Guard Cutter *Pontchartrain* wallows in a following sea in the North Atlantic. *(U.S. Coast Guard)*

Waves break against St. George Reef Lighthouse on Northwest Seal Rock near Crescent City, California. During a storm in 1923, waves actually broke over the lighthouse platform—70 feet above sea level. *(U.S. Coast Guard)*

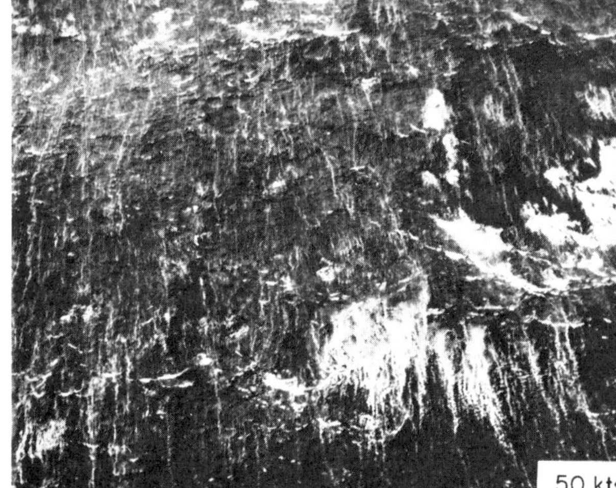

Surface (wind) waves are caused by the plucking, pulling, and pushing effect of the wind as it flows across the ocean surface. The ultimate height of the waves created in this manner depends upon the wind speed, as well as the time (duration) and distance (fetch) over which the reaction takes place. At a speed of about 40 kn (or kts) the wind begins to shear off the tops of the waves, thus reducing their rate of growth. It is the longer period waves from these generating areas which, after traveling up to several thousand miles, produce the long swells so sought after by surfers and swimmers. (*U.S. Navy*)

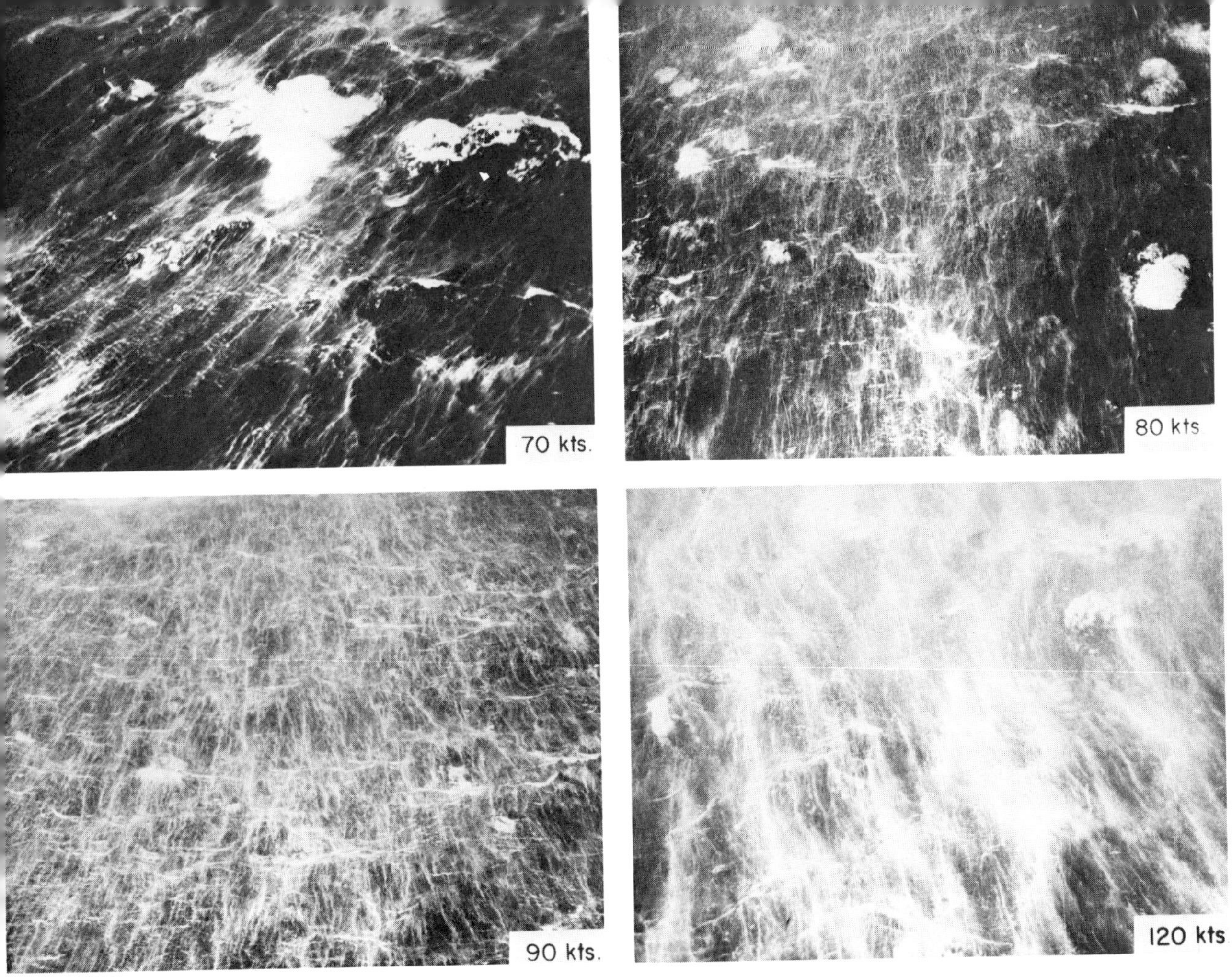

70 kts.

80 kts

90 kts.

120 kts

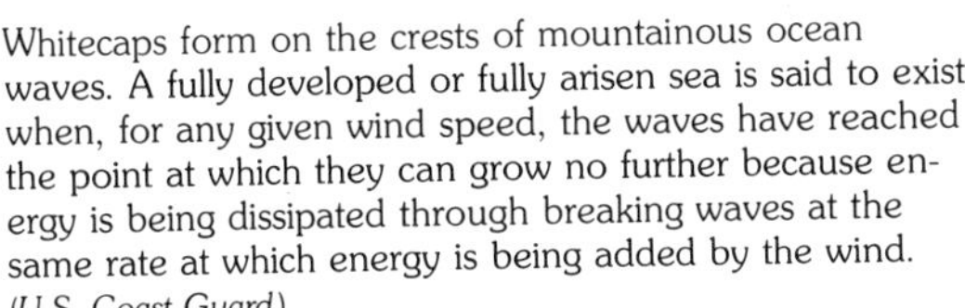

Whitecaps form on the crests of mountainous ocean waves. A fully developed or fully arisen sea is said to exist when, for any given wind speed, the waves have reached the point at which they can grow no further because energy is being dissipated through breaking waves at the same rate at which energy is being added by the wind. *(U.S. Coast Guard)*

ter at the surface approximates the height of the wave crest. The particle moves forward in its orbit (direction of wave travel) under the wave crest and returns to near its starting point under the trough. The orbit is not quite closed, so that a small net transport of water occurs in the direction of wave travel. The diameter of the particle orbit drops off exponentially with depth, so that at a depth equal to one-half the wavelength, the diameter of the orbit is only one-twenty-third that at the surface. This wave-induced motion is an important factor in mixing and aerating the surface layer of the ocean.

As swell begins to move into shallow water (waves begin to "touch" bottom when the depth is equal to half their wavelength), the orbital motion of the water particles becomes elliptical rather than circular, and the orbit becomes progressively flattened with depth. At the water-bottom interface the motion is simply back-and-forth in a current which is of some importance in the movement and winnowing of bottom material. This influence is somewhat enhanced by the fact that orbital speeds increase as the water depth decreases.

At this point of interface with the bottom waves have passed through the intermediate wave zone and are true shallow-water waves. They all travel at the same speed regardless of wavelength, and their speed and wave height are somewhat reduced by energy lost through interaction with the bottom. Very near the beach the combination of bottom friction and the growing tendency for the water particles to travel faster in their elliptical orbits than the wave profile causes a marked increase in wave steepness. When the crest achieves an angle of less than 120°, or when the wave height divided by the wavelength becomes greater than one-seventh, the wave rises dramatically in height (surf), becomes unstable, and begins to curl forward and break.

Due to a process known as diffraction, waves do not necessarily strike the beach at the angle of their original approach. For instance, in the case of waves approaching the beach at some large angle, say 45°, the part of the wave nearest the beach is in shallower water than that further out. The part of the wave that is in shallower water must travel more slowly because of its greater interaction with the bottom. This process, similar to a long line playing "crack-the-whip" on ice or roller skates, gradually turns the waves until their final approach to the beach is nearly parallel. This has the long-term effect of straightening shore lines. Offshore irregularities on the bottom may tend to focus wave energy by diffraction so that severe erosion is experienced along a localized section of beach, or it may disperse the energy that reaches the beach, thereby reducing local erosion. Diffraction also tends to focus wave energy on headlands and points so that they are eroded at a more rapid rate than that of the flanking coastline.

Where the coast is made up of rocky cliffs with deep water at their base, several interesting waves may be formed. For instance, if a train of waves with a single period is reflected back on itself in a process known as clapotis, a standing wave may occur. This results from the fact that there are now two wave trains with the same wavelength and period passing through each other in opposite directions. If perfectly tuned, the wave trains may give the appearance of being stationary, with each major segment simply oscillating between crest and trough. If, on the other hand, the wave train approaches the cliff at an angle of less than 20°, it is not reflected back to sea but tends to hug the cliff face. Under certain conditions such waves grow progressively in height to form the dangerous KING WAVE.

WEDDELL SEA is a sea of the South ATLANTIC OCEAN that occupies a large indentation in the coastline of Antarctica across Drake Passage from the tip of South America. The sea is flanked by the Antarctic Peninsula on the west, the 100 360 mi² (260 000 km²) Filchner Ice Shelf on the south, and Coats Land on the east. It is separated from the BELLINGSHAUSEN SEA on the west by the Antarctic Peninsula. To the north the seafloor plunges rapidly to depths that exceed 26 240 ft (8000 m) in the vicinity of the Scotia Ridge which partially blocks the Weddell Basin on the northwest. The Weddell Abyssal Plain is one of the largest yet discovered, measuring some 200 mi (322 km) wide (north-south) in the vicinity of 20° W longitude. The CONTINENTAL SHELF along the Antarctic Peninsula and Coats Land is strongly influenced by glaciation and glacial debris. Like most shelf areas of the Antarctic, it breaks at a depth somewhat below that characteristic of the world's shelves [600 ft (183 m)] as a result of isostatic adjustment to the weight of ice during the last Ice Age (11 000–8000 years ago).

The area of the Weddell Sea represents one of the more inhospitable environments on earth. Surface WATER temperatures are near 32° F (0° C) throughout the year, and air temperatures range from that of the surface water to −27° F (−33° C). The huge Filchner Ice Shelf sheds tablelike icebergs (see SEA ICE)—some over 60 mi (100 km) long—that pass out through the Weddell Sea into the South Atlantic. The Weddell itself is largely covered by ice throughout the year. The surface water circulation in the sea is clockwise as a result of the prevailing easterly winds. The combined action of the wind and current tends to pile the ice against the western edge of the sea.

The Weddell Sea takes its name from James Weddell, a veteran of the Royal Navy and a captain with the British firm of Enderby. In 1822, Weddell, blessed by exceptionally mild weather, established a record for "fartherest south." He was able to penetrate what is now the Weddell Sea as far south as latitude 74°15′.

Companies such as Enderby played an important role in exploring hitherto unknown regions of the world's oceans during the sixteenth to the nineteenth centuries. However, they operated at a time when the conservation of animal resources was a matter of little concern. For instance, when the Russian F.G. von Bellingshausen was exploring the South Shetland Islands on the northwestern rim of the Weddell Sea in 1820, he met an American sealing captain named Nathaniel Palmer (credited by some authorities with discovering the Palmer Peninsula—now called the Antarctic Peninsula). Palmer told Bellingshausen of the extraordinarily profitable seal hunting in the South Shetland area. At the time the seal population of these islands was estimated at about one million. Three years later (1822) the boom was over; the seal population had been reduced to the point that a voyage to the islands was no longer profitable.

WEST WIND DRIFT CURRENT See ANTARCTIC CIRCUMPOLAR CURRENT.

WHALES are any of about 37 species of large MARINE MAMMALS of the order Cetacea. The body of the whale is streamlined, the broad tail is used for propulsion, and the limbs, bony-skeleton flippers, are used for balancing. There are no external ears or nostrils—a small opening leads to the eardrum, and a blowhole (or two in the case of the baleen whales) on the top of the head leads to the large lungs.

Whales, porpoises, and other allied forms with a tapering body all belong to the order Cetacea of aquatic mammals. (See PORPOISE.) In this order the true whales are divided into two groups—the baleen whales (suborder Mysticeti) and the toothed whales (suborder Odontoceti).

Baleen Whales The baleen whales have no teeth. Instead, they have a peculiar but highly effective feeding apparatus in their gigantic mouths. This consists of a series of 300 or more horny plates which hang down into the cavity of the mouth. The substance of which these plates are composed is whalebone or baleen (actually keratin or a modified dermal derivative). This material is tough and more or less flexible, and each plate is coarsely frayed so that the thin baffle plates form a sieve through which gulps of water can be strained to collect PLANKTON, tiny CRUSTACEANS (e.g., KRILL), and small FISH.

There are about 10 species of baleen whales. The smallest, the pygmy right whale (*Caperea marginata*) grows to about 20 ft (6 m) in length. The largest living animal known to evolutionary history, the blue whale (*Balaenoptera musculus*), sometimes called SIBBALD'S RORQUAL or the sulfur-bottom because of the heavy growth of DIATOMS that cover its underside and make it look yellow, grows to be over 100 ft (30 m) and can weigh more than 160 tons. A baby blue whale is born at a length of 23 ft (7 m) having grown from an ovum in only 11 months. Such whale calves are said to need over a half-ton of milk per day. It has been estimated that a blue whale, during the first 5 years of its life, gains at the average rate of 90 lb (40.8 kg) per day. An adult weighing 90 tons consumes more than a ton of food a day (the blue whale feeds only half the year) in order to provide the energy needed for propulsion, maintenance of body temperature, and processes such as digestion and respiration. The heart of this animal weighs around 1200 lb (545 kg), and the tongue about 650 lb (294.8 kg).

Five other baleen whales are the finback whale (*Balaenoptera physalus*), the sei whale (*B. borealis*), and three smaller whales, Bryde's whale (*B. edeni*) and the two minke whales (*B. acutorostrata* and *B. bonaerensis*). The finback averages little more than 65 ft (20 m) in length, and the sei averages 55 ft(17 m). Bryde's whale and the minkes average 45 ft (14 m) and 30 ft (9 m), respectively. The humpback whale (*Megaptera novaeangliae,* a Latin term which roughly translated means "big-winged New Englander") is also shorter than the sei, averaging about 45 ft (14 m). Another interesting baleen whale is the gray (*Eschrichtius robustus*), an animal that makes long migratory journeys apparently with little rest or food. BARNACLES and skin pigments lend these whales a mottled appearance.

Blubber, the thick layer of fat which is protected on most species of whales by a thin outer skin, keeps the whales insulated from the cold water in which they swim. It also serves as a food reservoir similar to human fat. And, it yields most of the whale oil, the coveted commercial product of the whaling industry. Today, most of this oil is made into margarine and soap, some whale meat is eaten, and much of the animal is used as feed for domestic animals and for fertilizer.

The baleen whales are mostly animals of temperate and cold waters—primarily those of the Antarctic. They leave their feeding grounds in the Antarctic in the winter and migrate toward the equator into warmer water to reproduce and to nurture their young. For example, the southern right whales

(*Eubalaena glacialis* and *E. australis*) are believed to make a 4000-nmi (7408-km) migration, normally cruising at a rate of about 6 kn to and from their breeding grounds off Argentina to Antarctica every year. These animals average 55 ft (17 m) in length and have an enormous head about one-third as long as the rest of the body. They have no dorsal fin.

The northern right whale, or bowhead whale, *Balaena mysticetus* (length about 50 ft or 15 m), is found in the Arctic and northern subarctic waters. Its numbers were greatly reduced over a period of about 300 years, initially in the European Arctic, then in the eastern Canadian Arctic and the Okhotsk Sea. (See OKHOTSK, SEA OF.) Commercial whaling for bowheads existing in the CHUKCHI SEA and later in the BEAUFORT SEA has been going on for almost 100 years. These animals have been completely protected from commercial whaling by the International Convention for the Regulation of Whaling since 1947, and subsequently, by the Marine Mammal Protection Act (MMPA) of 1972 and the Endangered Species Act (ESA) of 1973.

The so-called right whales derive their name from the old-time whalers, who characterized these as the "right" ones to catch. To these whalers, they were right because they swim slowly, do not sink after they die, and possess a particularly rich store of oil and baleen. These whalebone plates, or baleen, were utilized by the whalers of old for a number of products: corset stays, umbrella ribs, clock springs, harpoon heads, knife handles, and scrimshaw (ornamental handwork—usually carving).

Unlike other whales, the right whales have callosities, or patches of thickened white skin growths several inches thick, on various parts of their body. The principal function of these callosities seems to be that of a splash deflector, preventing water from entering the whale's blowhole, especially when the animal breaches. Additionally, some whales represent floating islands for various plants and animals with, for example, some humpback whales carrying over a half a ton of barnacles on their flukes, flippers and bellies.

Toothed Whales The toothed whales have no baleen, and the tooth arrangement varies with the species. Such a variance ranges from 260 teeth in some to a single pair in others. In the narwhal the upper left canine may attain an 8-ft (2.4 m) length and has the appearance of a curved tusk. The primary species is the sperm whale (*Physeter catodon*), a deep-water animal which feeds at great depths in the ocean. These sperm whales breed in tropical coastal waters during the winter but seldom go any farther north than 40° LATITUDE for summer feeding. This is in contrast to the migratory habits of the baleen whales such as the fin, sei, and humpback which all breed in tropical waters during the winters but travel to the polar regions in the summers. Their principal food consists of giant SQUID, and it is known that sperm whales dive to at least 620 fathoms (1133 m) in search of squid. Giant squid grow to formidable proportions. Some about 60 ft (18 m) in length have been found, but it is believed that they grow much larger. Judging from the squid (*Architeuthis*) suction-disk scars found on the bodies of sone whales, some rather titanic battles take place thousands of feet beneath the surface between whales and squids. A cavity in the nasal passage of the sperm whale contains an oil which, on contact with air, forms spermaceti wax. This animal also forms an intestinal substance, ambergris, which is used as a base for perfumes.

Another form of the toothed whale is the grampus or KILLER WHALE (*Orcinus orca*). This predator, although a cetacean, is not really a whale, but a dolphin of the family Delphinidae. However, unlike other dolphins, it has a very high dorsal fin and will aggressively attack almost any creature with the possible exception of the adult WALRUS. Killer whales usually hunt in packs, and although they measure only 15–30 ft (5–10 m) in length, they often attack much larger whales. The grampus is characterized by a bluntly rounded snout, high black dorsal fin, white patch just behind the eye, a jet-black color above, and white underparts. These swift-swimming animals have large conical teeth which interlock when the jaw is enclosed. Their intelligence has been said to be on the level with that of a domestic dog. They are found in ocean waters from the BARENTS SEA to beyond the ANTARCTIC CIRCLE.

Other Characteristics Whales, the largest air-breathing animals of the world's oceans, have characteristics and habits that are still little known. For example, how they feed and raise their young, how they navigate, or how deep in the ocean they can submerge are all still mysteries. They have been observed in all the oceans of the world and even in fresh water, and most of the large whales, traveling in herds, continually migrate from ocean to ocean using a very sophisticated SONAR or echolocation system.

Shaped like a giant torpedo and driven forward by horizontal fins (flukes) that form the tail, the whale can achieve rather respectable speeds in the water. For example, finbacks are capable of making speeds in short bursts of more than 20 kn, although the usual cruising speed of a whale is about 6 kn.

Whales, like most other deep-sea cetaceans, have very large and complex brains. This suggests that to

WHALES. (*Top left*) A pilot whale performs at a California Marineland. (*U.S. Navy*) (*Top right*) Scientists are experimenting with the use of whales in deep-water work. Here, Morgan the pilot whale presses a grabber claw against a target, locking the lift device onto a torpedo on the sea bottom. As he pulls away, the mouthpiece separates, activating the hydrazine gas system which inflates a balloon and floats the torpedo to the surface. (*U.S. Navy*) (*Below*) A whale surfaces for air. When whales come to the surface to "blow," they exhale a cubic meter of air in the space of a second, producing a hoarse bellow and a water-vapor geyser. (*U.S. Navy*)

WOODS HOLE OCEANOGRAPHIC INSTITUTION. This world-renowned research center, on Massachusetts' Cape Cod, functions also as an educational organization, granting advanced degrees in oceanography and marine engineering. (*Woods Hole Oceanographic Institution*)

live in the sea, breathing air with a mammalian physiology and a mammalian skin, requires a large brain for a successful adaptation over the millennia. Water-breathing forms of comparable body sizes have very much smaller brains. The largest known brain is that of the sperm whale, *Physeter catodon,* ranging from 14–20 lb (6.4–9.2 kg) in animals whose body lengths range from 40 to 60 ft (12 to 18 m) and whose body weights can be up to 60 tons.

Whales have the ability to emit sounds at different frequencies, in the form of moans, moos, and chirps, while swimming. It is thought that the variations in the sounds are used either to communicate with one another or to chart their migration course. As opposed to the upper limit of human hearing (about 20 000 Hz), the whales' sounds are extremely shrill and may reach levels more than 12 times greater than the human ear can detect, or up to 256 000 Hz. Blue-whale voice signals are much longer than those recorded from other whales, and the signals are produced at standardized intervals. There is a precise duration of 100 s from the beginning of one voice sequence to the beginning of the next. The so-called right whales also have very characteristic phonations. They repeat a complicated 12-minute stanza of signals in exactly the same way, signal for signal. Another sound made by the whales is that produced by the "blow," or their method of breathing. This sound is often relatively loud and resembles a hoarse bellow or snort. In producing it, a whale comes to the surface, raises the blowhole, which is one nostril or two, depending on the type of whale, out of the water, exhales, and then inhales. The exhalation is rapid, and in less than a second, approximately 35 ft^3 (1 m^3) of turbulent air is forced out through the blowhole, which measures from about 1 to 5 in (2.5 to 12.7 cm) in diameter. A geyserlike water spout usually marks this blow since the exhaled air which has been under pressure, warmed and moistened, expands rapidly, and the water vapor condenses. During the intake cycle, which is also fast (usually less than a second), the blowhole is expanded more than during exhalation, and air is sucked into the animal's large lungs. Variations in the frequency of the whale's need to come to the surface for air are based upon differences of individuals and the types of activities engaged in, just as in the case of humans who increase their breathing cycle when under stress or during exercise. In the case of individual differences, a humpback whale, for instance, returns more often to the surface for air than does a sperm whale which has much shorter flippers than the humpback.

WHIRLWIND See HURRICANE.

WHITE SEA, one of the seas of the ARCTIC OCEAN, lies off the northwest coast of the U.S.S.R. and is largely confined by the Kola Peninsula to the west and the north, the Russian mainland to the south, and the Kanin Peninsula to the east and north. The sea is connected to the Arctic Ocean by the Gorlo Strait. There are three large bays (Mezen, Dvina, and Onega) along the eastern coast, and there is a long gulf (Kandalaksha) on the western coast. A number of rivers, three of which bear the names of the three large bays, flow into the White Sea, and the Sea itself is connected to the Caspian Sea and the BLACK SEA on the Turkish and Iranian borders by the Dvina, Volga, and Dnieper Rivers. The area of the White Sea is about 36 670 mi² (95 000 km²). It has a maximum depth of 1111 ft (340 m), with depths of 328–984 ft (100–200 m) predominating. The Norwegian explorer Ottar is credited with discovering the White Sea in the ninth century A.D.

WILLY-WILLY See HURRICANE.

WINKLER TITRATION See INSTRUMENTATION.

WOODS HOLE OCEANOGRAPHIC INSTITUTION is one of the world's foremost centers for oceanographic research and a leader in marine science education.

In 1930, the Woods Hole Oceanographic Institution (Woods Hole, Massachusetts) was formed as an independent nonprofit organization incorporated under the laws of the state of Massachusetts, "To prosecute the study of oceanography in all its branches; to maintain a laboratory or laboratories, together with boats and equipment and a school for instruction in oceanography and allied subjects. . . ." A $27 000 gift from the Carnegie Corporation provided land, and a $2.5 million ($\$2.5 \times 10^6$) grant from the Rockefeller Foundation provided funds for a building, for other construction, and for boats, equipment, expenses, and an endownment.

The Institution has grown from its founding as an oceanographic research center to become an educational organization for the marine sciences as well. All aspects of OCEANOGRAPHY are covered in the scope of the Woods Hole research program, and a fleet of several research vessels and the submersible *Alvin* are used in this oceanographic research. In addition, OCEAN ENGINEERING subjects are also under its purview as well as work in acoustics, deep submergence engineering and operations, information processing, INSTRUMENTATION, and ocean structure, moorings, and materials. It is a graduate degree granting institution, with programs of study and research leading to doctoral degrees in oceanography or the degree of ocean engineer.

WRASSE is the name for any of the small usually brilliantly colored bony fish in the family Labridae of the suborder Percoidei, which includes the typical Perciformes such as the perch and SNAPPER. The wrasses average about 4 in (10 cm) in length and are distinguished by the fact that the spiny part of the dorsal fin is longer than the soft-rayed part.

Throughout the ocean world, some 26 species of marine FISH, six species of SHRIMP, one CRAB, and one of the MARINE BIRDS are known to be cleaners of other fishes and marine animals. This SYMBIOSIS consists of one being cleaned of parasites and damaged or infected tissue while the other obtains food from the fish it "doctors" and earns immunity from attack. Wrasses are the most diligent of all the fish species which regularly clean and groom other marine fishes. (See BARBER FISHES.)

The wrasses, like most other fishes that perform such functions, primarily inhabit the tropical and sometimes temperate waters of the world's oceans. They have established a special relationship or mutualism with the BARRACUDA and feed on the BACTERIA that grows in the mouth of this fish.

The Pacific wrasse, *Lamroides dimidiatus,* found in the waters off Tahiti and Hawaii groom MORAY eels and various large fishes in complete safety.

Since parasitic infections take place in fish in all the world's oceans, it is thought that the service performed by the wrasses is invaluable to the health of marine populations.

WRECKFISH (*Polyprion americanus*) is the name for a serranid (a member of the family Serranidae) that is noted for inhabiting sunken ships and wrecks in the MEDITERRANEAN SEA and the tropical parts of the ATLANTIC OCEAN. It is also known as the stone bass.

YZ

YELLOW SEA (HWANG HAI) is a marginal sea of the western PACIFIC OCEAN nestled between the EAST CHINA SEA on the south, the Korean peninsula on the east, and the mainland of China on the north and west. The northern extremity of the sea is the Gulf of Po Hai with the demarcation between the two being a line running from the Liao-Tung Peninsula to the San-Tung Peninsula. The separation between the Yellow Sea and the East China Sea is taken to be a line running from the Korean mainland, southwest of Saishu Island, to the mainland of China north of the mouth of the Yangtze River. The Yellow Sea is about 621 mi (1000 km) long and 435 mi (700 km) wide at its widest point. It covers an area of 160 962 mi² (417 000 km²), occupies a volume of 4 078 470 mi³ (17 000 000 km³), and has a mean depth of 131 ft (40 m).

The Yellow Sea gets its name from the yellow loess (windblown) sediment (hwang-tu) of central China that is brought down to the sea by such great rivers as the Liao, White, Yellow, Yangtze, and Yalu. This fine-grained material gives the water over some parts of the sea a yellow color. Winds also carry this yellow dust to the sea, sometimes creating dust falls so thick that boat traffic must be stopped. The central part of the Yellow Sea basin is made up of muds and silts because of the lack of strong currents to flush them out. On the other hand the flanks of the seafloor are usually composed of sand owing to the winnowing effect of strong tidal currents. The entire floor of the sea was exposed during the drop in SEA LEVEL associated with the last Ice Age.

A branch of the warm Tsushima Current flowing northward toward Korea Strait is split off in the vicinity of Saishu Island and flows into the Yellow Sea. It is called the Yellow Sea Warm Current and has a speed of only 0.5 kn. Cool coastal currents flow southward along the coast of China and Korea. The sea is subjected to semidiurnal TIDES which measure 13–26 ft (4–8 m) along the coast of Korea and 3.2–10 ft (1–3 m) along the coast of China. The Gulf of Po-Hai experiences diurnal tides with a range of 10 ft (3 m).

The Yellow Sea region is under the climatic influence of a northerly monsoon during the winter which often brings severe blizzards, and a southwest monsoon during the summer which occasionally results in the severe dustfalls mentioned earlier. The air temperature ranges from 82 to 21° F (28 to −6° C), and the surface WATER temperature ranges from 32 to 82° F (0 to 28° C). Only the Gulf of Po Hai freezes over in winter.

YELLOW SEA. A marginal sea bordered by the East China Sea, the Korean peninsula, and mainland China, the Yellow Sea gets its name from the sediment brought from the interior by the great rivers of China. In this photograph taken from Skylab, the Chinese coast, with Kiangsu Province and North Kiangsu Canal, is discernible on the left. *(NASA)*

ZOOPLANKTON See PLANKTON.

INDEX

Boldface page numbers refer to article.